# Nanoscale Science and Engineering Forum 2014

Core Programming Area at the 2014 AIChE Annual Meeting

Atlanta, Georgia, USA
16-21 November 2014

ISBN: 978-1-5108-1265-9

**Printed from e-media with permission by:**

Curran Associates, Inc.
57 Morehouse Lane
Red Hook, NY 12571

**Some format issues inherent in the e-media version may also appear in this print version.**



Printed by Curran Associates, Inc. (2015)



AIChE
120 Wall Street, FL 23
New York, NY 10005-4020

Phone: (800) 242-4363
Fax: (203) 775-5177

www.aiche.org

**Additional copies of this publication are available from:**

Curran Associates, Inc.
57 Morehouse Lane
Red Hook, NY 12571 USA
Phone: 845-758-0400
Fax: 845-758-2634
Email: curran@proceedings.com
Web: www.proceedings.com

# TABLE OF CONTENTS

**(14a) Casting Solvent Effects on Block Ionomer Morphology and Corresponding Property Changes** ........ 1
*Kenneth Mineart, Richard Spontak*

**(14b) Phase Separation Kinetics in Pressure Sensitive Adhesive Melts Containing Styrene-Diene Based Block Copolymers** ........ 2
*Ninad Dixit, Alicia Pape, Stephen M. Martin, Eugene Joseph*

**(14c) Impact of Conformational and Chemical Correlations on Microphase Segregation in Random Copolymers** ........ 3
*Andrew J. Spakowitz, Shifan Mao, Steve He, Elyse Coletta, Michael Essien, Curtis W. Frank*

**(14d) Sequential Infiltration Synthesis into Native -OH Containing Block Copolymer** ........ 4
*Caleb Breaux, Clifford L. Henderson, Richard A. Lawson, Peter Ludovice*

**(14e) Kinetically-Controlled Dynamics in Block Polymer Micelles** ........ 5
*Thomas H. Epps III, Elizabeth G. Kelley, Ryan P. Murphy, Millicent O. Sullivan*

**(14f) Small Angle Scattering Studies of Nanoparticle Packing into Block Copolymer Micelles** ........ 6
*Gauri M. Nabar, Matthew S. Souva, Barbara E. Wyslouzil, Jessica O. Winter, Randall Winans, Sönke Siefert*

**(14g) Surface-Initiated Polymerization of Ionic Liquid Monomers to Achieve Films with Tunable Properties** ........ 7
*Ian G. Njoroge, Paul A. Kempler, G. Kane Jennings*

**(14h) Synthesis of Anion-Exchange Membranes for Applications Artificial Photosynthesis and in CO2 Capture** ........ 8
*Siwei Liang, Meenesh R. Singh, Daniel J. Miller, Nathaniel A Lynd*

**(14i) Polymerized Ionic Liquid Membranes for CO2 Capture: The Effect of Water Vapor** ........ 9
*Luca Ansaloni, Jacob Nykaza, Matteo Minelli, Yossef A. Elabd, Marco Giacinti Baschetti*

**(14j) Effect of the First Layer on Polyelectrolyte Multilayer Properties and Function** ........ 11
*Amy M. Peterson, Ramiro Magboo*

**(22a) Development of Chitosan/PVA Electrospinning Nanoporous Nanofibers for Improving Wound Healing** ........ 12
*Mian Wang, Thomas J. Webster*

**(22b) Fabrication of Polyaniline/Poly(ethylene oxide)/Graphene Nanofibers Via Electrospinning** ........ 13
*Ali Moayeri, Abdellah Ajji*

**(22c) Wafer-Scale Epitaxial Graphene on SiC: Synthesis, Transfer, and Devices** ........ 14
*Christos Dimitrakopoulos, Jeehwan Kim, Hongsik Park*

**(22d) In-Situ Gisaxs Study of Pore Orientation Effect on Thermal Transformation Mechanism of Surfactant Templated Mesoporous Titania Thin Films** ........ 15
*Saikat Das, Suraj Nagpure, Ravinder Garlapalli, Syed Islam, Qingliu Wu, Joseph Strzalka, Stephen E. Rankin*

**(22e) Scalable Directed Assembly of Electrosprayed Nanoparticles By Durotaxis** ........ 16
*Jonathan P. Singer, Robert W. Style, Hanqiong Hu, Rostislav Boltyanskiy, Eric R. Dufresne, Chinedum O. Osuji*

**(22f) Versatile Surface Functionalization and Micropatterning Enabled By Microcontact Printing of Poly-4-Aminostyrene** ........ 17
*Zhibin Wang, Junfei Xia, Sida Luo, Zhiwei Xiao, Tao Liu, Jingjiao Guan*

**(22g) Sub-10nm Graphoepitaxial Directed Self-Assembly of PS-PDMS By Laser Spike Annealing** ........ 24
*Jing Jiang, Brandon Wenning, Michael Thompson, Christopher K. Ober*

**(22h) Scalable Nanofabrication of Biomimetic Broadband Anti-Reflection Coatings** ........ 25
*Blayne Phillips, Peng Jiang*

**(25a) Selenium Nanoparticles and Nanocomposites As Scaffolds to Alter Cancer Proliferation and the Tumor Microenviroment** ........ 26
*Michelle Stolzoff, Thomas J. Webster*

**(25b) Preparation of a Nano-Patterned Polymer Replica for Reducing Catheter Associated Inflammation and Infection** ........ 28
*Luting Liu, Batur Ercan, Siyu Ni, Linlin Sun, Thomas J. Webster*

**(25d) MgO Nanoparticles Enhance Osteoblast Functions on Hydroxyapatite Nanocomposites for Antibacterial Orthopedic Tissue Engineering Applications** ........ 30
*Daniel J. Hickey, Batur Ercan, Linlin Sun, Thomas J. Webster*

**NSEF Young Investigator Award Presentation: Engineered Hydrogels for Regenerative Medicine Applications** ........ 43
*Ali Khademhosseini*

**NSEF Forum Award Presentation: Nanoscale Organic Hybrid Materials** ........ 44
*Lynden A. Archer*

**(107a) Lipidic Templates and Coatings for Designing Nanotheranostics** ........ 45
*Geoffrey D. Bothun*
**(107b) Invited: Engineering the Protein/Nanomaterial Interface for Biosensors and Biofuel Cells** ........ 46
*Scott A. Banta*
**(107c) Invited: Tailored Virus-like Particles for Biomedicine and Catalysis** ........ 47
*M.G. Finn*
**(133a) Characterization of Graphene, Graphene Oxide and Nanographite Dispersions Using NMR Solvent Relaxation** ........ 48
*David Fairhurst, Terence Cosgrove*
**(133b) Designer Stabilizer for Preparation of Pristine Graphene/Polysiloxane Films and Networks** ........ 58
*Dorsa Parviz, Micah Green, Ziniu Yu, Ronald Hedden*
**(133c) Award Submission - Molecular Order in Graphene-Oxide Suspensions and Graphene-Based Films: What Do We Learn from It?** ........ 59
*Rachel Tkacz, Rudolf Oldenbourg, Shalin B Mehta, Mainak Majumder*
**(133d) Microstructural Variations of Various Commercial Single-Walled Carbon Nanotubes Dispersed in Unsaturated Polyester Resin** ........ 61
*Joyanta Goswami, Virginia A. Davis*
**(133e) Behavior of Polymer – Carbon Nanotube Hybrid Particle in Porous Media** ........ 62
*Minh Vo, Dimitrios V. Papavassiliou*
**(133f) Merging Tertiary Amine with Palladium Complex on Graphene Oxide for Promoted Tsuji-Trost Allylation** ........ 64
*Qingshan Zhao, Yang Li, Guoliang Zhang, Fengbao Zhang, Xiaobin Fan*
**(133g) Novel Organometallic-Nanocarbon Complexes for Electrochemical Applications** ........ 76
*Zhongtao Zhang, Haining Liu, C. Heath Turner*
**(167a) Nanostructured and Dynamic Biomaterials for Controlling the Cell Microenvironment** ........ 77
*April M. Kloxin*
**(167b) DNA-Based Nanostructures As Versatile Delivery Vehicles** ........ 78
*Harry Bermudez*
**(167c) Designer Biomaterial Surfaces for Drug Delivery and Regenerative Medicine** ........ 79
*Anita Shukla*
**(189a) Phonon-Limited Transport Coefficients in Extrinsic Graphene** ........ 80
*Enrique Munoz*
**(189b) Study on the Surface Energy of Supported Graphene By Contact Angle Measurements** ........ 81
*Lei Li, Andrew Kazbial, Zhiting Li, Feng Zhou, Vahid Vahdat, Haitao Liu*
**(189c) A 2D Equation-of-State Model for Corona Phase Molecular Recognition on Single Walled Carbon Nanotube Surfaces** ........ 82
*Zachary Ulissi, Michael S. Strano, Jingqing Zhang, Vishnu Sresht, Daniel Blankschtein*
**(189d) Characterization of Carbon-Based Nanomaterials By Vapor Sorption Methods** ........ 83
*Daniel J. Burnett, Armando R. Garcia, Majid Naderi*
**(189e) Molecular Dynamics Simulations of Natural Gas Transport in Carbon Nano-Pore Structures** ........ 84
*Shuai He, Jacinta C. Conrad, Guan Qin*
**(189f) The Mobility Vs. Carrier Density Trend in Wafer-Scale Epitaxial Graphene on SiC, and How It Can be Defeated** ........ 85
*Christos Dimitrakopoulos, Timothy J. McArdle, Damon Farmer, Alfred Grill, Dirk Pfeiffer, Phaedon Avouris*
**(189g) Molecular Barrier Functions of Graphene Oxide in Liquid-Liquid Systems** ........ 86
*Megan A. Creighton, Finn van Krieken, Robert Hurt*
**(189h) Surface Properties of Carbon Nanotubes Adsorbed at an Air-Water Interface** ........ 87
*Sahil Vora, Brice Bognet, Huseini Patanwala, Francisco Chinesta, Anson W. K. Ma*
**(241b) Award Submission:Development of Peptide-Conjugated Iron Oxide Nanoparticles for Hyperthermia Treatment of Cancer** ........ 88
*Anastasia M. Kruse, Samantha A. Meenach, Kimberly W. Anderson, James Z. Hilt*
**(241c) Award Submission: Viscosity Increase in Concentrated Mab Solutions Due to Large Transient Clusters** ........ 89
*P. Douglas Godfrin, Peter Falus, Lionel Porcar, Jonathan Zarzar, Isidro (Dan) Zarraga, Norman J. Wagner, Yun Liu*
**(241d) Award Submission: Electronic Platform to Quantify Cellular Mechanisms Associated with Carbon Nanotubes Exposure in Real-Time** ........ 90
*Reem Eldawud, Chenbo Dong, Yon Rojanasakul, Cerasela Zoica Dinu*
**(241e) Award Submission:Poly(Lactide-co-Glycolide) Iron Oxide Microspheres for MRI-Monitored Delivery of Sorafenib to Liver Tumors in a Rabbit VX2 Model** ........ 91
*Jeane Chen*

**(241f) Award Submission: Identification of Kinase Inhibitors for Enhanced Polymer-Mediated Transgene Expression** ..... 92
*Matthew Christensen, Jacob Elmer, Seron Eaton, Laura Gonzalez, Joshua LaBaer, Kaushal Rege*

**(268a) Detecting Temperature-Dependent Desorption with Carbon Nanotube Cluster Coated Microsphere Sensors** ..... 93
*Maria V. Chistiakova, Andrea M. Armani*

**(268b) An Integrated Ionic Logic Circuit for Addressing/Reading Multiplex Sensor Platform** ..... 95
*Gongchen Sun, Zdenek Slouka, Hsueh-Chia Chang*

**(268c) Self-Referenced Detection of Cu2+ Using a Crossbar Array of Protein-Stabilized Gold Nanocluster and a Thermoplastic** ..... 96
*Junfei Xia, Zhibin Wang, Danting Huang, Peipei Zhang, Brett Kirkland, Jingjiao Guan*

**(268d) Low-Power Microflow Sensor Using Thermal Capacitance** ..... 103
*Wen-Chi Lin, Mark A. Burns*

**(268e) Poly (ethylene-glycol) Based Hydrogels for on-Chip Microsensors** ..... 104
*Paige Fischer, Bingzhao Xia, John Oakey*

**(268f) A Microfluidic Technique for Measuring the Wax Appearance Temperature (WAT) of Crude Oils** ..... 105
*Shahnawaz Molla, Farshid Mostowfi*

**(268g) Molecular Sensing at Graphene Grain Boundaries** ..... 106
*Poya Yasaei, Bijandra Kumar, Artem Baskin, Nikita Repnin, Petr Král, Amin Salehi-Khojin*

**(268h) Pt Nanowire Array Coated with Au Nanoparticles As a Novel Electrochemical Sensor** ..... 107
*Zhiyang Li, Fan Gao, Zhiyong Gu*

**(268i) Functionalization of Carbon Nanotube Thin-Film Transistors Fabricated By Material Recognition Property of Peptide Aptamer** ..... 108
*Yuichiro Shimada, Akane Hosokawa, Daisuke Kobayashi, Atsushi Shono, Katsuto Otake*

**(268j) Sequence-Specific DNA Detection at 10 Fm By Electromechanical Signal Transduction** ..... 109
*Leyla Esfandiari, Jacob Schmidt, Harold G. Monbouquette*

**(273a) Atomic Layer Deposited Highly Dispersed Platinum Nanoparticles Supported on Multiwalled Carbon Nanotubes for Liquid Phase Reactions** ..... 110
*Xinhua Liang, Chengjun Jiang*

**(273b) Temporal and Spatial Reactors for Atomic Layer Deposition on Particles** ..... 111
*J. Ruud van Ommen, Fabio Grillo, David Valdesueiro, Dirkjan Kooijman, Mojgan Talebi, Aristeidis Goulas*

**(273c) Metallic Co ALD Catalysts for Fuels from Syngas** ..... 113
*Staci A. Van Norman, J. Ruud van Ommen, John L. Falconer, Alan W. Weimer*

**(273d) Chemical Vapour Impregnation- a Versatile Route to High Redox Activity Nanoparticle Catalysts** ..... 114
*Michael Forde, Graham J. Hutchings, Chris K. Kiely, Qian He, Nikolaos Dimitratos, Ren Su, Lokesh Kesavan*

**(273e) Preparation of Carbon Supported Pt Nanoparticles By Atomic Layer Deposition** ..... 127
*Alia M. Lubers, Christopher L. Muhich, Kelly M. Anderson, Alan W. Weimer*

**(273f) Particle Size and Support Effect of Platinum Catalyst in a Proton Exchange Membrane Fuel Cell** ..... 128
*Shibely Saha, Dongmei (Katie) Li*

**(273g) Carboxyl Multi-Walled Carbon Nanotubes Stabilized Palladium Nanocatalysts for Fuel Cell Application** ..... 130
*Zhanhu Guo, Yiran Wang, Qingliang He, Jiang Guo, Huige Wei, Xingru Yan, Suying Wei*

**(290a) Impact of Support, Nanoparticle Size, and Carbon Precursor on the Nanotube Growth Mechanism: Viability of Controlling Nanotube Structure** ..... 131
*Diego A. Gomez Gualdron, Jenni M. Beetge, Juan C. Burgos, Perla B. Balbuena*

**(290b) Role and Stability of a Metal Carbide Catalyst in the Growth of Single-Walled Carbon Nanotubes** ..... 132
*Jose L. Gomez-Ballesteros, Pin Ann Lin, Matthieu Picher, Renu Sharma, Perla B. Balbuena*

**(290c) Diffusion Boundary Layer in the Growth of Centimeter-Tall Carbon Nanotube Forests** ..... 133
*Jaegeun Lee, Eugene Oh, Kun-Hong Lee*

**(290d) Architectural Design of Hollow Mesoporous Composite Nanotubes As Robust Catalysts for Benzyl Alcohol Oxidation** ..... 135
*Guowu Zhan, Hua Chun Zeng*

**(290e) Synthesis of Carbon Nanofibers from Light Liquid Paraffins Using the Liquid Pulse Injection Technique** ..... 136
*Shin R. Mukai, Ryoto Hirahashi, Yuusuke Rikima, Riku Furukawa, Shinichiro Iwamura, Isao Ogino*

**(290f) Stabilizing Effect of Sulfur Incorporation on Multi- and Single-Wall Carbon Nanotubes** ..... 138
*Magdalena M. Majewska*

**(290g) Synthesis of Holey Graphene Via Gold Nanoparticle-Catalyzed Hydroxyl Radical Attack** ........ 139
*James G. Radich, Prashant V. Kamat*
**(290h) Synthesizing Few Atom Thick Boron Based Nanostructures Isostructural to Chemically Modified Graphene Sheets** ........ 140
*Saroj Kumar Das, Amita Bedar, Asha Liza James, Kabeer Jasuja*
**(309b) Award Submission - Molecular Order in Graphene-Oxide Suspensions and Graphene-Based Films: What Do We Learn from It?** ........ 141
*Rachel Tkacz, Rudolf Oldenbourg, Shalin B Mehta, Mainak Majumder*
**(309c) Award Submission: Molecular Barrier Functions of Graphene Oxide in Liquid-Liquid Systems** ........ 143
*Megan A. Creighton, Finn van Krieken, Robert H. Hurt*
**(309d) Study on the Surface Energy of Supported Graphene By Contact Angle Measurements** ........ 144
*Lei Li, Andrew Kazbial, Zhiting Li, Feng Zhou, Vahid Vahdat*
**(309e) Strengthened Magnetic Epoxy Nanocomposites with Protruding Nanoparticles on the Graphene Nanosheets** ........ 145
*Xi Zhang, Ouassima Alloul, Suying Wei, Zhanhu Guo*
**(328a) A Model for Nanoscale Friction on Few Layer Graphene** ........ 146
*Sesha Hari Vemuri, Pil Seung Chung, Myung S. Jhon, Robert Smith*
**(328b) Crumpling and Unfolding Behavior of Spray Dried Pristine Graphene Sheets** ........ 147
*Micah Green, Dorsa Parviz, Sriya Das, Fahmida Irin*
**(328c) Quantitative Theory of Adsorptive Separation for the Electronic Sorting of Single-Walled Carbon Nanotubes** ........ 148
*Rishabh Jain, Kevin Tvrdy, Rebecca Han, Zachary Ulissi, Michael S. Strano*
**(328d) Evaluation of Critical Parameters in the Separation of Single-Wall Carbon Nanotubes through Selective Adsorption Onto Hydrogels** ........ 149
*Justin G. Clar, Tianyu Yuan, Yang Zhao, Jean-Claude J. Bonzongo, Kirk J. Ziegler*
**(328e) Two Phase Extraction Separation of Single-Wall Carbon Nanotubes** ........ 150
*Jeffrey A. Fagan, Erik Haroz, Ming Zheng*
**(328f) DNA-Controlled Carbon Nanotube Purification in Polymer Aqueous Two-Phase Systems** ........ 151
*Geyou Ao, Constantine Khripin, Ming Zheng*
**(328g) Separation Mechanism of Carbon Nanotubes Via Aqueous Two-Phase Method: Towards Simple Isolation** ........ 152
*Juan G. Duque, Navaneetha Subbaiyan, Sofie Cambre, Erik Haroz, Stephen Doorn, Nicholas Parra-Vasquez*
**(341a) Lifetime and Catalytic Reactivity of Fe-Ni Nanoparticles** ........ 153
*Lauren F. Greenlee, Nikki S. Rentz*
**(341b) Controlled Deposition of Iron Oxide Nanoparticles Using a Novel Gas-Expanded Liquid (GXL) Process to Generate Supported Fischer-Tropsch Catalysts** ........ 154
*Pranav S. Vengsarkar, Rui Xu, Christopher B. Roberts*
**(341c) Controlling Metal Nanoparticles in Fischer Tropsch Synthesis** ........ 155
*Freek Kapteijn, Jorge Gascon*
**(341d) Coke-Resistant Core-Shell Catalysts for Dry Reforming of Methane** ........ 156
*Junshe Zhang, Fanxing Li*
**(341e) Reactions Catalyzed By Metal Particles Supported on Functionalized Carbon Nanotubes in Biphasic Systems** ........ 157
*Nicholas Briggs, Javen Weston, Zheng Zhao, Deepika Venkataramani, Clint P. Aichele, Jeffrey Harwell, Daniel Resasco, Steven Crossley*
**(341f) Solid State Chemistry Mass Production of Platinum Group Metal Catalysts with Tailored Particle Morphology** ........ 158
*Zhenmeng Peng, Changlin Zhang, Sang Youp Hwang*
**(341g) Optimization of Green Synthesis of Gold Nanoparticles Using Delonix Regia Leaf Extract and Evaluation of Their Catalytic Activity** ........ 159
*Mausumi Mukhopadhyay, Preeti Dauthal*
**(383a) Elastic Properties and Fracture Mechanics of Graphene Nanomeshes** ........ 164
*Dimitrios Maroudas, Corinne Carpenter, Spencer Wyant, Lin Hu, Andre R. Muniz, Ashwin Ramasubramaniam*
**(383b) Mechanical Behavior of Interlayer-Bonded Nanostructures Obtained from Bilayer Graphene** ........ 165
*Andre R. Muniz, Alyne Machado, Dimitrios Maroudas*
**(383c) Award Submission: Encapsulation of Particle Ensembles in Graphene Nanosacks As a New Route to Multifunctional Materials** ........ 166
*Yantao Chen*
**(383d) Thermal Properties of Graphene Nanomeshes** ........ 167
*Lin Hu, Dimitrios Maroudas*
**(383e) Building a 3D Graphene/Graphene Oxide Matrix for Cooperative Catalysis** ........ 168
*DelRae Haag, Harold H. Kung*

**(383f) Carbon Black-Derived Graphene Quantum Dots As an Efficient and a Stable Catalyst for I3-/I- Redox Couple** ..... 169
*Chun-Chieh Wang, Shih-Yuan Lu*

**(383g) Strengthened Magnetic Epoxy Nanocomposites with Protruding Nanoparticles on the Graphene Nanosheets** ..... 170
*Xi Zhang, Ouassima Alloul, Suying Wei, John Zhanhu Guo*

**(383h) High Electrical Conductive Ethylene 1-Octene Copolymers Reinforced with Low Loading Carbon Nanotubes** ..... 171
*Xingru Yan, Xi Zhang, Huige Wei, Qingliang He, Suying Wei, Zhanhu Guo*

**(391a) Development and Characterization of β -Dicalcium Silicate Nanoparticles for Greener Cementitious Materials** ..... 172
*Scarlett Widgeon, Mariané Silva de Miranda, Rahul P. Sangodkar, Elizabeth Cisneros, Flávio A. Rodrigues, Bradley F. Chmelka*

**(391b) Multiscale Modeling for Nanoclearcoat Curing** ..... 173
*Hao Song, Jie Xiao, Yinlun Huang*

**(391c) Continuous Sizing and Characterization of Suspension-Based Nanomaterials** ..... 174
*Fanxu Meng, Victor M. Ugaz*

**(391e) Novel Exposure and Toxicological Methods to Define the End-of-Life Environmental Health and Safety Implications of Nano-Enabled Products** ..... 175
*Georgios A. Sotiriou, Dilpreet Singh, Fang Zhang, Wendel Wohlleben, Philip Demokritou*

**(404a) Self-Assembly of Aqueous Surfactants on Chemical and Morphological Heterogeneous Surfaces** ..... 176
*Manaswee Suttipong, Brian P. Grady, Alberto Striolo*

**(404b) Folding By Design** ..... 177
*Paul Dodd, Pablo F. Damasceno, Sharon C. Glotzer*

**(404c) Chiral Templating of Self-Assembling Nanostructures By Circularly Polarized Light** ..... 178
*Jihyeon Yeom*

**(404d) A Multiresolution Approach to Optimally Control the Dynamic Directed Self-Assembly of Nanostructures** ..... 179
*Sivaraman Ramaswamy, Paul I. Barton, George Stephanopoulos*

**(404e) Multiscale Modeling Approach to Determine the Role of Amphiphilic Building Block in the Stability of Paclitaxel Drug Delivery Nanocarriers** ..... 180
*Shikha Nangia, Wenjuan Jiang*

**(404f) Predicting Process Windows for Block Copolymer Directed Self-Assembly with Chemoepitaxial Guiding Underlayers with Pattern Density Multiplication** ..... 181
*Benjamin Nation, Andrew J. Peters, Richard A. Lawson, Peter J. Ludovice, Clifford L. Henderson*

**(404g) Worm-like Micelles of Polymerizable Surfactant As a Template for Polymerization** ..... 182
*Simonetta Rima, Marco Lattuada*

**(404h) Simultaneous Synthesis and Assembly of Noble Metal Nanoclusters with Variable Micellar Templates** ..... 183
*Yao Zhou*

**(420a) Nanocrystals and Nanostructures for Study of Catalysis** ..... 184
*Yijin Kang, Christopher B. Murray, Peidong Yang, Nenad Markovic, Vojislav Stamenkovic*

**(420aa) Scalable Process Control of Highly Luminescent and Environmentally Friendly ZnS Quantum Dot Synthesis By Anaerobic Bacteria in Aqueous Solutions** ..... 185
*Gyoung gug Jang, Ilia N. Ivanov, Harry M. Meyer III, Michelle K Kidder, David E. Graham, Ji-Won Moon*

**(420ab) Real-Time Analysis of the Cellular Mechanisms Associated with Carbon Nanotubes Exposure** ..... 186
*Reem Eldawud, Chenbo Dong, Yon Rojanasakul, Cerasela Zoica Dinu*

**(420ac) Thermal Management in MEMS/Nems** ..... 187
*Sesha Hari Vemuri, Myung S. Jhon*

**(420ad) Gold Nanoparticle Interactions with Model Biological Membranes** ..... 188
*Christina Bailey, Elaheh Kamaloo, Kellie Waterman, Kathleen Wang, Terri A. Camesano*

**(420ae) Preparation of High Performance Graphene Fibers By Wet-Spinning Method** ..... 189
*Chunyan Zhao, Songzhao Tong, Shenbin Mo, Jian Wang, Tianju Fan, Wei Tang, Chunqiu Yuan, Yidong Liu, Yong G. Min*

**(420af) Accelerated Generation of Free Radicals By Iron Oxide Nanoparticles in the Presence of an Alternating Magnetic Field** ..... 190
*Robert J. Wydra, Kimberly W. Anderson, Thomas D. Dziubla, J. Zach Hilt*

**(420ag) Mechanism of Nanoparticle Assembly in Polymer Thin Films from Quantitative SEM Image Analysis** ..... 191
*Chaitanya Murthy, Gaurav Arya*

**(420ah) Catalyst-Free Growth of Ultra-High Density ZnO Nanowires for Dye-Sensitized Solar Cells** ..................... 192
*Cheng Xu, Yang Zhao, Luping Li, Kirk J. Ziegler*

**(420ai) Steps Towards Achieving Uniform Electrodeposition in Nanoporous Aluminum Oxide Templates** ..................... 193
*Justin C. Wong, Kirk J. Ziegler*

**(420aj) Self-Assembly of Amphiphilic Peptide (AF)6H5K15: Roles of Hydrophobic and Hydrophilic Residues** ..................... 194
*Naresh Thota, Jianwen Jiang*

**(420ak) Self Assembly in of Soft Particle Fluids, and Hard-Soft Particle Mixtures** ..................... 195
*Mukta Tripathy*

**(420al) Different Ways of Looking at the Force Between Nano Particles** ..................... 196
*Alexander Lange, Gernot Bauer, Fabian Danecker, Joachim Gross*

**(420am) Preparation and Characterization of Curcurmin Loaded Polymeric Nanoparticles By Rapid Precipitation Method** ..................... 197
*Maulick Chopra, Sonal Mazumder*

**(420an) Rheological Study of Order to Disorder Transition (ODT) in Microphase Separated Block Copolymer Surfactant-Functional Small Molecule Additive Complexes** ..................... 198
*Rohit Kothari, H. Henning Winter, James J. Watkins*

**(420ao) Analysis of Lipid Transfer Rates of Phospholipid Nanodiscs (Bicelles) Using Time-Resolved Differential Scanning Calorimetry and Small Angle Neutron Scattering** ..................... 205
*Yan Xia, Kamil Charubin, Ying Liu, Ming Li, Frederick A. Herberle, Drew Marquardt, John Katsaras, Mu-Ping Nieh*

**(420ap) Mesoporous Magnetic Nano-Assemblies for Thermo-Chemotherapy** ..................... 206
*Sunil Kumar*

**(420aq) Structure and Transport Property of Sodium Dodecyl Surfate Micelle in Water Phase: Molecular Dynamics Simulation Approach** ..................... 207
*ByeongJae (Ben) Chun, Seung Soon Jang*

**(420ar) Linking the Electronic Character of Single-Wall Carbon Nanotubes to Their Biological Impacts Using a Model Aquatic Organism** ..................... 208
*Justin G. Clar, Jean-Claude J. Bonzongo, Kirk J. Ziegler, Carlos Silvera-Batista, Sejin Youn, Sarah Gustitus*

**(420au) Rheology of Nanodiamond Particles Aligned in DC Electric Fields** ..................... 209
*Carl McIntyre, Ericson Pereira*

**(420av) Structure, Dynamics and Rheology of Surfactant Micelle and Micelle-Nanoparticle Solutions from Molecular Dynamics Simulations** ..................... 210
*Abhinanden Sambasivam, Subas Dhakal, Radhakrishna Sureshkumar*

**(420aw) Equilibrium and Non-Equilibrium Self-Assembly of Nanostructured Materials** ..................... 211
*Bartosz A. Grzybowski*

**(420b) Synthesis FeCo Nanoparticles By NaBH4 Reduction in Oleylamine** ..................... 212
*Xia Xu, Jihoon Park, Yang-Ki Hong, Alan Lane*

**(420c) Water-Soluble RADA16-I Peptide Species within RADA16-I Designer Nanofiber Hydrogels** ..................... 213
*Benjamin C. Hudson, Anant K. Paravastu*

**(420d) The Impacts of pH and Tryptophan Concentration on Controllable Synthesis of Gold Nanoparticles** ..................... 214
*Leilei Dai, Fang Cai, Hao Lv, Jinsheng Sun, Jie He, Xijia Cao*

**(420f) Purification of DNA-Wrapped Carbon Nanotubes in Polymer Aqueous Two-Phase Systems** ..................... 228
*Geyou Ao, Ming Zheng*

**(420g) A Biofabrication Approach for Controlled Synthesis of Silver Nanoparticles with High Catalytic and Antibacterial Activities** ..................... 229
*Cuixian Yang, Sukwon Jung, Hyunmin Yi*

**(420h) Low Temperature Catalysts for Direct Remediation of H2S Malodor in Air** ..................... 230
*Kei Bo Cheung, Hao Chen, Wei Han, King Lun Yeung*

**(420i) Block Copolymer Templating of Hydrothermally-Grown ZnO Nanorod Arrays with Controlled Density for Photovoltaic Devices** ..................... 231
*Candice Pelligra, Su Huang, Noga Kornblum, Chinedum Osuji*

**(420j) Designing Composite Polymer Electrolyte Interfaces for Stable Electrodes** ..................... 232
*Guang Yang, Daniel T. Hallinan Jr.*

**(420k) Experimental Characterization and Multiphysics Modeling of Surface Tethered, Environmentally Responsive Poly(acrylic acid) Brushes on Thin Film Au and Cellulose Substrates** ..................... 233
*Vanessa Wood Braband, Steven Schneiderman, Navaraj Gurung, Suvarna Talluri, Todd Menkhaus*

**(420m) Triboelectric Nanogenerators from Carbon Nanotubes** ..................... 234
*Moses Oguntoye, Lawrence R. Pratt, Noshir S. Pesika*

**(420n) Fabrication of Graphene Nanofiber from Graphene Oxide By Ice Template** ..... 235
*Tianju Fan, Wei Tang, Chunqiu Yuan, Songzhao Tong, Shenbin Mo, Chunyan Zhao, Jian Wang, Yidong Liu, Yong Min*

**(420o) Preparation of Large Area Transparent Conductive Graphene Film Utilizing Langmuir-Blodgett Method** ..... 236
*Tianju Fan, Chunqiu Yuan, Wei Tang, Songzhao Tong, Yidong Liu, Yong Min, Arthur Epstein*

**(420p) Effectiveness and Toxicity of Magnetic Micelles for Use As Cancer Therapy Agents** ..... 237
*Rhythm R. Shah, Christopher S. Brazel*

**(420q) MODEL for the Synthesis of and Macrophage Responses to Nanoscale Hydroxyapatite** ..... 238
*Garima Bhardwaj, Thomas J. Webster, Arthur Gonzales*

**(420r) The Role of Structure and Surface Composition on the Contact Angle and Wetting Characteristics of Carbon Nanofiber Coatings** ..... 241
*Akanksha Visi, Aditya Sharma, Sonal Mazumder*

**(420s) The Effects of Synthesis Method on the Physical and Chemical Properties of Iron Oxide Nanoparticles** ..... 242
*Anastasia M. Kruse, Ronita Mathias, Kimberly W. Anderson, James Z. Hilt*

**(420t) Porous Carbon Microsphere-Encapsulated ZnO-Coo Nanoparticles Anode for High-Performance Lithium-Ion Batteries** ..... 243
*Lianjun Liu, Cunyu Zhao, Huilei Zhao, Qianyi Zhang, Ying Li*

**(420u) Characterization of Complex Solution-Nanotube Interfaces Using AUC and SANS** ..... 244
*Stephanie Lam, Jeffrey A. Fagan*

**(420v) Scalable Synthesis Methods to Produce Janus Nanoparticles** ..... 245
*Alan Hanley, Allan David*

**(420w) Evaluating the Potential of Dynamic Magnetic Susceptometry for Ricin B Detection** ..... 246
*Lorena Maldonado-Camargo, Srinand Sreevatsan, Carlos Rinaldi*

**(420x) Magnetic Nanoparticles for Non-Invasive Quantification of Prostate Cancer Aggressiveness** ..... 247
*Tareq Anani, Allan David, Young S. Choi*

**(420y) Simulations of Phenol-Capped Alkanethiol-Coated Gold Nanoparticles in Organic Solvents** ..... 248
*Albert L. Kwansa, Danielle S. Stallings, Yaroslava G. Yingling*

**(420z) Catalytic Conversion of Methane to Methanol on Metal Phthalocyanine Functionalized Graphene** ..... 249
*Sierra Headrick, Pabitra Choudhury*

**(449a) Poly(Lactide-co-Glycolide) Iron Oxide Microspheres for MRI-Monitored Delivery of Sorafenib to Liver Tumors in a Rabbit VX2 Model** ..... 250
*Jeane Chen, Sarah B. White, Kathleen R. Harris, Woon Teck Yap, Robert J. Lewandowski, Lonnie D. Shea, Andrew C. Larson*

**(449b) Controlling Ferrofluids Permeabilities to Cross an in Vitro Blood-Brain Barrier Model** ..... 251
*Di Shi, Linlin Sun, Gujie Mi, Soumya Bhattacharya, Suprabha Nayar, Thomas J. Webster*

**(449c) Development of Peptide-Conjugated Iron Oxide Nanoparticles for Hyperthermia Treatment of Cancer** ..... 254
*Anastasia M. Kruse, Samantha A. Meenach, Kimberly W. Anderson, James Z. Hilt*

**(449d) Preparation, Characterization and Properties of Janus Magnetic Liposomes** ..... 255
*Christophe Monnier, Marco Lattuada, Alke Fink*

**(449e) In Situ Evaluation of Nanoparticle–Protein Interactions By Dynamic Magnetic Susceptibility Measurements** ..... 268
*Ana Bohorquez, Carlos Rinaldi*

**(449f) Synthesis of a Polymer-Magnetic Particle Platform for "Tailored" Multimodal Materials for Imaging and Treatment** ..... 269
*O. Thompson Mefford*

**(455a) Nanostructured Size-Selective Catalysts Prepared By Molecular Layer Deposition** ..... 270
*Xinhua Liang, Zeyu Shang, Rajankumar Patel*

**(455b) Platinum Nanoparticles Supported on Multiwalled Carbon Nanotubes Encapsulated By Cerium Oxide: A Highly Reactive and Stable Catalytic System** ..... 271
*Chao Wan, Dangguo Cheng, Fengqiu Chen, Xiaoli Zhan*

**(455c) Using Active Site Confinement and Organic Ligand Approaches to Control Catalysis** ..... 279
*Michael M. Nigra, Alexander Katz, Marc-Olivier Coppens*

**(455d) Sinter-Resistant Metal Nanoparticles Encapsulated By Zeolite Nanoshell** ..... 280
*Shilpi Saxena*

**(455e) Controlling Metal Nanoparticle Dispersion with 'Nanocavity' Oxide Supports** ..... 282
*Zhenyu Bo, Cassie George, Gallagher James R., Jeffrey T. Miller, Peter C. Stair, Justin M. Notestein*

**(455f) Single-Walled Copper Silicate Nanotubes Doped with 3d Transition Metals for CO2 Hydrogenation** ........ 283
*Yuan Sheng, Hua Chun Zeng*
**(455g) Computational Design of Highly Selective Transition Metal Catalysts Encapsulated By Metal-Organic Frameworks for Butane Oxidation to 1-Butanol** ........ 284
*Sean T. Dix, Rachel B. Getman*
**(457a) Semiconductor Nanowire Fabric** ........ 285
*Brian A. Korgel*
**(457b) Piezotronics and Piezo-Phototronics in Semiconductor Nanowires for Active Electronics and Optoelectronics** ........ 286
*Zhong Lin (ZL.) Wang, Wenzhuo Wu*
**(457c) The Hydrothermal Synthesis of Palladium Nanorods on the Tobacco Mosaic Virus: A Study of Its Kinetic Mechanism and Nanorod Properties** ........ 287
*Oluwamayowa Adigun, Alexander Freer, Michael T. Harris*
**(457d) Kinetic Control of Catalyst-Free Growth of Ultra-High Density ZnO Nanowires** ........ 288
*Cheng Xu, Yang Zhao, Luping Li, Kirk J. Ziegler*
**(457e) Interplay Between Defect Propagation and Surface Hydrogen in Si Nanowire Kinking Superstructures** ........ 289
*Nae Chul Shin, Miaofang Chi, Michael A. Filler*
**(457f) Controlled Chirality in Nanoscale Semi-Conducting CdTe Nanowires** ........ 290
*Ryan L. Marson, Greg van Anders, Yunlong Zhou, Nicholas Kotov, Sharon C. Glotzer*
**(457g) Solid-Liquid-Vapor Anisotropic Etching of Semiconductor Nanowires** ........ 291
*Ho Yee Hui, Ildar Musin, Michael A. Filler*
**(457h) Characterization and Effects of Barrier Layer Thinning and Pore Opening in Porous Aluminum Oxide for Nanowire Growth** ........ 292
*Justin C. Wong, Kirk J. Ziegler*
**(503a) Optimizing Heat Generation in Magnetic Iron Oxide Nanoparticles for Use in Cancer Hyperthermia** ........ 293
*Rhythm R. Shah, Todd P. Davis, David E. Nikles, Christopher S. Brazel*
**(503b) Magnetic Nanoparticles for Measuring Prostate Cancer Aggressiveness** ........ 294
*Tareq Anani, Allan David, Young S. Choi*
**(503c) Controlled, Wireless Heating of Iron Oxide Nanoparticle Composites** ........ 295
*Joel Coffel, Eric Nuxoll*
**(503d) Shape-Controlled Iron Oxide Nanoparticles Fpr Bioimaging** ........ 296
*Yuping Bao, Jennifer Sherwood, Yaolin Yaolin, Thomas Macher, Ming Lee*
**(503e) Preparation and Characterization of Magnetic Gene Transfection Agents Consisting of Polyethyleneimine Coated Iron Oxide Nanoparticles** ........ 297
*Melissa Cruz-Acuña, Lorena Maldonado-Camargo, Jon Dobson, Carlos Rinaldi*
**(503f) Biocompatible Iron Oxide Nanoparticles Obtained Via the Non-Aqueous Synthesis in Hydrophilic Solvents for High Gradient Magnetic Protein Purifications** ........ 298
*Ingke-C. Masthoff, Mathias Kraken, Johannes Gädke, Rainer Krull, Jochen Litterst, Georg Garnweitner*
**(513a) Synthesis of p-Type ZnO Nanorods on an InP Substrate By Pulse Laser Deposition** ........ 300
*Yu Dongqi*
**(513b) Molecular Synthesis and Characterisation of Nanoparticles Using IR Spectrometry** ........ 307
*Muhammad Nuru Idris Sr.*
**(513c) Preparation of Drug Nanoparticles Stabilized By Polysaccharide Surfactant through Four-Inlet Vortex Mixer** ........ 329
*Mingwei Wang, Zhenyu Yuan, Jie Wang, Li Li, Xuhong Guo*
**(513d) Scale-up of Nanoparticle Manufacturing By Flame Spray Pyrolysis** ........ 330
*Karsten Wegner, Arto J. Gröhn, Sotiris E. Pratsinis*
**(513e) Numerical Analisys of Multicomponent Catalysts Production By Double Flame Spray Pyrolysis (DFSP)** ........ 331
*Dirceu Noriler, Maximilian J Hodapp, Florian Meierhofer, Henry F. Meier, Udo Fritsching*
**(582a) A Coarse Grain Model for Protein-Nanoparticle Interactions** ........ 332
*Shuai Wei, Charles L. Brooks III*
**(582b) Molecular Dynamics Simulations of Peptide-SWCNT Interactions: Role of Curvature** ........ 333
*Olukayode Karunwi, Saptarshi Chakraborty, Anthony Guiseppi-Elie*
**(582c) Enhanced Activity and Stability of Organophosphorus Hydrolase Via Interaction with an Amphiphilic Polymer** ........ 334
*Minkyu Kim, Manos Gkikas, Aaron Huang, Jeon Woong Kang, Nisaraporn Suthiwangcharoen, Ramanathan Nagarajan, Bradley D. Olsen*

**(582d) Enabling Multi-Enzyme Biocatalysis Using Coaxial Electrospun Hollow Nanofibers: Design and Applications** ........ 335
*Songping Zhang, Xiaoyuan Ji, Ping Wang, Zhiguo Su, Guanghui Ma*
**(582e) Confinement Protection Effects of Mesoporous Silica, Inspired By Chaperonin Complexes** ........ 336
*Michele Lynch, Marc-Olivier Coppens*
**(582f) A Combinatorial Strategy Towards a Better Sugar-Powered Enzymatic Fuel Cell** ........ 337
*Zhiguang Zhu, Y.-H. Percival Zhang*
**(583a) Invited: Packing and Assembly of Particles: Nanowires and Beyond** ........ 338
*Sharon C. Glotzer*
**(583b) Forming Aligned Micellar Nanowires By De-Wetting on Micro-Pillars** ........ 339
*Pouyan E. Boukany*
**(583c) The New Liquid Crystals: Self-Assembly and Processing of Nanocylinder Dispersions** ........ 340
*Virginia Davis*
**(583d) Flow-Guided Template Manufacturing of Nanowire-Array Sensors** ........ 341
*Yingbo Zu, Juan Chen, kartik Kumar Rajagopalan, Shengnian Wang*
**(583e) Synthesis and Characterization of Silver Nanowire for Printable Conductive Media** ........ 342
*Shohreh Hemmati, Dale P. Barkey, Xiaowei Teng, Ryan Banfield*
**(583f) In Situ Formation of Silver Nanowire Networks on Paper and Plastic Substrates Using a Desktop Inkjet Printer** ........ 343
*Pushkaraj Joshi, Shravan Kumar Parmar, Venugopal Santhanam*
**(583g) Phase Transformation As a Strategy for the Synthesis and Assembly Via Welding of Single-Crystalline Metal Silicide Nanowires** ........ 344
*Sreeram Vaddiraju, Yongmin Kang, Venkata Vasiraju*
**(583h) Nanofiber-Based Dye Sensitized Solar Cells with a Novel Nano-Particle Scattering Layer** ........ 345
*Lijun Yang, Wallace Woon-Fong Leung*
**(618a) Hydrophobically Modified Chitosan Coated Liposomes for Gene Delivery** ........ 346
*Jaspreet S. Arora, Dibyadyuti Datta, Nirbhay Kumar, Vijay T. John*
**(618b) Local Sustained Knockdown for Accelerated Diabetic Wound Healing** ........ 347
*Steven Castleberry, Wei Li, Benjamin D. Almquist, Paula T. Hammond*
**(618c) Identification of Kinase Inhibitors for Enhanced Polymer-Mediated Transgene Expression** ........ 348
*Matthew Christensen, Jacob Elmer, Seron Eaton, Laura Gonzalez, Joshua LaBaer, Kaushal Rege*
**(618d) Protease-Activatable Viruses Based on Adeno-Associated Virus for Cardiovascular Disease Therapy** ........ 349
*Michelle L. Ho, Michael Lam, Momona Yamagami, Caitlin Guenther, Junghae Suh*
**(618e) Gold Nanoparticle Electroporation Enhanced Polyplex Delivery** ........ 350
*Shuyan Huang, Yingbo Zu, kartik Kumar Rajagopalan, Shengnian Wang*
**(618f) Near-Infrared–Actuated Devices for Remotely Controlled Drug Delivery** ........ 351
*Brian P. Timko, Daniel S. Kohane*
**(618g) Degradable Lipid Nanoparticles with Predictable in Vivo siRNA Delivery Activity** ........ 352
*Kathryn A. Whitehead, Daniel G. Anderson*
**(618h) Poly (propargyl L-glutamate) Based Linear-Dendritic Block Copolymers As pH-Responsive Hydrophobic Drug Carrier** ........ 353
*Mohiuddin Quadir, Stephen Morton, Lawrence Mensah, Kevin Shopsowitz, Paula T. Hammond*
**(618i) PLGA-Peg-Based pH Sensitive Nanoparticles for Targeted and Controlled Drug Delivery: In Vitro Evaluation** ........ 354
*Zilan Zhou, Apurva Badkas, Joo-Youp Lee*
**(646a) Molecular Bioelectronics: Enzyme-Swnt Conjugates for Bioanalytical Biochips** ........ 355
*Anthony Guiseppi-Elie, Olukayode Karunwi*
**(646b) Binding Between DNA and Carbon Nanotubes Strongly Depends upon Sequence and Chirality** ........ 356
*Akshaya Shankar, Jeetain Mittal, Anand Jagota*
**(646c) Thermodynamics of DNA Strands Encapsulated into Electrically Charged Nanotubes** ........ 357
*Fernando J.A.L. Cruz, José P.B. Mota*
**(646d) Effect of Organic Acids on Nucleation and Growth of Calcium Phosphate Crystals on Glutamic Acid Functionalized Nanofibers** ........ 362
*Danial Barati, Seyedsina Moeinzadeh, Esmaiel Jabbari*
**(646e) Biophysical Characterization of Immobilized, Self-Assembled Phospholipid Bilayer Based Structures in Sol-Gel Derived Silica** ........ 364
*Wade F. Zeno, Silvia L. Hilt, Kannan K. Aravagiri, Subhash H. Risbud, John C. Voss, Atul N. Parikh, Marjorie L. Longo*
**(646f) Computational Modelling of Functionalized Nanoparticles Interacting with Bio-Inspired Membranes and Vesicles** ........ 365
*Fikret Aydin, Michael Sebastiano, Xiaolei Chu, Meenakshi Dutt*

**(646g) Morphological Characterization of Detergent-Mediated Photosystem I (PS I)-Proteoliposome Formation** ........ 366
*Hanieh Niroomand, Bamin Khomami, Dibyendu Mukherjee*
**(646h) Multi-Site Functionalization of Protein Scaffolds for Bimetallic Nanoparticle Templating** ........ 367
*Kelly Huggins, Sarah C. Heilshorn*
**(646i) Nature-Inspired Microfluidic Template As a Substrate for Plasmonic Nanoparticles** ........ 368
*Karthik Pushpavanam, Sanjitarani Santra, Kaushal Rege*
**(674a) Diffusing Colloidal Probes of Cancer Cells** ........ 369
*Gregg Duncan, Michael A. Bevan*
**(674b) Enhancement of Cancer Cellular Uptake By the Morphology of Lipid-Based Nanodiscs** ........ 370
*Wafa Aresh, Ying Liu, Jessica Sine, Derek Thayer, Anu Puri, Yike Huang, Yong Wang, Mu-Ping Nieh*
**(674c) Mitochondrial Targeting with Triphenylphosphonium (TPP)-Conjugated and TPP-Peg-Conjugated PAMAM Dendrimer Nanocarriers** ........ 371
*Elizabeth Bielski, Qian Zhong, Lin Yang, Matthew Brown, Vladimir Cabral, Sandro R.P. da Rocha*
**(674d) Effect of Incorporation of Lysolipid on Dipalmitoylphosphatidylcholine Bilayer Membrane: Study on the Permeability Enhancement of Liposomes at the Phase Transition Using Molecular Dynamics Simulations** ........ 372
*Hyea Hennim Hwang, Jae Hyung Choi, Keewon Lee, Young Kyoung Kim, Seung Soon Jang*
**(674e) Self-Assembled Polymer Nanomedicines for Synergistic and Synthetic Lethal Drugging of Breast and Ovarian Tumors** ........ 373
*Erik Dreaden, Yi Wen Kong, Stephen Morton, Michael Yaffe, Paula T. Hammond*
**(674f) Dendrimer Nanodevices for Pediatric Brain Injury** ........ 374
*Elizabeth Nance, Michael Porambo, Fan Zhang, Manoj Mishra, Zhi Zhang, Michael Johnston, Ali Fatemi, Rangaramanujam Kannan, Sujatha Kannan*
**(674g) Dry Powder Aerosols Based on Mucus-Penetrating Nanocomposite Microparticles for Pulmonary Antibiotics Delivery** ........ 375
*Zimeng Wang*
**(674h) Multi-Modal Gold Nanoparticles for Simultaneous Lineage-Specific Targeted Delivery and Tracking of Chemotherapeutics** ........ 376
*Shunji Egusa, Yogen Saunthararajah*
**(674i) Controlled Uptake of Therapeutic Protein Nanoparticles** ........ 377
*Kevin Ling, Julie A. Champion*
**(674j) Improved Entrapment of N-Acetyl Cysteine (NAC) in Nanoparticles of Poly[Lactic-Co-Glycolic Acid] (PLGA) Prepared By Double Emulsion Method By Using Bovine Serum Albumin (BSA)** ........ 378
*Ruth J. Lancheros Salas, José A. Beleño, Carlos Arturo Guerrero, Rubén D. Godoy*
**(703a) DNA Nanotubes and Nanotapes Formed Via Self-Assembly of ssDNA-Amphiphiles** ........ 379
*Timothy R. Pearce, Efrosini Kokkoli*
**(703b) Targeted Binding of Nanolipoprotein Particles to Phase Separated Lipid Domains** ........ 380
*Wade F. Zeno, Marjorie L. Longo*
**(703c) Self-Assembly of Functionalized Virus Capsids** ........ 381
*Leebyn Chong, Vyshnavi Karra, Meenakshi Dutt*
**(703d) Design of Sterically Stable Bio-Inspired Hybrid Soft Biomaterials Via Computational Modelling** ........ 382
*Fikret Aydin, Geetartha Uppaladadium, Meenakshi Dutt*
**Programming Colloidal Phase Transitions with DNA Strand Displacement** ........ 383
*William Benjamin Rogers, Vinothan Monaharan*
**(703f) Ibuprofen Loading and Release in Micelles Formed By Amphiphilic Peptide (AF)6H5K15: A Coarse-Grained Molecular Dynamics Simulation** ........ 384
*Naresh Thota, Zhongqiao Hu, Jianwen Jiang*
**(732a) Measuring How the Local Glass Transition Temperature Shifts Across a Glassy-Rubbery Polymer-Polymer Interface** ........ 385
*Roman R. Baglay, Connie B. Roth*
**(732b) Tuning the Wall Thickness of Templated Polystyrene Nanotubes Produced from Melt Infiltration** ........ 387
*Anthony Tan, John M. Torkelson*
**(732c) Depletion, Bridging, and Ordering in Bare and Grafted Nanorod-Polymer Mixtures** ........ 389
*Uma K. Shankar, Anirudh Mantri, Mukta Tripathy*
**(732d) Ordered Polymer-Grafted Nanoparticle Networks for Ionic Conductivity** ........ 390
*Pinar Akcora*
**(732e) Preparation of Conductive Polymer Nanofibers with a Coaxial Nozzle Electrospinning Method** ........ 391
*Takashi Uruma, Yuichiro Shimada, Daisuke Kobayashi, Atsushi Shono, Katsuto Otake*

**(732f) Novel Pressure-Responsive Shape Memory Polymers** .......... 392
*Yin Fang, Peng Jiang*

**(732g) Synthesis of Polymer Brush in Oil-Phase through Photo-Emulsion Polymerization** .......... 393
*Xiaochi Liu, Xiaohan Wang, Yisheng Xu, Li Li, Rui Zhang, Jun Xu, Xuhong Guo*

**(732h) Thermodynamic Interactions and Tunable Properties of Thermoplastic Elastomers Derived from Vegetable Oils** .......... 394
*Megan L. Robertson, Shu Wang, Sameer Vajjala Kesava, Enrique D. Gomez*

**(732i) Invited Talk: Using Light to Locally Tune the Properties of Polymers** .......... 395
*Jan Genzer, Ying Liu, Robin Mays, Michael D. Dickey*

**(733a) Neurotransmitter Detection Using Corona Phase Molecular Recognition on Fluorescent Single-Walled Carbon Nanotube Sensors** .......... 396
*Markita Landry, Sebastian Kruss, Emma Vander Ende, Barbara Lima, Nigel F. Reuel, Jingqing Zhang, Justin Nelson, Bin Mu, Andrew J. Hilmer, Michael S. Strano*

**(733b) Photo-Switchable Quantum Dots Based on Reversible FRET and Azobenzene DNA** .......... 397
*Qirui Fan, Jessica O. Winter*

**(733c) Selective Intracellular Labeling Using Microfluidic Electroporation-Delivered Quantum Dots** .......... 398
*Chen Sun, Zhenning Cao, Tao Geng, Chang Lu*

**(733d) Biochemical Ratiometric Sensing By Single Chirality Carbon Nanotubes in Living Tissues** .......... 399
*Juan Pablo Giraldo, Markita Landry, Seonyeong Kwak, Rishabh Jain, Nicole Iverson, Min Hao Wong, Michael S. Strano*

**(733e) Si Nanocrystal Surface Chemistry: Room Temperature Hydrosilylation and Thermally-Induced Thiolation** .......... 400
*Yixuan Yu, Colin Hessel, Timothy Bogart, Matthew Panthani, Michael Rasch, Brian A. Korgel*

**(733f) Detection of a Peanut Allergen, Ara h1, on a Novel, Biodegradable Biosensor Platform Using Surface Enhanced Raman Spectroscopy** .......... 401
*Pervin Gizem Gezer, G Logan Liu, Jozef Kokini*

**(733g) Spatiotemporal Intracellular Nitric Oxide Signaling Captured Using Internalized, Near Infrared Fluorescent Carbon Nanotube Nanosensors** .......... 402
*Zachary Ulissi, Xun Gong, Michael S. Strano, Debabrata Mukhopadhyay, Fatih Sen, Nicole Iverson, Gerald Wogan, Luiz Godoy, Ardemis A. Boghossian, Selda Sen*

**(733h) A Facile Colorimetric Sensor of Ionizing Radiation Using Polypeptide Mediated Nanoparticle Formation** .......... 403
*Karthik Pushpavanam, Kaushal Rege, Stephen Sapareto, John Chang*

**(734a) Self-Assembled Polymeric Nanoparticles Encapsulating Hydrophobic Curcumin to Attenuate Morphine Tolerance** .......... 404
*Magdalena Szymusiak, Hao Shen, Xiaoyu Hu, Zaijie Jim Wang, Ying Liu*

**(734b) Silver Nanoparticle-Embedded Polymersome Nanocarriers for the Treatment of Antibiotic-Resistant Infections** .......... 405
*Benjamin M Geilich, Anne L. van de Ven, Thomas J. Webster*

**(734c) Tumor-Penetrating Nanoparticles Capable of Tandem Delivery of Agents for the Treatment of Lung Cancer** .......... 408
*Qihua Sun, Samantha A. Meenach*

**(734d) Design and Synthesis of Aminoglycoside Antibiotic-Derived Polymeric Micelles for Drug Delivery to Cancer Cells** .......... 416
*Bhavani Miryala, Taraka Sai Pavan Grandhi, Matthew Christensen, Yanqing Tian, Kaushal Rege*

**(734e) Size Effect of Hollow Gold Nanoparticles on Photo-Activated Drug Release of a Liposomal Carrier** .......... 418
*Maria O. Ogunyankin, Joseph A. Zasadzinski, Dmitri Lapotko*

**(734f) Single Lipoplex Fluorescent Nanoparticle Tracking Analysis and Applications for the Study of Exosomes** .......... 419
*Clayton Deighan, Scott Baldwin, Michael Paulaitis, Jeffrey J. Chalmers*

**(734g) Electronic Platform to Quantify Cellular Mechanisms Associated with Carbon Nanotubes Exposure in Real-Time** .......... 420
*Reem Eldawud, Chenbo Dong, Yon Rojanasakul, Cerasela Zoica Dinu*

**(734h) Viscosity Increase in Concentrated Mab Solutions Due to Large Transient Clusters** .......... 421
*P. Douglas Godfrin, Peter Falus, Lionel Porcar, Jonathan Zarzar, Isidro (Dan) Zarraga, Norman J. Wagner, Yun Liu*

**(734i) Protein Confinement in Mesoporous Materials** .......... 422
*Justin Siefker, Marc-Olivier Coppens*

**Author Index**

# 386436 Casting Solvent Effects on Block Ionomer Morphology and Corresponding Property Changes


**Abstract:**

The role of casting solvent has been shown to play an important role in the microstructure of block copolymers that have not been, or are not able to be, annealed. Casting solvent plays a stronger influence in block ionomer ordering because these materials can possess strongly incompatible blocks that are easily templated in various solvent environments. Here, this principle is utilized to produce different morphologies from a sulfonated pentablock ionomer. The judicious selection of polar and nonpolar solvents, and their mixtures, translates to different solution structures that in turn create different solid film morphologies. Transmission electron micrographs and tomograms provide morphological information amongst samples varying in sulfonate concentration and casting solvent(s). The obtained data shows that casting solvent identity does, in fact, alter the ordering of the pentablock copolymer into vastly different morphologies. Following identification, the effect of the produced morphologies on swelling (water uptake), water vapor transmission rate (WVTR), and water-based semi-permanent deformation will be presented. Preliminary results indicate that geometric restriction of the sulfonated, hydrophilic phases reduces water uptake. Additionally, systems that exhibit more long range order have a lower apparent sulfonate-domain percolation threshold that results in maintained, high WVTR at low sulfonate concentration. Lastly, more disordered sulfonate domains translate to more easily deformed microstructures under the stress of liquid water and heat.

## Phase Separation Kinetics in Pressure Sensitive Adhesive Melts Containing Styrene-Diene Based Block Copolymers

**Ninad Dixit, Alicia Pape, Eugene Joseph and Stephen Martin**

*Virginia Polytechnic Institute and State University, Macromolecules and Interfaces Institute, Blacksburg, VA, ninad22@vt.edu.*

Pressure sensitive adhesives are often processed by coating from the melt, and the final adhesive properties depend on the morphology of the resulting film. We report on the phase separation kinetics in melts of commercially available styrene-diene based block copolymers and adhesive blends containing various amounts of tackifier. A modified Percus-Yevick hard sphere model was used to determine morphological parameters from real-time small-angle X-ray scattering (SAXS) analysis. The Avrami kinetics model was used to study ordering of the polystyrene domains based on rheological analysis. The addition of tackifier in the adhesive blends led to a gradual decrease in the extent of ordering of polystyrene domains. The phase separation mechanism also shifted from a 'Nucleation and Growth' to a 'Spinodal Decomposition' type process with increasing amounts of tackifier in the blend. Isoprene chain mobility and tackifier content in the blend were found to be the most dominant factors affecting the ordering kinetics. At higher tackifier content, polystyrene domains remained in a liquid-like disordered state. We also concluded that the concentration fluctuations in the disordered melts drive the initial stage of phase separation process in adhesive blends.

# 379646 Impact of Conformational and Chemical Correlations on Microphase Segregation in Random Copolymers

**Abstract:**

Random copolymers play an important role in a range of soft materials applications and biological phenomena. For example, polymer membranes for fuel cells are frequently composed of random copolymers synthesized with hydrophobic and hydrophilic monomers. Microphase segregation of these monomers results in a solid material whose mechanical integrity arises from a space spanning random arrangement of hydrophobic domains, and ion transport occurs through the corresponding hydrophilic domains. The structure of the microphase segregated domains dictates the pathways for ion transport and ultimately the macroscopic ion mobility and material performance. An individual monomer is typically a single chemical unit whose length is comparable to a Kuhn length, resulting in a monomer segment that is structurally rigid at length scales of a segregated domain. Previous work on random copolymer phase segregation addresses the impact of correlations between the chemical identities along the chains for flexible polymers; thus, a monomer block behaves as a random walk without structural correlation associated with semiflexibility. In our work, we develop a model of semiflexible random copolymers to address the thermodynamics of microphase segregation and the structure of the segregated domains. Using a random phase approximation up to quartic order in density fluctuations, we show the onset of phase segregation and the correlation length of domains is extremely sensitive to chain rigidity. We then leverage particle-field Monte Carlo simulations to address the structure of the domains and discuss the role of density fluctuations in the thermodynamic behavior.

# Sequential Infiltration Synthesis into Native –OH containing Block Copolymer

Caleb Breaux, Richard Lawson, Peter Ludovice, Clifford Henderson

cbreaux6@gatech.edumailto:Ben.Nation@gatech.edu

School of Chemical and Biomolecular Engineering

Georgia Institute of Technology, Atlanta, GA

Microphase separation of block copolymers into structures with characteristic length scales on the order of 10 nm is a promising alternative to conventional lithography. After aligning a block copolymer to a pre-patterned underlayer, one block of the copolymer must be removed, typically through an etch step. However, if the etch resistance between the two blocks are similar, it can be difficult to retain the phase separated pattern. Sequential Infiltration Synthesis (SIS) offers a route by which the etch resistance of one block may be increased by the infiltration of organometallic into one block. Extensive study of SIS has been done regarding PS-b-PMMA due to its ease of phase separation via thermal annealing, and because the PMMA block's carbonyl moieties readily act as sites for SIS of Al using the precursors trimethyl-aluminum (TMA) and water. For the SIS of other metals (i.e. Ti, W, Zn), a pretreatment of one cycle of TMA/Water is needed for the PMMA domains to have the –OH groups for their growth. Here, a high $\chi$-block copolymer (PS-b-PHOST) with a domain that has native –OH groups is used in SIS to form perpendicular and parallel morphologies.

# 387881 Kinetically-Controlled Dynamics in Block Polymer Micelles

**Abstract:**

Amphiphilic block polymers consist of covalently-bonded hydrophobic and hydrophilic polymer segments. Akin to small molecule amphiphiles, these macromolecules self-assemble into a variety of well-defined structures in aqueous solution such as micelles and vesicles. However, the macromolecular hydrophobic block leads to very slow and seemingly nonexistent micelle dynamics. This kinetic entrapment is an attractive advantage of amphiphilic block polymers for applications in drug delivery; however, it also leads to path-dependent solution assembly. Consequently, careful optimization of preparation conditions are necessary to produce well defined, uniform, and reproducible macromolecular solution assemblies. We are interested in understanding the effects of common processing conditions on the structure, dynamics, and long-term stability of block polymer micelles. In particular, we found that organic cosolvent removal led to an unexpected micelle growth, in which the sizes evolved through a distinct bimodal distribution separated by multiple fusion events. These studies revealed that the micelle stability was critically dependent on stirring the solution, which motivated a detailed investigation into the effects of solution agitation on the chain exchange dynamics in polymer micelles. The results presented here demonstrate the influence of kinetically-controlled processes on block polymer micelle dynamics, and emphasize the importance of selecting and controlling processing conditions when preparing these assemblies.

## Small Angle Scattering studies of nanoparticle packing into block copolymer micelles

**Gauri Nabar[1], Mathew Souva[1], Sönke Siefert[4], Randall Winans[4], Jessica Winter[1,2], Barbara Wyslouzil[1,3]**

The Ohio State University [1]William G. Lowrie Department of Chemical and Biomolecular Engineering, [2]Department of Biomedical Engineering, [3]Department of Chemistry and Biochemistry

Argonne National Laboratory [4] X-ray science division

Micellar nanocomposites are finding increasing applications in medical nanotechnology because of their ability to encapsulate a variety of nanoparticles. In our work, we synthesize micelles composed of poly(styrene-b-ethyleneoxide) (PSPEO) co-encapsulating quantum dots (QDs) and superparamagnetic iron oxide nanoparticles (SPIONS) for in-*vitro* cancer cell detection and manipulation. We have extended this approach to introduce poly(lactic-co-glycolic acid) PLGA into PSPEO micelles either by solubilization or a mixed micelle formation mechanism. Successful application of these nanocomposites requires an understanding of how their morphology evolves with changing process conditions and the nature of polymer-encapsulant interactions. To understand the multifunctional properties of the micelles, it is also important to know the number of nanoparticles encapsulated within the micelle and their distribution inside the heterogeneous micellar micro-environment. Here, we evaluate nanoparticle packing as a function of number and shape, in PSPEO micelles using Small Angle X-ray Scattering.

Our multifunctional micelles are synthesized using the interfacial instability approach, which involves combining the polymer and nanoparticles initially present in a common solvent such as chloroform with a selective solvent such as water. Micelles form, trapping the hydrophobic nanoparticles. An emulsifier, poly vinyl alcohol (PVA) is used in this process. We investigated the micelle encapsulation behavior of differently shaped nanoparticles (spheres and cylinders) at different concentrations. At sufficiently high nanoparticle concentration, crystalline behavior is exhibited. Using small angle x-ray scattering (SAXS), we observed diffraction peaks in the scattering profiles at high nanoparticle loadings. The inter-nanoparticle spacings calculated for the QDs and the SPIONS are consistent with their size and geometry. Our future efforts are directed towards studying the onset of packing behavior within micelles for different classes of nanoparticles. Further, by changing the original polymer used to a longer polymer known to form cylindrical micelles, we observed nodular loading behavior of the nanoparticles inside the micelles. The dynamic behavior of the nodes within the cylindrical micelle was studied using fluorescence microscopy.

Nanoparticle loading was also compared to that of a hydrophobic polymer, PLGA. PLGA was found to exist in two phases, solubilized in micelles or as independent microspheres, based on initial conditions. We were able to observe the scattering behavior of each of these phases to determine PLGA distribution on a volumetric basis.

Nanostructure investigation using SAXS provides an in-*situ* exploration of the sample. Combined with other analytical tools such as Transmission Electron Microscopy (TEM) and fluorescence microscopy, scattering methods yield vital structural information about the sample probed. This forms the basis for successful implementation of these nanocomposites in a wide variety fields in nanotechnology and nanomedicine.

Ionic liquids are organic salts that are liquid at or near room temperature (i.e. below 100°C) and possess unique properties that are tunable based on cation and anion composition. ILs can be attached as side chains to polymerizable monomers to create poly(ionic liquid)s, which have applications as surface coatings, anti-bacterial surfaces, and lubrication. We report the surface-initiated ring-opening metathesis polymerization (SI-ROMP) of norbornene-functionalized ionic liquids to grow poly(ionic liquid) films on gold and silicon substrates. The ionic liquids consist of the cation 1-norbornylmethylene-3-methylimidazolium with the anions bromine ($Br^-$), hexafluorophosphate ($PF_6^-$) and trifluoromethanesulfonate ($TFO^-$) that are easily ion exchanged into the film after SI-ROMP. The properties of the poly(ionic liquid) films depend on the selection of anion and swelling solvent, as characterized by reflectance absorption infrared spectroscopy, ellipsometry, electrochemical impedance spectroscopy, and tribometry.

# 391107 Synthesis of Anion-Exchange Membranes for Applications Artificial Photosynthesis and in CO2 Capture

**Abstract:**

Anion exchange materials are fundamental components of fuel cells and solar fuel generators, as polymer electrolyte and polymer membrane, attributed to its flexibility and ease of fabrication. Herein we report the synthesis and characterization of anion conducting membranes derived from poly(phenylene oxide) (PPO) and poly(epichlorohydrin) (PECH) precursors. A variety of ionic groups including quaternized ammonium with different alky substituents, and imidazolium were attached to the PPO and PECH backbone. A range of compositions were synthesized, and gas permeabilities, and ionic conductivities were measured as a function of ionic content. The stability of the ionomers under harsh alkaline conditions was confirmed by thermogravimetric analysis and infrared spectroscopy. The properties of the PPO- and PECH-derived ionic membranes are presented within the context of solar fuels generation, and as materials for $CO_2$ capture.

---

# *Polymerized ionic liquid membranes for $CO_2$ capture: the effect of water vapor*

*Luca Ansaloni[1], Jacob R. Nykaza[2], Matteo Minelli[1,3], Yossef A. Elabd[2], Marco Giacinti Baschetti[1,3]*

[1]Dipartimento di Ingegneria Civile Chimica, Ambientale e dei Materiali (DICAM)
Alma Mater Studiorum - Università di Bologna, Bologna, Italy

[2]Department of Chemical and Biological Engineering,
Drexel University, Philadelphia, Pennsylvania

[3]Centro Interdipartimentale per Ricerca Industriale - Meccanica Avanzata e Materiali (CIRI-MAM)
Alma Mater Studiorum - Università di Bologna, Bologna, Italy

Polymerized ionic liquids (PILs) represent an attractive material for membrane-based gas separation technology, thanks to the remarkable properties of ionic liquids, such as the large $CO_2$ solubility at room temperature[1,2].

Water is a common impurity present in several gas separation processes and the investigation of its effect on the membrane transport properties is required for a reliable and comprehensive material characterization. The determination of the $H_2O$ influence is even more important when hydrophilic moieties, such as side-chain charged ions, are present in the membrane matrix.

As previously reported[3] for perfluorosulfonated ionomers (Nafion and Aquivion), which are of interest for some gas separation applications, the presence of water enhances significantly the gas transport rate of the various gaseous species inspected, with up to a two orders of magnitude increase in the permeability and a simultaneous improvement of $CO_2$ selectivity versus other gases.

Accordingly, gas transport properties of imidazolium-based PIL homopolymer and block copolymer membranes are investigated under humid conditions in order to characterize the water vapor influence on the separation performance. In particular, $CO_2$ and $CH_4$ are considered as gaseous probes and their permeability is measured in two different PILs membranes: the p(MEBIm-TFSI) homopolymer and the p(MMA-b-MUBIm-TFSI) block copolymer. The synthesis of these two polymers are detailed elsewhere[2,4,5].

First, water sorption experiments are carried out by means of a quartz-spring microbalance apparatus[6] at different temperatures (25, 35 and 45°C). Experimental results show a rather low solubility of $H_2O$ in both PILs membranes, likely due to the large hydrophobic character of the polymeric chains supporting the imidazolium moiety. A slight increase in water uptake is observed with temperature, although the solubility value remains rather limited (≤ wt. 2.5%).

The permeability of $CO_2$ and $CH_4$ in both PILs is then investigated in a humid permeometer apparatus[3], which allows the membrane characterization maintaining the polymer film at a fixed water activity, which can span in a rather wide range (0 - 75%), and at different operative temperatures (25, 35 and 45°C).

The results in terms of gas permeability shows a monotonous increasing behavior with the water activity, regardless to the low water content in the membrane, both in the homopolymer and the block copolymer, although the extent of the variation is different for the two materials. In case of p(MMA-b-MUBIm-TFSI), at 75% R.H. a 30% increase in the transport rate is observed, whereas in case of the p(MEBIm-TFSI), at the same water vapor pressure, the gas permeability is 70% larger than the values measured in dry conditions. Interestingly, the water vapor presence affects in a similar way both penetrants, and the selectivity is barely influenced by the R.H..

Differently from what is commonly observed for other membrane materials with similar water solubility (e.g. polyimides[7]), the effect of water vapor on the gas permeation process is in an overall increase of the performance of PILs membranes, making them attractive materials for gas separation applications since no dehydration step is required before the membrane module.

***References:***


1. Scovazzo, *J. Memb. Sci.* **343,** 199–211 (2009).
2. Chen, H *et al., Macromoleculs* **42,** 4809–4816 (2009).
3. Giacinti Baschetti *et al., Int. J. Hydrogen Energy* **38,** 11973–11982 (2013).
4. Ye *et al., Macromolecules* **45,** 7027 – 7035 (2012).
5. Nykaza *et al., Polymer* (2014). doi:10.1016/j.polymer.2014.04.003
6. Davis *et al., Macromolecules* **45,** 7486–7494 (2012).
7. Piroux *et al., J. Memb. Sci.* **232,** 115–122 (2004).

# 378496 Effect of the First Layer on Polyelectrolyte Multilayer Properties and Function

**Abstract:**

Polyelectrolyte multilayers have been attracting attention due to the ease of preparation and potential for tailorable properties. We are interested in polyelectrolyte multilayers as coatings for orthopedic implants. Polyelectrolyte multilayers were prepared with bone morphogenetic protein 2 as the first layer and control systems were prepared with a polycation as the first layer. Significantly different surface roughness was observed between the two, which impacted the proliferation and differentiation of pre-osteoblasts cultured on both coatings. Furthermore, it was observed that processing conditions such as deposition solution pH and and salt concentration impacted greatly this difference in surface roughness and also affected other physical properties. Techniques including quartz crystal microbalance with dissipation, atomic force microscopy and goniometry were employed to try to understand the observed phenomena. We will discuss our findings and present thought regarding potential mechanisms.

## Development of Chitosan/PVA Electrospinning Nanoporous Nanofibers for Improving Wound Healing

Mian Wang[1], Thomas J. Webster[1, 2]

[1]Department of Chemical Engineering, Northeastern University, Boston, Massachusetts, USA 02115

[2]Center of Excellence for Advanced Materials Research, King Abdulaziz University, Jeddah, Saudi Arabia

**Abstract:** Chronic dermal wounds, such as venous ulcers and diabetic foot ulcers, represent a major health problem that affects millions of people worldwide and induces billions of dollars in social and economic costs, explaining the large research efforts now focused on developing new therapeutic approaches to improve wound healing [1]. The entire process of normal cutaneous wound healing requires the formulation of scaffolds with high regeneration properties. The wet scaffold plays a very important role in wound healing since water swells in the scaffold and is of great help to reduce necrotic tissue. That's one of the reasons why chitosan was chosen for wound healing because of its rich number of hydrogen bonds between chitosan chains. Previous research revealed that the swelling ratio of chitosan nanofibers is 70% more than chitosan particles. It is believed that the swelling ratio of nanoporous nanofiber scaffolds is much higher than this [2]. Besides, chitosan is a cheap resource, is biocompatable, and is antibacterial.In this research, a nanoporous nanofiber scaffold was designed by electrospinning chitosan blended with poly(vinyl alcohol) (PVA). Individual chitosan can not easily be incorporated into any fiber via electrospinning due to its high molecular weight resulting in high surface tension. Therefore, alkali hydrolysis was used to decrease its molecular weight by soaking for 24h and 48h. In order to achieve nanoporous nanofibers, PVA was blended with low molecular weight chitosan. After fabrication, scaffolds were soaked in water since PVA can dissolve in water. Thus, nanoporous nanofiber scaffolds were fabricated when water takes the place of PVA. SEM images revealed that both non-treated chitosan and 24h hydrolysis chitosan could not be electrospinned individually. Although after 24h of hydrolysis, chitosan blended with PVA were electrospuninto nanofibers, they are were not uniform and small beads existed. SEM images of the 48h hydrolysis chitosan indicated that the individual chitosan had nanofibers although a few nanobeads still existed. When blended with PVA at different volume ratios (chitosan/PVA, 75: 25, 50: 50), uniformed nanofibers were obtained without nanobeads. In addition, when observed under high magnification, the diameter of the 75: 25 volume ratio of chitosan/PVA was around 70 nm, which was smaller than the 50: 50 volume ratio of chitosan/PVA (around 110nm). In conclusion, 48h hydrolysis chitosan blended with different volume ratios with PVA is promising nanoporous nanofiber scaffold for the cell viability experiments.

**Acknowledgement:** We thank Northeastern University for funding this research.

**References:** [1] Tchemtchoua VT, Atanasova G, Aqil A, et al. Development of a Chitosan Nanofibrillar Scaffold for Skin Repair and Regeneration. Biomacromolecules 2011, 12, 3194–3204.

[2] Ashleigh C, Rachael O, Hongyan M, et al. Chitosan-based nanofibrous membranes for antibacterial filter applications. Carbohydrate Polymers. 2013, 1, 92(1): 254–259.

Electrospinning is a convenient method to produce polymer nanofibers with controlled diameters on the order of tens of nanometers to micrometers. The resulting non-woven fiber mats have high specific surface areas of around 1−100 $m^2/g$. Combining these properties with the high electrical conductivity of intrinsically conductive polymers, conductive electrospun fiber mats are promising for a variety of applications such as electronic, magnetic, biomedical, sensor and optical fields.

Polyaniline (PANi) is one of the most studied intrinsically conductive polymers (ICPs) and it is unique due to its ease of synthesis, environmental stability, and simple doping/dedoping chemistry, yet it is relatively hard to process compared to most other polymers. As is common among ICPs, it has a fairly rigid backbone due to its high aromaticity. Thus, the elasticity of its solutions is generally insufficient for it to be electrospun directly into fibers. Moreover, PANi has poor solubility in common solvents, which further complicates it electrospinnability. To circumvent these problems, it is possible to electrospun PANi nanofibers by adding an easily electrospinable polymer to the organic solutions such as poly(ethylene oxide) (PEO). However, the presence of an insulator copolymer (PEO) decreases the fibers conductivity due to dilution of the conducting component. A good strategy for improving the electrical properties of this electrospun fibers blend is to incorporate graphene, which has unique electrical conductivity, into the nanofibers.

Since the discovery of graphene in 2004, it has attracted tremendous research interest. Graphene is a single-atom-thick, two-dimensional sheet of $sp^2$-hybrized carbon atoms arranged in a honeycomb crystal structure with exceptionally high strength, surface area, thermal conductivity, and electronic conductivity. Therefore, considering the excellent properties of graphene and PANi, a highly conductive PANi/PEO composite with graphene as nanofillers may be obtained.

In this work, we report the fabrication of electrospun PANi/PEO nanofibers filled with non-covalently functionalized graphene sheets. The fabricated graphene/polymer nanofibers have the potential to serve in a variety of applications, including conductive wires and sensors.

## Wafer-Scale Epitaxial Graphene on SiC: Synthesis, Transfer, and Devices

Christos Dimitrakopoulos,[1] Jeehwan Kim,[2] Hongsik Park.[2]

1. Chemical Engineering Department, University of Massachusetts, Amherst, MA 01003.
2. IBM Thomas J. Watson Research Center, Yorktown Heights, NY 10598.

Graphene's exceptional properties make it attractive for technological applications in many areas, including high-speed electronics.[1,2] The establishment of processes for producing high quality, large-scale graphene is necessary for such applications. Large area growth of epitaxial graphene on the Si-face of semi-insulating (SI) hexagonal SiC wafers [e.g. 4H(0001) or 6H(0001)] exhibits manageable growth kinetics, and most importantly, its azimuthal orientation is fixed, as it is determined by the structure of the single crystal substrate. Therefore, this is a viable method for producing graphene with uniform coverage and structural coherence at wafer-scale.

While SI SiC is a good substrate for graphene RF transistors, its high cost limits epitaxial graphene grown on it to few niche applications. Furthermore, to enable hybrid electronics, where standard digital cirquits built on Si perform digital logic functions, while graphene, which does not exhibit a band gap, is used for ultrafast analog devices, we would need to transfer epitaxial graphene onto Si substrates. To address these issues, we have recently developed a method in which a graphene film grown on a 4-inch SiC wafer is exfoliated via the stress induced by a Ni film and transferred to other substrates, resulting in a wafer-scale monolayer of graphene that is continuous and has a single azimuthal orientation.[3] This growth and transfer process in principle can be repeated on the same SiC wafer hundreds to thousands of times, dramatically reducing the cost per wafer-scale graphene layer.

A 300 mm Si wafer, the standard size used in modern microelectronics fabrication facilities, could in principle be covered almost completely by a hexagonal close-packed (*hcp*) arrangement of seven 4-inch diameter, almost circular graphene layers (discs). Since all of these graphene discs can be grown on and transferred from the same SiC wafer, which has a straight edge (flat), each graphene disc also exhibits an identical straight edge. By alligning the straight edges of the transferred and *hcp* graphene discs, the whole 300 mm Si wafer could in principle be covered by graphene having a fixed, specific lattice orientation. This has not been achieved by CVD graphene grown at large scale on Cu or other catalytic substrates and then transferred, or by any other method, todate.

*Support by DARPA Contracts FA8650-08-C-7838 & HR0011-12-C-0038 is gratefully acknowledged by CD (while he was at IBM Research).*

[1] "100-GHz Transistors from Wafer-Scale Epitaxial Graphene" Y.-M. Lin, C. Dimitrakopoulos, K. A. Jenkins, D. B. Farmer, H.-Y. Chiu, Ph. Avouris ***Science* 327**, 662 (2010).

[2] "Wafer-Scale Graphene Integrated Circuit" Y.-M. Lin, A. Valdes-Garcia, S.-J. Han, D. B. Farmer, I. Meric, Y. Sun, Y. Wu, C. Dimitrakopoulos, A. Grill, Ph. Avouris, K. A. Jenkins ***Science* 332**, 1294 (2011).

[3] "Layer-Resolved Graphene Transfer via Engineered Strain Layers" J. Kim, H. Park, J. B. Hannon, S. W. Bedell, K. Fogel, D. K. Sadana, C. Dimitrakopoulos ***Science 342,*** *833* (2013).

# 384095 In-Situ Gisaxs Study of Pore Orientation Effect on Thermal Transformation Mechanism of Surfactant Templated Mesoporous Titania Thin Films

**Abstract:**

Titania thin films with orthogonally oriented cylindrical mesopores have numerous potential applications in photovoltaic cells, electrochemical devices and photocatalytic technologies due to their accessible cylindrical straight pores which provide well-defined short diffusion paths for reactants and charge carriers[1]. The Rankin group at the University Of Kentucky has established a technique based on Pluronic P123 surfactant templating to synthesize titania thin films with hexagonally ordered cylindrical mesopores tilted orthogonal to the substrate [2] and also showed that these orthogonal structures have better thermal stability than films with mesostructure oriented parallel to the substrate.[3] However, the mechanisms of formation and thermal transformation of this ideal mesostructure are in need of further investigation, since the ability to orient pores in these films may allow the films to better withstand stress during heating and crystallization. This study addresses the detailed kinetics of orthogonally oriented mesostructure transformation, and compares them with films oriented parallel to the substrate using in-situ grazing incidence small angle x-ray scattering (GISAXS). Complementary supporting information is provided by ex situ transmission and electron microscopy.

The pores in the films are template using Pluronic surfactants P123, and are oriented by using a crosslinked P123 modified substrate. Prior to calcination, the films are aged at 4 °C, and GISAXS indicates that the orthogonal orientation of ordered mesoporous films develops during this time. In-situ GISAXS measurements dueing thermal treatment show that the orthogonally oriented films have higher activation energy (155 KJ/mole) for mesostructurdeformation loss during crystallization than those with pores oriented parallel to the substrate (126 KJ/mole), and that this makes the orthogonally oriented films more stable during thermal treatment at higher temperatures (> 600 °C).The reasons for this difference are ascribed to greater resistance to anisotropic stress during heating of the orthogonally oriented pores and to nucleation of titania crystallites at the top surface of the films with orthogonally oriented pores rather than throughout the thickness of the film.

**References**

1. Coakley et al., MRS Bulletin, 2005, 30-37.

2. Koganti et al., Nano Letters, 2006, 6, 2567-2570.

3. Das et al., The Journal of Physical Chemistry C, 2014, 118, 968-976.

# 388500 Scalable Directed Assembly of Electrosprayed Nanoparticles By Durotaxis

**Abstract:**

Despite the ever-growing library of fabrication techniques, achieving scalability in the deliberate deposition of nanoparticles remains a challenge. We have demonstrated a method for the positioning of nanoparticles though a combination of liquid durotaxis and electrospray. Durotaxis describes the phenomenon wherein liquid droplets on flat, soft surfaces spontaneously move towards softer, or deeper, regions of the surface due to changes in the effective contact angle. By using a patterned hard substrate (here, a rigid plastic) coated by a soft overlayer (here, a silicone elastomer), the motion of microdroplets generated by an electrospray process is directed on the sub-micron scale during their evaporation. As a result, materials (*e.g.* dye, polymer, and nanoparticles) transported within the droplets are deposited in a spatially controlled manner. The final resulting pattern requires no direct patterning of the particles beyond the large-area electrospray and therefore represents a high-throughput technique. Additionally, the particle assembly has the benefit of being generated on a flat, conformable surface on a flexible patterned substrate. This latter factor is advantageous as it both allows for subsequent transfer printing of the materials and protects the underlying pattern from potentially damaging mechanical contact with the particles and transferred substrate, increasing its usable lifetime. This greatly increases the potential for scaling to a reel-to-reel process.

# Versatile Surface Micropatterning and Functionalization Enabled by Microcontact Printing of Poly-4-Aminostyrene

Zhibin Wang
11/16/2014

Department of Chemical and Biomedical Engineering
FAMU-FSU College of Engineering
Florida State University

1

# Micro/nanocontact printing of polyelectrolytes

Schematic of contact printing process

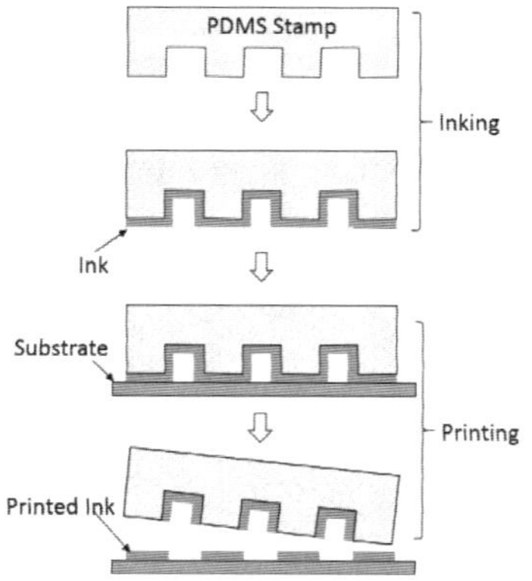


AFM image of DNA nano-array

Cha et al. *ACS Nano.* 2009

Phase contrast image of single cell pattern

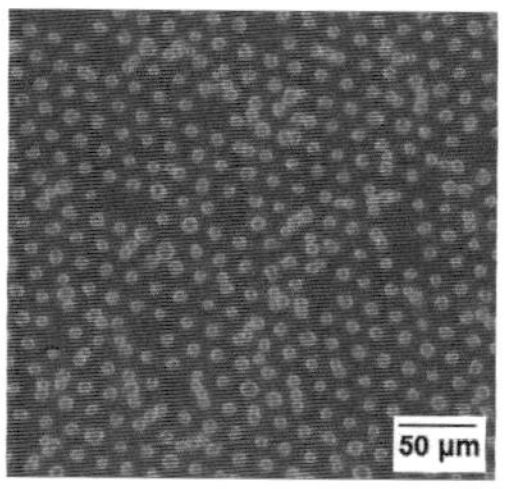


Z Wang et al. *Soft Matter* 2012

# Diazonium salts

Diazotization process

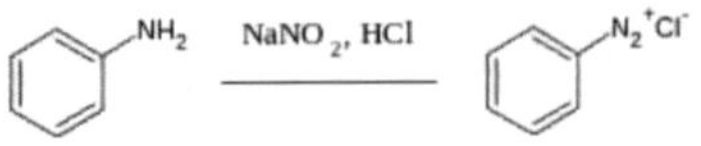


- Carbon
- Metals (gold, iron & stainless steel)
- Semiconductors
- Biomolecules (DNA, proteins)

- ➢ Molecular electronics
- ➢ Biosensing
- ➢ DNA microarrays
- ➢ Tissue engineering

Microcontact printing of aryldiazonium salts

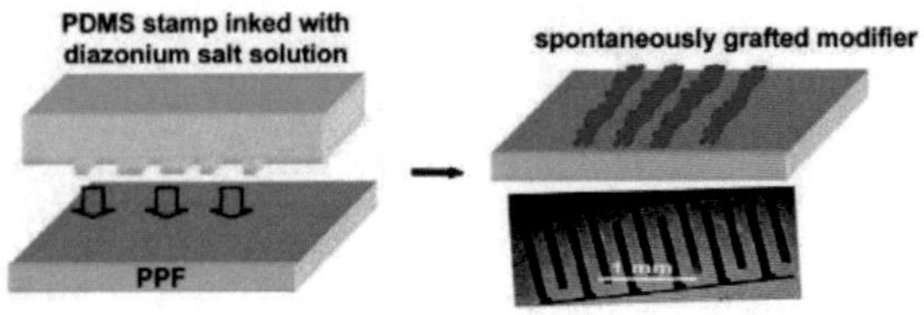


Downard et al. *J. Am. Chem. SOC. 2007.*

# Functional ink: poly-4-aminostyrene (PAS)

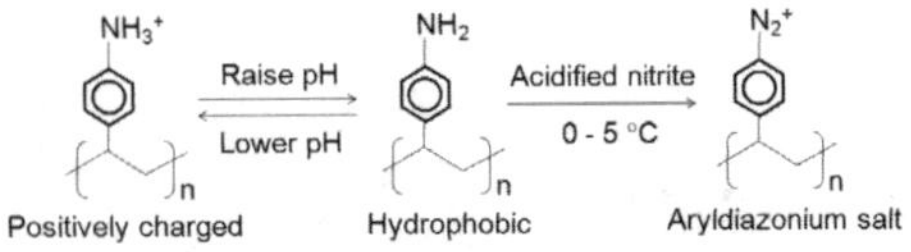


- Polymer ink
- Diazotion (aryldiazonium salt)
- Intrinsic fluorescence
- Hydrophobicity

Fluorescence image of printed PAS array

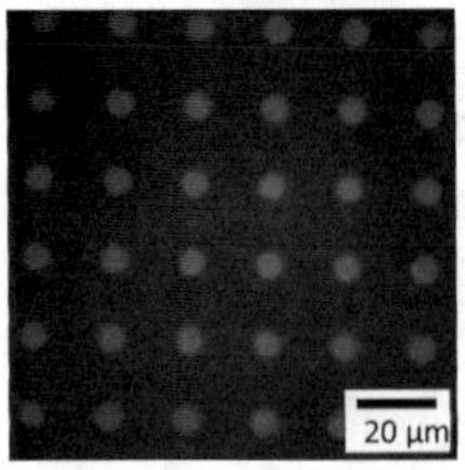


AFM image of printed PAS array

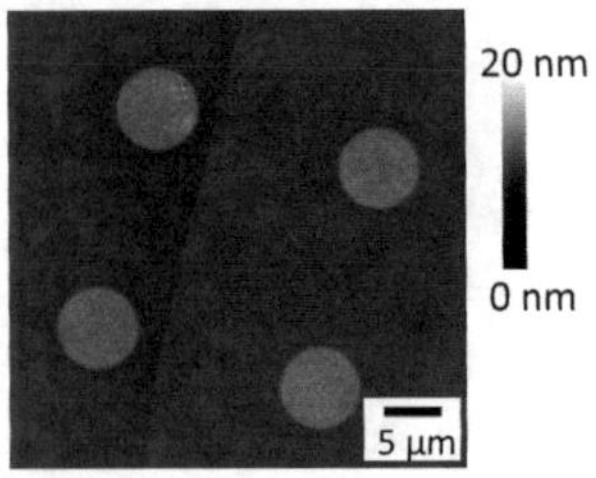


Z Wang et al. *Langmuir* 2014, 30: 13483

# Printed PAS for micropatterning bio & nanomaterials

Fluorescence image of co-micropatterned avidin stripes (green) and DNA stripes (red)

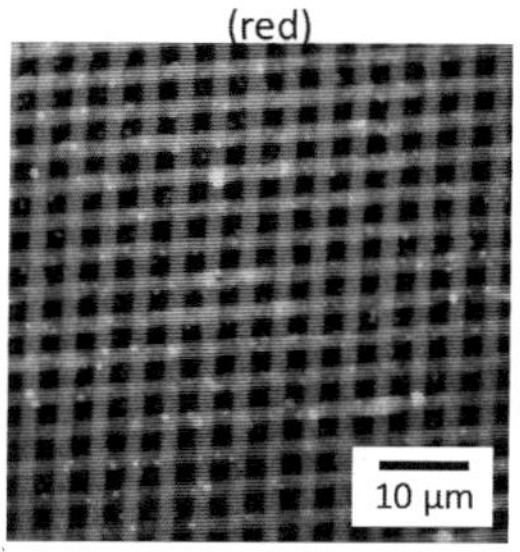


Avidin: stained by fluorescein-biotin
DNA: stained by propidium iodide

AFM image and Raman spectrum of SWCNTs immobilized on diazotized PAS stripes

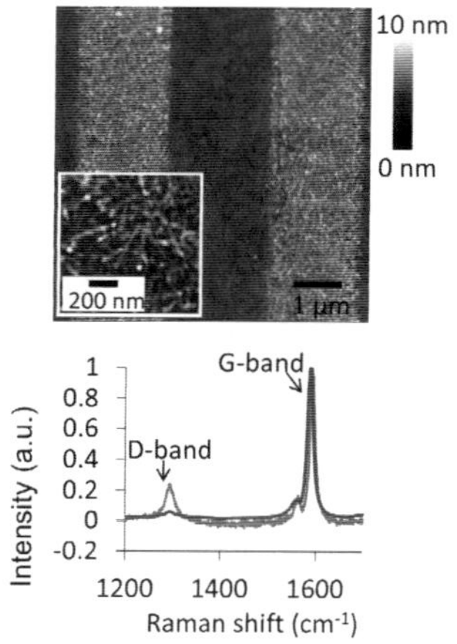


AFM image and UV-vis of Au NPs immobilized on diazotized PAS stripes

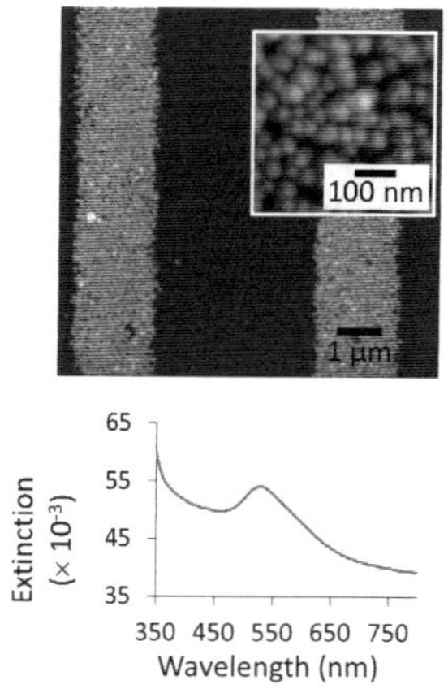


# µCP of a metallic structure on a carbon surface

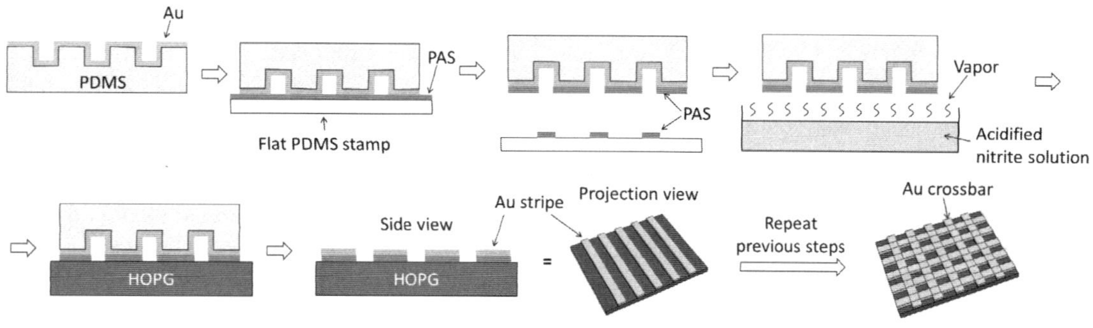


Schematic of fabrication procedure

Z Wang et al. *Langmuir* 2014, 30, 13483

# μCP of a metallic structure on a carbon surface

AFM image of μCP of a metallic structure on a carbon surface without PAS

AFM image of μCP of a metallic structure on a carbon surface

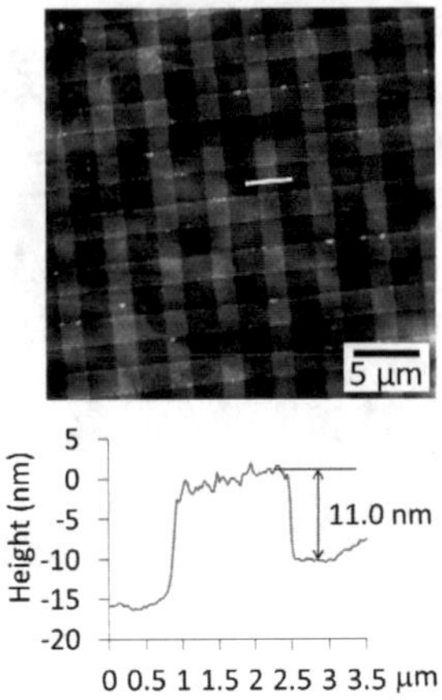


SEM image of μCP of a metallic structure on a carbon surface

Z Wang et al. *Langmuir* 2014, 30, 13483

# Printed PAS-based polyelectrolyte multilayer as a mask for wet etching

Contact angle of PAS/PSS/PAS coated surface

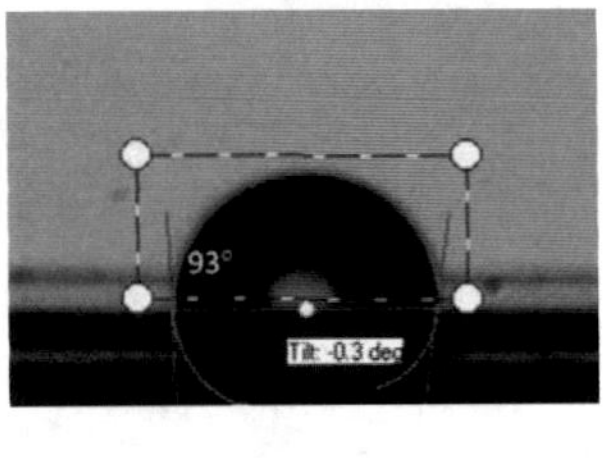


Schematic of PAS based etching mask fabrication

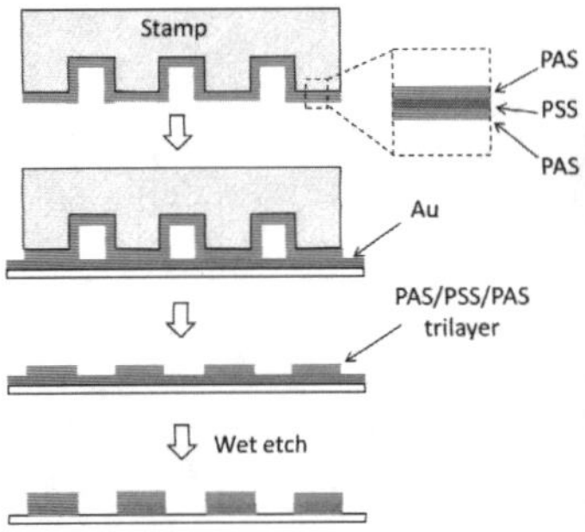


Height AFM image of gold surface before etching (top) and after etching (bottom)

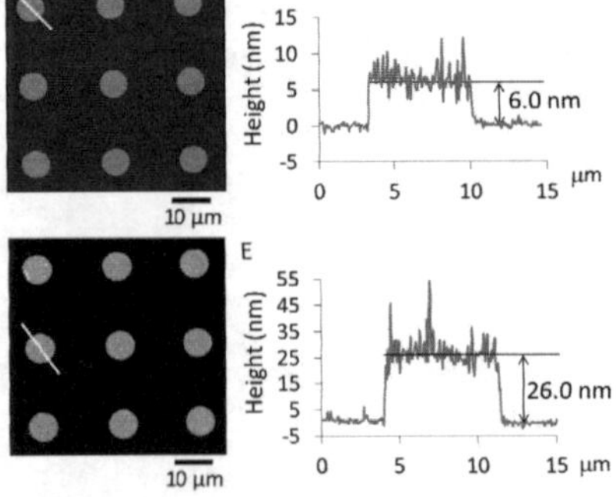


Z Wang et al. *Langmuir* 2014, 30, 13483

# Fabrication of highly engineered microparticles

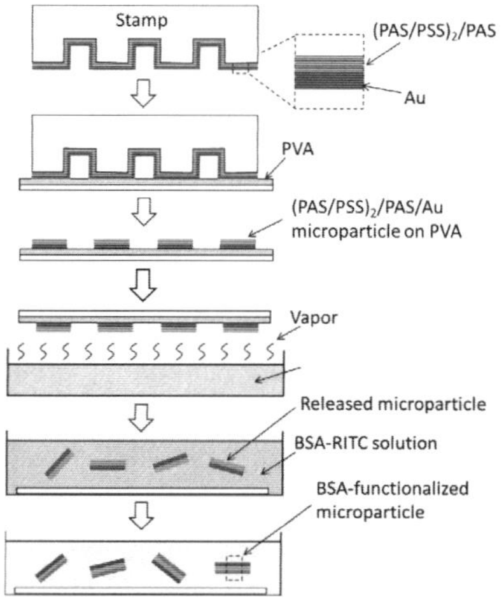


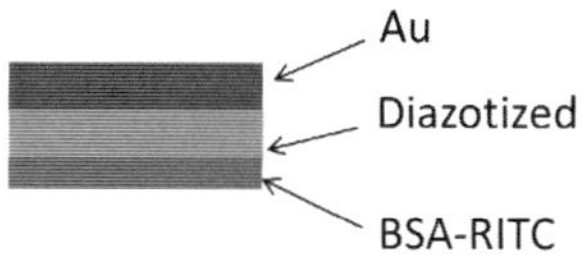


Merged image of released engineering particles

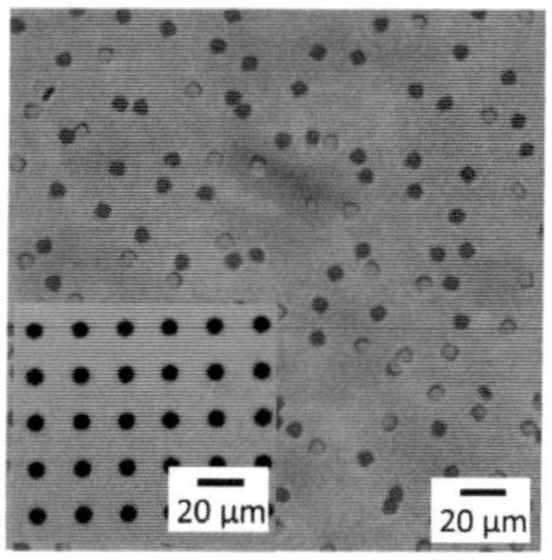


Z Wang et al. *Langmuir* 2014, 30, 13483

# Summary

- The microcontact-printed PAS can be diazotized to micropattern biomolecules including DNA and protein, and nanomaterials including single-walled carbon nanotubes and gold nanoparticles.
- The diazotized PAS enables μCP of a metallic structure on a carbon surface.
- The hydrophobic nature of PAS at the neutral pH allows the microcontact-printed PAS-based polyelectrolyte multilayer to be used as masks for wet etching.
- This technique allows fabrication of highly engineered microparticles with a unique structure.

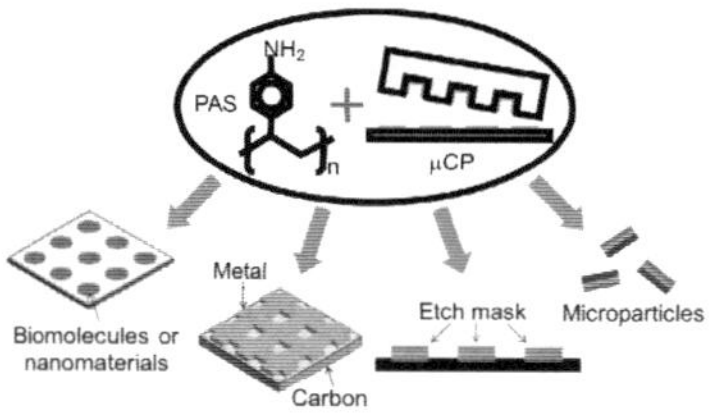

## Acknowledgement

Dr. Guan's group:

Dr. Jingjiao Guan (Chemical & Biomedical Engineering, FSU)

Junfei Xia (Chemical & Biomedical Engineering, FSU)

Dr. Peipei Zhang (University of Maryland)

Dr. Tao Liu (Industrial & Manufacturing Eng., FSU)

Dr. Sida Luo (Industrial & Manufacturing Eng., FSU)

Dr. Zhiwei Xiao (CGG)

Dr. Eric Lochner (Physics, FSU)

Dr. Peng Xiong (Physics, FSU)


# Thank you

PAS/PSS/PAS
etching mask

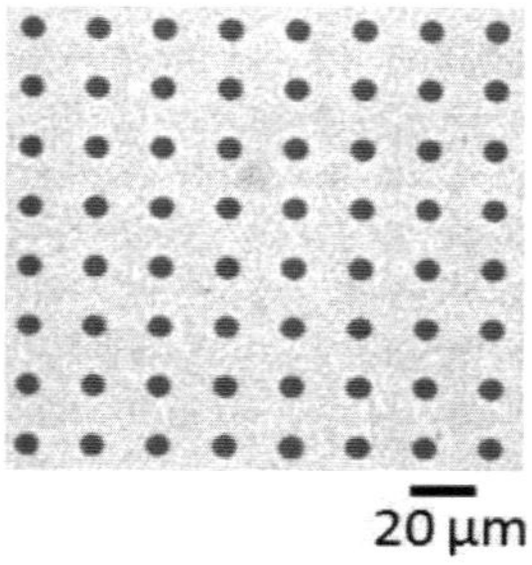


PAH/PSS/PAH
etching mask

PAH: poly(allylamine hydrochloride)

Abstract: Directed self-assembly of blcok copolymer is becoming a practical alternative choice for the next generation nanolithography. High chi block copolymers like PS-PDMS are capable of forming sub 10nm half pitch with the help the some annealing methods to pass the order disorder transition. Previous studies indicate that the thermal annealing temperature and rates of PS-PMMA follows the arrehenus law. In this studies, we used diode laser to decrease the thermal annealing time to less than 10ms which is $10^{-4}$ to $10^{-6}$ order of the conventional oven annealing. With the design of patterned substrate and underlayer to tune the surface energy, PS-PDMS is able to assembly in long range order in milliseconds. We synthesized PS-PDMS of three different molecular weight with the same PS content and compared the effect of molecular weight on the pattern size. We also applied additional solvent annealing prior to laser annealing to help the directed self-assembly. In summary, we are able to achieve sub-10nm structure using graphoepitaxy directed Self-Assembly of PS-PDMS by laser spike annealing.

# 383574 Scalable Nanofabrication of Biomimetic Broadband Anti-Reflection Coatings

**Abstract:**

Millions of years before people began to fabricate functional nanostructures, biological systems were using nanometer-scale architectures to produce unique functionalities. Some nocturnal moths use hexagonal arrays of subwavelength nipples as anti-reflection (AR) coatings to reduce reflection from their compound eyes for the purpose of night camouflage. Similar periodic arrays of nanopillars have also been observed on the wings of cicada to render superhydrophobic surfaces for self-cleaning functionality. Inspired by these natural nanostructures, we have developed a simple and scalable templating technique for fabricating self-cleaning, broadband AR coatings on a large variety of technologically important substrates, such as single crystalline Si, multicrystalline Si, GaAs, GaSb, and glass. The technique is based on two colloidal self-assembly technologies that combines the simplicity and cost benefits of bottom-up self-assembly with the scalability and compatibility of standard top-down microfabrication. The resulting subwavelength-structured AR coatings exhibit improved broadband anti-reflection properties than traditional multilayer dielectric coatings, and show superhydrophobic surface states, promising for applications ranging from highly efficient solar cells and photodetectors to self-cleaning flat-panel displays.

**Selenium Nanoparticles and Nanocomposites as Scaffolds to Alter Cancer Proliferation and the Tumor Microenviroment**

Michelle Stolzoff, Thomas J. Webster.
Departments of Bioengineering and Chemical Engineering, Northeastern University, Boston, MA 02115

**Introduction:** Nanoparticle-based drug delivery has been explored to circumvent the often-toxic chemotherapy treatments used today by providing a more efficient and specific delivery to diseased tissues. Specifically, cancer nanotherapeutics have generated therapies with more specific targeting, larger therapeutic windows and most importantly of all, the possibility for greater survival rates. The development of a monodisperse, colloidally stable, and bioactive nanoparticle which is effective for cancer therapy, among other applications, without the use of nonspecific, often-toxic drugs is thus a worthwhile venture. A candidate material for the core of such a nanoparticle is elemental selenium. Selenium has become a popular dietary supplement for its anti-cancer, anti-inflammatory and other properties,[1,2] and nano-selenium has been demonstrated to be cytotoxic to cancer cells and bacteria when applied directly.[3-5] Specifically, selenium is a major component of several anti-oxidant proteins. Here, we describe the synthesis, characterization and efficacies of different selenium nanoparticle (SeNP) formulations in cancer cell cytotoxicity and modulation in ROS generation and gene expression.

**Materials and Methods:** Both free and nanocomposite SeNP were explored in this study. SeNP were generated by mixing equal parts $DIH_2O$, 0.1M glutathione (GSH) and 0.025M $Na_2SeO_3$. To precipitate SeNP, 2M NaOH was added to bring the solution to the alkaline range and diluted in $DiH_2O$ to stop the reaction. The SeNPs were spun down and rinsed to further prevent further SeNP growth. For the nanocomposites, poly(l-lactic acid) (PLLA) was used as a foundation substrate, with SeNP either grown onto or mixed throughout the substrate. To generate "nanosmooth" films, the PLLA was dissolved in chloroform and cast onto glass dishes. The films were cut into discs and coated with selenium nanoclusters using the same steps for generating SeNP, resulting in an even coating of selenium nanoclusters. To modify the SeNP dosage, the PLLA films were incubated in $GSH/Na_2SeO_3$ or NaOH for varied development times of 0, 5, 30 or 60 s. The discs were prepared for *in vitro* viability assays by sterilizing in EtOH for 1 hr and rinsing thoroughly with sterile PBS prior to cell seeding. Each disc was seeded with either 5,000 MG-63 osteosarcoma cells (ATCC CRL-1427) in EMEM with 10% FBS and 1% PS, or human osteoblasts (ATCC CRL-11372) in PromoCell Osteoblast Media with 10% Supplementmix and 1% PS. The plates were incubated for two days, and cell viability was determined via an MTS assay. Additionally, cell morphology and spreading was observed via SEM. Cells were grown onto the films as before and fixed following a previously described. All experiments were conducted in triplicate and repeated three times each.

**Results and Discussion:** Increases in both the $GSH/Na_2SeO_3$ (G/Se) and NaOH exposure steps resulted in a general increase in SeNP deposition without altering the sizes of the SeNP (ranging from 90-110nm). Notably, the longest times of 60 s (60-60) for each step did not significantly increase the coverage. Comparing the 60-60 coverage of ~14 $NP/\mu m^3$ to 60 s in G/Se and 30 s in NaOH (30-30) with ~13 $NP/\mu m^3$ suggests that there is a saturation point in SeNP coverage after a certain point. As more SeNP were deposited onto the PLLA films, greater nanoroughness was present on the surface. In Fig. 1, it is shown that this increase of nanoroughness (and Se) decreased MG-63 proliferation, most notably after 30 s of development times in each step.

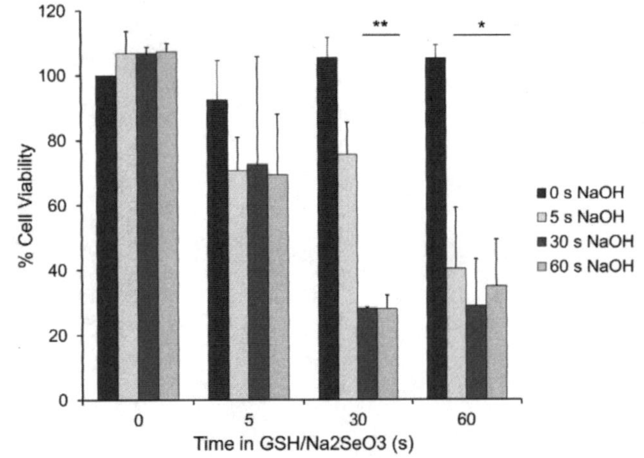


**Figure 1. MG-63 osteosarcoma cells were grown on PLLA films with varied amounts of SeNP dictated by the length of precipitation times in each solution. Mild increases in nanoroughness from the NaOH (blue) did not significantly alter cell proliferation but the inclusion of SeNP (as well as higher resulting roughness) greatly decreased proliferation (grey, red, gold). n=2, data represented as SEM. * = $p < 0.05$, ** = $p < 0.01$.**

**Conclusions:** The precipitation of SeNP is nearly instantaneous, but coating a substrate with SeNP can be controlled by varying the development times for each of the protocol steps. This increase in coverage does not affect the SeNP size, however, and there was an apparent saturation limit to create stable coatings of higher concentrations of SeNP using this specific precipitation method. We demonstrated the anti-cancer properties of selenium when coated onto PLLA films, resulting in the decreased proliferation of osteosarcoma cells. With longer development times, the films collected greater amounts of SeNP and, thus, a rougher nano-scale topography. The isolation of the effects of nano-selenium dosages and nanoroughness will further characterize the cell-material interactions we have demonstrated here.

**Acknowledgements:** The authors thank Northeastern University and the College of Engineering for funding.

**References**

1 Huang, Z., Rose, A. H. & Hoffmann, P. R. The role of selenium in inflammation and immunity: from molecular mechanisms to therapeutic opportunities. *Antioxidants & redox signaling* **16**, 705-743 (2012).

2 Combs Jr, G. F. & Gray, W. P. Chemopreventive agents: selenium. *Pharmacology & therapeutics* **79**, 179-192 (1998).

3 Tran, P. A., Sarin, L., Hurt, R. H. & Webster, T. J. Differential effects of nanoselenium doping on healthy and cancerous osteoblasts in coculture on titanium. *International journal of nanomedicine* **5**, 351 (2010).

4 Kong, L. *et al.* The suppression of prostate LNCaP cancer cells growth by Selenium nanoparticles through Akt/Mdm2/AR controlled apoptosis. *Biomaterials* **32**, 6515-6522 (2011).

5 Tran, P. A. & Webster, T. J. Selenium nanoparticles inhibit Staphylococcus aureus growth. *International journal of nanomedicine* **6**, 1553 (2011).

**Introduction:** Catheters are medical devices that are inserted into the body to remove fluid, create an opening, or deliver a drug, and have become an indispensible part in almost all fields of medicine. However, nowadays catheter associated infections are the most serious and costly healthcare-associated infections. Moreover, shunt catheter obstruction most often caused by macrophages (or inflammatory cells) is also very problematic, especially in the treatment of hydrocephalus. Recent studies have suggested a sensitivity of cell and bacteria to nano-scale topographical properties as a potential tool for selectively increasing desirable cell functions while simutaneously decreasing competitive cell functions. It is the hypothesis of this study that Polydimethylsiloxane (PDMS), a commonly used polymeric catheter material, which mimics the nano-patterned topography of natural tissue, will decrease inflammation and infection while remaining non-toxic to mammalian cells. Here we present a simpe and cheap methodology to prepare a nano-pattened PDMS replica by using highly ordered nanotubular anodized titanium (Ti) as the template.

**Materials and Methods:** 99.5% pure titanium foils (Alfa Aesar) were cut into 2.5cm×2.5cm squares and cleaned with acetone, 70% ethanol, deionized water. The samples were etched for 1min with a solution of 1.5% nitric acid and 1.5% hydrofluoric acid (HF) to remove the thin oxidized layer. The cleaned Ti sample was used as an anode, while a platinum (Pt) mesh served as a cathode. Both were immersed in an electrolyte solution consisting of 1.5% HF. The Ti and Pt samples were connected to a DC power supply. After anodization Ti samples were rinsed with large amounts of deionized water immediately and air dried. Next, the PDMS precursor and liquid stabilizer (Sylgard 184, Dow Chemical Co.) were mixed at 10:1 and were then placed in a vacuum chamber for 1h. The mixture was cast onto the anodized Ti template and then was placed into a vacuum chamber for another 1h. The PDMS sample was cured at 60 °C for 2h followed by cooling and was gently peeled away from the Ti template.

*Staphylococcus aureus* (ATCC 25923) was used in this study. Sterile samples were treated with the prepared bacterial solution ($\times 10^7$) and cultured for either 24h or 48h in an incubator (37°C, 5% $CO_2$, humidified). Afterwards, the samples were rinsed with a 10mg/mL PBS solution twice and placed into 15 mL tubes with 4 mL PBS. These tubes were shaken at 3000 rpm for 15 minutes on a vortex mixer to release the bacteria attached on the surface into the solution. Solutions with bacteria were spread on agar plates and bacteria colonies were counted after 18 hours of incubation. Bacterial tests were conducted in triplicate and repeated three times.

Human fibroblasts (ATCC, CCL-110) and macrophages (TIB-186; ATCC) at population numbers less than seven were used for all cell experiments. Cells were seeded onto the substrates at a density of 5000 cells/cm$^2$ and were allowed to adhere for 4 h in a 37°C, humidified 5% $CO_2$ atmosphere. The medium was removed from the wells four hours later and the cells were washed twice with PBS. Sterile samples were transferred to fresh 24-well tissue culture plates, and 1 mL of fresh growth medium was added to each well. Next, 150 mL of a dye solution from the cell proliferation assay kit (3-(4, 5- dimethylthiazol-2-yl)-2, 5-diphenyltetrazolium (MTT) (Promega) was added to each well and the plates were incubated for 4h. Afterwards, 1 mL of the Solubilization solution/Stop MixV1R (Promega) was added to each well and the plates were allowed to stand overnight. The contents of the wells were then mixed to obtain a uniformly colored solution, and then transferred to 96-well plates (200 mL per well). The absorbance at 570 nm was recorded using a microplate reader (Molecular Devices, SpectraMax® M3). The test was performed in triplicate and repeated three times.

**Results and Discussion:** As expected, the nano-sized tubes were distributed uniformly on the titanium surface after anodization. Moreover, nano-patterned structures were fabricated successfully on the surface of PDMS by using anodized Ti as the template. The contact angles for distilled water over plain and nanostructured PDMS surface were about 100° and 76°, respectively. These values highlighted that upon nanostructuring the PDMS surface became slightly more hydrophilic. It was found that nano-patterned PDMS templates increased fibroblast adhesion compared with plain PDMS and, thus, were non-toxic to fibroblasts. Results also showed that nano-patterned PDMS inhibited *S.aureus* growth by 70.5% and 79.2% after 24h and 48h, respectively (Fig. 1).

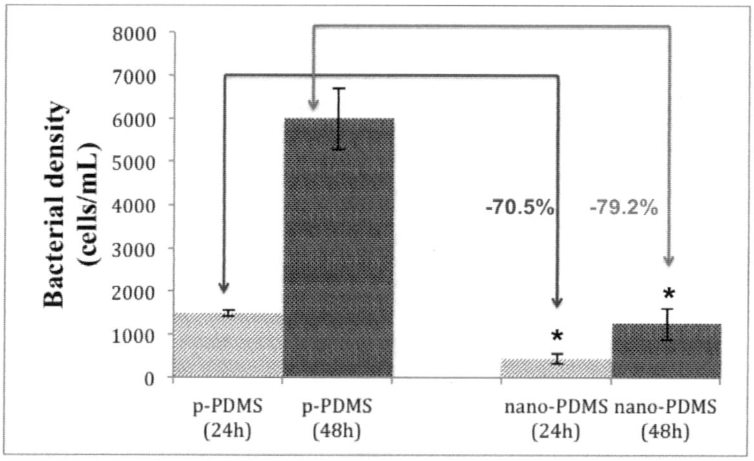


**Figure 1.** *S. aureus* growth on the surface of nano-patterned PDMS (nano-PDMS) and plain PDMS (p-PDMS).
Data represents mean ± SD, n=3. *$p<0.05$ compared with p-PDMS

**Conclusions:** Nano-patterned structures were successfully fabricated on the surface of PDMS through an anodized titanium template method. The direct consequence of this modification was that the nanostructured surface became slightly more hydrophilic. This increase in the hydrophilicity could in part be responsible for the little increase of the fibroblast adhesion on the nano-patterned PDMS. Further more, the nano topography on PDMS could inhibit *S.aureus* growth and also demonstrated promise to inhibit inflammatory cells potentially, and thus should strongly be considered for reducing catheter associated inflammation and infection.

**Acknowledgements:** The author would like to thank the Webster Nanomedicine Lab and the Department of Chemical Engineering, Northeastern University for help and funding.

# Antibacterial MgO Nanomaterials for Orthopedic Tissue Engineering

Daniel J. Hickey[1] and Thomas J. Webster[1,2]

[1]Department of Chemical Engineering, Northeastern University, Boston, MA 02115, USA
[2]Center of Excellence for Advanced Materials Research, King Abdulaziz University, Jeddah, Saudi Arabia

AIChE Annual Meeting
November 16, 2014

14AIChE
Annual Meeting, Atlanta, GA

Northeastern University

## Research Motivation

- More than 3 million musculoskeletal procedures are performed annually in the US
- About half involve bone grafting
- Poor tissue integration leading to instability is the leading cause of revision surgeries

Biomaterials with improved chemical and mechanical properties are badly needed to facilitate material integration and tissue regeneration

Jahangir AA *et al.*, *AAOS Now*, Jan. 2008.

2

# Research Motivation

Additionally, bacterial infection of implanted biomaterials is a recurring problem that can lead to implant failure, revision surgeries, and the spread of infection

- For total knee arthroplasty, infection is reported as the second main cause of revision (0.5-5%), after instability
- Average 5% of orthopedic fixation devices become infected (out of ~2 million implanted annually in U.S.)

1. Campoccia, D., et al., Biomaterials, 27(11), 2006.
2. Trampuz, A., and W. Zimmerli, Injury 37, (Suppl 2), 2006.

3

# Objective

The objective of this study was to implement MgO nanomaterials into scaffolds for bone tissue engineering to enhance tissue integration and impart bactericidal efficacy

4

# Magnesium Oxide (MgO) Nanoparticles

Biocompatible, biodegradable, low-cost,
environmentally friendly,
and present naturally in the human body

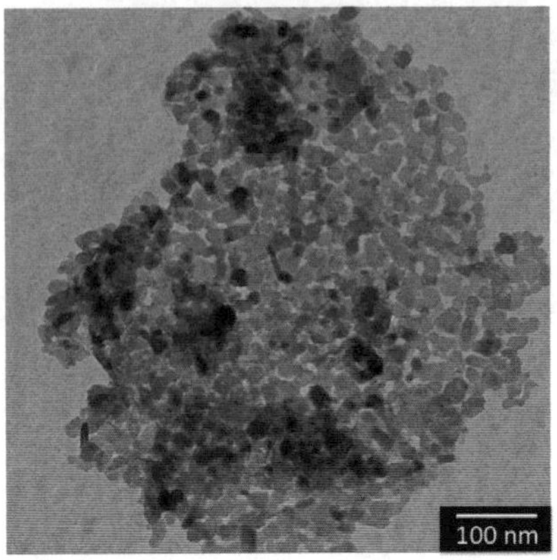


5

# Magnesium for Bone

- $Mg^{2+}$ cations coat the edges of bone apatite (HA) crystals and directly influence crystal size and density
- Indirectly influence mineral metabolism through activation of alkaline phosphatase (ALP)

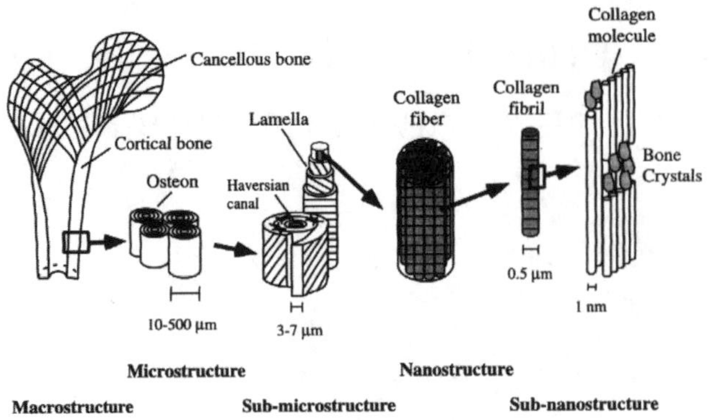


1. Serre, C.M., et al., J Biomed Mater Res., 42, 1998.
2. Rho, J.Y., et al., Medical Engineering & Physics, 20, 1998.

6

# Osteoblast Proliferation in Conditioned Medium

Osteoblasts grown on tissue culture polystyrene in supernatant from pH experiment

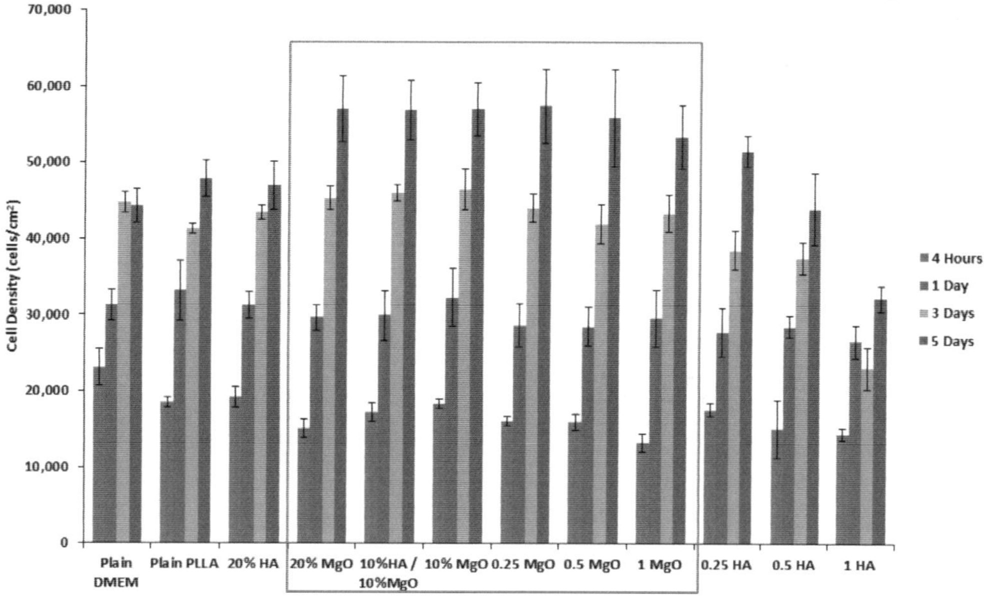


Data represents mean +/- SD, N=1, n=4

7

# Magnesium for Cell Functions

$Mg^{2+}$ (and $Ca^{2+}$) cations initiate conformational activation of integrin receptors to bind ligands

Integrins mediate cell-ECM interactions, which direct cell functions

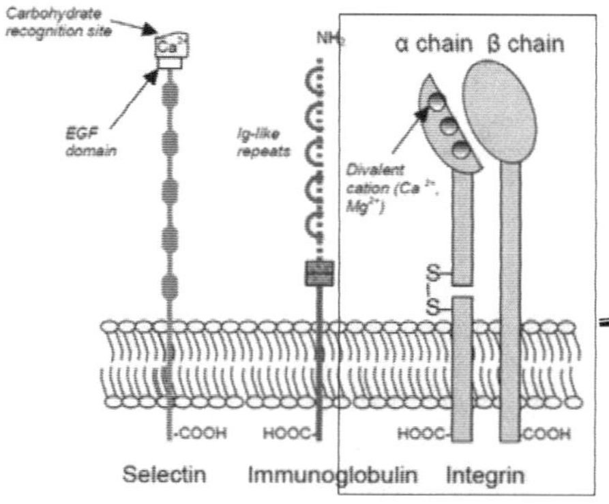


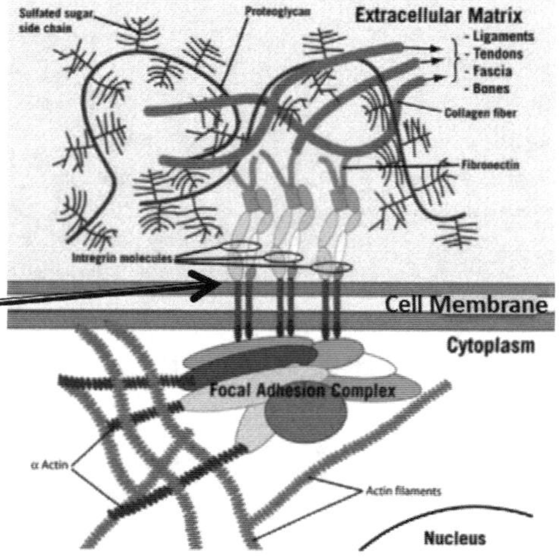



1. Benoit, D.S.W., Peppas N.A., Taylor and Francis, London, 2007.
2. Ravin, T., Journal of Prolotherapy, 3, 2011.


8

# Fibroblast Proliferation in Nanoparticle-Supplemented Media

HDFs grown in complete DMEM containing indicated nanoparticle concentrations [mg/mL]

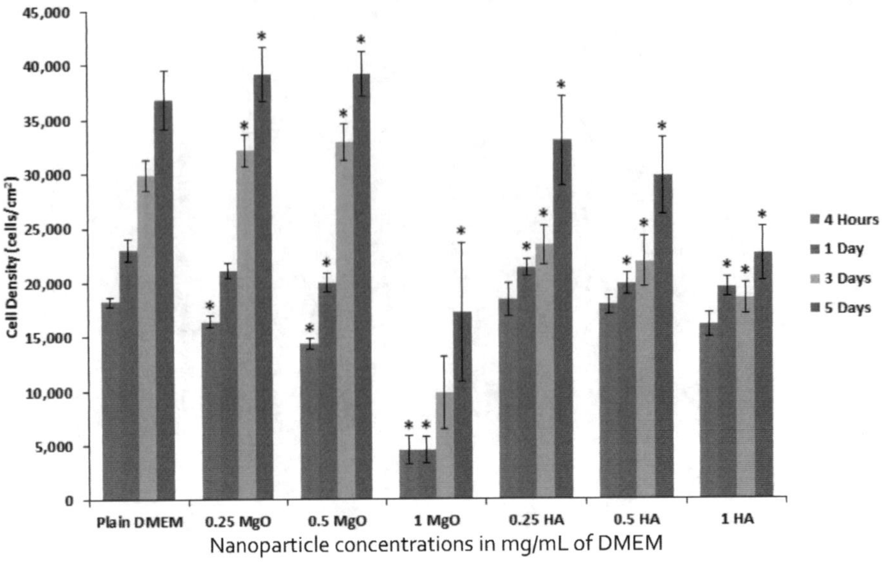


Data represents mean +/- SEM, N=3. *P<0.05 compared to plain DMEM at the same time point 9

# MgO NPs are Highly Antibacterial

## MgO (metal oxide) nanoparticles exhibit strong broad-range antibacterial activity

- Mechanism: **Not** reactive oxygen species (ROS) production
  - Direct NP-bacteria contact is required
  - Initiate bacterial membrane damage and eventual cell death

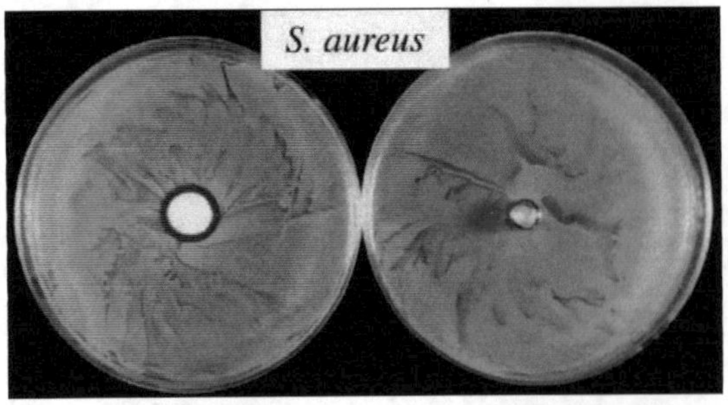


MgO NP    w/ penicillin cap


1. Sawai, J., et al., World J Microbiol Biotechnol, 16, 2000.
2. Leung, Y.H., Small, 13, 2013.


10

# Antibacterial Efficacy of Free MgO

Free MgO NPs show outstanding toxicity towards ***S. aureus***

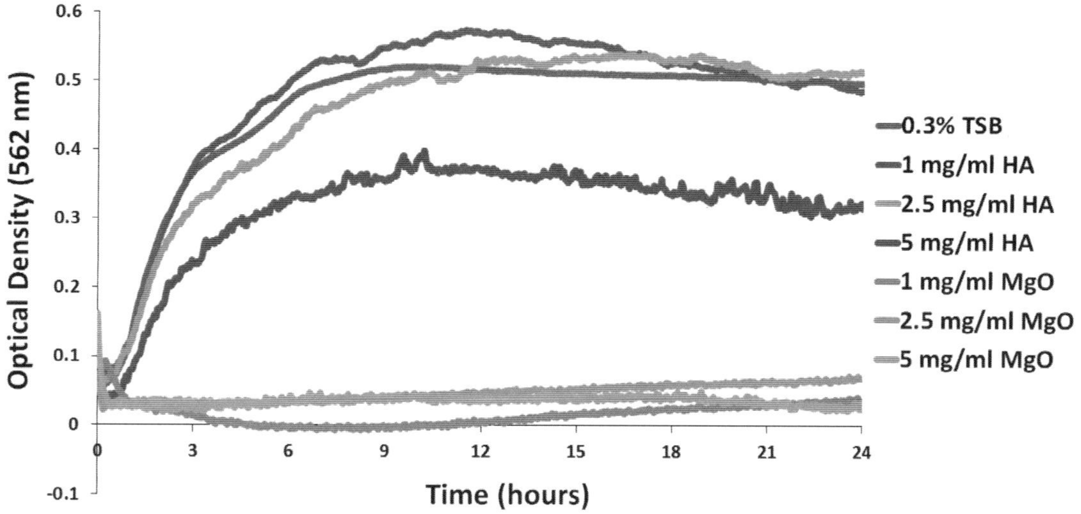


11

# Nanocomposite Fabrication

- Poly (l-lactic acid) (PLLA)
- Hydroxyapatite (HA) NPs
- Magnesium oxide (MgO) NPs

- Combine in chloroform
- Sonicate
- Cast to petri dish
- Evaporate solvent

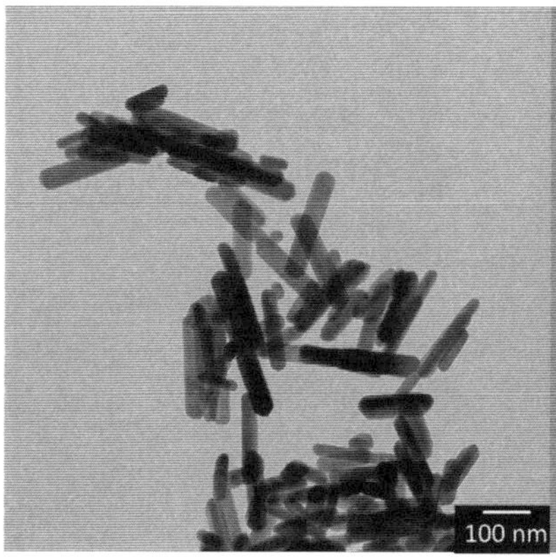


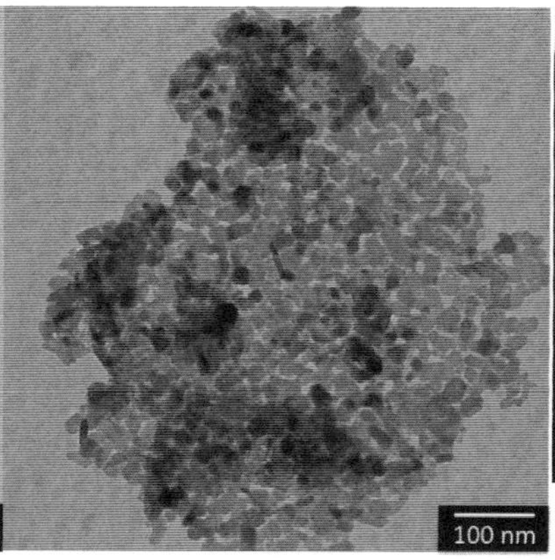


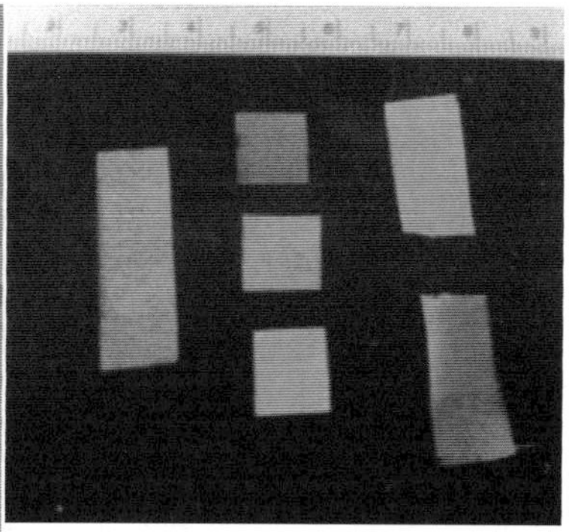

12

# Nanocomposite Surface Topography

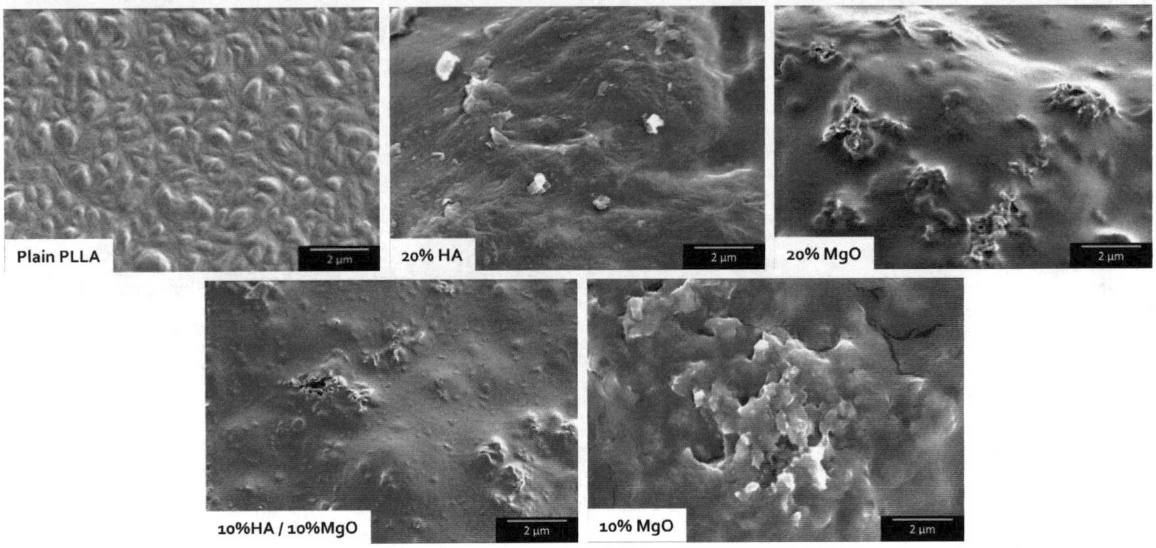


SEM images (mag = 10k) show variations in surface topography and particle exposure

13

# Osteoblast Adhesion

MgO-NPs increase **osteoblast adhesion** on PLLA

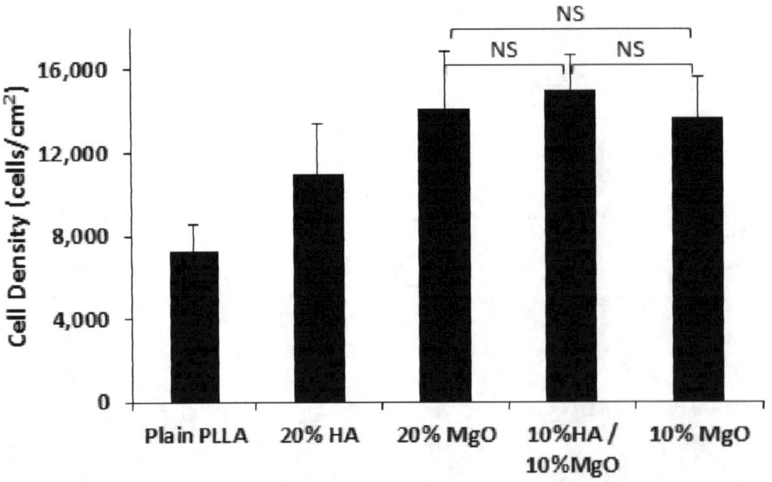


Data is average +/- SEM, N=3. NS signifies no statistical difference between samples

14

# Fibroblast Adhesion

MgO-NPs increase **fibroblast adhesion** on PLLA

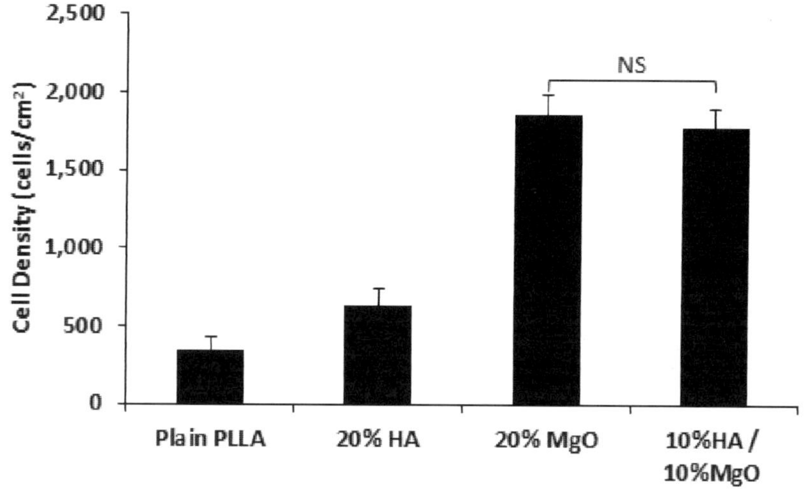


Data is average +/- SEM, N=3. NS signifies no statistical difference between samples

15

# Osteoblast Cell Proliferation

MgO-NPs enhance **OB proliferation** on HA-PLLA n-comp's

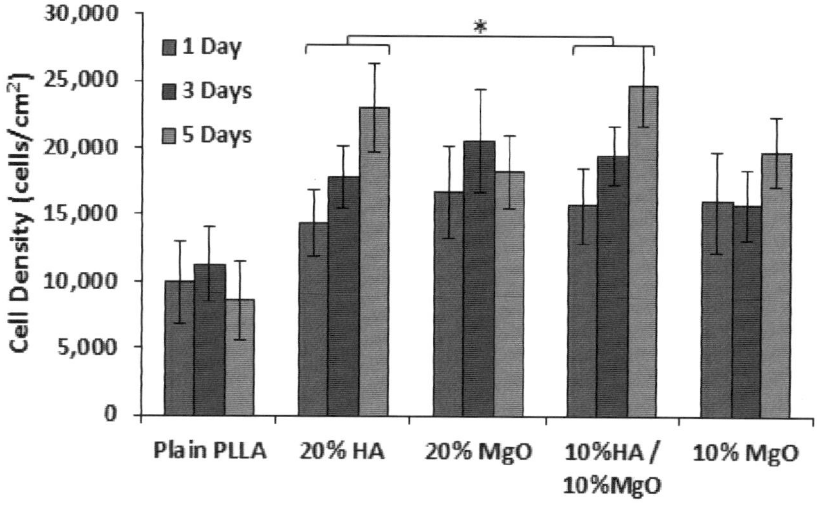


Data is average +/- SEM, N=3. *P<0.05.
Cell densities on all nanocomposites were statistically higher than on plain PLLA at all time points

16

# Mechanical Tensile Testing

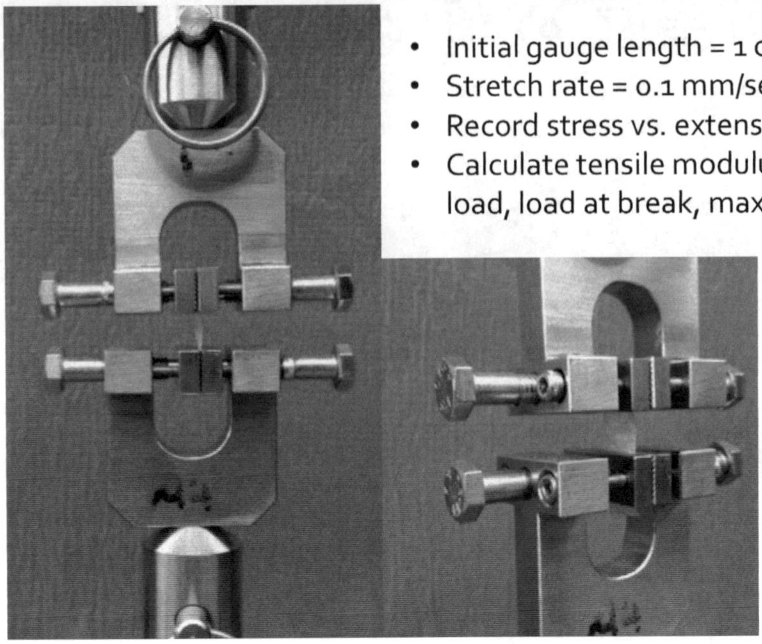

- Initial gauge length = 1 cm
- Stretch rate = 0.1 mm/sec, to sample break
- Record stress vs. extension
- Calculate tensile modulus, extension, max load, load at break, max stress, stress at break

17

# Mechanical Tensile Testing

Nano secondary phase increases the elastic modulus of PLLA

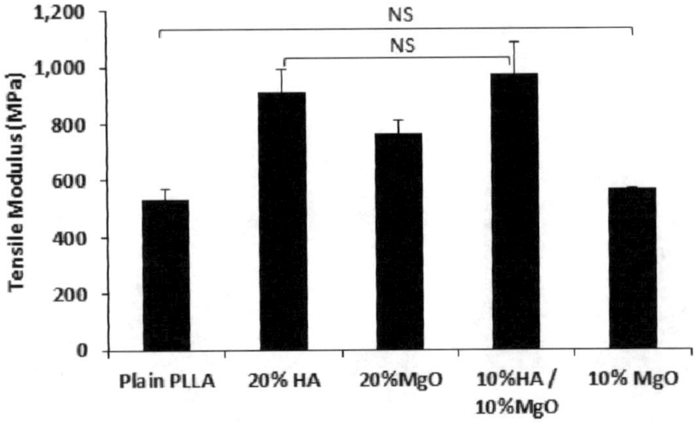


Data is average +/- SEM, N=3. NS signifies no statistical difference between samples

**Typical Young's Moduli:** [1,2]

| | |
|---|---|
| Ligament material | ~300 MPa |
| Tendon material | 1 – 2 GPa |
| Cortical bone | ~17 GPa |
| Cancellous bone | 0.1 – 4.5 GPa |

Modulus of all samples was within acceptable range for cancellous bone


1. Turner *et al.*, J. Biomech. 23, 549-561, 1990.
2. http://wings.buffalo.edu/eng/mae/courses/417-517/Orthopaedic%20Biomechanics/Lecture%203u.pdf


18

# Material Elongation Decreases with Addition of Secondary Phase

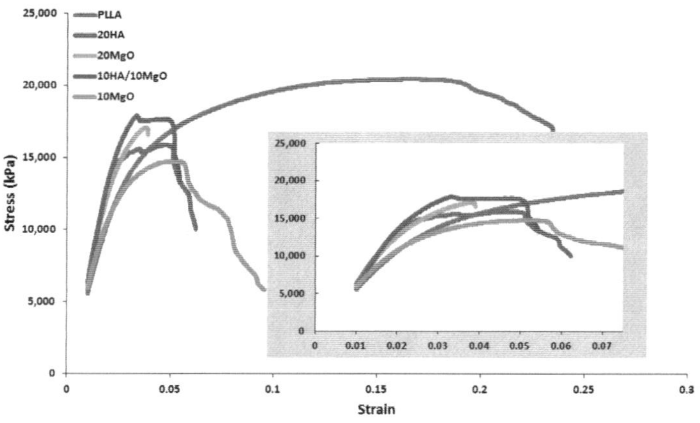


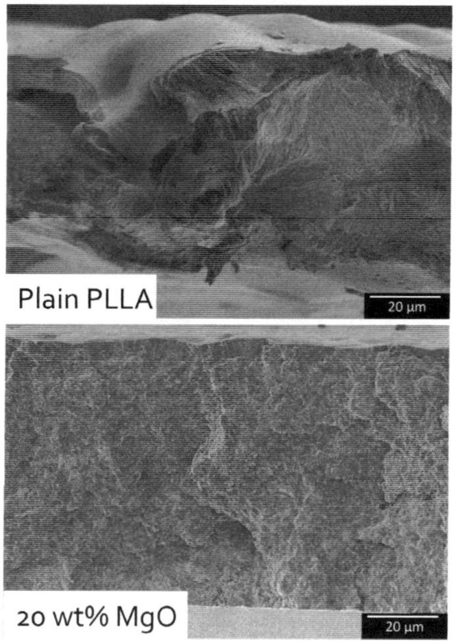


Analysis of SEM images after tensile testing gives insight into the modes of failure

19

# Nanocomposite Degradation

Percent weight reduction of 1-$cm^2$ nanocomposite samples incubated at standard conditions for 7 days in 2 mL of complete osteoblast media

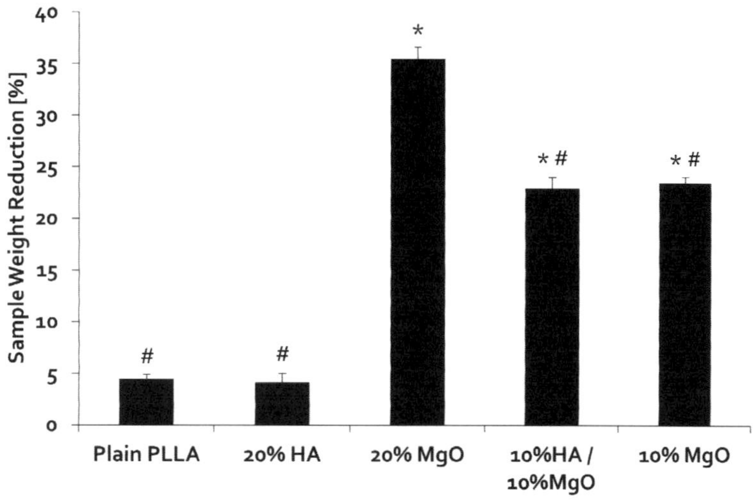


Data represents mean +/- SEM, N=3. *$p<0.05$ compared to plain PLLA, #$p<0.05$ compared to 20% MgO

20

# Degraded Nanocomposites

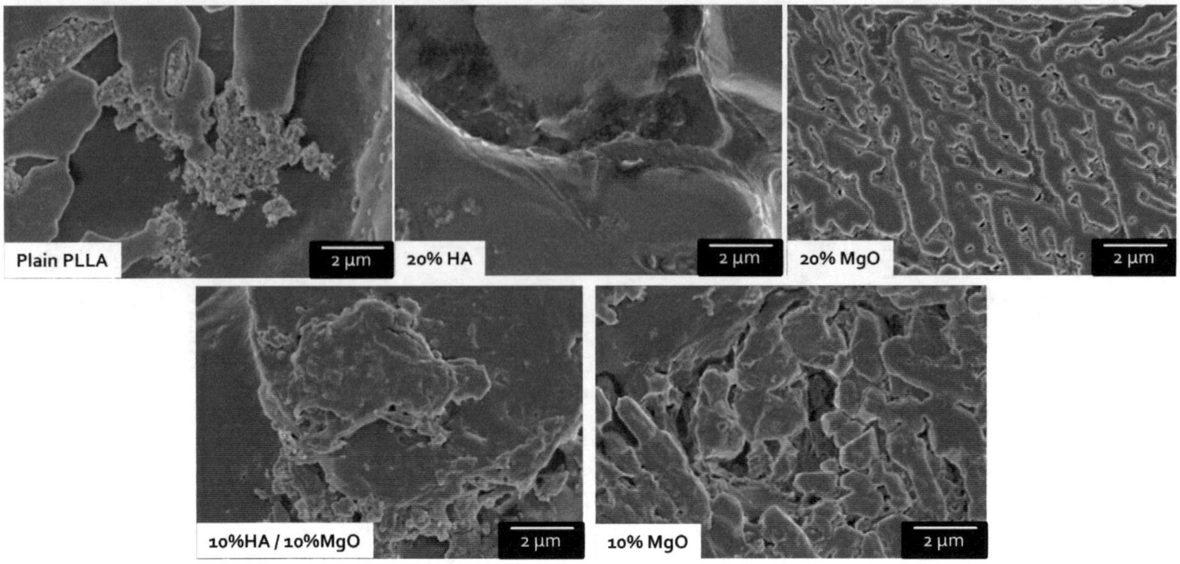


SEM images (mag = 10k) show degraded nanocomposite surfaces

21

# Nanocomposite pH Variations

pH variations of 2 mL of complete osteoblast media containing 1-$cm^2$ nanocomposite samples incubated at standard conditions for 7 days

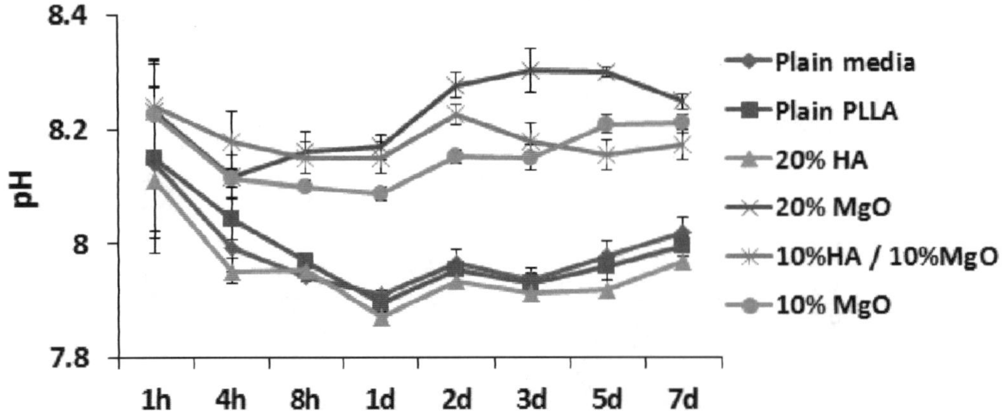


$$MgO + H_2O \rightarrow Mg(OH)_2$$
$$Mg(OH)_2 \leftrightarrow Mg^{2+} + 2\,OH^-$$

22

# Nanoparticle pH Variations

pH variations of 2 mL of complete osteoblast media containing indicated nanoparticle concentrations [mg/mL] incubated at standard conditions for 7 days

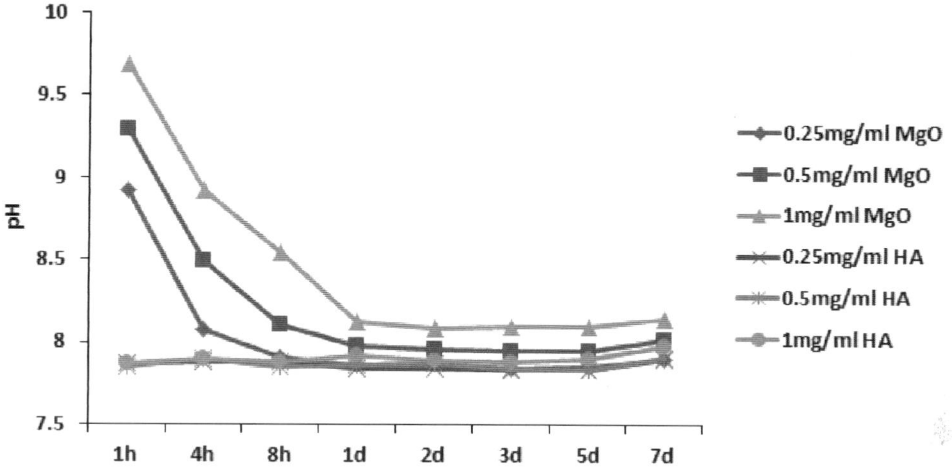


23

# Osteoblast Proliferation in Conditioned Medium

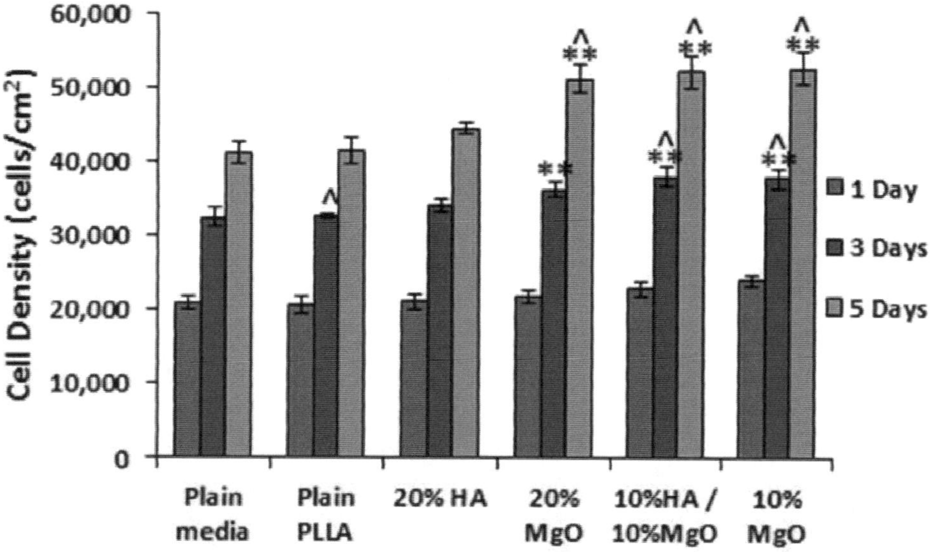


Data represents mean +/- SEM, N=3. $^{**}p<0.05$ compared to plain media, ^$p<0.05$ compared to 20% HA
All cell densities under the same treatment were statistically greater with time

24

# Conclusions

- MgO nanoparticles improve osteoblast adhesion and proliferation
  - Enhanced fibroblast adhesion also
- MgO degradation products improve osteoblast proliferation
- Adding MgO nanoparticles to HA composites could improve bone integration
- Antibacterial properties of MgO NPs could reduce biomaterial infections

25

# Acknowledgements


- Webster Nanomedicine Lab
- Rob Eagan, Bill Fowle, Linlin Sun, Batur Ercan
- NEU Department of Chemical Engineering
- NSF-IGERT Grant No. 0965843

# 394697 NSEF Young Investigator Award Presentation: Engineered Hydrogels for Regenerative Medicine Applications

**Abstract:**

Engineered materials that integrate advances in polymer chemistry, nanotechnology, and biological sciences have the potential to create powerful medical therapies. Our group aims to engineer tissue regenerative therapies using water-containing polymer networks, called hydrogels, that can regulate cell behavior. Specifically, we have developed photocrosslinkable hybrid hydrogels that combine natural biomolecules with nanoparticles to regulate the chemical, biological, mechanical and electrical properties of gels. These functional scaffolds induce the differentiation of stem cells to desired cell types and direct the formation of vascularized heart or bone tissues. Since tissue function is highly dependent on architecture, we have also used microfabrication methods, such as microfluidics, photolithography, bioprinting, and molding, to regulate the architecture of these materials. We have employed these strategies to generate miniaturized tissues. To create tissue complexity, we have also developed directed assembly techniques to compile small tissue modules into larger constructs. It is anticipated that such approaches will lead to the development of next-generation regenerative therapeutics and biomedical devices.

# 394699 NSEF Forum Award Presentation: Nanoscale Organic Hybrid Materials

**Abstract:**

Polymer-nanoparticle composites are used in virtually every field of science and technology. When the particles approach nanometer dimensions, large interfacial regions are created, which present opportunities and challenges for both basic science and applications. In favorable situations, both the size and spatial distribution of these interfaces can be controlled to create new hybrid and nanoporous materials with physical and transport properties inaccessible in their pure constituents or in their conventionally prepared mixtures. This talk focuses on a particluar class of hybrid materials created through bottom-up synthesis of nanoparticles densely grafted with short polymer chains. At high grafting densities, these systems spontaneously form self-suspended suspensions in which each nanoparticle carries a share of the suspending polymeric fluid tethered to its surface. Consequences of this intimate coupling of a suspended and suspending material on nanocomposite structure, phase stability, and transport properties are explored in the talk. Because each hairy nanoparticle building block that constitutes a self-suspended suspension is itself a nanoscale hybrid, the materials provide convenient systems for studying confinement dynamics of polymers and flexible platforms for synthesis of nanoporous materials. Towards application of these hybrid materials in next-generation energy storage technologies, the talk will consider the stability of electrodeposition of metals in a structured electrolyte created by internalizing charges tethered to nanoparticles in the pore-space between self-suspended particles. Generalization of these ideas to create novel nanoporous electrolytes that stabilize electrodeposition of metals will be briefly discussed in the context of next-generation rechargeable battery technologies.

# 370762 Lipidic Templates and Coatings for Designing Nanotheranostics

**Abstract:**

Lipidic platforms, such as liposomes, have been used for decades as drug delivery vehicles and remain as one of the most effective platforms. This stems in part from their biocompatibility, their ability to encapsulate hydrophilic and hydrophilic cargo, and their chemical and biophysical diversity. In the quest to improve the detection and treatment of diseases, there has been significant recent interest in capitalizing on these properties and creating nanoscale assemblies capable of combined diagnostics and therapy (i.e. nanotheranostics) by merging lipidic platforms with active or functional nanotechnologies. This presentation will focus on two examples where lipidic approaches are being used to design multifunctional nanotheranostics based on gold or iron oxide that respond to external stimuli. The first example is a templating technique where small hydrophobic nanoparticles (ca. 2-5 nm) are embedded within liposomal bilayers. The liposome templates the formation of a 'loose' nanoparticle shell and the embedded nanoparticles control the release of encapsulated molecules from within. This control stems from the coupled phase and transport behavior of the bilayers, which is dependent upon nanoparticle concentration and size. By using iron oxide nanoparticles, intracellular release of doxorubicin was achieved *in vivo* using an alternating electromagnetic field. The second example also involves iron oxide nanoparticles, but in this case the nanoparticles are larger (ca. 30 nm) and susceptible to permanent and alternating magnetic fields. A lipid monolayer shell is formed on the surface through a solvent exchange technique that yields monodispersed particles. The particles achieve extremely high uptake *in vivo* and are able to control the loading and intracellular delivery of small interfering ribonucleic acid (siRNA). These examples highlight the role lipidic platforms can play in designing stimuli-responsive nanotheranostics that are capable of achieving multiple therapeutic objectives.

# 392528 Invited: Engineering the Protein/Nanomaterial Interface for Biosensors and Biofuel Cells


**Abstract:**

The optimization of the protein/nanomaterial interface is a critical in the development of bioelectrocatalytic systems such as biosensors and biofuel cells. We have been developing protein engineering strategies to produce enzymes optimized for performance in these systems. In this presentation, I will review some recent accomplishments in this endeavor.

One key challenge in producing bioelectrocatalytic systems is the immobilization of enzymes and mediators on an electrode with high density while maximizing activity. We have developed self-assembling proteinaceous hydrogels for this purpose, and we can incorporate mediators in this system as needed. We have recently expanded this approach to make a biocatalytic hydrogel comprised of a 3-enzyme metabolic network that can oxidize methanol to $CO_2$.

Some enzymes require soluble cofactors, and this can limit performance. We have recently engineered an alcohol dehydrogenase to have both altered substrate and cofactor specificities. Now we have explored its ability to utilize a smaller non-natural cofactor and we have demonstrated the advantages of using artificial cofactors on biofuel cell performance.

In some applications, mediatorless systems are preferred, and we have explored the appending of nanomaterials on enzymes to serve as redox relays. We have been able to functionalize glucose oxidase with a gold nanoparticle and demonstrated direct electron transfer with this system.

We are currently exploring protein assemblies for improved biocatalytic performance. We are reverse engineering a classic metabolon structure from the mammalian TCA cycle. We are also using computational protein design to create new self-assembling enzymatic complexes. The most recent results of these new projects will also be discussed.

# 392527 Invited: Tailored Virus-like Particles for Biomedicine and Catalysis

**Abstract:**

Viruses and virus-like particles are the largest scaffolds conveniently available to the molecular scientist with structures known to atomic resolution. They therefore represent unique tools with which to explore a variety of questions in nanoscience, biology, and materials research. Our work in recent years has focused on the particle derived from the bacteriophage Q-beta capsid as a model system. Methods for the chemical and genetic modification of the coat protein, its targeting to particular cell types, and its use in immunology and as a container for functional biological molecules will be described.

# Characterization of Graphene, Graphene Oxide and Nanographite Dispersions using NMR Solvent Relaxation

**David Fairhurst, PhD.**
XiGo Nanotools, LLC., Bethlehem, PA 18015, USA
and
**Terence Cosgrove, DSc.**
University of Bristol, Bristol BS8 1TS, UK

AIChE 2014
November 17th 2014
Atlanta, GA



# Nanocarbon Materials

**Enhanced performance capabilities**
stable physicochemical properties, low cost, excellent chemical stability, wide operating temperature range and long cycle life

**Formulations**
opaque, (concentrated) dispersions in aqueous and non- aqueous media

**Physical Characterization**
poses challenges

XiGo Nanotools 2014

## NMR Solvent Relaxation Measurements

NMR experiments not only useful for identification of unknown compounds

*Solvent* relaxation measurements can be used to probe details of molecular structure and dynamics at particle-liquid interfaces

Particulate Analysis

- ❑ wetted surface area
- ❑ influence of morphology/porosity
- ❑ impact of surface chemistry

Molecular Analysis

- ❑ surfactant and polymer adsorption
- ❑ polymer complexation
- ❑ competitive adsorption/desorption

CL.Glaves, PJ Davis and DM Smith, *Surface Area determination via NMR: Fluid and Frequency Effects*, Powder Technology 54 (4) 261-269. (1988)

XiGo Nanotools 2014

## Practical Advantages of NMR Solvent Relaxation

- ❑ any type of particle
- ❑ any liquid (with at least one H atom)
- ❑ includes mixtures of liquids (miscible)
- ❑ high concentrations
- ❑ no dilution, no sample preparation
- ❑ rapid measurements
- ❑ small samples
- ❑ non-invasive, non-destructive

XiGo Nanotools 2014

## How it works...

- H nuclei in liquids behave like little "magnets"

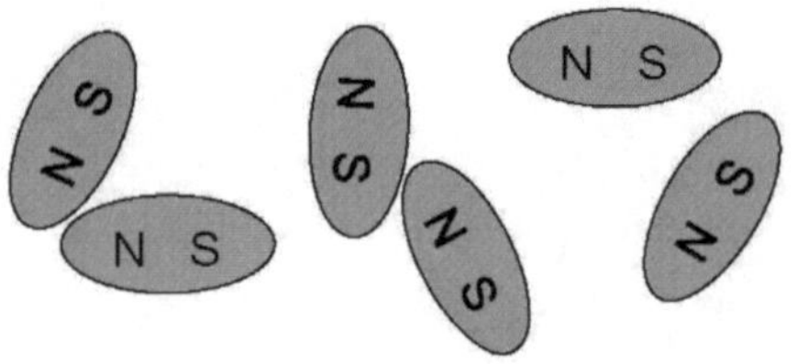


- without any external field → random orientation

XiGo Nanotools 2014

## How it works...

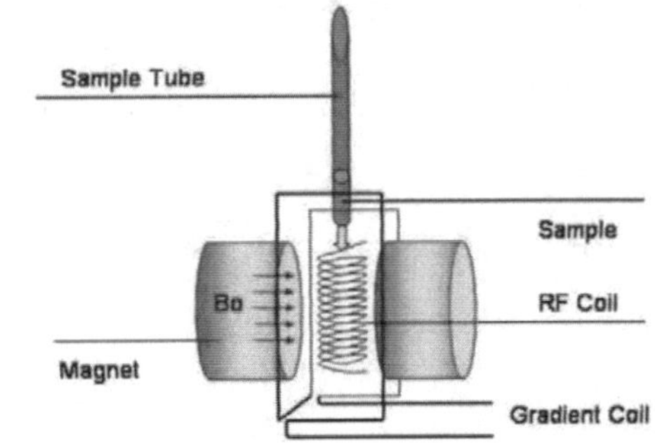


- An external field, $B_0$, aligns the nuclear spins

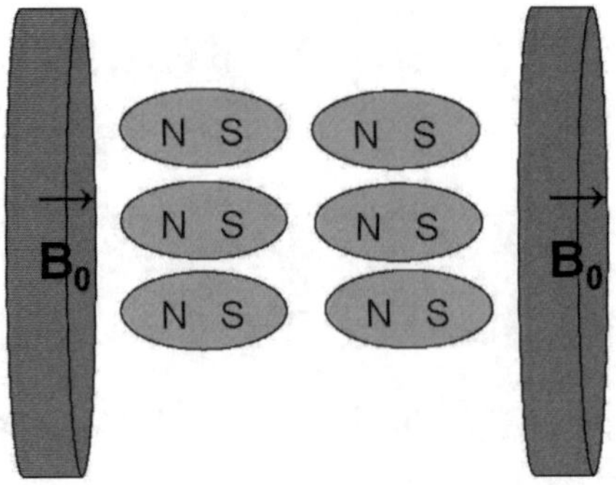


XiGo Nanotools 2014

# How it works…

- A short RF Pulse, $B_1$, changes the orientation of the H atoms

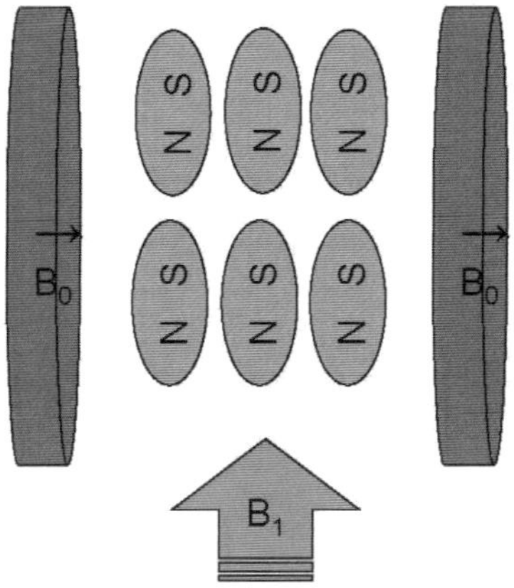


- When the field is switched off $B_1$ disappears and the H atoms then realign with $B_0$ producing a signal

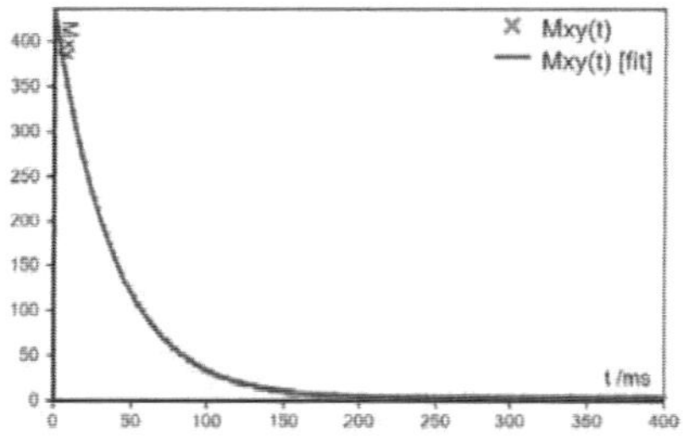


Mxy fit for $T_2$ measurement

XiGo Nanotools 2014

# Proton NMR relaxation

Liquid molecules adsorbed at an interface have a much shorter relaxation time compared with that in the bulk fluid: difference can be orders of magnitude

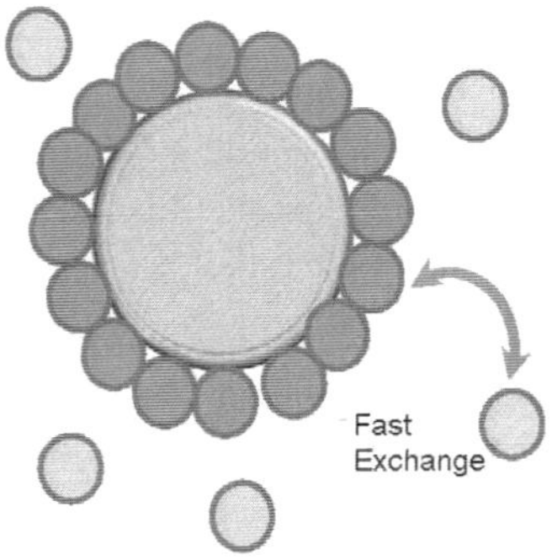


We distinguish between two liquid phases:

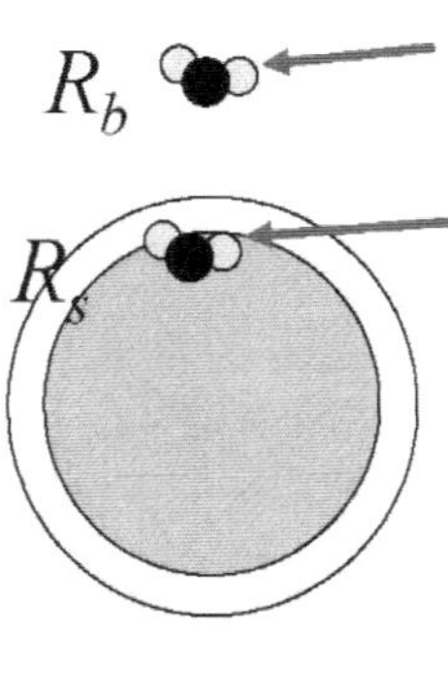


**Changes in the extent of the surface, or the rate of exchange, can be measured with NMR**

XiGo Nanotools 2014

## NMR Relaxation: Surface, Liquid and Suspension

xigo nanotools

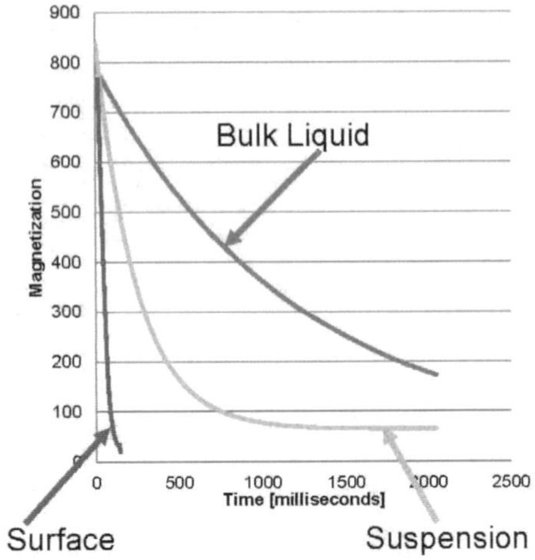


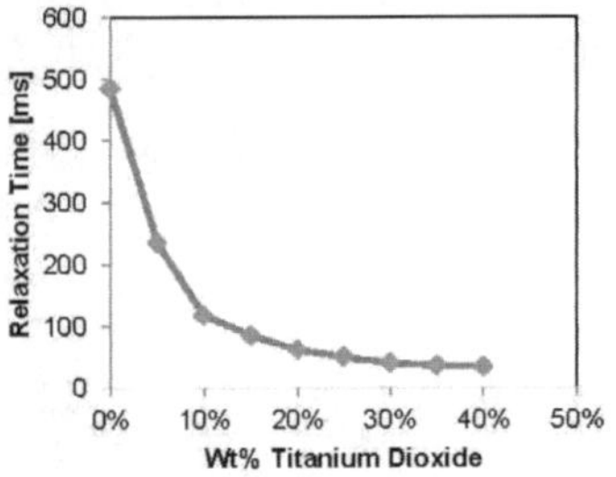


Relaxation rate, R = 1/relaxation time, T

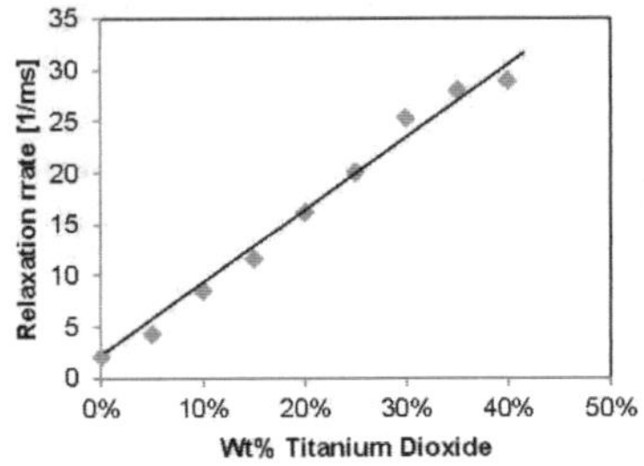


XiGo Nanotools 2014

## Proton NMR Relaxation Rate

xigo nanotools

Each phase, bulk (or "free") liquid and adsorbed (or "bound") liquid has a characteristic relaxation time, T and rate, R (=1/T)

The average relaxation rate of any suspension, $R_{d(av)}$, is:

$$R_{d(av)} = p_s R_{ns} + p_b R_{nb}$$

n=1 for spin-lattice relaxation ($T_1$ method)
n=2 for spin-spin (or spin-echo) relaxation ($T_2$ method)
$p_b$ is the fraction of solvent in the bulk phase
$p_s$ is the fraction of solvent at the surface

Relaxation time ↓ as wetted particle surface area↑

XiGo Nanotools 2014

# $T_1$ or $T_2$ relaxation?

- $T_2$ is more sensitive to changes in surface area
  - $T_2$ is better for more dilute suspensions
  - $T_2$ can become very small at high concentrations
- $T_1$ works across the largest range of concentration
  - $T_1$ is best for highly concentrated suspensions
- $T_2$ is faster, so less affected by settling
- $T_1$ and $T_2$ have a different relaxivity, $k_a$

Choice of $T_1$ or $T_2$ depends upon specific application
The XiGo Acorn Area™ provides both measurements*

* Patents issued in the US, China, Japan and EU

XiGo Nanotools 2014

# NMR Relaxation: Non-porous Materials

Relation between NMR relaxation and the wetted surface area of suspensions given by:

$$R = k_A S \varphi + R_b$$

Where,
R is the average relaxation rate constant
$\varphi$ is the particle volume-to-liquid volume ratio
S is the total surface area per unit weight
$R_b$ is the relaxation rate constant for the bulk suspension liquid

Using a standard reference material we can define the constant, $k_A$, the specific surface relaxivity for the solid/liquid system

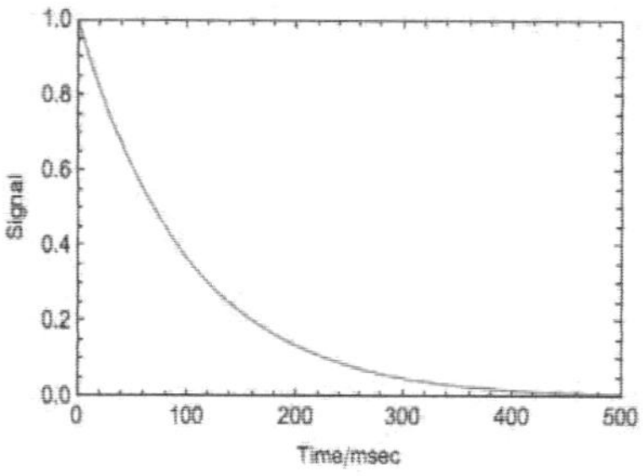

Schematic of single exponential

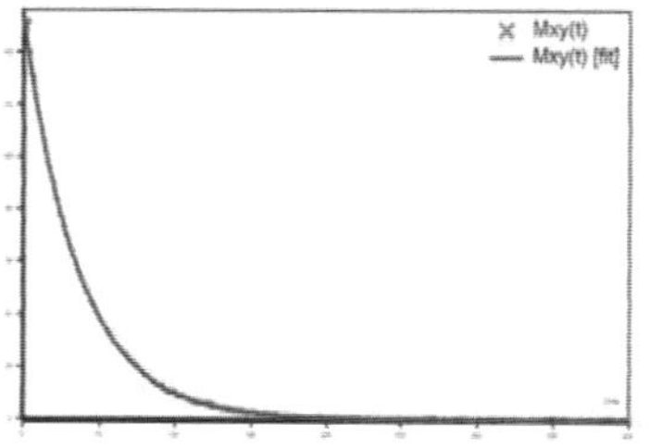

Experimental relaxation plot for a sample of a 25% 100nm PU latex

XiGo Nanotools 2014

# Nanocarbon

5w/w% suspension in THF

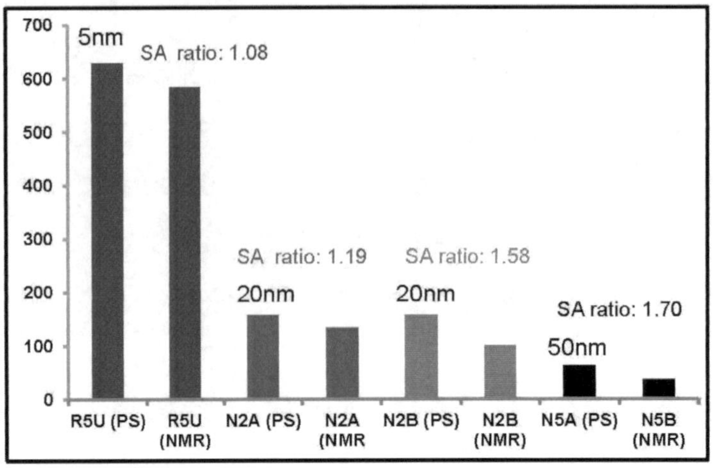


NMR relaxation provides a simple and fast measurement of wetted SA; it is sensitive to the degree of aggregation in a dispersion

# Graphene Oxide

Known to contain a range of reactive oxygen functional groups

→ presence of surface hydroxyl groups.

→ increase in hydroxyl content → increased wettability

Two dilute aqueous dispersions

→ different oxidation states: sample 1 > sample 2*

→ measured as received

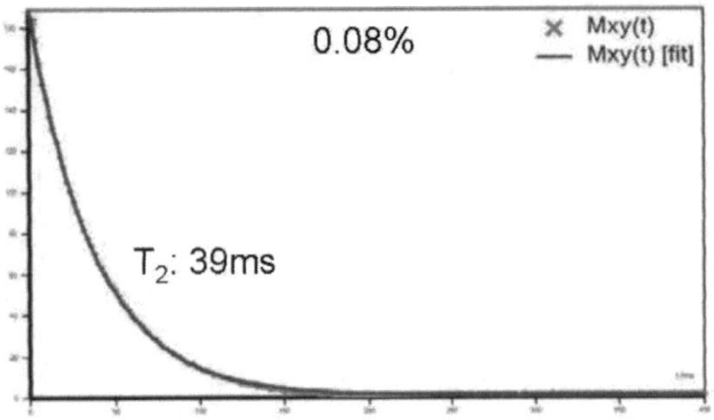


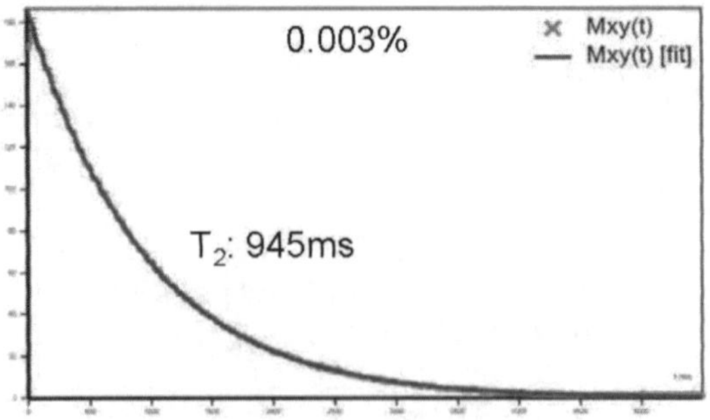


Sample 1: SSA = 1326$m^2g^{-1}$

Sample 2: SSA = 786$m^2g^{-1}$

NMR relaxation can be used to probe the physico-chemical nature of the particle-liquid interface (surface chemistry)

* Luna Imperiali, PhD Thesis, The Catholic University of Leuven, Belgium

XiGo Nanotools 2014

# Nanographite

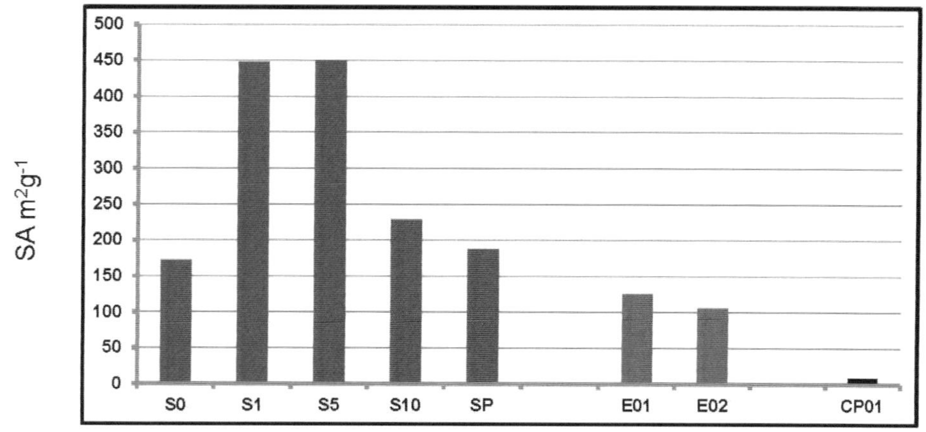


NMR relaxation measurements can be used to optimize dispersibility

XiGo Nanotools 2014

# NMR Relaxation: Porous Materials

NMR relaxation behavior of a liquid inside a porous material different depending upon specific (morphological and chemical) characteristics:
→ pore volume, pore throat diameter and the interfacial chemistry

Two separate magnetisation decays may be observed:

$$M(t) = M_e \exp[-R_e t] + M_i \exp[-R_i t]$$

Where,
$R_i$ and $R_e$ are the average relaxation rates for internal(i) and external(e) environments
M(t) is the total signal
$M_e$ and $M_i$ are the signals arising from the total liquid in the external and internal system

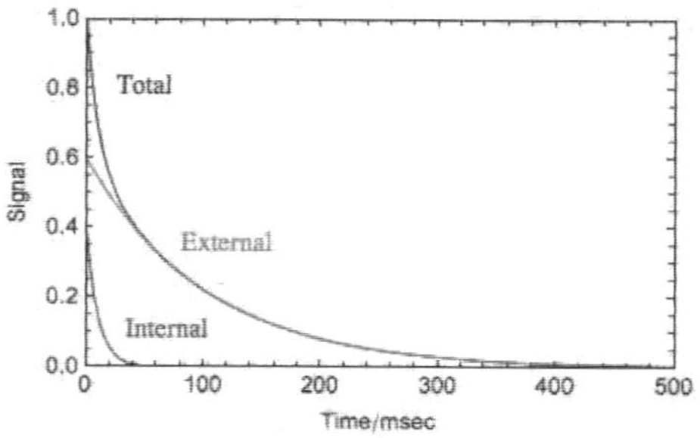


Schematic of double exponential

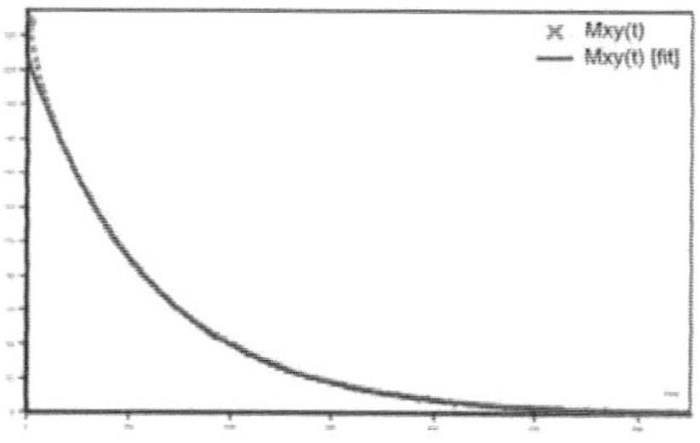


Experimental relaxation plot for a sample of a 10wt% 1μm porous silica

XiGo Nanotools 2014

# Porous Graphene

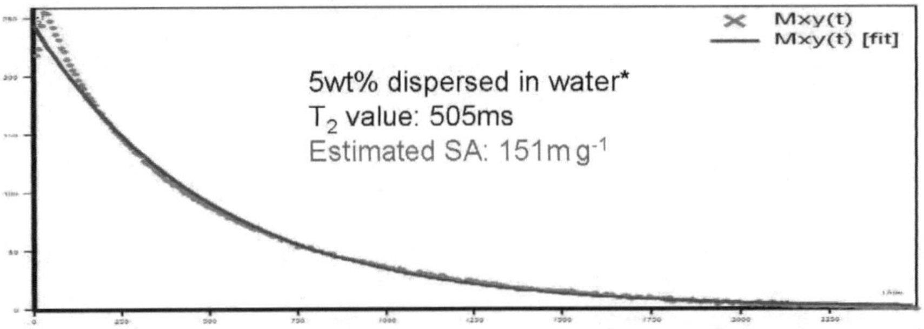


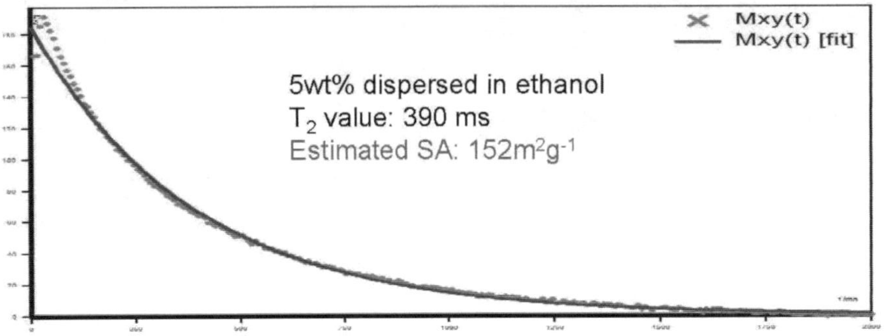


* With 1% surfactant (a naphthalene polycondensate)

XiGo Nanotools 2014

# Porous Graphene

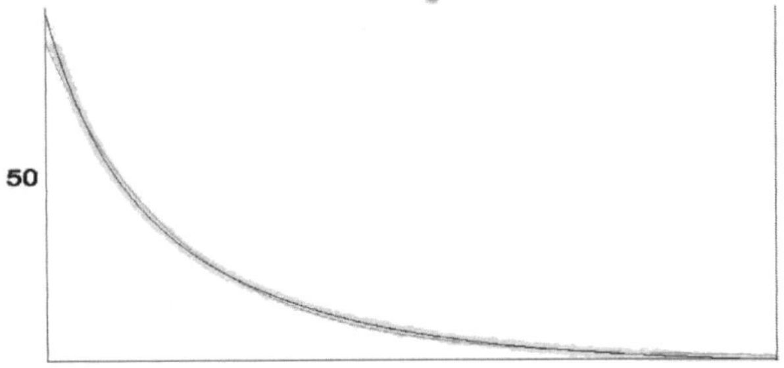


Intensity (1): 101.2
Relaxation (short): 177.4 ms
Estimated SA: 803m$^2$g$^{-1}$
Intensity (2): 186.8
Relaxation (long) 639.5 ms

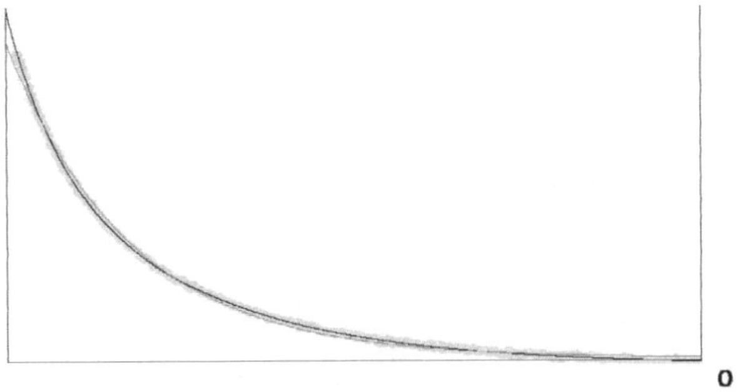


Intensity (1): 80.3
Relaxation (short) 137.7 ms
Estimated SA: 1200m$^2$g$^{-1}$
Intensity (2): 140.7
Relaxation (long) 492.3 ms

# In Conclusion

- NMR relaxation is a powerful technique for probing events and interactions at any particle-liquid interface
- Optimize Formulation Development of Nanocarbon Materials
  - wetted surface area
  - adsorption of surfactants/polymers/dispersants
  - effects of milling
- Almost any liquid can be studied
- Wide solids concentration range (ca 0.1% to 60+%); no dilution and no sample preparation
- The technique is non-destructive; sample can be stored and re-measured
- Measurements in a few minutes
- Useful for R&D and QC



# The Acorn Area and Flow

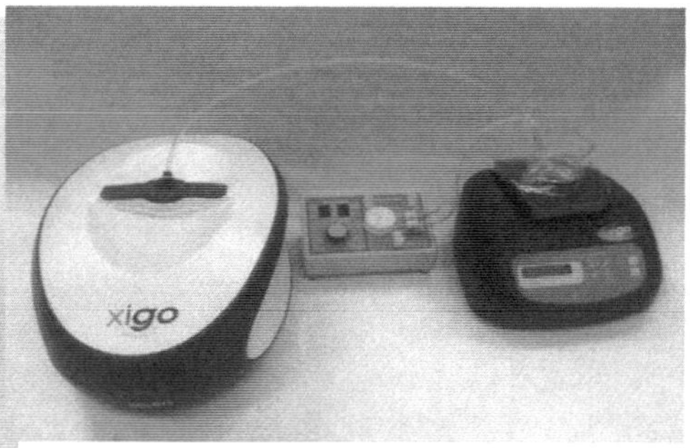


www.xigonanotools.com

# 366398 Designer Stabilizer for Preparation of Pristine Graphene/Polysiloxane Films and Networks

**Abstract:**

A designer polysiloxane-based stabilizer for graphene was synthesized and applied to prepare a highly conductive polymer film. This stabilizer was simultaneously used as the polymer matrix. To synthesize the stabilizer, 1-ethynylpyrene was grafted to the backbone of a poly(dimethylsiloxane)-*co*-(methylhydrosiloxane) random copolymer by Pt-catalyzed hydrosilylation with a SiH: ethynyl ratio of 1.0:1.3. Dispersion of graphene in chloroform was prepared through the $\pi$-$\pi$ interactions between the graphene sheets and the pyrene groups of the resulting copolymer. A graphene/polymer film was cast from this dispersion. SEM and TEM images confirmed the homogeneous distribution of the graphene sheets in the film. The conductivity of this film with 4 wt% loading of graphene was measured to be 220 S/m. This is the first case of a melt-processable, conductive graphene/polymer film reported in the literature. Later the ratio of SiH:ethynyl was changed to 1.7:1.0, which led to self-crosslinking of the polymer at 110 C and resulted in a direct production of a conductive graphene/silicon elastomeric composite. The crosslinking reaction was observed by FT-IR spectroscopy and the network formation was confirmed by swelling and extraction of the product.

**Molecular order in graphene-oxide suspensions and graphene-based Films: What do we learn from it?**

Rachel Tkacz[a], Rudolf Oldenbourg [b], Shalin Mehta[b], Mainak Majumder,[a]

a. Nanoscale Science and Engineering Laboratory (NSEL), Mechanical and Aerospace Engineering, Monash University, VIC 3800, Australia

b. Marine Biological Laboratory, Woods Hole, MA 02543, USA

Graphene-based materials are poised to play a significant role in several technological applications including transparent conducting films, filtration membranes, electrochemical energy storage technologies to corrosion-resistant films. This arises because graphene is a multifunctional material with combination of properties such as large surface area, high electrical conductivity and mechanical strength, low light absorption and chemical inertness. It is becoming increasingly apparent that graphene produced by exfoliation of graphite in liquids possess advantages regarding production scalability and cost-efficiency. While research efforts on achieving applications are burgeoning, correlating processing and property relationships of these materials, in general, is under explored, yet vital to developing high performance materials. To this end, we use orientation-mapped polarized light imaging to determine the fine-structure, orientation of graphene sheets, molecular order and dynamic behaviour in the liquid state and dry state by diattenuation and birefringence imaging (see figure 1 for images of GO sheets around an air bubble and highly ordered films). In doing so, we have recently discovered a confined liquid crystalline droplet phase of graphene oxide with ($20 \pm 2.8$ μm diameter) and an isotropic interior. [1] This novel droplet-phase could find potential application in molecular sensing and drug delivery applications. Additionally, we have identified capillary-force-assisted self-assembly (CAS) as a method for preparation of thin films of chemically reduced graphene oxide (rGO) with unidirectional organization of rGO sheets. The slow axis of the rGO platelets within the CAS films displayed a narrow angular distribution (±3°) within a film area of 1 $mm^2$, resulting in the highest possible order parameter (*S*) of ~1 with 8-fold enhancement of electrical conductivity compared to films with disordered structure. [2] In summary, our studies will enable to deliver high-performing materials from graphene-based assemblies while also develop novel applications by discovery of novel self-assembled architectures.

1. R.Tkacz, R. Oldenbourg, S. B. Mehta, M. Miansari, A. Verma and M. Majumder, “pH dependent isotropic to nematic phase transitions in graphene oxide dispersions reveal droplet liquid crystalline phases”, Chem. Commun, 2014, DOI:10.1039/C4CC00970C

2. R.Tkacz, R.Oldenbourg, A.Fulcher, M.Miansari, M.Majumder, "Capillary-Force Assisted Self-Assembly (CAS) of highly Ordered and Anisotropic Graphene-Based Thin Films", *J.Phys.Chem.C,* 2014, 118 (1), 259–267

0251658240251659264

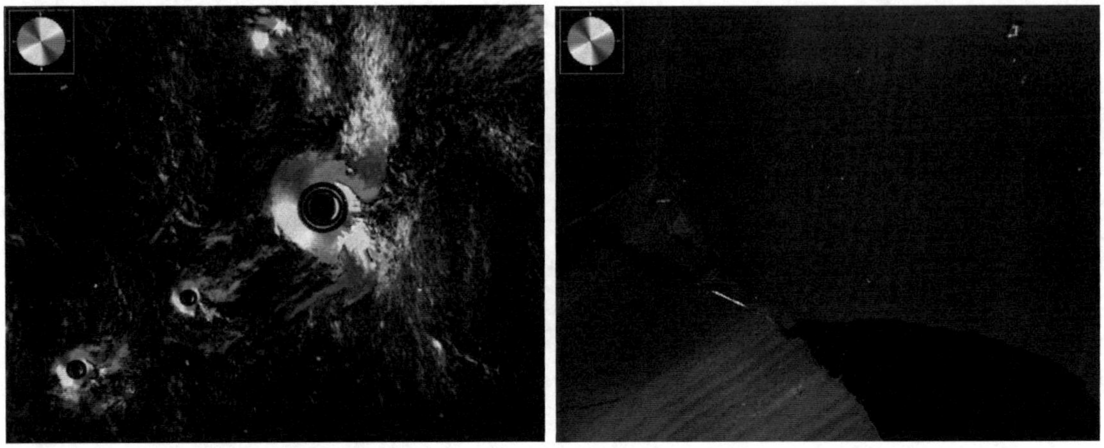

Figure 1: The polarized light LC-PolScope cross-polar processed images displaying retardance by brightness and azimuth of the slow axis by color (as shown by the circular legend) of (right) an air bubble trapped in GO suspension, triggers anchoring forces on the nematics phase, and (left) rGO film formed with the assistance of capillary forces at the meniscus of three phases interface.

# 371095 Microstructural Variations of Various Commercial Single-Walled Carbon Nanotubes Dispersed in Unsaturated Polyester Resin

**Abstract:**

In this work, the dispersion states of different single-walled carbon nanotube (SWNT) grades in unsaturated polyester resin (UPR) were investigated using rheology and microscopy. SG65 and CG200 SWNT from Southwest Nanotechnologies were dispersed in UPR by a previously reported high shear mixing method. The SG65 grade has a dominant (6, 5) chiral content and lower pure SWNT composition; the CG200 grade is dominant in metallic chirality with higher pure SWNT content. Atomic force microscopy (AFM) and x-ray photoelectron spectroscopy (XPS) showed similar SWNT aspect ratios and surface chemistries. Oscillatory and steady shear rheology of the dispersions at similar volume fractions of pure SWNT showed noticeable variations in storage and viscous moduli, crossover frequencies and low shear viscosities. These differences indicate differences in the dispersion microstructure. Since properties of polymer nanocomposites depend on the SWNT dispersion state, the observed rheological variations are expected to correlate to differences in cured composite properties. This work demonstrates that actual SWNT content and chirality can significantly contribute to dispersion microstructure. Thus, the availability of more commercial SWNT grades has made SWNT type selection an important criterion for achieving nanocomposites with enhanced properties.

# Behavior of Polymer – carbon nanotube hybrid particle in porous media

Minh D. Vo, Dimitrios V. Papavassiliou

School of Chemical, Biological, and Materials Engineering

The University of Oklahoma,

Norman, OK -73019

**Abstract**

Nanohybrid particles display many interesting properties for applications in oil and gas exploration and production. They could be used as nanoscale vehicles for catalyst as well as emulsion stabilizers in oil recovery process. In experiments, carbon nanotubes (CNT) are usually dispersed and stabilized successfully in the presence of commercial polymers (like PVP and HEC-10). In addition, the propagation of polymer - CNT particles (nanohybrid) in porous media has been achieved experimentally [1]. However, the exact configuration of the polymers as they wrap around individual CNTs the effects of polymers on the behavior of the nanoparticles in flow in tight spaces are not fully clarified. This study uses the dissipative particle dynamics (DPD) method in order to figure out some properties of nanohybrid particles in porous media. DPD is a coarse grained method, which is promising for studying systems with polymer, surfactants and colloidal particles in the mesoscopic scale [2]. The appropriate DPD parameters for polymers are validated in our simulation by comparing end-to-end distance and radius of gyration obtained from the calculations to those available in the literature. After the system with cylindrical CNTs and polymers reach equilibrium, the final conformation of the nanohybrid particle is obtained. The steric repulsion effects will be discussed. Additionally, different shear rates are investigate to examine the conditions that could lead to an unstable nanohybrid particle in the flow. Finally, the interaction among nanohybrid particles is also considered and the drag, orientation and trajectory of these particles in flow in microchannels will be discussed.

## ACKNOWLEDGEMENTS

The financial support of the Advanced Energy Consortium (AEC BEG08-022) and the computational support of XSEDE (CTS090017) are acknowledged.

## REFERENCES

1. Mohannad J. Kadhum, Daniel P. Swatske, Jeffrey H. Harwell, Ben Shiau, Daniel E. Resasco, "Propagation of interfacially active carbon nanohybrids in porous media", Energy & Fuels, 27, 6518-6527, 2013.
2. Robert D. Groot, Patrick B. Warren, "Dissipative particle dynamics: Bridging the gap between atomistic and mesoscopic simulation," *J. Chem. Phys.,* **107**(11), 4423-4435, 1997.

# Merging Tertiary Amine with Palladium Complex on Graphene Oxide for Promoted Tsuji-Trost Allylation

• **Speaker: Prof. Xiaobin Fan**

School of Chemical Engineering & Technology,
Tianjin University, China
Tel: +86 22-27890090
Email: xiaobinfan@tju.edu.cn

## Introduction

### Education

- **B.S.**: 2003, Chemistry, Nankai University, China
- **M.S.**: 2006, Chemical Engineering, Tianjin University, China
- **Ph.D.**: 2009, Chemical Engineering, Tianjin University, China

### Professional Experience

- **Professor**, Department of Chemical Engineering, Tianjin University, China, 2013- Present
- **Associate Professor**, Department of Chemical Engineering, Tianjin University, China, 2011- 2013
- **Lecturer**, Department of Chemical Engineering, Tianjin University, China, 2009-2011

## Outline

1. Background
2. Research Methodology
3. Experimental Results
4. Conclusions

## 1. Background

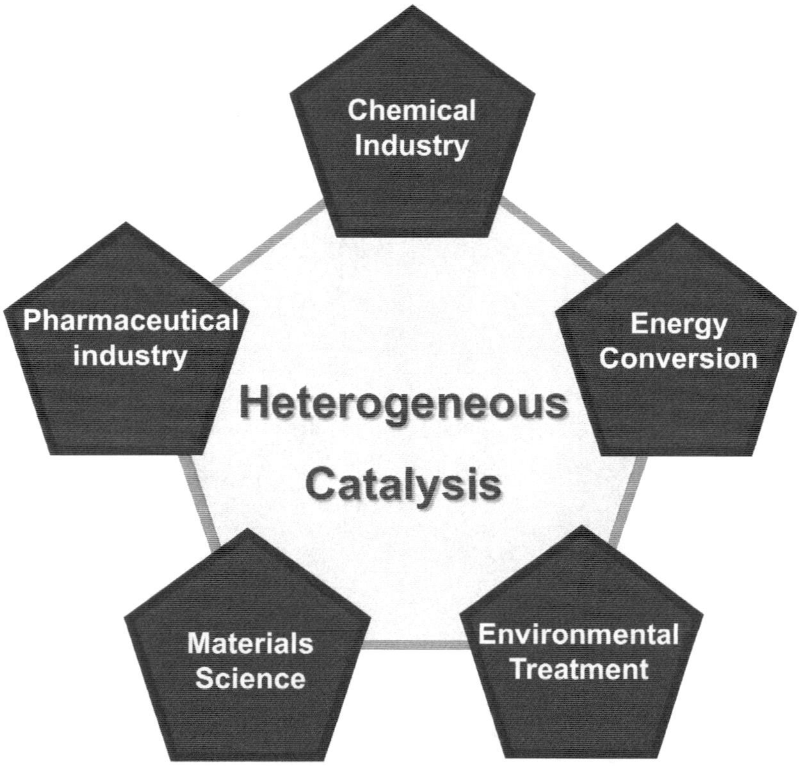

## 1. Background

※ **The classical seven steps for a heterogeneous catalytic reaction**

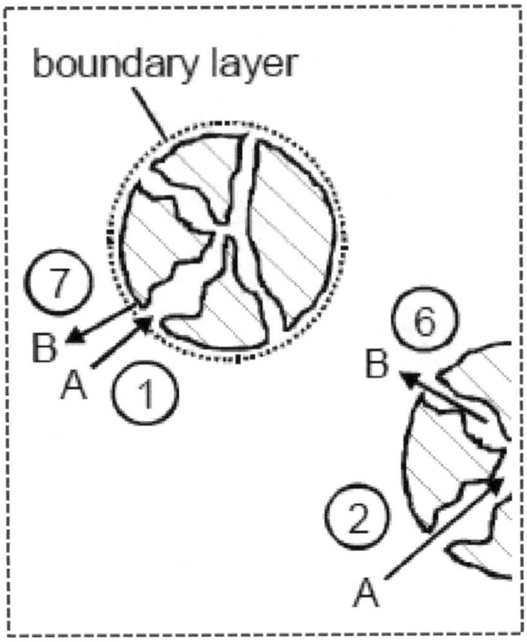


(1) Diffusion of the reactants to the surface of catalyst (External-diffusion);

(2) Diffusion of the reactant into the catalyst pores (Internal-diffusion);

(3) Adsorption of reactants;

(4) Reaction at specific active sites;

(5) Desorption of the products;

(6) Diffusion of the products from the catalyst pores (Internal-diffusion);

(7) Diffusion of the products from the surface of catalyst (External-diffusion).

## 1. Background

※ **Problems of traditional supports**

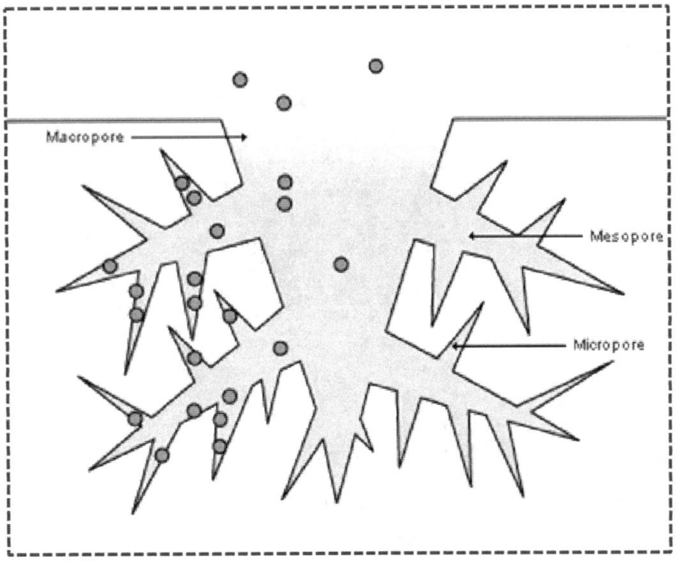


(1) Heterogeneous reaction needs large surface area;

(2) Larger surface area = Porous structure with internal mass transfer resistance;

(3) Many fast reactions are controlled by internal mass transfer.

# 1. Background

## ※ Why graphene?

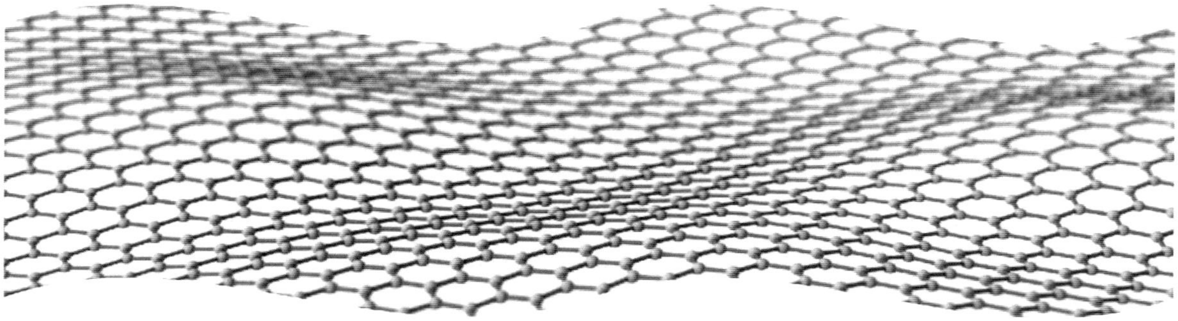

# 1. Background

## ➢ Organocatalysts

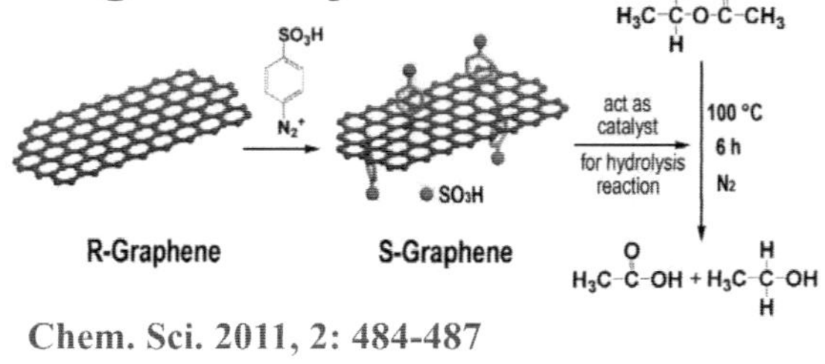


Chem. Sci. 2011, 2: 484-487

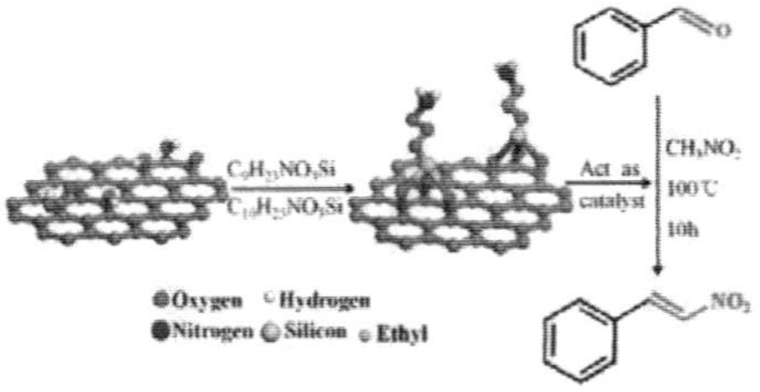


Nanoscale 2013, 5: 6030–6033

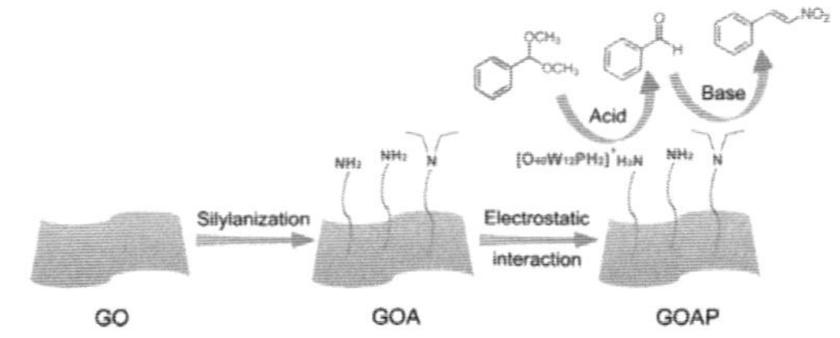


Ind. Eng. Chem. Res. 2014, 53: 1437-1441

## ➢ Organometallic catalysts

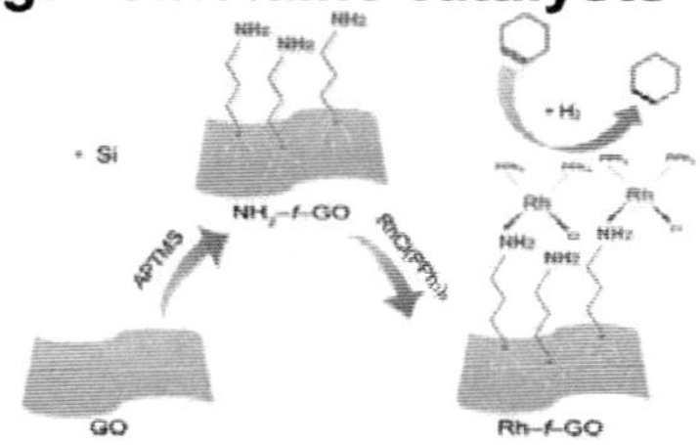


Nanoscale 2013, 5: 882-885

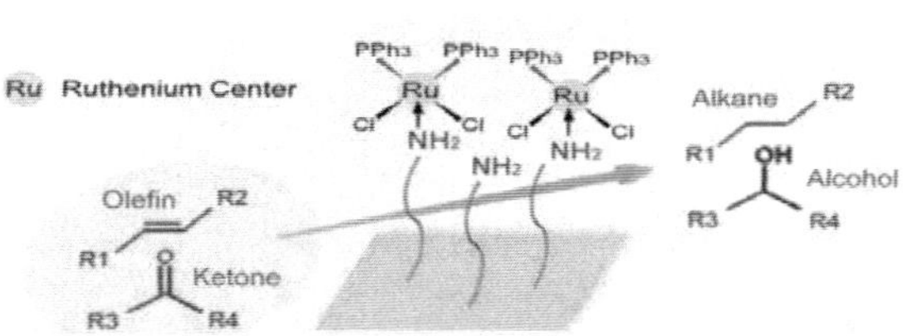


J. Mater. Chem. A 2013, 1: 15039-15045

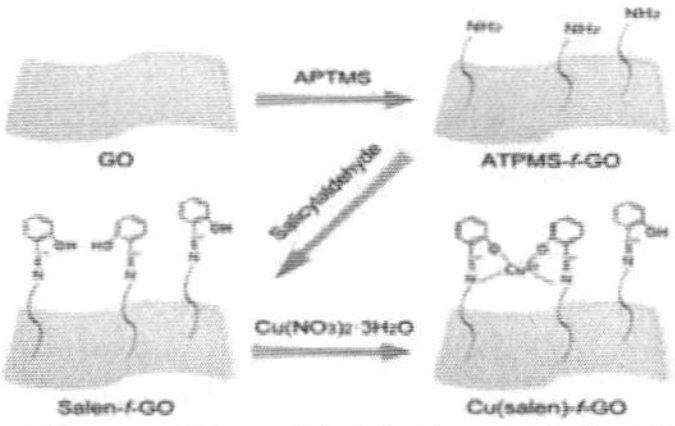


Ind. Eng. Chem. Res. 2014, 53: 4232-4238

# 1. Background

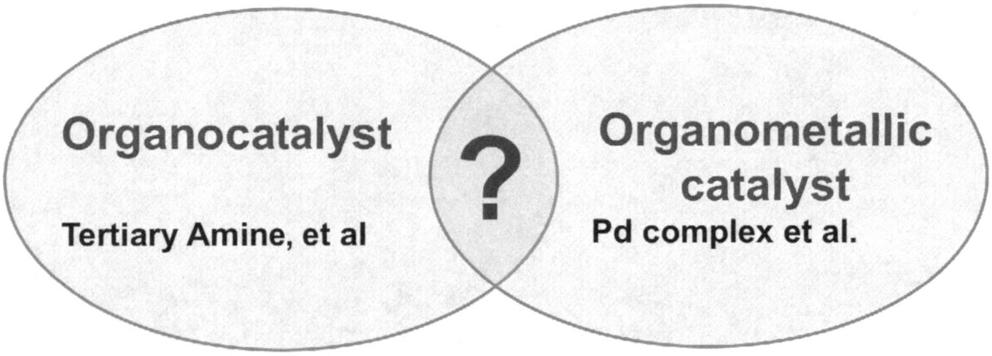


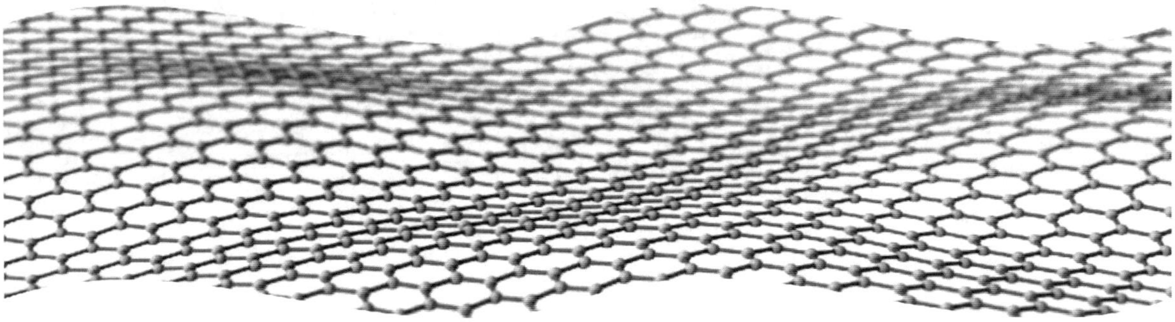

# 2. Research Methodology

**Strategy**

**Step 1**

Chemical modification of graphene oxide **( Silylanization)**

**Step 2**

In situ synthesis of palladium complex **(PdCl2 as the precursor )**

**Step 3**

Characterization & performance test **(FTIR, XPS, SEM, TEM, EDS, GC)**

# 2. Research Methodology

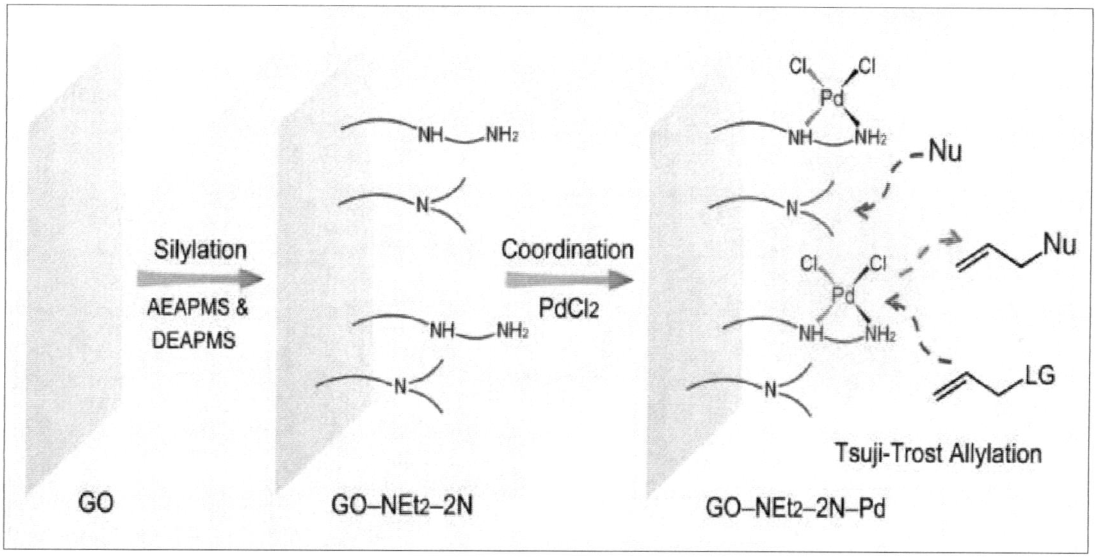


**Illustration for the synthetic methodology**

# 3. Experimental Results

- **FTIR spectra**

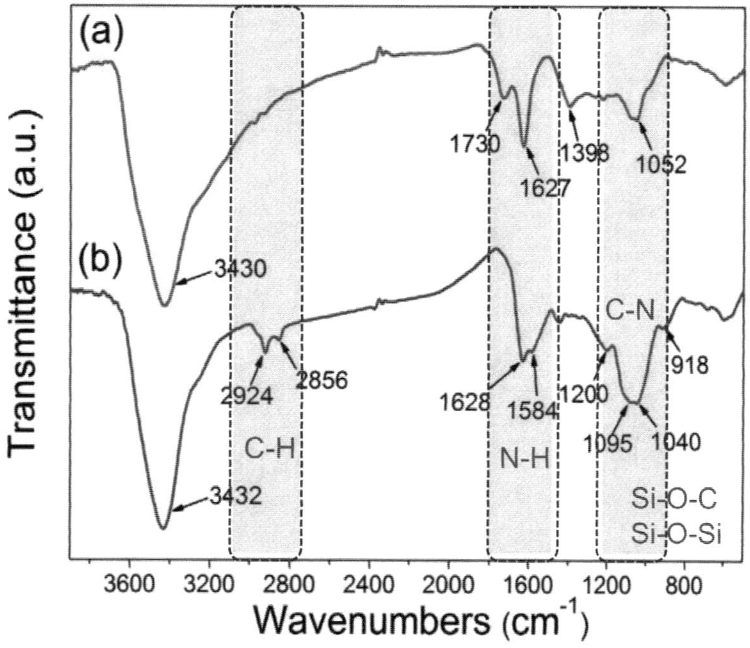

## 3. Experimental Results

➢ **XPS spectra**

**Pd Loading: 1% wt**

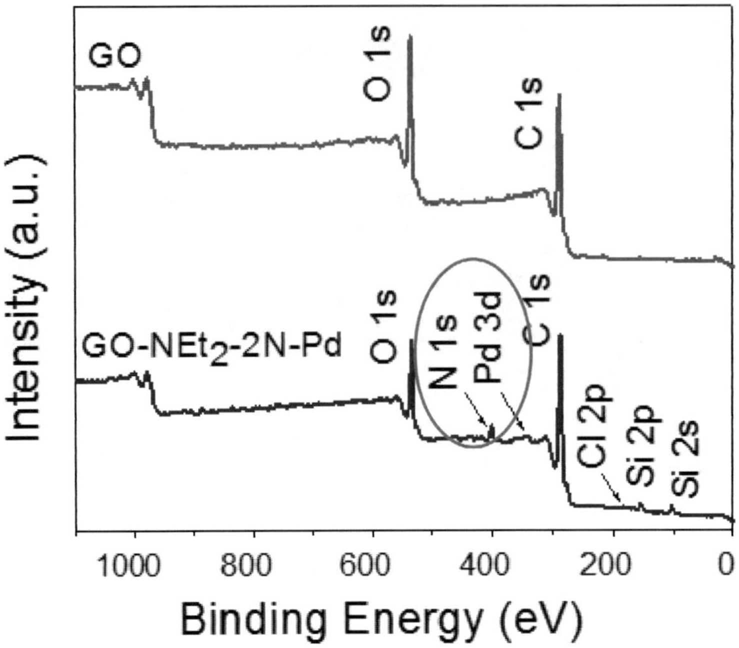


## 3. Experimental Results

➢ **Pd 3d XPS spectra**

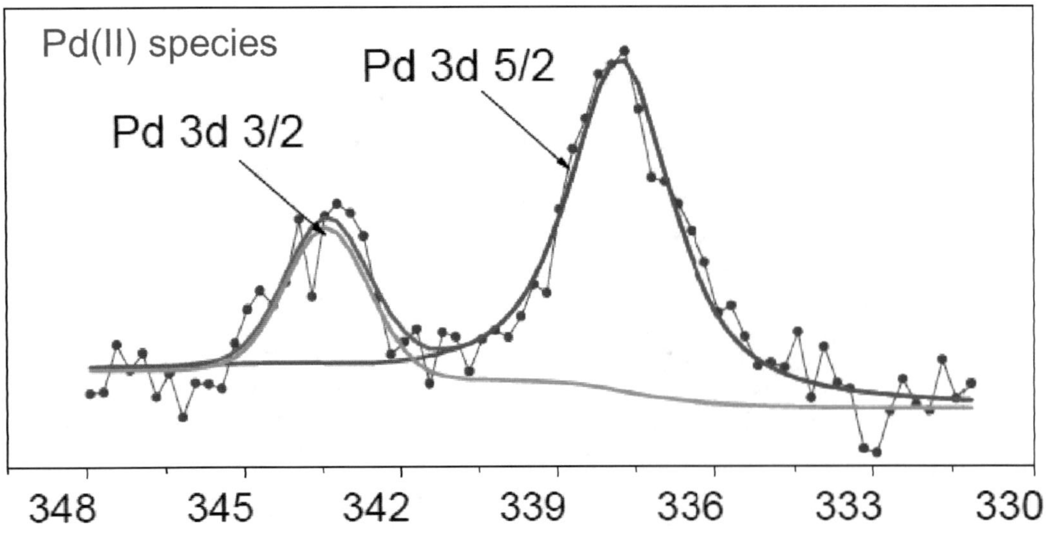

# 3. Experimental Results

- **N 1s XPS spectra**

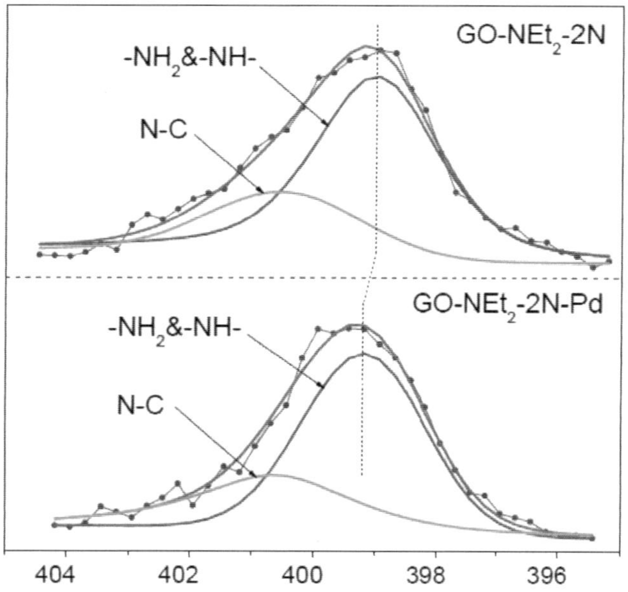


# 3. Experimental Results

- **Morphology**

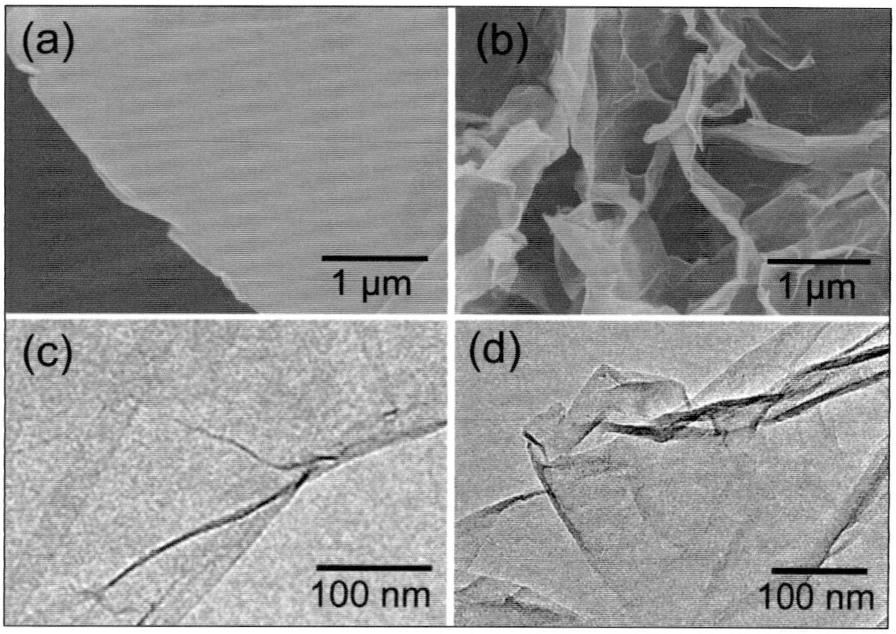


SEM and TEM images of (a) (c) GO and (b) (d) GO-NEt$_2$-2N-Pd.

# 3. Experimental Results

## ➢ Distribution

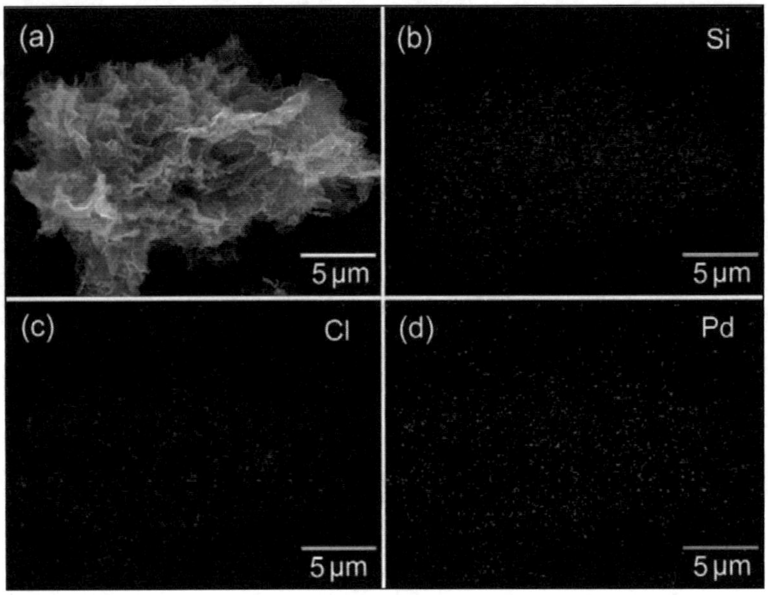


SEM image and corresponding EDS element mapping.

# 3. Experimental Results

## ➢ Performance test

The model reaction: Tsuji–Trost allylation.

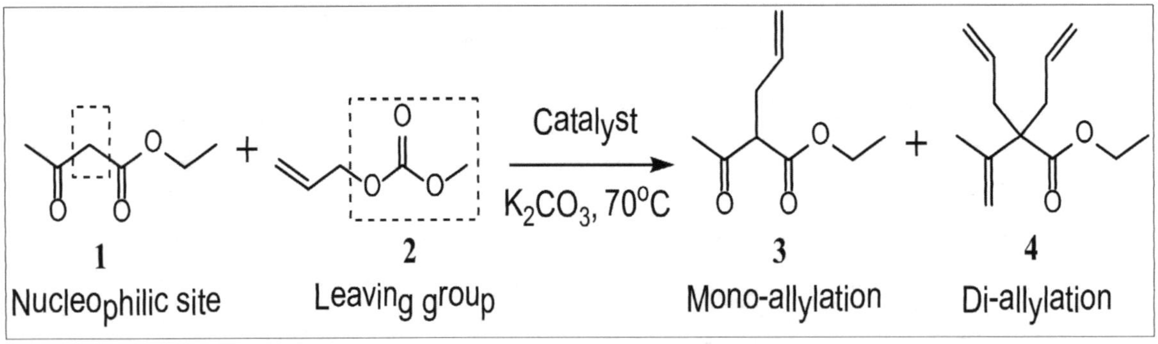


Experimental conditions: 1 mmol of **1**, 2.5 mmol of **2**, Catalyst: low Pd-loading of 0.4 mol %, $K_2CO_3$ (3.0 mmol), 70 C, 5 h.

# 3. Experimental Results

## ➢ Catalytic Performance

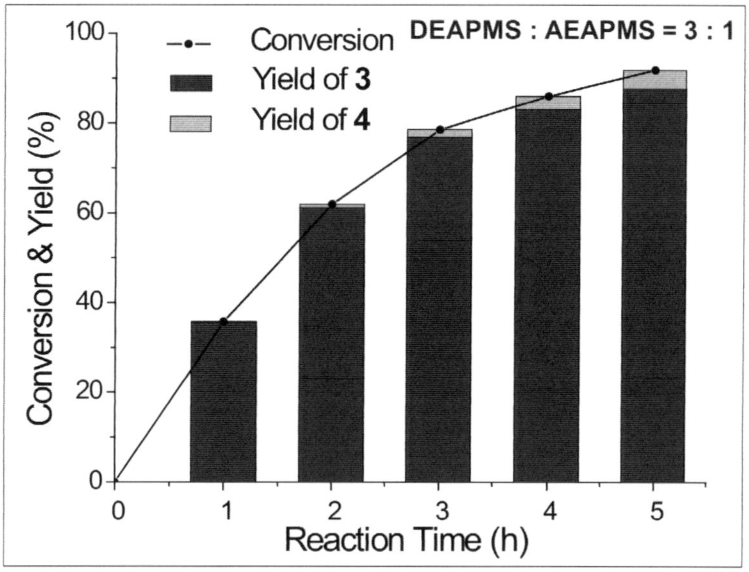


**Conversion: ~92%; TOF: ~46 /h**

***TOF: <10 /h, Arne T. Dickschat et al, Chem. Commun., 2013, 49, 2195**

# 3. Experimental Results

## ➢ Recycle Experiments

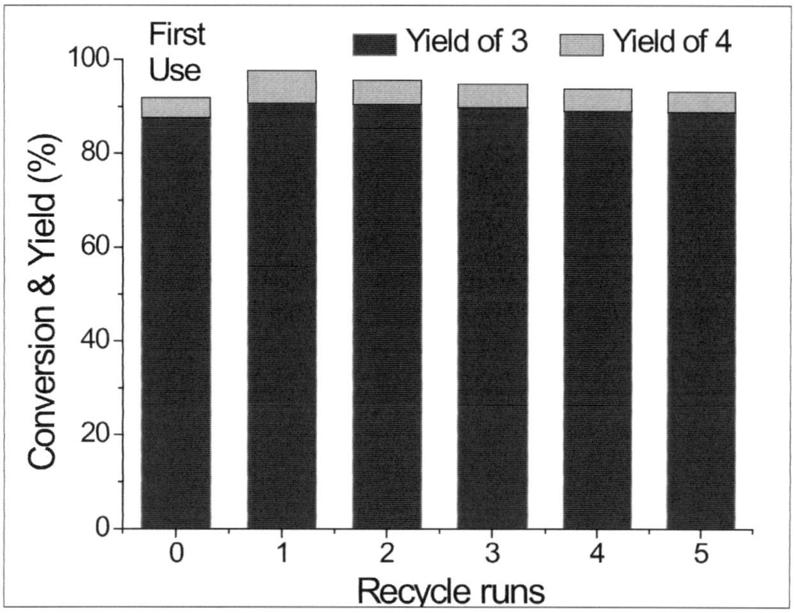

# 3. Experimental Results

## ➢ Control Experiments

| Entry | Catalyst | Conversion (%) [b] | Yield (%) [b] 3 | Yield (%) [b] 4 |
|---|---|---|---|---|
| 1 | None | 2.70 | 0 | 0 |
| 2 | GO-NEt$_2$-2N [c] | 4.03 | 0 | 0 |
| 3 | $PdCl_2$ | 6.43 | 2.82 | 0 |
| 4 | GO-2N-Pd | 44.52 | 44.03 | 0.49 |
| 5 | $PdCl_2$ + Triethylamine [d] | 10.44 | 9.20 | 0 |
| 6 | GO-2N-Pd + Triethylamine [d] | 50.83 | 50.11 | 0.72 |
| 7 | GO-NEt$_2$-2N-Pd | 91.85 | 87.67 | 4.18 |
| 8 | Pd/C [e] | 54.93 | 54.34 | 0.59 |

**Entry**

3 vs 7

4 vs 7

5 vs 7

8 vs 7

**Efficient** and **Promoted**

# 3. Experimental Results

**Excellent performance of the combined catalyst**

**First**

Cooperation of the palladium and amine

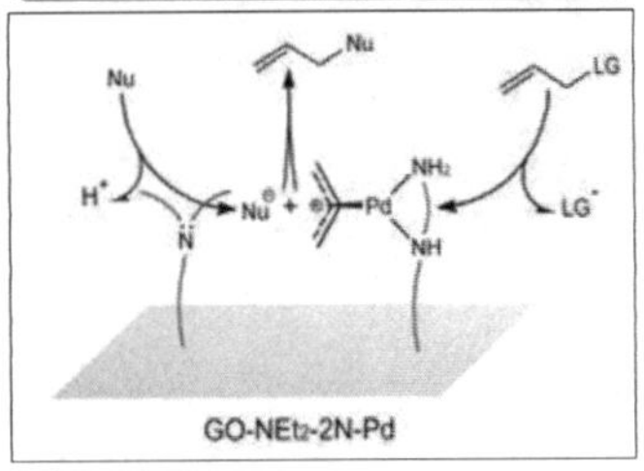


**Second**

Facial mass transfer of the system

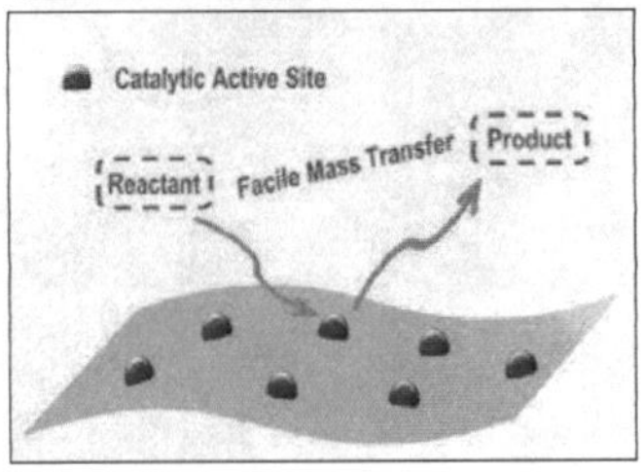


**Third**

Immobilization and site-isolation effect

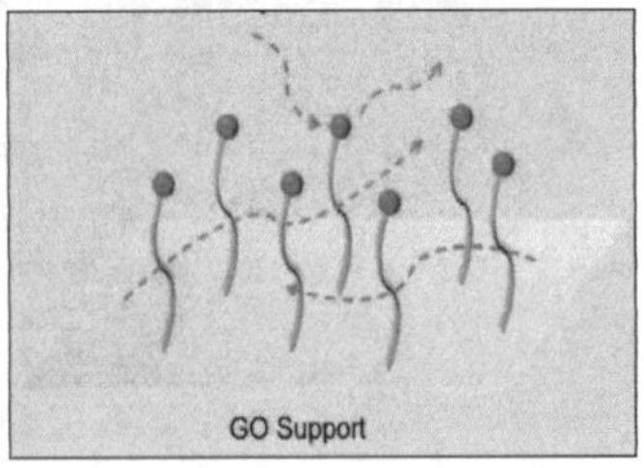

# 4. Conclusions

➤ **Developing an efficient strategy to combine a palladium complex and an organic amine on graphene oxide.**

➤ **The obtained bifunctional catalyst can be used as a promoted and recyclable catalyst for Tsuji–Trost allylation.**

天津大学
TIANJIN UNIVERSITY

Qingshan Zhao (PHD)

NSFC

国家自然科学
基金委员会
National Natural Science
Foundation of China

**Thanks for your attention !**

# 378104 Novel Organometallic-Nanocarbon Complexes for Electrochemical Applications

**Abstract:**

In this work, we report a theoretical study on novel organometallic-nanocarbon complexes for electrochemical and catalytic applications. These materials include the N-heterocyclic-transition metal(NhTM)-nanocarbon complexes (pyrrole, pyrazole, and triazole) and porphyrin-transition metal-nanocarbon complexes (PORTM), with TM = Fe, Zn, and Mn. The geometries, binding energies, electronic structures, and magnetic properties of these materials are investigated using density functional theory (DFT) methods.

These complexes are similar to our previous work on cyclopentadienyl-transition metal (CpTM) functionalized nanocarbons, which included both pristine and B-doped carbon substrates. In comparison, the binding energies of the NhTM-nanocarbon complexes with the carbon substrates are predicted to be even greater than those of CpTM (which surpassed the intramolecular binding energy within ferrocene, a molecule with a similar structure and chemical characteristics). The electronic structures and magnetic properties of the NhTM-nanocarbon complexes are also predicted from band structure analyses, natural bond order partial charges, and deformation charge density analyses. These complexes exhibit metallic band structures with strong magnetic momenta, indicating good conductivity for electrochemical applications and potential magnetic applications. In addition, the redox potentials and the charge transfer mechanisms in different solvents are calculated using DFT, combined with the conductor-like polarizable continuum model (CPCM) solvation technique.

# 392532 Nanostructured and Dynamic Biomaterials for Controlling the Cell Microenvironment

**Abstract:**

The nano- to macro-scale biophysical and biochemical properties of the environment that surrounds a cell are known to play an important role in cell function and fate. However, less is known about how changes in these properties influence biological functions. For example, driven by transient bidirectional crosstalk between cells and the extracellular matrix (ECM), cell differentiation and tissue regeneration are complex processes that often involve the presentation of multiple cues that are tightly regulated over multiple time and size scales[1]. Studying such complex, dynamic processes in vitro remains challenging. Biomaterials, particularly hydrogels, have emerged as synthetic mimics of the ECM to probe how changes in the microenvironment regulate cell behavior in regeneration or disease[2,3]. This talk will focus on strategies to impart highly-regulated property control by synthesizing monomers capable of forming hydrogels in the presence of cells and subsequently allowing triggered modification (e.g., light, enzymes, or reducing conditions) to tune the network's biophysical or biochemical properties. Toward mimicking the hierarchical nanostructure of the native ECM, monomers that assemble and allow in situ modification will be presented. In particular, we will highlight recent results in using these materials to understand the critical cues that direct progenitor cell fate within collagenous tissues during regeneration or disease.

1 Rehmann, M. S. & Kloxin, A. M. Tunable and dynamic soft materials for three-dimensional cell culture. *Soft Matter* **9**, 6737-6746 (2013).

2 Kharkar, P. M., Kiick, K. L. & Kloxin, A. M. Designing degradable hydrogels for orthogonal control of cell microenvironments. *Chem. Soc. Rev.* **42**, 7335-7372 (2013).

3 Smithmyer, M. E., Sawicki, L. A. & Kloxin, A. M. Hydrogel scaffolds as in vitro models to study fibroblast activation in wound healing and disease. *Biomaterials Science* **2**, 634-650 (2014).

# 392533 DNA-Based Nanostructures As Versatile Delivery Vehicles

**Abstract:**

The ability of DNA to form predictable nanostructures through sequence-directed hybridization has allowed the design of complex supramolecular materials. Our discovery of the nuclease resistance of DNA nanostructures led us to explore their potential as delivery vehicles; a potential "third way" alternative to the fields of viral and non-viral delivery. To this end we successfully demonstrated the ability of DNA nanostructures to deliver functional antisense cargo in vitro. We have also demonstrated the use of non-canonical Watson-Crick base pairing to achieve controlled release from DNA nanostructures through sensitivity to solution conditions (i.e., pH). More recently, we have incorporated aptamer sequences into DNA nanostructures to achieve both targeting and uptake without the need for transfection reagents. Thus DNA-based nanostructures form the basis for a versatile platform with great potential as an alternative approach to molecular delivery.

# 388727 Designer Biomaterial Surfaces for Drug Delivery and Regenerative Medicine

**Abstract:**

Research in biomaterials is continuing to lead advances in treatments for a variety of critical medical conditions. Our research group designs and develops optimized biomaterials to address several challenges in the areas of drug delivery and regenerative medicine. In this presentation, I will focus on our recent research in the area of developing drug delivery coatings aimed at treating aspects of traumatic injury including infection. Current treatments for microbial infection often result in significant systemic toxicity and exacerbate problems such as antibiotic-resistance. We have developed controlled release antimicrobial delivery coatings that can avoid many of these complications. We use techniques including layer-by-layer (LbL) self-assembly to develop coatings that exhibit a range of favorable drug release profiles of a variety of therapeutics, ranging from antibacterial agents, to anti-biofilm drugs, and antifungal agents. These drugs cover a wide range of chemical and structural properties including hydrophilic and hydrophobic small molecules and proteins. Several of these LbL film architectures have successfully been applied to a range of medical device surfaces and have demonstrated efficacy against their microbial targets. I will discuss our focus on increasing drug loading in these coatings and developing target-activated drug releasing systems. Our work in developing functional surfaces for stem cell engineering will also be discussed.

# Phonon-limited transport coefficients in extrinsic graphene

**Enrique Muñoz**, Instituto de Fisica, Pontificia Universidad Catolica de Chile, Vicuña Mackenna 4860, Santiago, Chile.

## Abstract

The effect of electron-phonon scattering processes over the thermoelectric properties of extrinsic graphene was studied. Electron-phonon interaction is described within the deformation potential approximation. Electrical and thermal resistivity, as well as the thermopower, were calculated within the Bloch theory approximations. Analytical expressions for the different transport coefficients were obtained from a variational solution of the Boltzmann transport equation. The phonon-limited electrical resistivity $\rho_{e-ph}$ shows a linear in temperature dependence at high temperatures, and follows a $\rho_{e-ph} \sim T^4$ at low temperatures, in agreement with experiments. The phonon-limited thermal resistivity at low temperatures exhibits a $\sim T$ dependence and achieves a nearly constant value at high temperatures. The predicted Seebeck coefficient at very low temperature is $Q(T) \sim \pi^2 k_B T/(3eE_F)$, which shows a $n^{-1/2}$ dependence with the carrier density $n$, in agreement with experiments.

[1] E. Muñoz, Journal of Physics: Condensed Matter 24 (2012), 195302.

[2] E. Muñoz, J. Lu and B. I. Yakobson, Nano Letters 10 (2010), 1652.

# 369820 Study on the Surface Energy of Supported Graphene By Contact Angle Measurements


**Abstract:**

**Study on the surface energy of supported graphene by contact angle measurements**

Due to the atomic thinness of graphene, its integration into a device will always involve its interaction with at least one supporting substrate, making the surface energy of graphene critical to its fabrication and real-life applications. In the current work, the contact angle of graphene synthesized by chemical vapor deposition (CVD) was monitored temporally after synthesis using water, diiodomethane, ethylene glycol, and glycerol. The surface energy was then calculated based on the contact angle data by Fowkes, Owens-Wendt (extended Fowkes), and Neumann models. The surface energy of fresh CVD graphene grown on a copper substrate (G/Cu) immediately after synthesis was determined to be 62.2 $mJ/m^2$ (Fowkes), 53.0 $mJ/m^2$ (Owens-Wendt) and 63.8 $mJ/m^2$ (Neumann), which decreased to 45.6 $mJ/m^2$, 37.5 $mJ/m^2$ and 57.4 $mJ/m^2$, respectively, after 24 hours of air exposure. The ellipsometry and ATR-FTIR characterization indicates that the surface energy of G/Cu is affected by airborne hydrocarbon contamination. G/Cu exhibits the highest surface energy immediately after synthesis and the surface energy decreases gradually after airborne hydrocarbon contamination occurs. The intrinsically mild polarity of G/Cu surface has been discussed and could be attributed to high-energy surface defects, pi-hydrogen bonding, and partial wetting transparency.

# 389955 A 2D Equation-of-State Model for Corona Phase Molecular Recognition on Single Walled Carbon Nanotube Surfaces

**Abstract:**

Corona Phase Molecular Recognition (CoPhMoRe) has been recently introduced as a means of generating synthetic molecular recognition sites onto nanoparticle surfaces. A synthetic heteropolymer is adsorbed and confined to the surface of a nanoparticle, forming a corona phase capable of highly selective molecular recognition due to the conformational imposition of the particle surface on the polymer. In this work, we develop a computationally predictive model for analytes adsorbing onto a polymer corona phase around a single walled carbon nanotube surface using a 2D equation of state that takes into consideration the analye-polymer, analyte-nanoparticle, and polymer-nanoparticle interactions using parameters determined independently from molecular simulation. The curvature dependence of the interaction energies is determined to be a weak contribution to the overall interaction energy. Overall, the resulting model is able to correctly predict 83% of a 374 analyte-polymer library generating experimental fluorescence responses within 20% error of the experimental values. This modeling framework is an important step to the design of such polymers to target a specific analyte.

Characterization of Carbon-Based Nanomaterials by Vapor Sorption Methods

Daniel J. Burnett[1], Majid Naderi[1], and Armando R. Garcia[1]

1. Surface Measurement Systems, Ltd., Allentown, PA USA

Surface and bulk properties are important to the successful development of carbon-based nanomaterials used for real-world applications. For instance, surface energy of nanoparticles can determine initial wetting behavior (i.e. hydrophilic, hydrophobic, or amphiphilic surfaces). Also, solvent chemical selectivity can be strongly dependent on surface energy and surface acid-base properties. Additionally, bulk properties like vapor diffusion/flux coefficients can be useful in determining permeability of different solvents. For the above reasons, vapor sorption techniques like Inverse Gas Chromatography (IGC) and Dynamic Vapor Sorption (DVS) are ideal methods for characterizing membranes. Both techniques allow a wide range of experiment temperatures, solvents, and sample morphologies to be studied.

This presentation will outline several studies where IGC and DVS have been used to characterize the physico-chemical properties of various nanomaterials. For instance, the surface energy and wettability of three carbon-based candidates for aqueous filtration of polyaromatic compounds (PACs): activated carbon, single-walled carbon nanotube paper, and graphene pellets. . DVS results indicate that water sorption capacity obeyed the following trend: Activated Carbon > Graphene Film > Graphene Pellets. Finite Dilution-IGC results show that all three samples are energetically heterogeneous. Surface polarity was determined by taking the ratio of the acid-base component with the total surface energy. This measurement gives an indication of the density of polar surface groups, thus can be an indication of surface wettability.

Further, the surface energetic heterogeneity of commercial multi-walled cabon nanotubes (CNTs) was measured, relating to the effects of different modification treatments (i.e. annealing and oxidation). Dispersive surface energy profiles in show that all MWCNT samples were energetically. However the degree of energetic heterogeneity was found to depend on the modification treatment, with oxidized MWCNTs found to be energetically most active and most heterogeneous. Regarding the acid-base component of the surface energy, oxidized MWCNTs had relatively higher concentration of Lewis acid-base functional groups and have the highest surface polarity, followed by the as-received and annealed samples.

Other examples and their implications on performance, formulation, and stability will be discussed.

# 387836 Molecular Dynamics Simulations of Natural Gas Transport in Carbon Nano-Pore Structures

**Abstract:**

Shale gas is attracting increasing attention as a rising energy supply and numerous successful production plays have been achieved. However, accurate reserve estimation and production projection are still challenging and require a sound understanding about mechanisms of natural gas transport. In contrast to conventional reservoirs, more than 80% of pores in shale formation are micro- or meso-pores, where the length scale of pores is smaller than 30 nm. Gas permeability through nano-porous media is ultra-low and the transport behavior remains poorly understood because continuum theory breaks down at these length scales. Moreover, the majority of shale gas is stored as adsorbed gas in organic matter. Consequently, the interactions between natural gas and the organic matter molecules may affect the flow of natural gas through nano-pore structures. Modeling the transport of gas through shale requires fundamental understanding of the interaction between molecules, for which molecular dynamics simulation provides the theoretical foundation.

There are two main challenges for modeling natural gas transport through nano-pore structure. First, organic matter molecules found in shale formation exhibit complex molecular structures. As one alternative, the molecular structure can be replaced by a simpler existing carbon structure. Although carbon structures cannot reproduce the detailed atomic-scale interactions, such structures are computationally inexpensive and are easily implemented. Therefore, we model the organic matter molecules using carbon structure. Second, shale exhibits a complex pore network structure. Using digital rock reconstruction techniques, we statistically model 3-D nano-pore structure in organic matter from scanning electron micrographs of shale rock sample. Subsequently, the complex pore structure is discretized into orthogonal solid and pore grids whose length is determined by the resolution of the SEM figure. By inserting a particular carbon structure into solid grid, we generate a different complex 3-D carbon nano-pore structure for each simulation.

To study gas transport behavior through shale nano-porous media, we employ methane to model natural gas. Two geometries are investigated to model the organic nano-slit pore: two parallel plates with FCC (001) carbon structure and a carbon nanotube with zigzag structure. Lennard-Jones intermolecular potential is applied to model the methane-methane and carbon-carbon interaction. Methane-carbon interaction is modelled by taking the average of the two interactions based on Lorentz-Berthelot mixing rule. All simulations are conducted using LAMMPS (Large-scale Atomic/Molecular Massively Parallel Simulator). We first use equilibrium molecular dynamics simulations to determine the mechanism of storage of methane. Density profiles of methane confined between two parallel plates and inside a nanotube reveal that the density of methane near the walls is much higher than that in the center. It confirms that the methane adsorbs on the carbon. We measure the amount of methane molecules in the adsorbed layer as a function of the free gas pressure to obtain the adsorption isotherm, which obeys the Langmuir isotherm and is in good agreement with experimental data from literatures; this finding indicates that adsorption is single-layer adsorption and adsorbed sites on the surface are energetically homogeneous. To characterize the flow behavior, the velocity profile and the relation between the average velocity and external body force are simulated using non-equilibrium molecular dynamics simulation based on the intermolecular interaction from equilibrium molecular dynamics simulation. Different velocity profiles through two geometries are observed. The velocity profile through the two parallel plates is nearly uniform, whereas the velocity profile in carbon nanotube depends on the external body force. If the external body force is small, gas transport near the boundary is much slower than gas transport in the center and the velocity profile is nearly parabolic. If the external body force is large, the velocity profile in the nanotube is again nearly constant. A linear relationship between average velocity and external body force is observed for flow between parallel plates, whereas a nonlinear relationship is observed for flow in a carbon nanotube. This deviation between two geometries indicates that surface interaction affects gas transport in nano-scale systems because the specific surface areas of two geometries are different. Different numbers of carbon atoms exposed in two geometries dramatically change the intermolecular force between methane and carbon molecule, so that the methane-carbon interactions dominate the flow behavior. Finally, we model flow in a 3-D FCC (001) carbon nanopore structure and find a linear relationship between the average flux and the external body force. These results represent the flow dynamics in realistic nano-porous media and yield new insights into gas transport in carbon nano-porous media.

# The mobility vs. carrier density trend in wafer-scale epitaxial graphene on SiC, and how it can be defeated

C. Dimitrakopoulos,[1] T. J. McArdle,[2] D. Farmer,[3] A. Grill,[3] D. Pfeiffer,[3] Ph. Avouris[3]

1. Chemical Engineering Dept. University of Massachusetts, Amherst, MA 01003.
2. IBM Microelectronics, E. Fishkill, NY 12533.
3. IBM Thomas J. Watson Research Center, Yorktown Heights, NY 10598.

Graphene's unique and exceptional properties make it attractive for technological applications in many areas, including high-speed electronics. The establishment of processes for producing high quality, large-scale graphene is necessary for such applications to be enabled. Large area growth of graphene on the Si-terminated face of hexagonal SiC wafers exhibits manageable growth kinetics – contrary to the C-face – allowing good control over the number of graphene layers. Most importantly, the azimuthal orientation of epi-graphene on the Si-face of a SiC wafer is fixed. Therefore, it is a viable method for producing graphene with uniform coverage and structural coherence at wafer-scale.

Graphene was grown on SiC with two methods: i) By combining SiC surface preparation steps in disilane/He with a higher temperature (T=1550 °C) graphenization step in Ar and ii) By $H_2$-etching SiC wafers at T>1500 °C to produce a vicinal, defect-free surface and then annealing that surface at T~1575 °C in Ar atmosphere (P~ 100 mbar).

After optimizing the process parameters for graphene growth on 4H(0001) SiC with 0.00° miscut, we have measured average Hall mobilities μ up to 4900 $cm^2\ V^{-1}\ s^{-1}$ at sheet carrier densities N ~ $4\times10^{11}\ cm^{-2}$ after a long exposure to ambient atmosphere. This mobility is higher and the electron density lower than the freshly made devices. The strong n-type character of pristine graphene on SiC (0001) is due to the dangling bonds below the $\left(6\sqrt{3}\times6\sqrt{3}\right)R30°$ reconstruction (buffer layer), which act as positively charged traps. The reduction in N by ambient exposure is linked to the oxygen and water adsorbed onto the graphene surface. We can explain the large improvement in mobility with reduced N by taking into account several factors, including long-range Coulomb scattering and short-range scattering. A plot of μ vs. N will be introduced as a benchmark of epi-graphene quality. After $H_2$ intercalation, our epi-graphene becomes p-type and exhibits a record combination of high mobility and high hole density (far above the benchmark curve).

*Support by DARPA Contracts FA8650-08-C-7838 & HR0011-12-C-0038 is gratefully acknowledged.*

# 382588 Molecular Barrier Functions of Graphene Oxide in Liquid-Liquid Systems


**Abstract:**

Graphene oxide (GO) is an effective stabilizer for a range of organic-in-water emulsions. The high stability of GO-based emulsions is due to in part to the formation of robust multilayers at the interface that effectively cover the droplet surface, allowing for full passivation of droplet surface patches left uncovered by packing inefficiencies in the first monolayer. In this work, we show that these GO interfacial films demonstrate unique molecular barrier properties at the liquid− liquid interface, a phenomena that is not observed when non-platelike stabilizers are used. These films are able to inhibit transport from the droplet interior and cause a significant suppression of dispersed-phase evaporation rates with potential applications in controlled release. They are also able to protect the dispersed phase from attack by reactive oxygen species, which typically react at diffusion-limited rates. The mechanism for protection is hypothesized to be a combination of hydroxyl radical scavenging and creation of tortuous pathlengths that hinder diffusion to the interior phase.

# 387550 Surface Properties of Carbon Nanotubes Adsorbed at an Air-Water Interface

**Abstract:**

Particles of appropriate size and wettability adsorb strongly at fluid-fluid liquid interfaces, lowering the interfacial energy and thereby stabilizing emulsions and foams. In the present work, we investigated the surface properties of both functionalized and non-functionalized carbon nanotubes (CNTs) at an air-water interface. Our findings indicated that both types of CNTs adsorbed strongly at the air-water interface, leading to an increase in surface pressure. The surface elastic modulus was calculated upon compression, and the Volmer equation of states has been applied to model and understand the surface pressure data. The CNT-decorated interface was studied both optically (*in situ*) and with electron microscopy after performing Langmuir-Schaefer deposition. Both types of CNTs showed some degree of orientational ordering upon compression. At high surface coverage, CNTs formed a layer which wrinkled upon further compression. The corresponding compressional modulus was calculated from the wavelength of the undulations. The findings may have implications on the development of more stable emulsions and the production of CNT thin films with controllable packing density and orientational ordering. This work is supported by NSF CAREER award #1253613.

# 384000 Award Submission:Development of Peptide-Conjugated Iron Oxide Nanoparticles for Hyperthermia Treatment of Cancer

**Abstract:**

Hyperthermia, the heating of tissue between 42 and 45°C, has been shown to enhance the effects of radiation and chemotherapeutics, but current methods of hyperthermia often result in severe side effects due to lack of localization and overheating of tissue. Magnetically mediated hyperthermia provides the opportunity for localized heating, however, this method is currently limited by the lack of particle penetration into tumor tissue. In this research, a nanoparticle system composed of an iron oxide core with a crosslinked dextran coating functionalized with the tumor homing peptide, CREKA, has been developed to overcome this limitation by homing to and penetrating into tumor tissue. The iron oxide core allows for particle heating upon exposure to an alternating magnetic field while the dextran coating stabilizes the particles in suspension and decreases the cytotoxicity. The overall goal of this study was to optimize CREKA-conjugated iron oxide nanoparticles for enhanced tumor homing and penetration for effective hyperthermia treatment applications. This specific particle system can enhance particle accumulation at the tumor site, providing a high enough concentration to induce hyperthermia conditions. The particles were characterized for size, stability, biocompatibility, and heating capabilities. The particles were stable in PBS and media over at least twelve hours, have a hydrated diameter of 50 nm, and can generate a significant amount of heat to raise solution temperatures well into the hyperthermia range. The cytotoxicity of the particles was analyzed through studies on A549 lung cancer cells for low particle concentrations with high exposure time and results determined that the particles have low cytotoxicity over both time frames and concentrations. Fibrinogen clots were used to determine the binding affinity of CREKA conjugated iron oxide nanoparticles compared to non-targeting nanoparticles. The binding and uptake of CREKA conjugated iron oxide nanoparticles into multicellular spheroids and monolayer cell cultures was also evaluated. This particle system is novel in that it incorporates the heating properties of iron oxide with the tumor homing properties of CREKA, to enhance particle homing to tumor sites.

# 372785 Award Submission: Viscosity Increase in Concentrated Mab Solutions Due to Large Transient Clusters

**Abstract:**

This work addresses a fundamental challenge in controlling properties and stability in concentrated protein solutions, which is both of basic scientific interest and industrial importance. Proteins in densely packed environments are abundant in our bodies and in commercially produced biopharmaceutical products. In addition to product stability during lyophilization and long-term storage, successful drug product formulation is essential for the processing and drug delivery stages of a product's life cycle. In particular, an increase in a therapeutic's viscosity could introduce challenges to material handling, such as during filling, and also adversely limit the variety of available drug delivery methods. Recent reports on monoclonal antibodies (mAbs) have highlighted their tendency to form small, strongly bound, reversible clusters, which produce an undesirable significant increase in viscosity at elevated concentrations (Biophys. J. 105:720, 2013). Clusters in fluids are important structural features formed by the balance of different interaction potential features that are especially prominent in concentrated protein solutions. Several types of short-range attractive interactions contribute to effective inter-protein interactions that may serve as the origin of clustering. While previous studies have discussed clustering in mAb solutions via electrostatic driven interactions, here we highlight a more hydrophobic mAb where the addition of salt induces the formation of large transient clusters. Cluster formation becomes more extensive with increasing salt concentration and reduced temperature. At the lowest temperatures studied, signs of slow kinetic arrest are present. However, despite the formation of a gel, the local dynamics of individual proteins in these transient clusters measured by neutron spin echo are surprisingly found to be very fast. This is dramatically different from that of globular proteins and previously studied mAb systems. Despite the fast motion of these mAbs, the short range interaction is still strong enough to produce a very large weakly linked protein network causing the increase of solution viscosity or gelation. These results highlight the difference between transient and strongly bound dynamic clusters and the sensitivity of mAb therapeutic material properties to the origin of inter-protein interactions and the purity of product formulation.

# 388410 Award Submission: Electronic Platform to Quantify Cellular Mechanisms Associated with Carbon Nanotubes Exposure in Real-Time

**Abstract:**

Despite the intriguing potential of carbon nanotubes (CNTs) for biomedical applications, their implementation is currently being hindered by many uncertainties regarding their toxicity and fate inside the biological systems. Several reports have shown that CNTs toxicity can be attributed to their length, surface chemistry, aggregation and metal impurities, just to name a few. However, due to the various types of CNTs being synthesized every day along with the different surface functionalization techniques, there is no fundamental understanding of the toxicological and pharmacological profiles of cellular systems exposed to CNTs with different physico-chemical properties. In this research, we provide a comprehensive analysis of the cellular behavior of human lung cells post exposure to CNTs using a combination of electronic and cell-based techniques. Our analyses rely on a non-invasive electronic cell impedance sensing (ECIS) platform to provide real-time measurements of cell adhesion, cell-cell interactions as well as changes in cellular morphology upon exposure to CNTs with user-defined physical and chemical properties. The approach is complemented by standard microscopy techniques as well as conventional *in vitro* cellular-based assays, such as cell cycle analysis, cell proliferation and viability, in order to derive structure-function relations associated with the cytotoxic and apoptotic events induced by exposure to CNTs. Our results correlated the different physico-chemical properties of CNTs with the various biological responses and provided mechanistic insights into their Cyto- and genotoxic effects. Out study help define a novel platform for nanomaterials toxicity analysis and thus could expand the biomedical applications of CNTs and other nanomaterials.

Introduction: Sorafenib is a multikinase inhibitor that was approved by the FDA in 2007 for the treatment of hepatocellular carcinoma (HCC). Since then, it has quickly become the standard of care for patients with advanced stage disease. However, patient tolerance is an issue due to the side effects resulting from systemic delivery. To avoid systemic delivery of chemotherapies for HCC, a procedure called transarterial chemoembolization (TACE) has been developed to selectively place a catheter in tumor feeding arteries to locally deliver the chemotherapy. This is then followed by or combined with embolic agents to contain the chemotherapy and to further starve the tumor from nutrients and oxygen in the blood. Thus, poly(lactide-*co*-glycolide) (PLG) microspheres have been developed that can encapsulate sorafenib for local delivery to liver tumors via TACE as well as co-encapsulating iron oxide nanoparticles to enable MRI-monitoring of the delivery. Specifically, the purpose of this study aims to use MRI to correlate to microsphere concentrations at the tumor to determine estimated sorafenib doses delivered.

Methods: PLG sorafenib iron oxide microspheres were fabricated via a double emulsion/solvent evaporation method. The microspheres were then characterized for microsphere size, sorafenib and iron oxide loading and release, as well as MRI contrast effect properties *in vitro* (7T MRI, Bruker Clinscan, Billerica, MA, USA). From these *in vitro* MRI studies, the microsphere concentrations were plotted to correlate against changes in MRI signal changes. Afterwards, liver tumors were inoculated in 4 New Zealand White Rabbits in which a VX2 tumor fragment was injected through a biopsy needle under ultrasound guidance. The tumor was then allowed to grow to roughly 1 cm in diameter after 2-3 week after which the rabbits underwent TACE procedures and microsphere infusions (50 mg). The rabbits were also scanned pre-and post procedurally at the same 7T MRI Bruker Clinscan to compare tumor signals before and after microsphere infusion with $T_2$ and $T_2$* map protocols. Afterwards, the tumor $T_2$ and $T_2$* values were extracted from the MRI images via MATLAB analyses to determine changes in the $T_2$ and $T_2$*. These changes were then used to estimate the microsphere concentrations at the tumor using the correlations previously obtained following the *in vitro* MRI studies. The rabbits were then survived for 24 hours before sacrificing and harvesting the tissue for Prussian blue staining for iron.

Results: The resulting microspheres were of an average diameter of 13 microns. Loading and release studies indicated that the weight percentage of sorafenib and iron oxide in the microspheres were 8.8% and 0.89%, respectively and that 21% and 28% of the loaded sorafenib and iron oxide released in 72 hours, respectively. *In vitro* MRI studies showed that microsphere concentration was correlated to $1/T_2^*$ via the equation: $\mathbf{(1/T_2^*)}$ = 0.0205(**Microsphere Concentration**) + 0.1262. For subsequent *in vivo* studies, MRI indicated that the microspheres primarily deposited in the tumor periphery and the estimated microsphere concentrations were on average 45 mg/mL in those regions. After the rabbits were sacrificed, Prussian blue staining confirmed proper tumor targeting and that primary microsphere deposition was at the tumor periphery.

Conclusion: The developed PLG sorafenib iron oxide microspheres allow for targeted delivery of sorafenib to liver tumors with the advantage of MRI monitoring and proper estimations of the delivered dose. Further studies are evaluating biomarkers for tumor responses and possible correlations of the tumor responses to the microsphere doses that were effectively delivered.

**Award Submission: Identification of Kinase Inhibitors for Enhanced Polymer-Mediated Transgene Expression**

Matthew Christensen[1], Jacob Elmer[1], Seron Eaton[2], Laura Gonzalez[2], Joshua LaBaer[2], Kaushal Rege[1]

Department of Chemical Engineering[1], Biological Design Institute, Personalized Medicine[2]
Arizona State University

Human cells contain hundreds of kinase enzymes which regulate several cellular processes, including those that can influence endogenous gene and transgene expression. We carried out a kinome-level screen of nearly 200 kinase inhibitors, using which, we identified small molecule inhibitors that enhance polymer-mediated transgene expression in prostate and bladder cancer cells. In particular, our lead targets include kinases associated with cell cycle regulation as well as transcription. Dose response studies revealed optimal concentrations of kinase inhibitors for enhancing transgene expression in different cell lines. Furthermore, these dose response studies demonstrated that the use of different polymer carriers, or different polymer to DNA ratios, led to different levels of transgene expression enhancement induced by the drug. This suggests that different polymer carriers may assist in overcoming particular barriers to gene delivery, in some cases perhaps overlapping with the inhibitor's mechanism.

The role of cancer cell cycle and intracellular trafficking were investigated in order to elucidate mechanisms by which kinase inhibitors enhance transgene expression. Pairwise combination of small molecule inhibitors of epigenetic targets and kinases resulted in significant enhancement of polymer-mediated transgene expression, although the most dramatic enhancements were seen with a cocktail of two different kinase inhibitors and the one epigenetic modulator. The observed enhancement in transgene expression with the addition of each inhibitor suggests that the mechanisms by which these inhibitors gene expression are independent of each other.With combination treatments, minimal effects on cell viability were observed with each drug addition, indicating that inhibiting specific intracellular biomarkers is a powerful strategy for enhancing non-viral transgene expression. Our results on combination kinase inhibition for enhancing transgene expression have transformative implications for biotechnology and medicine.

# Detecting Temperature-Dependent Desorption with Carbon Nanotube Cluster Coated Microsphere Sensors

**Maria V. Chistiakova[1] and Andrea M. Armani[1,2*]**

[1]*Mork Family Dept of Chemical Engineering and Materials Science*

[2]*Ming Hsieh Dept of Electrical Engineering-Electrophysics*

*University of Southern California, 3651 Watt Way, Los Angeles, California 90089*

*Corresponding author: armani@usc.edu; http://armani.usc.edu

Detecting gas-surface interactions is critical in understanding material behavior and in developing improved thin films and coatings. There are many parameters that can control an interfacial interaction, including the chemical composition of the two materials and the pressure and temperature of the environment. To enable high sensitivity detection of trace species, many gas desorption experiments are performed using thermal desorption spectroscopy (TDS), which requires high vacuum to function. Therefore, it is not possible to study gas desorption under ambient conditions. In the present work, we have developed a new technique, which uses less complex experimental methods and which operates under ambient conditions..

The technique relies on evanescent label-free detection. Specifically, the optical field interrogates the surface, detecting the desorption of gas from the surface of the sensor. In our work, we use an optical whispering gallery mode (WGM) microresonator sensor coated with carbon nanotube clusters. Although the optical field is confined within the cavity, it is able to detect the desorption of gas from the CNT surfaces. The silica microsphere sensor is fabricated by melting an optical fiber tip with a $CO_2$ laser. To deposit the CNTs on the surface of the microsphere, we first suspend the CNTs in a solution of ethanol and sonicate it to disperse the CNTs. Then, we dip the spheres into the solution, achieving a 1-2% surface coverage with CNT clusters. The coverage is verified with scanning electron microscopy.

The microsphere sensor confines light of a particular wavelength, called the resonance wavelength, which not only re-circulates inside the device, but also exists outside the periphery, and is thus sensitive to environmental changes around the sphere. Since we can monitor the resonance wavelength, we can also detect when changes have occurred as the resonance shifts.

We perform experiments by first regenerating the CNTs on the surface of the microspheres with heat, then saturating with a particular gas (in this case, either CO or $CO_2$) at room temperature, andfinally recording the desorption response as we heat the CNT-covered microspheres from room temperature to 100°C. We record the change in resonance wavelength as a function of temperature change. Because resonance wavelength changes as the refractive index is changed, we convert the wavelength shift to changes in the refractive index. Plotting the change in refractive index vs. the change in temperature gives a slope equivalent to the thermo-optic coefficient (*dn*/*dT*). By comparing the *dn*/*dT* values, we can differentiate between different gases based on their boiling point and liquid-phase density.

# 373640 An Integrated Ionic Logic Circuit for Addressing/Reading Multiplex Sensor Platform

**Abstract:**

We present an integrated ionic logic circuit based on a novel fluid-based memristor and crossbar logic. The designed ionic logic circuit is used to address and read individual sensors in a large array with a single input. The circuit stores and retrieves intermediate signals in the circuit that come from an ion-selective membrane sensor array developed by our group (Annual Review of Analytical Chemistry, DOI: 10.1146/annurev-anchem-071213-020155; Biosensors and Bioelectronics, doi.org/10.1016/j.bios.2014.04.008.) These membrane sensors exhibit nonlinear I-V and rectification features, which allow optimum integration with the memristor to produce nonlinear circuit components with binary states, i.e. on-off logic components. A binary sequence at the input activates the various sensors and the sequence at the output reports the reading. The fluid-based memristor used for constructing circuit is achieved by a memristive phenomenon between silicon microelectrode and aqueous solution, due to a continuous formation of an oxide film on the silicon electrode under anodic polarization. This memristor can be simply fabricated by conventional silicon technology. The confined silicon microelectrode is defined by using thick silicon dioxide spacer, while the metal contact and on-chip counter electrode are made by lift-off technique. An optimal design that allows an on/off ratio larger than 10 and a short switching time less than 100ms of this memristor, accompanied with an operating current in microfluidic range will be presented. Employing crossbar logic, ionic logic circuit units, such as microfluidic latches and NAND/NOR gates, are built by embedding these fluid-based memristors on the microfluidic channel sidewalls. Under this logic configuration, all ionic signals from input and sensor units is delivered through ion migration in the microfluidic channels, while all logic computing operations are completed at the sidewall silicon microelectrodes. This fabrication strategy allows easy integration within our membrane sensor platform, without increasing fabrication complexity and system size. The microfluidic coder and decoder are built subsequently by this logic configuration for addressing and reading each sensor followed a binary sequence signal. The optimization and fabrication details of each logic units will be presented.

FLORIDA STATE UNIVERSITY

# Self-Referenced Detection of $Cu^{2+}$ Using a Crossbar Array of Protein-Stabilized Gold Nanocluster and a Thermoplastic

Junfei Xia, Zhibin Wang, Danting Huang, Peipei Zhang, Yingru Liu, Brett Kirkland, Jingjiao Guan

Department of Chemical and Biomedical Engineering
Florida State University

1

FLORIDA STATE UNIVERSITY

## Importance of $Cu^{2+}$ in Biological System

- Essential to biological processes
  - Enzyme cofactor
    - Structural property
    - Catalytic property

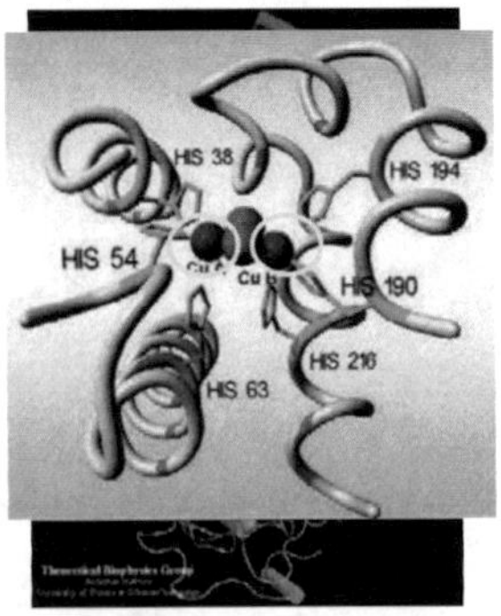


- Toxicity
  - Excess amount of $Cu^{2+}$
  - Damage liver, kidney, brain, etc.
  - $Cu^{2+}$ → Generate Reactive Oxygen Species (ROS) → Damage protein, lipid, DNA

L. M. Gaetke, C. K. Chow, Toxicology 189 (2003) 147.
V. Desai, S. G. Kaler, Am. J. Clin. Nutr. 88 (2008) 855.
http://www.ks.uiuc.edu, http://i.imgur.com

2

FLORIDA STATE UNIVERSITY

## Detection of $Cu^{2+}$

- Conventional detection approaches
  - Atomic adsorption/emission spectroscopy
  - Mass spectroscopy
  - Colorimetric assay
  - Fluorescent assay

- Lab-on-a-chip/Microarray-based sensing
  - Low fluid volume consumption
  - Fast analysis
  - Massive parallelization

Microarray detection of DNA

http://commons.wikimedia.org/wiki

3

FLORIDA STATE UNIVERSITY

## Protein Stabilized Gold Nanocluster (AuNC)

- Synthesis route

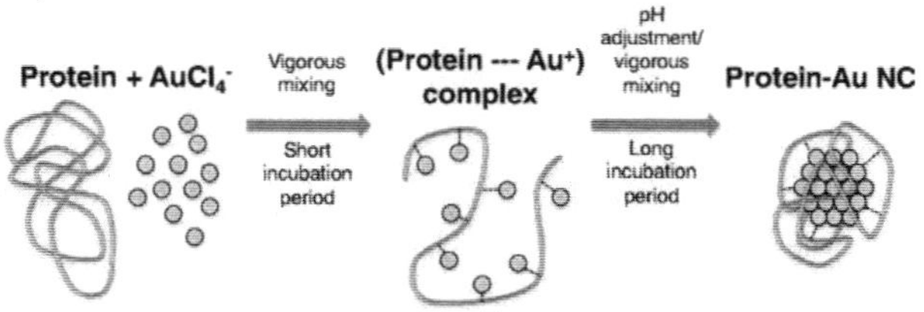


Protein:
BSA (Bovine Serum Albumin)

- Sensing ability
  - Fluorescence quenching by $Cu^{2+}$
  - Mechanism: $Cu^{2+}$ binding with BSA

D. Chevrier et. al, *J. Nanophoton.* 6(1), 064504

4

# Preparation of BSA-AuNC

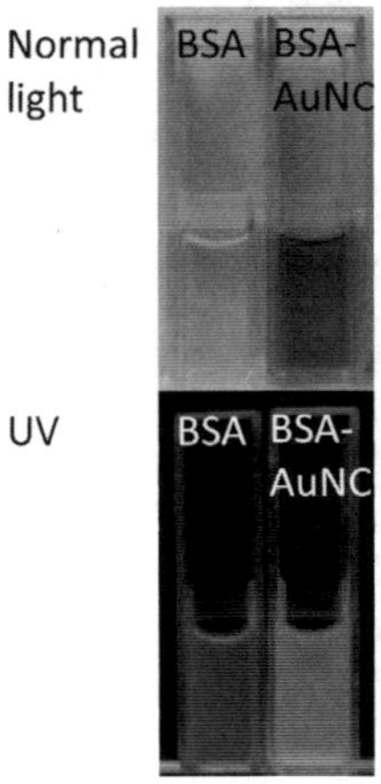


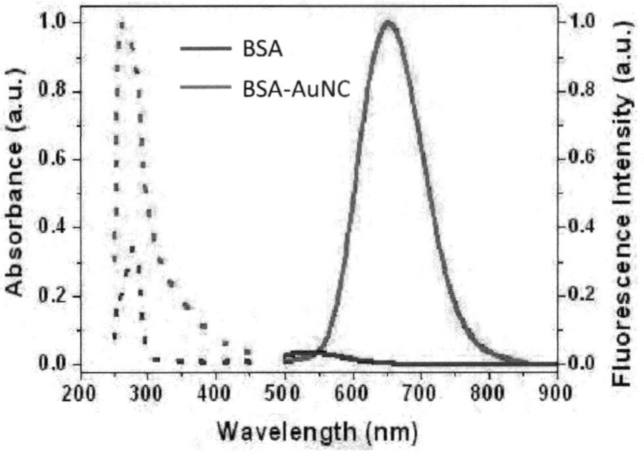


J. P. Xie, Y. G. Zheng, J. Y. Ying, J. Am. Chem. Soc. 131, 19 (2009) 888
Junfei Xia *et al.*: *J. Nanosci. Lett.* 2014, 4:19

5

FLORIDA STATE UNIVERSITY

# Fabrication of Microarray Sensor

**Microcontact Printing**

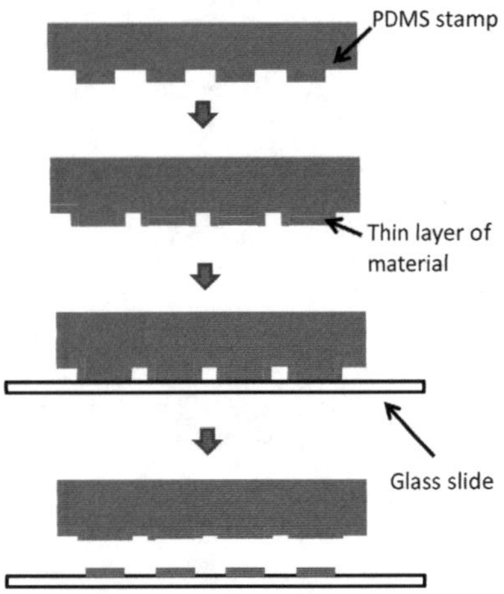


PDMS: polydimethysiloxane

**Fabrication Process**

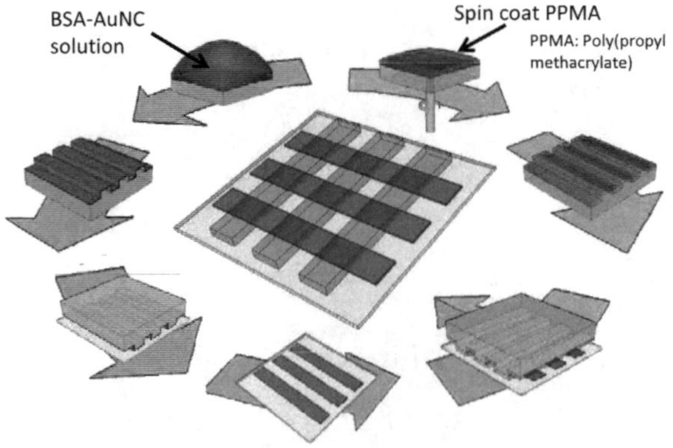


Unit cell (4 regions):
1. BSA-AuNC 2. Bare glass
3. PPMA 4. BSA-AuNC covered by PPMA

6

FLORIDA STATE UNIVERSITY

## Characterization of the Sensor Array

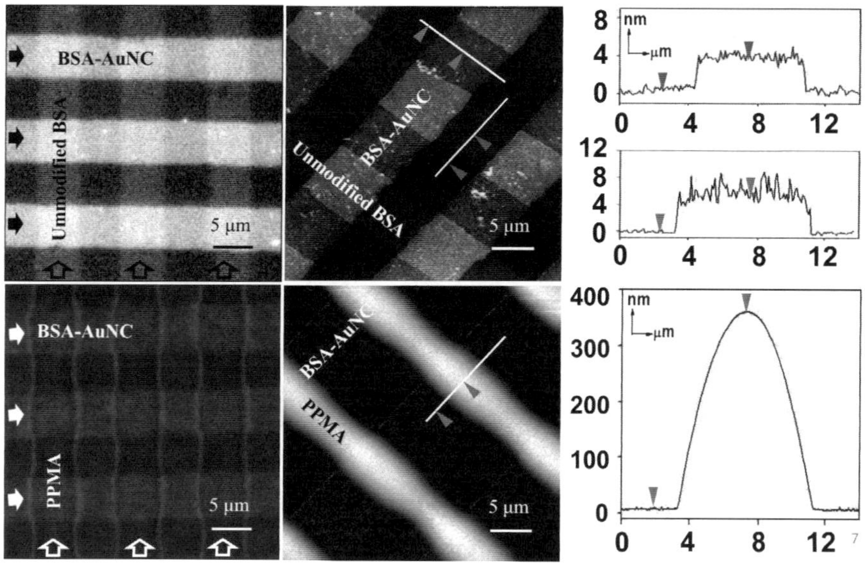


7

FLORIDA STATE UNIVERSITY

## Sensor Detection of $Cu^{2+}$

0.1 µM $Cu^{2+}$ 0.5 µM $Cu^{2+}$ 5 µM $Cu^{2+}$

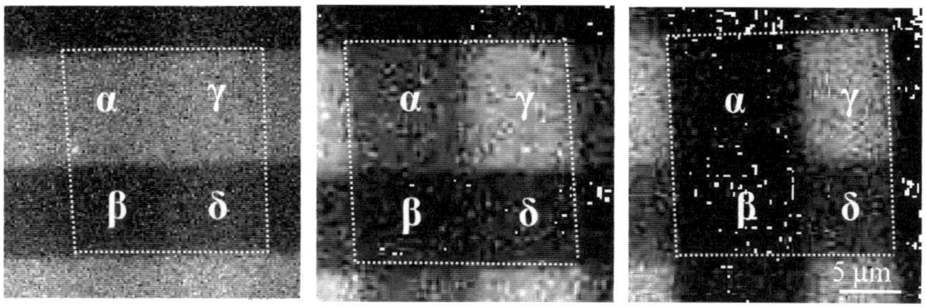


α: BSA-AuNC
β: Bare glass
γ: BSA-AuNC covered by PPMA
δ: PPMA

$$\frac{I}{I_{ref}} \equiv \frac{(I_\alpha - I_\beta)}{(I_\gamma - I_\delta)}$$

8

FLORIDA STATE UNIVERSITY

## Sensor Detection of $Cu^{2+}$

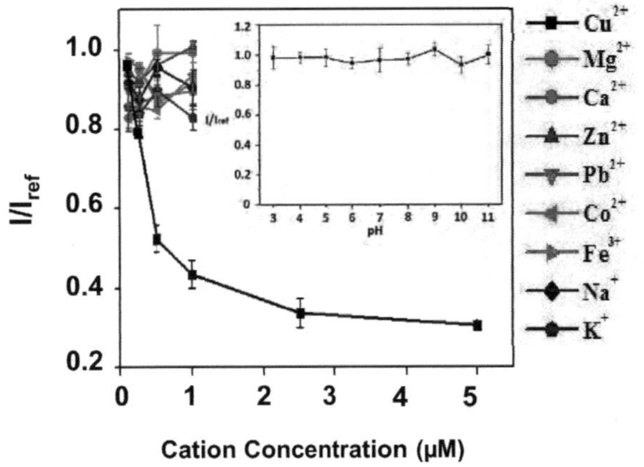


$$\frac{I}{I_{ref}} \equiv \frac{(I_\alpha - I_\beta)}{(I_\gamma - I_\delta)}$$

Detection of $Cu^{2+}$ level in tap water:

Significance: High $Cu^{2+}$ intake will cause damage to the body

Results:
$I/I_{ref} = 0.18$
$Cu^{2+}$ level > 5 μM

Local annual water quality report for $Cu^{2+}$ level in tap water: ~ 10 μM

9

FLORIDA STATE UNIVERSITY

## Reusability

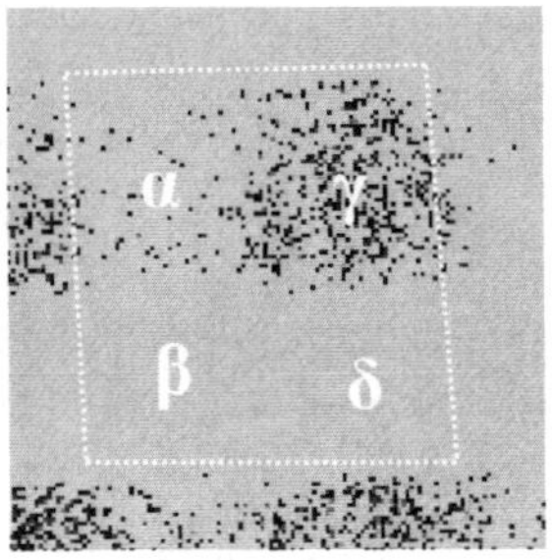


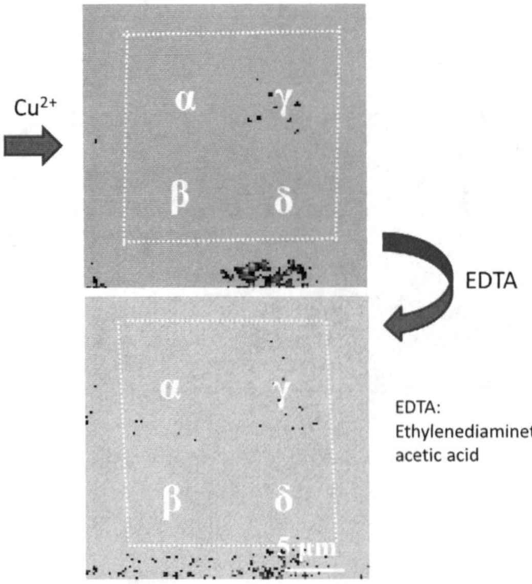


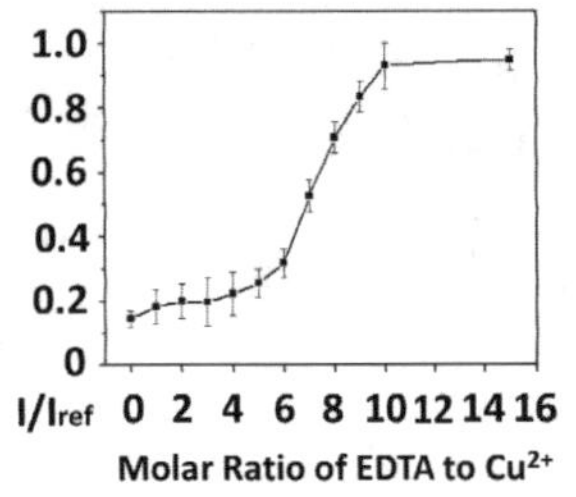


EDTA: Ethylenediaminetetra acetic acid

10

FLORIDA STATE UNIVERSITY

## Conclusion

- We fabricated a micrometer-sized $Cu^{2+}$ sensor based on microcontact printing BSA-AuNC and a thermoplastic polymer
- The sensor allows sensitive, selective, and self-referenced detection of $Cu^{2+}$
- This method is simple, inexpensive, extendable to other sensory materials such as small-molecule fluorophores and quantum dots

11

FLORIDA STATE UNIVERSITY

## Acknowledgement

Florida State University

*Department of Chemical & Biomedical Engineering*

**Dr. Jingjiao Guan**
Dr. Brett Kirkland
Dr. Yingru Liu
Peipei Zhang
Zhibin Wang
Danting Huang

*Department of Physics*
Dr. Eric Lochner

*Department of Chemistry & Biochemistry*
Dr. Bert van de Burgt

**This works was supported by FSU Startup Fund**

12

FLORIDA STATE UNIVERSITY

# THANKS FOR YOUR ATTENTIONS! QUESTIONS?

13

# 376829 Low-Power Microflow Sensor Using Thermal Capacitance

**Abstract:**

There are many different flow sensors commercially available, and the application of each depends on the properties of the fluid such as viscousity, pressure, particle concentration, and density. For example, high viscousity fluid or slurries are usually measured by turbine and fluid trapped meters, and can be cost effective. Other meters, such as ultrasonic and magnetic meters, are designed to be contactless, usually for the purpose of medical or food safety and can be relatively expensive. Cheap meters such as float gauges are typically difficult to integrate into control strategies.

A micro thermal-capacitance flow sensor is relatively simple and, if mass-produced, can cost less than a dollar to produce. These sensors measure the energy transfer change associated with flow over a solid surface. Thermal flow meters fall into three categories: time of flight, hot wire, and calorimetric flow meter. Hot wire meters sense the flow rate directly from the heater resistance change. Time of flight methods sense the time for a heated bolus of fluid to travel a set distance. Calorimetric sensors measure the temperature distribution around the heater and previous research has focused on low flow rates (e.g., microliter per minute). Though flow rates of several gallons per minute of liquid are much more common in industrial and home use, microflow sensors are rarely studied in this region.

Therefore, thermal flow rate sensor with higher sensitivity is required. Etching substrate for thermal isolation or high TCR resistance are two chief methods to increase calorimetric and hot wire sensor sensitivity. Nevertheless, thin substrate or cantilever design entail increasing fabrication cost. Most importantly, both can not sustain flushing water. In contrast, glass substrate is much cheaper and has excellent thermal isolation property. With higher yield strength, it is also more robust than silicon. Glass substrate can also be used without any pre-deposition, which silicon wafer needs for surface electrical isolation. Our goal is to fabricate a cheap but robust flow rate sensor suitable in water of high velocity. Glass wafer is therefore chosen for its convenient properties.

In this talk, we will describe the design and construction of microflow sensors for laminar and turbulent water flow rates ranging from 0.5 to 2.0 GPM -- more than one thousand times the typical current use. The sensor is easy to manufacture, requiring only two masks for fabrication. Influence of the sensor surface area, water temperature, and sensor substrate is analyzed experimentally and theoretically (i.e., COMSOL simulation). Thermal capacitance sensors from 0.5 to 2.0 gallon per minute (GPM) in a one inch diameter pipe is demonstrated.. Reynolds number range from 1600 to 6300, thus spanning laminar, transition, and turbulent flow. We found that the sensitivity rises with decreasing sensor surface area and increasing power input. Further, the sensor can function with only 2.75mW of applied power. The sensor can work at different temperatures, and the temperature shift is normalized in both simulation and experiment results.

**Poly (ethylene-glycol) Based Hydrogels for On-chip Microsensors**

*Paige Fischer, Bingzhao Xia, and John Oakey*
*University of Wyoming, Department of Chemical and Petroleum Engineering*

Poly (ethylene-glycol) (PEG) hydrogel-based cellular and molecular microsensors have been fabricated within microfluidic devices via polymer micromolding and photopolymerization. The fidelity of micromolding and photolithography allows various hydrogel surface topologies to be created easily and accurately. We demonstrate two implementations of hydrogel photopolymerization in microfluidic chambers: herringbone capture surfaces by micromolding and and post arrays by in situ photolithographic patterning. A primary advantage of PEG is its ease of modification, allowing different polymerization schemes and hydrogel properties to be tailored for custom control over hydrogel characteristics. PEG with an incorporated photodegradable group has been used to create cell capture surfaces that allow for the specific release of a desired cell population. PEG modified with thiol-ene end groups eliminates oxygen inhibition during the polymerization process that is seen in acrylate chemistry, thereby enhancing photopolymerization fidelity. Thiol-ene modified PEG has been used to create discrete or continuous structures for immunoassay platforms. By increasing or decreasing the amount of hydrogel forming monomer in solution the the hydrogel surface mechanical properties can be altered to effect cell-surface interactions or modification of the surface. A large array of on-chip microsensors can be created by PEG modification and easy manipulation of micromolding and photo polymerization techniques.

# A Microfluidic Technique for Measuring the Wax Appearance Temperature (WAT) of Crude Oils

Shahnawaz Molla and Farshid Mostowfi, Schlumberger

We present a novel technique for rapid measurement of the wax appearance temperature (WAT) of crude oil using a microfluidic device. The device comprises of a long (1 m) capillary with a cross-section of 50 μm x 50 μm, etched on to a silicon substrate. The technique is based on accurate pressure measurements across the device to detect changes in cross-sectional area of capillary due to wax deposition. The Poiseuille law dictates that when a fluid travels through a capillary at low Reynolds numbers, the flow rate correlates strongly with the capillary radius: $Q \square R^4$. Therefore, small changes in radius due to deposition of wax can be accurately detected by monitoring the changes in flow rate or the pressure across the capillary. Due to the small thermal mass of the device, the temperature of the system can be changed quickly. By monitoring the differential pressure, a rapid pressure increase was observed as sample temperature was systematically decreased to wax appearance temperature. The relationship between the rate of pressure changes and the sample temperature was used as the working principle in the proposed WAT measurement technique.

This technique was validated systematically by measuring the solid-liquid phase boundaries for single-component and multi-component synthetic samples, as well as crude oils. The measurements agreed well with literature data and cross-polar microscopy measurements. The crude oils used in this study were stock-tank samples with wax content ranging from 2.5-11 wt. %. The microfluidic device offers a simple and rapid measurement technique that is suitable for field applications. The technique can be extended to study rheology of live waxy oil and also to simulate flow assurance issues such as restart pressure.

# Molecular Sensing at Graphene Grain Boundaries

Poya Yasaei[1], Bijandra Kumar[1], Artem Baskin[2], Nikita Repnin[2], Petr Král[2], Amin Salehi-Khojin[1,*]

[1]Department of Mechanical and Industrial Engineering, University of Illinois at Chicago, Chicago, IL, 60607, USA.
[2]Department of Chemistry, University of Illinois at Chicago, Chicago, IL, 60607, USA.
[*] Corresponding author, salehikh@uic.edu,

**Abstract:**

In this study, we introduce the graphene GBs as an ideal structure for detection of toxic analytes. To this objective, we fabricated individual graphene GB based sensor and study their gas sensing characteristics by exposing against the dimethyl methylphosphonate (DMMP) or 1,2-dichlorobenzene (DCB) gas molecules. DMMP is an electron donor while DCB is electron acceptor molecules. We uncovered that an isolated graphene GB has ~300 times higher sensitivity to gas molecules than a single crystalline graphene grain. We also investigated the sensing response of a sequence of multiple GBs, naturally formed by the merging of the grains in the CVD growth. Our results show a clear sensing cross-over from a single GB toward a polycrystalline sensor, as the number of GBs are increased from 1 to more than 5. Our hybrid modeling of an electronic structure and transport in realistic GBs reveals that the ultra-sensitivity in GBs is caused by a synergetic combination of gas molecules accumulation at GB, together with the existence of ultrasensitive switchable transport channels in GBs. The discovered sensing platform opens up new pathways for the design of nanometer scale highly sensitive chemical detectors.

**Keywords:** Grain boundary, Graphene, Sensors, Transmission spectrum

AIChE 2014 abstract

# Pt Nanowire Array Coated with Au Nanoparticles as A Novel Electrochemical Sensor

Zhiyang Li (student presenter), Fan Gao, and Zhiyong Gu*
Department of Chemical Engineering
University of Massachusetts Lowell, Lowell, MA 01854, USA

A lot of different types of nanomaterials have been developed as new electrochemical sensors in recent years. The high surface to volume ratio associated with unique surface properties makes them sensitive to species adsorbed on surfaces and results in good sensitivity and activity. However, most of the sensors developed don't have satisfactory electrochemical properties because of the limit of their structure and electrode modification method. In our research, a highly sensitive electrochemical sensor based on vertical nanowire array has been developed. The vertical Pt nanowire array is prepared by electrodeposition method using AAO membranes; after then a large amount of Au nanoparticles are grown onto the surface of the vertical Pt nanowire array by electroless plating. This new sensor structure can overcome several disadvantages of traditional nanowire electrode modification method with nanowire. The well aligned array makes it possible that every nanowire can have full contact with the analyte without the nanowires-overlap; also, there is no serious bubble blocking problem because no coating layer needs to be used for nanowire immobilization, which makes this structure more stable and durable. Because of the large surface area including the extremely high density of Au NPs, this new structure shows good electrochemical performance and can be widely used for chemical sensing, such as $H_2O_2$, glucose, or other biomolecule. The results indicate that this novel structure, as a new platform, has a great potential for various kinds of electrochemical detection.

# 390246 Functionalization of Carbon Nanotube Thin-Film Transistors Fabricated By Material Recognition Property of Peptide Aptamer


**Abstract:**

Carbon nanotubes (CNTs) are an attractive material possessing unique electrical, mechanical and optical properties which are the capability of metal or semiconductor derived from its structure difference, high electrical and thermal conductivity, and high-tensile strength and flexibility. Those properties hold great promise for development of micro/nano-electronic components, such as transparent and flexible electrodes, wirings, field-effect transistors, sensors, and so on. Therefore, handling techniques of the CNTs have been becoming important. Especially, isolation of conducting or semiconducting CNTs, site-selective alignment of the each CNT, and surface modification of the CNTs are key techniques for creating devices with high energy efficiency or sensors with high-sensitivity and selectivity. Actually, the electron field-effect mobility in excess of $10^6$ $cm^2$ /V·s was achieved on the transistor with applying an individual semiconducting CNT, and various sensors complexed with nanoparticles or biomolecules were reported as the application of detecting against a specific gas or cancer, respectively. However, many complicated processes were required for constructing those device structures. Biomolecules with recognition against specific material surfaces, known as aptamer, have a potential to achieve them by simple process. In this study, we demonstrated autonomous immobilization of CNTs on a solid surface by utilizing the property of the peptide aptamer, and fabricated the CNT thin-film transistors (CNT-TFTs) with applicative field-effect mobility. Furthermore, we examined the functionalization process of the CNT surface by metal nanoparticles toward the construction of the sensors.

The CNT-TFT was fabricated by following procedure. Au as source and drain electrodes were patterned on the silicon substrate as a back gate (phosphorus-doped, 0.001 Ω·cm resistivity) with thermal oxide film with a thickness of 100 nm. For immobilizing the CNTs as a channel, the peptide aptamer (CNTBP) of amino-acid sequence with [HMGLTKIHYSAL] which was previously identified for single-walled CNTs (SWCNTs) were introduced via silane coupling reagent (3-Aminopropyl triethoxysilane). Subsequently, the droplet of the SWCNT distributed with the single-stranded DNA was applied between the electrodes, and the substrate was washed with enough deionized water. To functionalize the CNT surface by Pd nanoparticles, sodium tetrachloropalladate (II) and the following reducing agent, sodium borohydride or L(+)-ascorbic acid, was applied. Morphological observation was investigated of the CNT surface was investigated through a scanning electron microscope (SEM) or transmission electron microscope (TEM), and the component analysis was performed by energy dispersive X-ray *spectroscopy* (EDX). Electrical characteristics of the CNT-TFT device were evaluated through a semiconductor parameter analyzer.

The SWCNTs immobilized surface was observed through optical microscopy and SEM. Although the deposited CNTs were invisible through optical microscopy, high-density and thin layered CNTs network was found only in the case with the CNTBP through SEM observation. Furthermore, the drain current-voltage characteristics of the CNT-TFT were investigated. In the case of using semiconducting CNTs, the drain current was increased in response to the gate voltage and the field-effect mobility reached 0.23  $6.7\times10^{-1}$ $cm^2$ /V·s which value was not ideal but applicative. In contrast, ohmic properties were observed in the case with conducting CNTs. These results showed transparent CNTs electronic components could be formed by utilizing the peptide aptamer. Furthermore, Pd nanoparticles deposited on the CNT surface was evaluated through TEM and EDX. In the preparation using ascorbic acid as a reducing agent, the particles with an average diameter of 5 nm were found on the CNT networks successfully. In the case using sodium borohydride, aggregated particles were observed relatively. In the case of both, the depositing particles of Pd were confirmed. As evidenced here, this simple process has a widespread potential for the fabrication of sensors with high-sensitivity and selectivity using CNT-based nanostructures.

# 390755 Sequence-Specific DNA Detection at 10 Fm By Electromechanical Signal Transduction

**Abstract:**

The development of a rapid, sensitive, and cost-effective nucleic acid detection platform is highly desired for a range of diverse applications including in patient screening during epidemics, oncological status assessment during surgery, detection of food contaminants, and biowarfare agent detection. We previously described a simple, PCR-free, optics-free, and potentially low-cost device for sequence-specific nucleic acid (NA) detection. The key element of the system is a peptide nucleic acid (PNA) capture probe conjugated to polymer spheres. PNA oligomers are uncharged analogs to DNA and RNA that share the same base chemistry and hybridize strongly to complementary NA sequences. Since the sphere-PNA conjugates carry little or no charge, they do not exhibit electrophoretic movement in response to a steady, DC field imposed through a pore. However, the substantial negative charge acquired upon capture of a target NA sequence makes the hybridized conjugate electrophoretically mobile. If the pore diameter is smaller than that of the spheres, the charged conjugate carrying the hybridized PNA and target NA would be expected to block the pore and significantly decrease its conductance, thereby causing a very strong, sustained drop in measured current. In such a way, the selective PNA-NA hybridization signal is transduced electromechanically, and the system would be expected to give an essentially binary response signaling the absence or presence of a target NA. In proof of principle work with our device, we detected 20-mer DNA oligomers with a 10 pM detection limit. Here, we demonstrate the testing of our device with longer DNA strands, and we describe the resulting improvement in the limit of detection (LOD). We examined the detection of target DNA oligomers of 110, 235, 419, and 1613 nucleotides at 1 pM–1 fM concentrations, finding that the LOD decreased as DNA length increased, with 419 and 1613 nucleotide oligomers detectable down to 10 fM. In addition, no false positive responses were obtained with non-complementary, control DNA fragments of similar length. Also, we successfully detected target fragments against a DNA background simulating the non-complementary genomic DNA present in real samples. The 10 fM LOD achieved with a ~1600 base analyte suggests our device would be useful clinically for identification of bacteria using similar size, species specific 16S rRNA as the target. Results to date have been acquired with a micro-scale device; we currently are creating and testing nano-scale devices that should provide further performance enhancements.

# 364286 Atomic Layer Deposited Highly Dispersed Platinum Nanoparticles Supported on Multiwalled Carbon Nanotubes for Liquid Phase Reactions

**Abstract:**

Highly dispersed platinum nanoparticles were deposited on gram quantities of non-functionalized multiwalled carbon nanotubes (MWCNTs) by atomic layer deposition (ALD) in a fluidized bed reactor at 300 °C. (Methylcyclopentadienyl) trimethylplatinum and oxygen were used as precursors. The results of TEM analysis showed that ~1.3 nm Pt nanoparticles were highly dispersed on non-functionalized MWCNTs. The porous structures of MWCNTs did not change with the deposition of Pt nanoparticles. For comparison, the commercial 3 wt.% Pt/C catalyst was also characterized. The ALD-prepared Pt/MWCNT was used for the hydrogenation of xylose to xylitol and hydrogen transfer of ketones. The ALD-prepared Pt/MWCNT showed the best catalytic performance with 100 % conversion of xylose and 99.3 % selectivity to xylitol, compared to commercially available Pt/C, Ru/C, and Raney Ni catalysts. The heterogeneous hydrogen-transfer reactions of acetophenone indicated that an acetophenone conversion of 100% and a 1-phenylethanol selectivity of 99.0% could be obtained with a ketone to Pt mass ratio of 24,690 and a ketone to KOH mass ratio of 22 at 150 °C for 5 h. The stability of ALD produced Pt/MWCNT catalyst was much higher than that of the commercial Pt/C, due to the presence of surface defects on the MWCNTs and the strong metal–support interaction for the ALD-prepared Pt/MWCNT catalyst.

# Temporal and Spatial Reactors for Atomic Layer Deposition on Particles

J. Ruud van Ommen, Fabio Grillo, David Valdesueiro,
Dirkjan Kooijman, Mojgan Talebi, Aris Goulas

Delft University of Technology, ChemE, Julianalaan 136, 2628 BL Delft, the Netherlands
j.r.vanommen@tudelft.nl

Whereas most research on atomic layer deposition (ALD) is aimed at depositing ultrathin films on wafers and other flat substrates, ALD can also be used to provide the surface of particles with either an ultrathin continuous coating or a decoration of nanoclusters [1,2,3]. When carried out in a so-called fluidized bed, ALD is an attractive way of producing nanostructured particles, for example for catalytic applications, with excellent scale-up potential. In such a system, the particles are suspended in an upward gas flow. Nanoparticles – contrary to what is typically observed for micron-sized particles – are fluidized as very dilute agglomerates with distinctive fluidization characteristics. We will discuss the challenges related to fluidization and coating of nanoparticles.

Like most conventional ALD reactors, fluidized-bed ALD reactors are typically operated in a temporal mode: the pulses of the different gaseous reactants are delivered subsequently in time. An alternative is to separate the administering of the reactants in space [4]. We recently developed a spatial ALD reactor for particles. We will discuss spatial ALD of Pt on $TiO_2$ nanoparticles that are pneumatically transported through the reactor with nitrogen at atmospheric pressure. These nanoparticles are blown through a tube as very porous agglomerates, ensuring good accessibility to the precursor/co-reactant ($MeCpPtMe_3$ and $O_2$), which are injected along the tube. Optimizing the operating conditions, we obtained an even spread of 1-2 nm nanoclusters of Pt over the $TiO_2$ nanoparticles (see Fig. 1). This material showed to be a very active photocatalyst. Reaction tests demonstrated an increase of the photocatalytic activity with a factor 5, compared to native $TiO_2$ nanoparticles. With the current setup, we are able to produce nanostructured particles via ALD at a production rate of about 1 g/min. However, the experience with pneumatic transport from other fields opens up promising scale-up prospects for continuous particle ALD.

*References*


[1] D. M. King, X. Liang, A. W. Weimer, Powder Technol. 221 (2012) 13-25.

[2] D. Longrie, D. Deduytsche, C. Detavernier, J. Vac. Sci. Tech. A 32 (2014) 010802.

[3] A. Goulas, J.R. van Ommen, KONA Powder and Particle J. 31 (2014) 234-246.

[4] W.M.M. Kessels, M. Putkonen, MRS Bull. 36 (2011) 907-913.

**Figure 1.** TEM image of the Pt nanoclusters, deposited on $TiO_2$ nanoparticles by ALD in a spatial reactor.

# 377799 Metallic Co ALD Catalysts for Fuels from Syngas

**Abstract:**

Direct deposition of a metallic Co by ALD has been demonstrated using low-cost cobaltocene precursor and $H_2$; this catalyst preparation method eliminates reduction of oxide and improves catalytic productivity. Eliminating high-temperature treatment of catalyst avoids sintering and incorporation of metal into the support (aluminate). Two methods of depositing metallic Co are discussed: on $Al_2O_3$and on W ALD. Both surfaces have exhibited deposition of metallic Co; however, the deposition reactions are vastly different.

Application of the Co catalyst deposited by metallic ALD has multiple advantages. First, cobaltocene is a less-expensive reactant than other proposed chemistries for metallic Co ALD. Second, metallic Co ALD will be less likely to sinter than incipient wetness catalysts because reduction temperatures are avoided. Third, the deposition of metallic Co compared to $Co_3O_4$ ALD provides higher productivity of Co catalyst because the metal does not form a mixed oxide with the support.

LEHIGH UNIVERSITY

Center for Advanced Materials and Nanotechnology

CardiffCatalysisInstitute

# Chemical vapour impregnation- a versatile route to high redox activity nanoparticle catalysts

Michael M Forde PhD.

Graham J. Hutchings, Chris K. Kiely, Qian He, Nikolaos Dimitratos, Ren Su and Lokesh Kesavan

AIChE Meeting (Atlanta) November 2014

# Talk Outline

- Overview of our catalyst preparation techniques
- Characterisation of CVI materials- examples
- Catalysis with CVI materials
  - Toluene oxidation
  - Levulinic acid hydrogenation
  - $H_2$ evolution by photocatalysts

# Sol-immobilisation (SI)

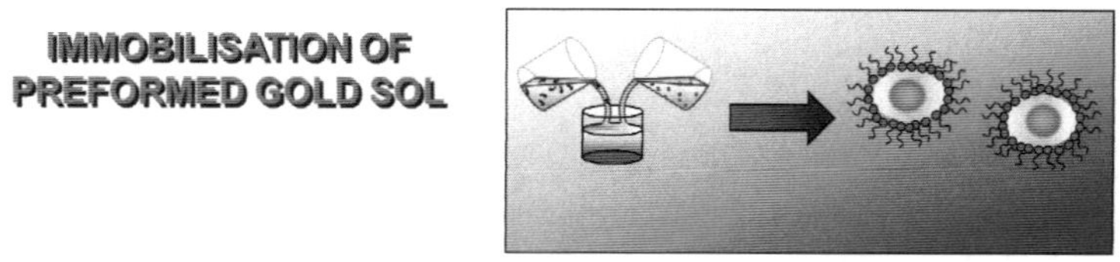


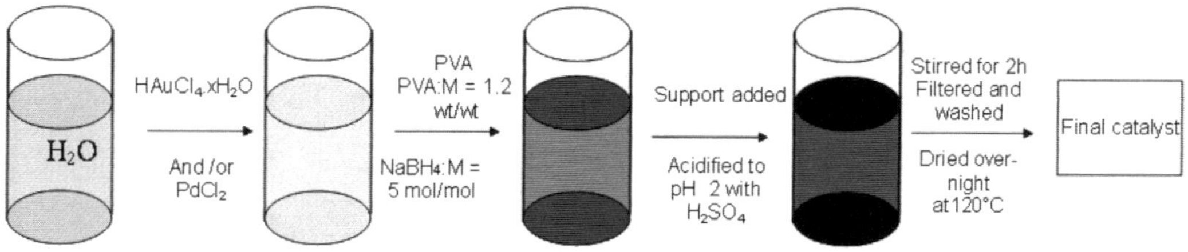


- Final catalyst not heat treated
- Can prepare bi and tri metallics in this way
- Order of metal precursor addition and reduction affects morphology

# Core-shell structure by SI

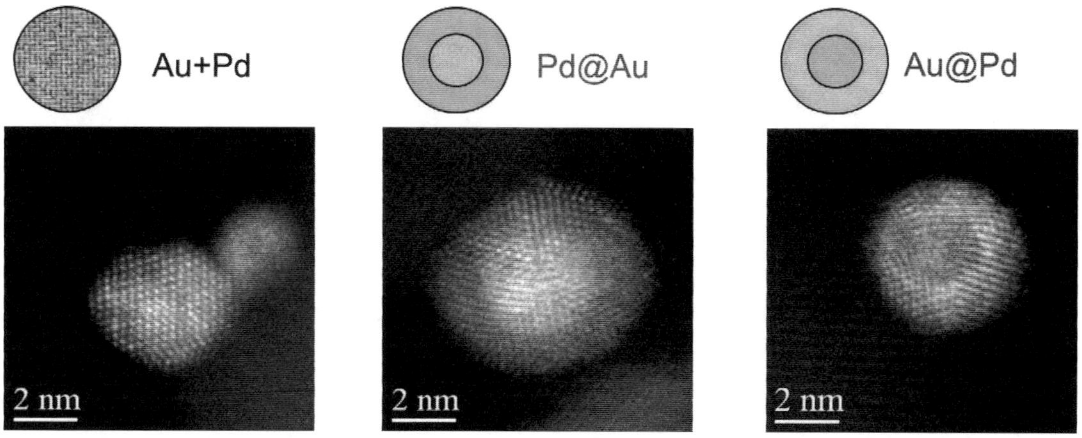


**Varying precursor addition order and reduction can be used to change morphology**

**Go to method for making small nanoparticles**

# Issues

- ➢ Surface ligands present
- ➢ Spectator species present (chloride etc)
- ➢ Materials not heat stable
- ➢ Lengthy synthesis
- ➢ Reproducibility and general application

Issues common to many techniques!

# Chemical Vapour Impregnation (CVI)

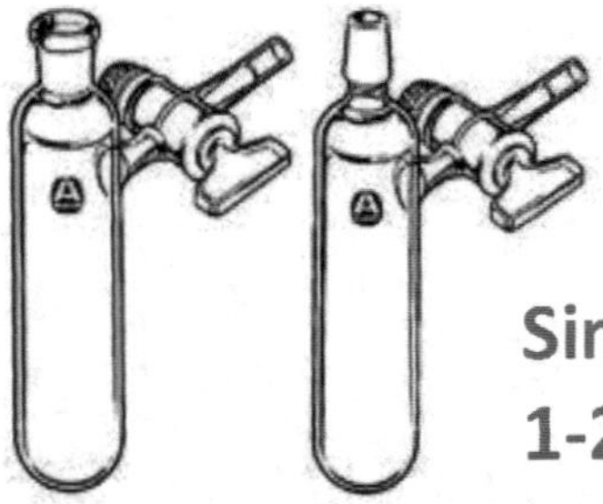

**Simple, scalable, reproducible technique**
**1-2h for deposition**

- ✓ Physically mix precursor and organometallic precursor
- ✓ Evacuate tube, heat under constant vacuum
- ✓ Remove vacuum, cool and remove material
- ✓ Heat treat to produce fine dispersion of nanoparticles on
- ✓ support

# Features of CVI

- ✓ Can deposit wide range of metals on almost any support
  - ✓ We've done Fe, Cu, Pt, Pd, Rh, Ru, Ag, Zn, Ga
- ✓ Very fine metal dispersion
  - ✓ Size depends on nature of metal, treatment conditions, loading
- ✓ Can be used to prepare bi and trimetallics
- ✓ Can be used to modify one material by selective deposition of a second metal.. Core shell structures
- ✓ Contamination free support
- ✓ Clean surface to NP's

# Pd and Pt - examples

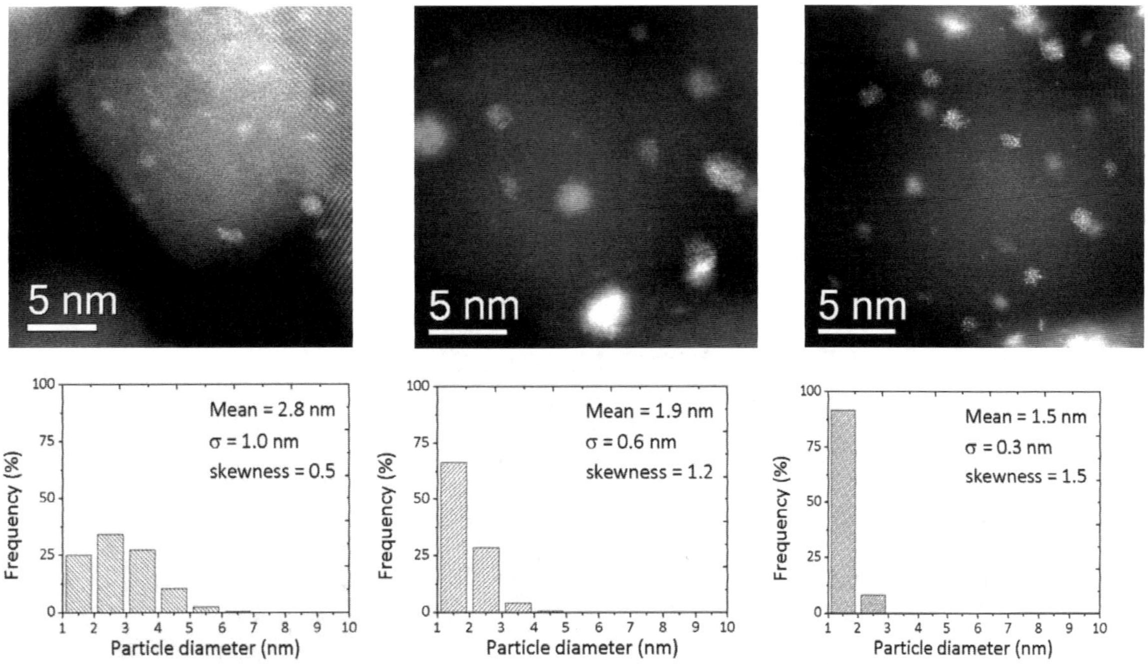


**2.5%Pd/$TiO_2$** **1.25%Pd-1.25%Pt/$TiO_2$** **2.5%Pt/$TiO_2$**

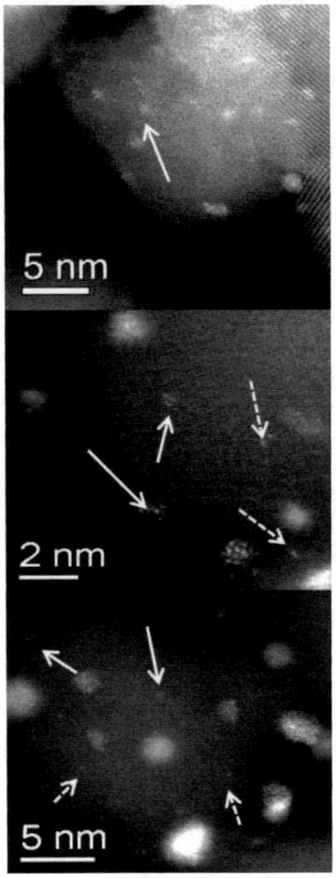


# Pd and Pt examples

**Sub-nm clusters**

**Single atoms**

**Still observed after 400°C treatment**

## Characterisation CVI - examples

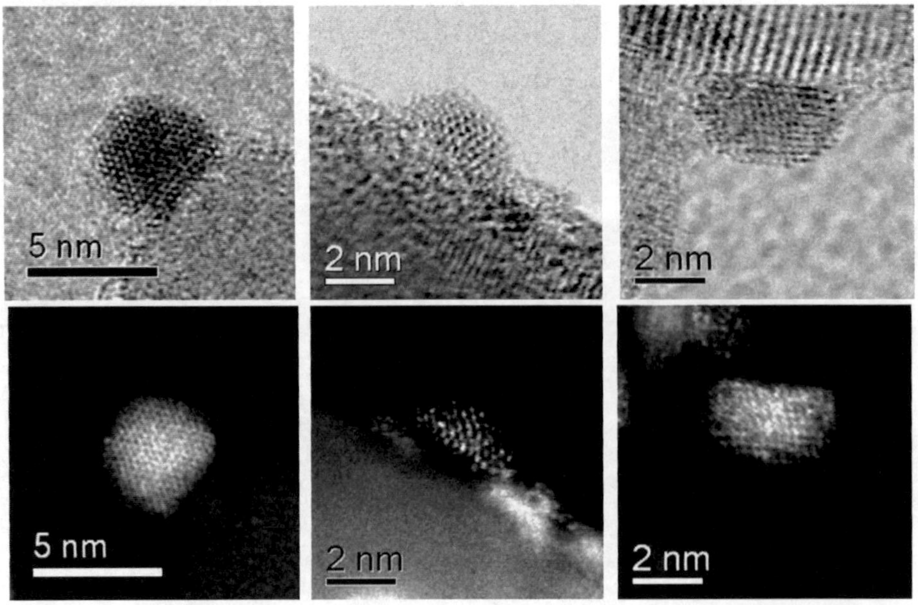


**Clean nanoparticles in all cases**

# Mixed Pd-Pt - example

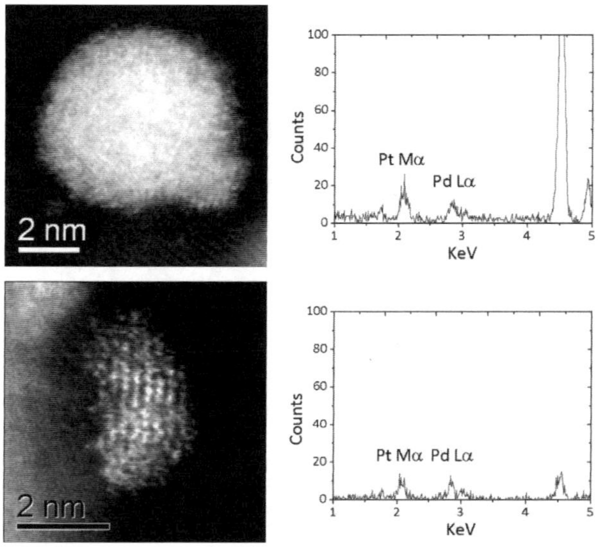


$1.25\%Pd\text{-}1.25\%Pt/TiO_2$

One step deposition

Homogeneous alloy expected

Most particles found to be **homogeneous alloys**

Occasional Pt and Pd only particle observed

# Modifying Au catalysts

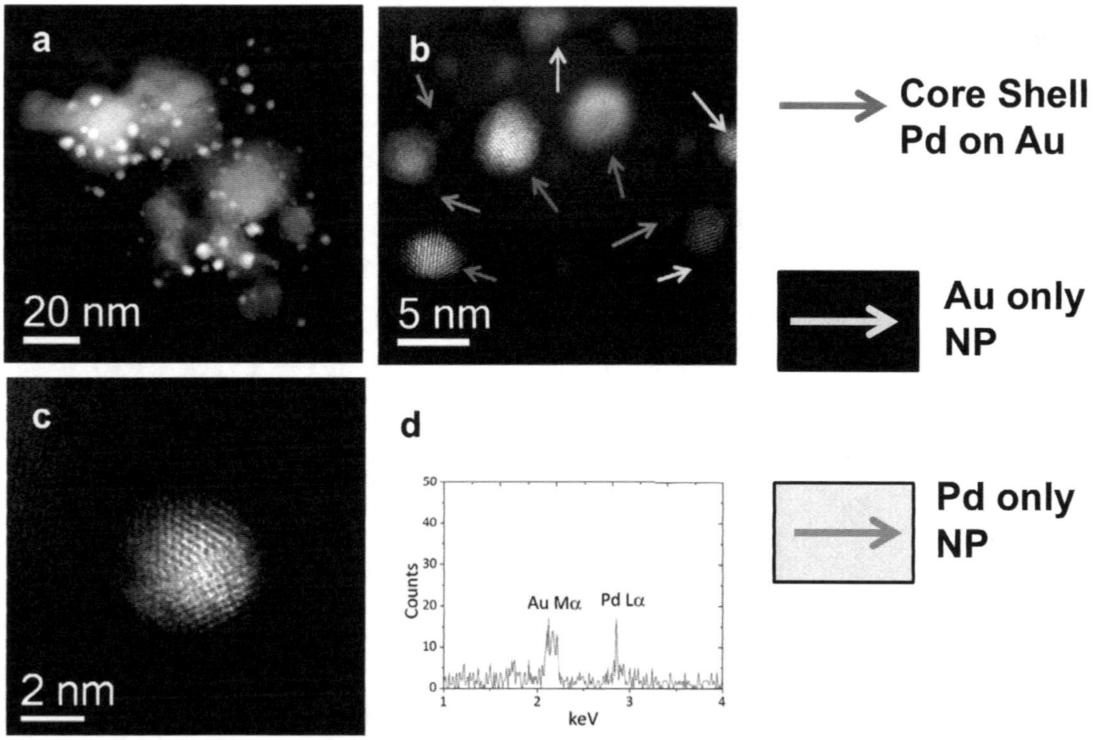

# Modifying Au catalysts

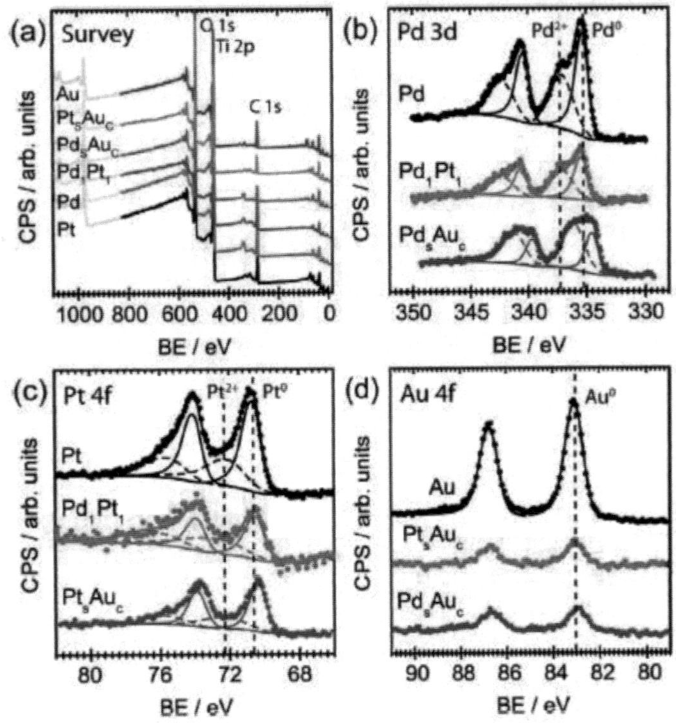


**In hybrid Au peak is reduced severely... Evidence of core shell structure**

**Pd can be found in +2/0 oxidation state (age)**

**Pt always found in 0 oxidation state**

# Characterisation- Fe/$SiO_2$ by CVI

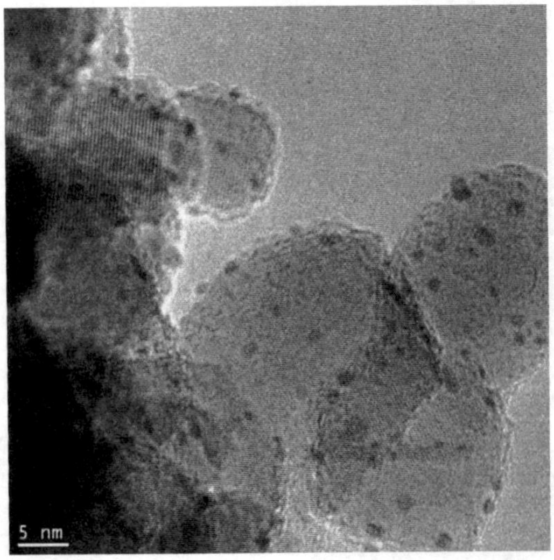


Uniform distribution of FexOy Particles ~ 3nm

Size affected by metal loading

If heat treatment is done under $H_2$ All surface Fe disappears

# TOLUENE OXIDATION

ACS Nano, 2014, 8, 957-969
Angew. Chem. Int. Ed., 2012, 51, 5981–5985
Phys. Chem. Chem. Phys., 2013,15, 10636-10644
Science, 2011, 331, 195

## Catalysis- Toluene Oxidation

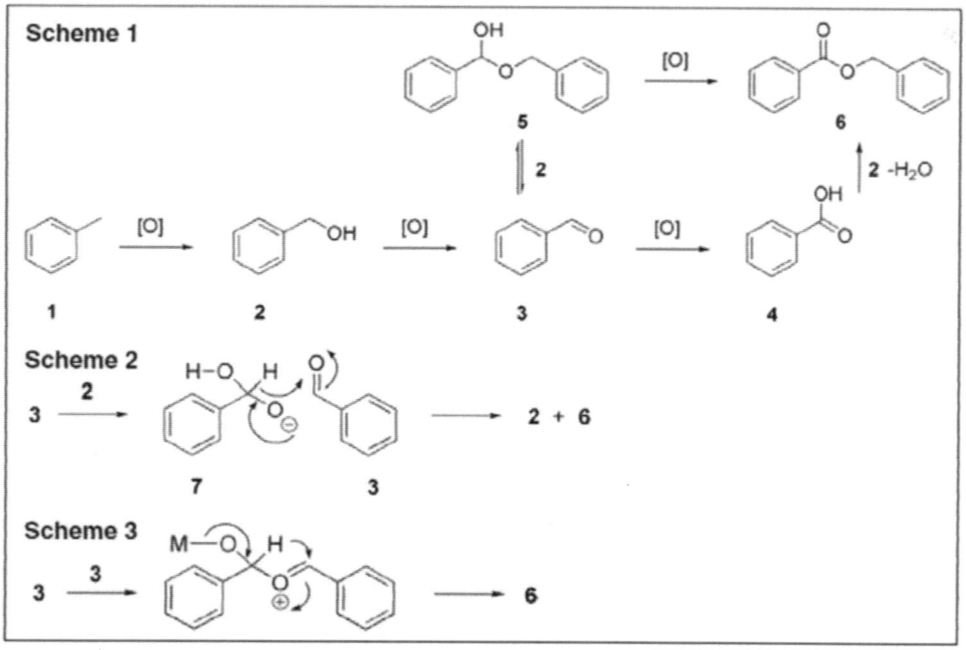

## Toluene oxidation- 1wt% (AuPd)/C by SI

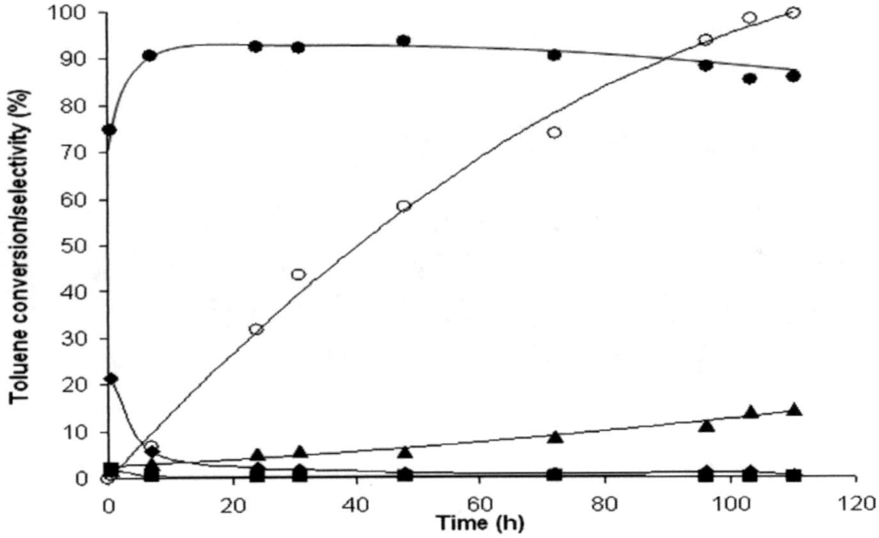


**Key result- High conversion to benzylbenzoate**

**Solvent free, base free, 160°C, 10bar $O_2$ (continuous)**

## Toluene oxidation- 1.25%Pd-1.25%Pt/$TiO_2$ CVI

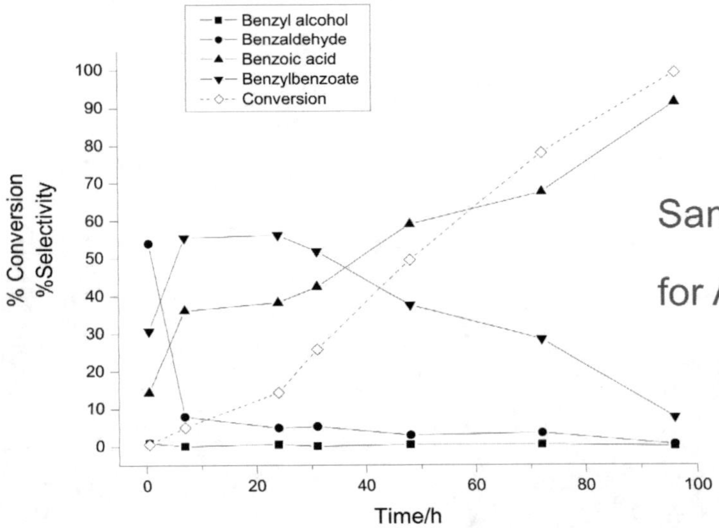


Same conditions as

for AuPd/C (SI) catalyst

**Excellent activity**

**Different product distribution**

**Pd-Pt/$TiO_2$ or Pd-Pt/C by SI does not achieve high conversion**

# Toluene oxidation with TBHP and Pd-Pt/$TiO_2$

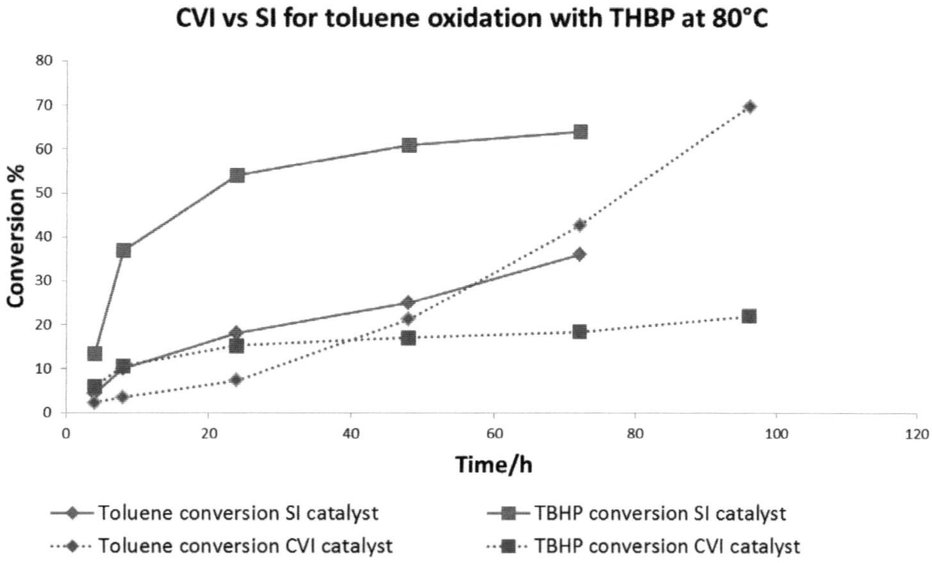


CVI catalyst is activating $O_2$...under $N_2$ toluene conversion drops, TBHP conversion increases

Higher conversion achieved with CVI catalyst

# Levulinic Acid Hydrogenation

**ACS Nano, 2014, 8, 957-969**

# Hydrogenation Catalysis- Levulinic Acid

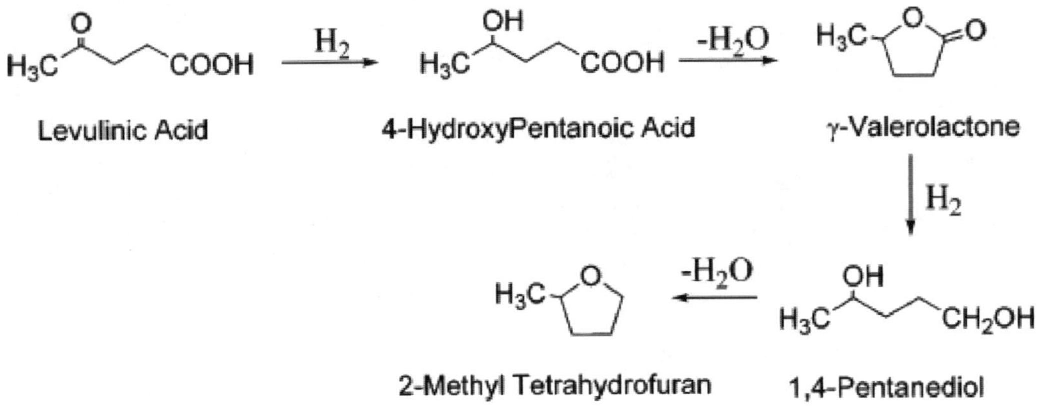


**High rate, milder conditions, Lower excess of $H_2$**

# Levulinic Acid Hydrogenation

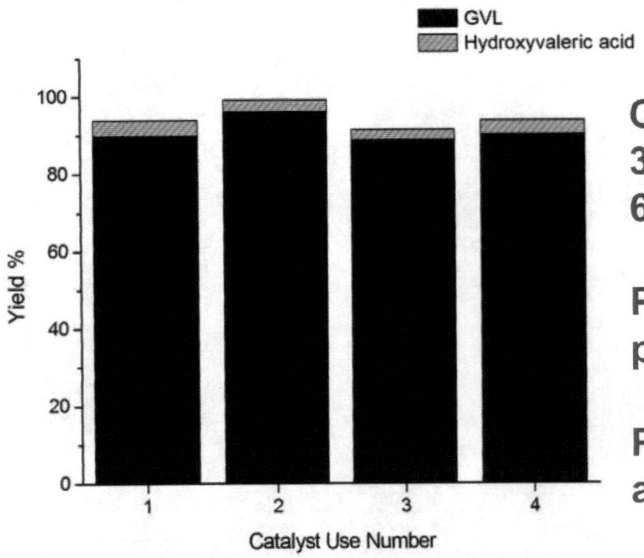


**Catalyst even works at 30°C !**
**60% yield GVL in 20h**

**Pd-Pt has good performance at 150°C**

**Pd only performs well above 180°C**

**Pt/$TiO_2$ catalyst, 100 °C, $H_2$:LA= 2:1, 4 h**

# Photocatalysis

Dalton Trans., 2014,43, 14976-14982

## $H_2$ from ethanol-water

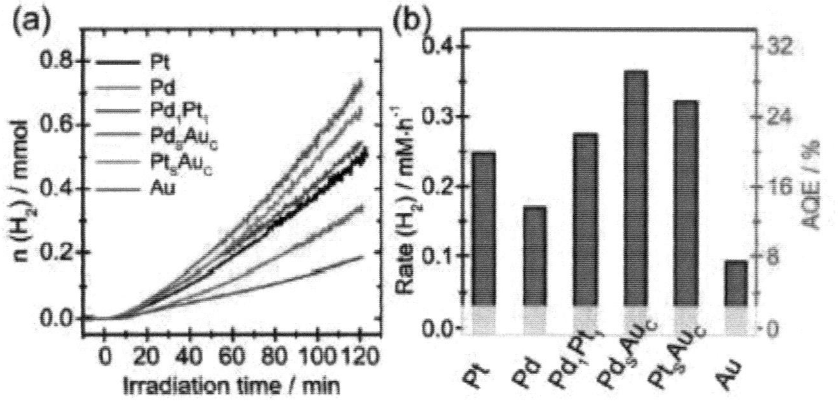


**Beneficial effect of core shell structure observed**
**UV led used as light source**
**No dark reaction observed**

## Summary

- Simple, scalable techniques
- Well defined nanoparticles
- Flexibility to make exotic nano-architectures
- Clean, stable nanoparticles
- Wide activity for oxidation, hydrogenation, photo-catalysis

# 388051 Preparation of Carbon Supported Pt Nanoparticles By Atomic Layer Deposition

**Abstract:**

Hydrogen fuel cells offer an efficient use of a more energy dense fuel compared to combustion engines using gasoline. However, adaptation of fuel cells to power vehicles depends on several factors, one of which is economic synthesis of each component. Atomic layer deposition (ALD) using a fluidized bed reactor is an economic and scalable method for manufacturing the noble metal nanoparticle catalyst. This work furthers understanding of reported mechanisms of platinum ALD by investigating reactions on a common carbon support, XC72-R. By-products of reactions during ALD using trimethyl(methylcyclopentadienyl)platinum(IV) ($MeCpPtMe_3$) and either oxygen or hydrogen as reactants on un-modified carbon and carbon functionalized using nitric acid are inspected using *in situ* mass spectrometry. This understanding will help fine-tune parameters during a commercial catalyst synthesis process. Platinum weight loading, average particle size and particle dispersion depend on the chemistry of the pre-processing and precursor reactions. Oxygen becomes integrated into the carbon during functionalization and binds platinum tighter than just carbon; therefore particles aggregate less during deposition on functionalized carbon. Combustion of the ligand of $MeCpPtMe_3$ attached to carbon, as well as substrate oxidation during the oxygen dose, contribute to more particle aggregation versus when hydrogen is utilized for ligand removal. Mechanisms for each ALD condition are proposed.

# Particle Size and Support Effect of Platinum catalyst in a Proton Exchange Membrane Fuel Cell

Shibely R. Saha[1] and Dongmei Li[1]*
[1]*University of Wyoming, 1000 E. University Ave. Laramie, WY 82071*
*dli1@uwyo.edu

Platinum (Pt) loading in proton exchange membrane fuel cells (PEMFC) is reduced and utilization increased by studying particle size and support effect on Pt catalysis. Generally, smaller Pt particle size in the PEMFC electrode catalyst layer provides increased reaction sites, thus maximizing the utilization and minimizing the loading. Another important factor that contributes to efficient catalyst use is its distribution upon the support, as well as how supported catalysts are dispersed in the gas diffusion layer (GDL). Catalysts that are well-distributed on its support and well-dispersed on the electrode provide desired contact for the reactants and improved mass transport of the reactants to the active catalytic sites [1].

To investigate the impact of support on Pt catalyst, transition metal carbides (TMCs), such as Molybdenum Carbide ($Mo_2C$), were selected as a support for Pt, due to their similarity in electronic and geometric structure to Pt [2] . Although previous research has used TMCs, including $Mo_2C$, as Pt supports, low TMC surface area has limited the catalytic performance[3]. To address this challenge, nanosize Pt has been discretely deposited upon the surface of a $Mo_2C$ hollow nanotube support (confirmed by HRTEM) via island growth mechanisms of Atomic Layer Deposition (ALD). Nanosize Pt particles supported onto hollow nanotube $Mo_2C$ (also referred as co-catalyst) with sub-monolayer coverage are presumed to further increase the reactivity of the catalyst, possibly resulting from both increased active Pt sites and enhanced interaction between the Pt particle and support. The reactivity of the co-catalyst towards Hydrogen Evolution Reaction (HER) and Hydrogen Oxidation Reaction (HOR) has been evaluated and compared to that of commercial 20% Pt supported on carbon black powder by electrochemical characterization using a three-electrode cell. This electrochemical characterization showed significantly higher HOR and HER activity from the co-catalyst with 2.4% Pt loading (confirmed by Atomic Absorption Spectroscopy (AAS)), compared to 20% Pt/C, demonstrating its potential applicability as PEMFC anode.

To evaluate co-catalyst performance in a device setting, PEMFC membrane electrode assembly (MEA) was fabricated using the co-catalyst as PEMFC anode. Device performance demonstrated a significant increase in maximum power per mg of Pt, compared with commercial 20% Pt supported on carbon black powder. As such, both electrochemistry cell and device testing results revealed that the sub-monolayer Pt coverage (a.k.a. discrete islands) on nanoscale $Mo_2C$ tubes generated a synergetic effect that requires further systematic theoretical and experimental study to understand.

To investigate how particle size and distribution of the co-catalyst on electrode affects the device performance of PEMFC MEA, ultrasonic AccuMist (Sono-Tek) spray technique was applied, where the catalyst can be atomized, minimizing co-catalyst agglomeration, often observed in conventional brush painting approach. Scanning Electron Microscopy (SEM) confirmed that smaller particle size was generated on the electrode with ultrasonic AccuMist spray approach, with 37% increase in maximum power per mg of Pt in the electrode compared to that by conventional brush painting technique.

**References**


1. Lister, S; McLeanG., *Journal of Power sources*, p.61-76, 130(2004)
2. Levy, *R.B., Science*, p. 547-549, 181(1973)
3. Esposito, D.V.; Hunt, S.T.; Kimmel, Y.C.; and Chen, J.G.; *J. Am. Chem.Soc.,* p. 3025–3033, 134(2012)

# 390688 Carboxyl Multi-Walled Carbon Nanotubes Stabilized Palladium Nanocatalysts for Fuel Cell Application


**Abstract:**

Carboxyl Multi-walled carbon nanotubes (MWNTs-COOH) supported palladium (Pd) nanoparticles (NPs) catalysts (Pd/MWNTs-COOH) were prepared by a one-pot solution-based method, i.e., direct decomposing palladium acetylacetonate ($Pd(acac)_2$) in a refluxing xylene solution in the presence of MWNTs-COOH without any additives. Pd/MWNTs-COOH with different Pd loadings were prepared by altering the amount of $Pd(acac)_2$. Different annealing temperature effect was also investigated on Pd/MWNTs-COOH in a reducing atmosphere. X-ray diffraction (XRD) results indicate a successful deposition of Pd NPs on MWNTs-COOH. Transmission electron microscope (TEM) micrographs revealed that Pd NPs with increasing size were uniformly grown onto MWNTs surface with the increase of Pd loadings, and varying Pd particle sizes were also observed implying varying steps of Pd NPs as diffusing, contacting and coercing on tube wall surface with the increase of annealing temperature. The loading of Pd was calculated from the thermal gravimetric analysis (TGA), the variation of –COOH was investigated by Raman spectroscopy and X-ray photoelectron spectroscopy (XPS). The electrocatalytic activities of the Pd/MWNTs-COOH catalysts were evaluated for methanol oxidation and ethanol oxidation in KOH solutions. For methanol oxidation, bare Pd NPs and Pd/MWNTs (MWNTs without -COOH) was also synthesized and compared, parameters including the concentration of methanol and KOH were investigated, showing that the best catalyst performance was achieved by the balance between the adsorption of hydroxyl and methanol. A pronounced influence of temperature on methanol oxidation has been manifested, which indicates the positive role of increasing temperature. For ethanol oxidation, Cyclic voltammetry (CV), chronoamperometry (CA), electrochemical impedance spectroscopy (EIS) and Tafel were introduced to characterize the electro-performances, Pd/MWNTs-COOH with initial precursor ratio of 2:1 gave the best performance, 100 °C annealed Pd/MWNTs-COOH also gave best performance compared with other temperature annealed ones

# 386479 Impact of Support, Nanoparticle Size, and Carbon Precursor on the Nanotube Growth Mechanism: Viability of Controlling Nanotube Structure

**Abstract:**

Single-walled carbon nanotubes are tubular crystalline materials that can be grown via chemical vapor deposition (CVD) and have outstanding electronic and optical properties that are potentially useful for a myriad of applications. However, such properties can vary significantly with the individual nanotube structure, which can be defined by the chirality (*n,m*). Therefore, designing a highly chiral-selective and large-scale nanotube synthesis process has been highly sought-after to avoid costly post-synthesis separation processes. Such level of control, however, demands a clear understanding of the nanotube growth mechanism, which is still under debate due to seemingly conflicting experimental observations. In this contribution, we use reactive molecular dynamics to simulate nanotube growth at 1000 K under different conditions with variations of i) nanocatalyst particle size, ii) carbon feedstock, and iii) interaction strength between the nanocatalyst and the support. We used computational tools to analyze the carbon diffusion mechanism and nanotube assembly, and the structural and dynamic state of the nanoparticle (and the possibility of carbide formation), and discuss our findings in the context of the viability of using the nanocatalyst particles as a template to control nanotube chirality. For instance, dissolution of carbon during the induction/nucleation period increases the dynamics of the nanocatalyst particle (surface), which is unfavorable to create a template effecting on the emerging carbon network. However, we observed that carbon dissolution and the formation of the carbon network happen as parallel processes, indicating that the latter can occur with or without the former. This finding reconciles the observation of nanotube growth in both carbon-dissolving and non-carbon-dissolving materials, and suggests that the latter scenario can be potentially exploited to achieve templated nanotube growth.

# 389539 Role and Stability of a Metal Carbide Catalyst in the Growth of Single-Walled Carbon Nanotubes

**Abstract:**

Achieving a high degree of selectivity in the synthesis of single-walled carbon nanotubes in a systematic and efficient way is one major barrier to be overcome before the exceptional properties of these materials can be exploited commercially. Many studies seem to indicate that the state and structure of the catalyst in CCVD play a fundamental role in determining structural features of the nanotubes such as chirality. Selective growth has been reported in both monometallic and carbide nanoparticles; however the contribution that either of them brings to this process has not been completely elucidated. The stability and the influence of the latter in the growing nanotube structure is the focus of this study. We determine the stability and relevant features of carbide nanoparticles and the interactions between carbide facets and the nascent carbon structure using ab initio molecular dynamics simulations. We use 55-metal-atom Ni and Co clusters to model the catalyst and run our simulations at three different temperatures to study the stability of the clusters. Nanotube caps of metallic and semiconducting nanotubes are placed in contact with the equilibrated nanoparticles to observe changes in the carbide due to the presence of the growing nanotube. Analyses of the electronic and geometric structure allow us to elucidate the nature of the carbide-nanotube interactions. We observe that the catalyst accommodates its overall shape to the nanotube cap according to an "inverse template effect" observed for unsupported nanoparticles. Although the original carbide structure is not completely maintained due to fluctuations in the structure of the solid, we observed that C atoms remained without forming bonds among themselves throughout the simulation time. There is also an apparent migration of the C atoms to the subsurface but no segregation is observed. We also performed simulations with surface slabs resembling different facets of the carbide catalyst placed in contact with a graphene fragment. Estimations of the work of adhesion of grahene and structural analysis reveal that the differences in the affinity of the graphene and the catalyst determine the mechanism of anchoring and liftoff of the nascent nanotube to the catalyst in accordance with electron microscopy observations.

# Diffusion Boundary Layer in the Growth of Centimeter-Tall Carbon Nanotube Forests

Jaegeun Lee, Eugene Oh, Kun-Hong Lee*

Department of Chemical Engineering, Pohang University of Science and Technology (POSTECH), 77 Cheongam-Ro, Nam-Gu, Pohang, Gyungbuk, South Korea 790-784.

*ce20047@postech.ac.kr

Despite their excellent mechanical properties, the application of carbon nanotubes (CNTs) has been much limited mainly due to their short length. The development of CNT fiber provides the opportunity of enabling the mechanical application of CNTs. To fully utilize the ultimate mechanical strength of individual CNTs, it is important to compose CNT fiber with long CNTs. However, the technique to mass produce long CNTs (centimeter scale) is still lacking.

In this talk, growth of centimeter-tall multi-walled carbon nanotube forests up to 1.9 cm is reported using chemical vapor deposition process with a typical tube furnace reactor. The total gas flow rate was changed from 340 to 1700 sccm, which corresponds to flow velocity of 1.25 cm/s to 6.25 cm/s. The growth rate was increased from 1.8 mm/hr to 3.9 mm/hr with increasing total flow rate.

The increased growth rate is attributed to the decreased diffusion boundary layer thickness. As the velocity of bulk fluid increases, the boundary layer thickness is decreased. The boundary layer thickness is proportional to the square root of bulk fluid velocity as described in figure 1. Under mass transport limited growth regime, the growth rate is proportional to the diffusion flux, and thus the growth rate should be proportional to the square root of the bulk fluid velocity. As shown in figure 2, a clear proportionality was observed from our experimental data. Further discussion on fostering growth rate will be given in the respect of transport phenomena.

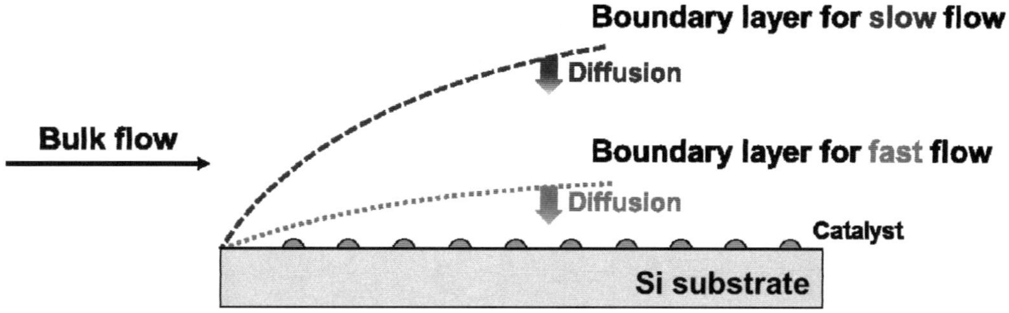


Figure 1. Boundary layer thickness under different flow velocity.

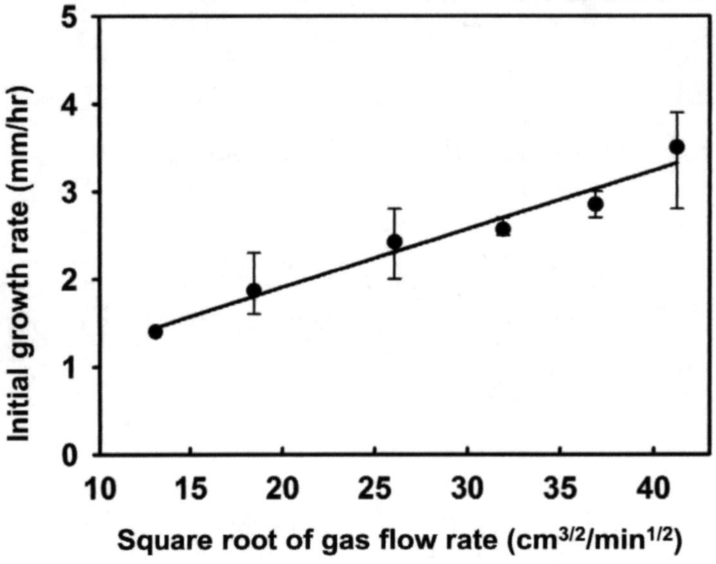


Figure 2. Initial growth rate of CNTs with increasing gas flow rate.

# Architectural Design of Hollow Mesoporous Composite Nanotubes as Robust Catalysts for Benzyl Alcohol Oxidation

Guowu Zhan, Hua Chun Zeng*

NUS Graduate School for Integrative Sciences and Engineering, and Department of Chemical and Biomolecular Engineering, Faculty of Engineering, National University of Singapore, 10 Kent Ridge Crescent, Singapore 119260. chezhc@nus.edu.sg

*KEYWORDS carbon nanotubes, catalysts, assemble, hollow, capsules, mesoporous*

**ABSTRACT:** Compared with the industrial physical mixture catalysts of poorly organized structure, the architecturally-designed integrative catalysts with multi-composition and well-defined morphology can bring new opportunities for high efficiency catalysis. Advanced hybrid materials with carbon nanotubes (CNTs) and metal or metal oxide or metal-organic frameworks (MOFs) enable catalysts possible to achieve cooperatively enhanced performances from the respective component. Herein, we report, for the first time, a facile and reliable synthetic method for the preparation of a series of complex and highly integrative materials possessing a central nanoscale cavity, a permeable silica shell and various catalytically active nanoparticles (e.g. Au, Pd, Au-Pd alloy, ZnO, $TiO_2$ and $Co_3O_4$) anchoring on the CNTs core. Integration of single or double layered mesoporous silica, microporous ZIF-8, or $Nb_2O_5$ onto the surface of CNTs was also demonstrated. Concerning their workability, we further demonstrated that the as-prepared integrative catalysts have higher catalytic performance for the selective oxidation of benzyl alcohol to benzaldehyde, compared with the conventional surface supported catalysts. Most importantly, the hollow CNT@Au-Pd@$mSiO_2$ catalyst was very robust and could be used in six cycles without any loss of activity. In addition, efforts were also made to further integrate the one-dimensional (1-D) CNT@$mSiO_2$ into 3-D superstructured spheres (ca. 64 μm) based on the polymerization of styrene using Pickering emulsion droplets (o/w type) as reaction vessels.

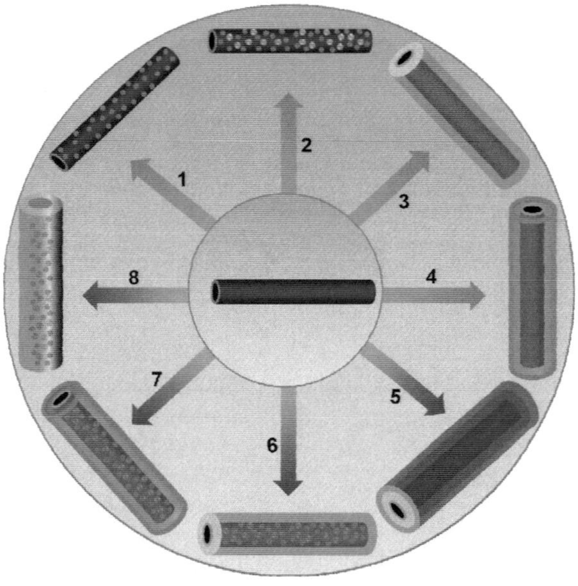

# 372138 Synthesis of Carbon Nanofibers from Light Liquid Paraffins Using the Liquid Pulse Injection Technique

**Abstract:**

INTRODUCTION

Carbon nanofibers (CNFs) show many unique properties, most of them common with carbon nanotubes (CNTs). As the mass production of CNFs is thought to be much easier than CNTs, the demand of CNFs is growing at a quicker pace. However, the production cost of CNFs is still high, therefore applications in which the usage of them is thought to be feasible is still quite limited.

Previously, we introduced a new method to efficiently produce CNFs, the Liquid Pulse Injection (LPI) technique. The main feature of this method is the introduction of the fiber source into the reactor as liquid pulses. Liquid pulse introduction leads to the generation of extremely active catalyst particles in a highly dense state. Therefore, CNFs can be easily obtained at high carbon yields, up to 85%, through this method. We also experimentally showed that CNFs can be obtained at high carbon yields even from an inexpensive mixture of benzene, toluene and xylene.

Aromatics are generally used as the carbon source for CNF production. However, such hydrocarbons are highly toxic, and many regulations exist for the usage of them. As the LPI technique allows the generation of extremely active catalyst particles, CNFs are thought to be attainable at high yields even from hydrocarbons which were thought not to be suitable for CNF production. Up till now, we have shown that CNFs can be efficiently produced from various alcohols. In this work we verified the possibility of efficiently producing CNFs from another category of inexpensive and abundant hydrocarbons, light liquid paraffins.

EXPERIMENTAL

An apparatus equipped with a tube reactor, an electric furnace and two mass flow controller systems, one to control the flow rate of hydrogen, and the other to control the flow rate of nitrogen was used for CNF production. Hydrogen was used as the carrier gas, and a mixture of ferrocene, the carbon source and thiophene (catalyst promoter) was used as the starting material. Hexane was selected to represent light liquid paraffins, and was used as the carbon source in production experiments. First the reactor was thoroughly purged with nitrogen, and then hydrogen was introduced to the reactor. Next the temperature of the reactor was raised to 1,473 K using the electric furnace. After the temperature distribution in the reactor reached a steady state, 20 liquid pulses of the starting material were intermittently introduced into it at an interval of 30 s or 60 s. In some experiments, this production sequence was modified in order to collect CNFs at different growth stages. After the starting material was injected into the reactor, the generated CNFs were quickly blown out from the hot temperature reaction zone by rapidly increasing the flow rate of the carrier gas when they reached the position where they were intended to be sampled. Through this sequence, fibers representing the CNFs at this position can be collected. By varying the blowing out timing of the CNFs, a series of fibers can be obtained, and the growth behavior of the fibers can be clarified by analysing the collected fibers. During CNF production, the composition of the reactor outlet gas was also monitored using a micro GC. After 60 s from the injection of the final liquid pulse, the reactor was quickly purged with nitrogen, and the electric furnace was turned off. The reactor was cooled down to room temperature, and the produced CNFs located at the bottom of the reactor were collected and weighed. The carbon yield was calculated as the ratio between the amount of carbon collected as CNFs, and the amount of carbon introduced into the reactor. The amount of pyrolytic carbon formed on the reactor wall was evaluated by oxidizing it using air, and measuring the amount of $CO_2$ formation. Finally, the carbon balance was calculated.

## RESULTS AND DISCUSSION

Although lower than the values achievable when aromatics are used as the carbon source, CNFs could be obtained at moderate carbon yields even when hexane was used as the carbon source. The obtained CNFs were extremely long, a feature common to typical CNFs produced using the LPI technique.

Through the observation of CNFs collected at various growth stages, it was found that when hexane is used as the carbon source, fiber growth initiates when the catalyst particles reach the region in the reactor where the temperature is around 1073 K. This temperature is about 200 K lower than that observed when benzene is used as the carbon source, and is thought to be due to the lower thermal stability of hexane. Interestingly, fiber growth was found to terminate when the growing fibers reach and reside in the region in the reactor where the temperature is in the range of about 1273 K to 1373 K and then growth resumes when the fibers reach the region where the temperature increases to about 1423 K. Gas analysis revealed that fiber growth termination occurs where methane dominates the gas phase and resumes when ethylene starts to appear in the gas phase through the coupling of methane.

Next we attempted to increase the productivity of CNFs by adjusting parameters of the LPI technique. As expected, decreasing the carrier gas flow rate to increase the residence time of the growing CNFs within the reactor led to the increase of carbon yield, but was accompanied by the decrease in the purity of the obtained CNFs. So next we increased the volume of the liquid pulses and shortened the pulse intervals. When benzene is used as the carbon source, such modifications lead to a significant decrease in product purity, but this was not the case when hexane was used as the carbon source. Such results can be explained by the fact that non-catalytic carbon formation in the gas phase, an undesirable reaction which leads to the generation of impurities during CNF production, proceeds much easier when benzene is used as the carbon source. By this modification of process conditions, a high productivity of 2.5 kg($m^3$-reaction zone volume)$^{-1}$$h^{-1}$ could be achieved even when hexane was used as the carbon source. This value is higher than the productivity achieved when CNFs are produced from benzene using standard process conditions.

# 389408 Stabilizing Effect of Sulfur Incorporation on Multi- and Single-Wall Carbon Nanotubes

**Abstract:**


## Abstract:

We have successfully incorporated sulfur into both multi-wall and single-wall carbon nanotubes. Lawesson reagent has been used as a sulfurization agent. Post synthesis characterization revealed that sulfur incorporation stabilizes carbon nanotubes against temperature treatment and improves conductivity without damaging the structure. Controlled heteroatom incorporation alters electronic and physicochemical properties of carbon nanotubes facilitating nanodevice, catalytic or electrode applications. For example sulfur doped carbon nanotubes might be used as material for lithium ion cathodes or low-cost electrocatalyst in fuel cells. Raman spectroscopy data have shown that D band representing defects in the graphene structure of carbon nanotubes decreases significantly after incorporation, the Raman breathing modes (RBM) though shifted due to electronic change increase in intensity compared to nitric acid treated sample. TEM images confirm that carbon nanotubes stay intact after Lawesson treatment. The amount of incorporated sulfur has been investigated using temperature programmed oxidation and is approximately 6.3wt. %. Near edge X-ray absorption fine structure spectroscopy was used to study the local bonding energy of sulfur to understand the mechanism of sulfur incorporation into carbon nanotube structure.

Reduced graphene oxide (RGO) is commonly employed as a conductive carbon support for active nanostructured materials in applications ranging from storage batteries, photovoltaics, and photocatalysis. The conjugated carbon network inherent to graphene derivatives offers excellent conductivity while the 2-D structure supports high surface area upon which active materials employed in these applications can be anchored. In recent work, we have demonstrated a new solution-based process to synthesize a porous RGO morphology via hydroxyl radical attack. In this talk, I will outline the motivation and approach in the development of the "holey graphene" material while also exploring potential applications.

# 378589 Synthesizing Few Atom Thick Boron Based Nanostructures Isostructural to Chemically Modified Graphene Sheets

**Abstract:**

In this talk, we demonstrate a chemical synthetic route that yields a colloidal dispersion of nanosheets constituted from chemically modified boron honeycomb lattice. Such boron-based nanosheets have not been synthesized earlier and thus making available a method that yields such nanosheets forms a foundational step in addition of a new member to the growing family of nanosheets. Here we report a sonochemical method that facilitates functionalization aided layer-by-layer exfoliation of a parent crystalline material and results in an aqueous dispersion of chemically modified boron-based nanosheets. We show that the as-synthesized nanosheets are analogous to the graphene-oxide nanosheets that are formed as a result of the oxidative exfoliation process of graphite. We have utilized this dispersion to lay down the fundamental physico-chemical properties of boron-based nanosheets. This talk will present details of the ultrasonication assisted exfoliation route; morphological characteristics of the nanosheets (average lateral dimensions of 10-50 μm, average thickness of <10 nm, crumpled appearance); optical properties of the colloidal dispersion (maximum extinction coefficient of 1.08 ml $mg^{-1}cm^{-1}$), and the nature of surface chemical functionalization (presence of oxygen containing functional groups). We also use these fundamental properties to divulge the exfoliation mechanism and show that the pH-dependent ionization of chemical functional groups facilitate a net negative charge on the surface of nanosheets to help stabilize the colloidal dispersion. We further show that these chemically modified boron-based nanosheets exhibit an excitation wavelength dependent fluorescence with emission peaks at ~358 nm and ~411 nm and attribute this to the heterogeneity created by the oxygen-based functional groups. This research pursuit is institutional in its merit as it presents to the scientific community an entirely new and unexplored class of nanosheets. We expect that the quasi-planar nature of these nanosheets, the capability to easily process their aqueous dispersion and the useful properties of boron atoms will open up several avenues for basic and applied research.

**Molecular order in graphene-oxide suspensions and graphene-based Films: What do we learn from it?**

Rachel Tkacz[a], Rudolf Oldenbourg [b], Shalin Mehta[b], Mainak Majumder,[a]

a. Nanoscale Science and Engineering Laboratory (NSEL), Mechanical and Aerospace Engineering, Monash University, VIC 3800, Australia

b. Marine Biological Laboratory, Woods Hole, MA 02543, USA

Graphene-based materials are poised to play a significant role in several technological applications including transparent conducting films, filtration membranes, electrochemical energy storage technologies to corrosion-resistant films. This arises because graphene is a multifunctional material with combination of properties such as large surface area, high electrical conductivity and mechanical strength, low light absorption and chemical inertness. It is becoming increasingly apparent that graphene produced by exfoliation of graphite in liquids possess advantages regarding production scalability and cost-efficiency. While research efforts on achieving applications are burgeoning, correlating processing and property relationships of these materials, in general, is under explored, yet vital to developing high performance materials. To this end, we use orientation-mapped polarized light imaging to determine the fine-structure, orientation of graphene sheets, molecular order and dynamic behaviour in the liquid state and dry state by diattenuation and birefringence imaging (see figure 1 for images of GO sheets around an air bubble and highly ordered films). In doing so, we have recently discovered a confined liquid crystalline droplet phase of graphene oxide with ($20 \pm 2.8$ μm diameter) and an isotropic interior. [1] This novel droplet-phase could find potential application in molecular sensing and drug delivery applications. Additionally, we have identified capillary-force-assisted self-assembly (CAS) as a method for preparation of thin films of chemically reduced graphene oxide (rGO) with unidirectional organization of rGO sheets. The slow axis of the rGO platelets within the CAS films displayed a narrow angular distribution (±3°) within a film area of 1 $mm^2$, resulting in the highest possible order parameter ($S$) of ~1 with 8-fold enhancement of electrical conductivity compared to films with disordered structure. [2] In summary, our studies will enable to deliver high-performing materials from graphene-based assemblies while also develop novel applications by discovery of novel self-assembled architectures.

1. R.Tkacz, R. Oldenbourg, S. B. Mehta, M. Miansari, A. Verma and M. Majumder, “pH dependent isotropic to nematic phase transitions in graphene oxide dispersions reveal droplet liquid crystalline phases”, Chem. Commun, 2014, DOI:10.1039/C4CC00970C

2. R.Tkacz, R.Oldenbourg, A.Fulcher, M.Miansari, M.Majumder, "Capillary-Force Assisted Self-Assembly (CAS) of highly Ordered and Anisotropic Graphene-Based Thin Films", *J.Phys.Chem.C,* 2014, 118 (1), 259–267

0251658240251659264

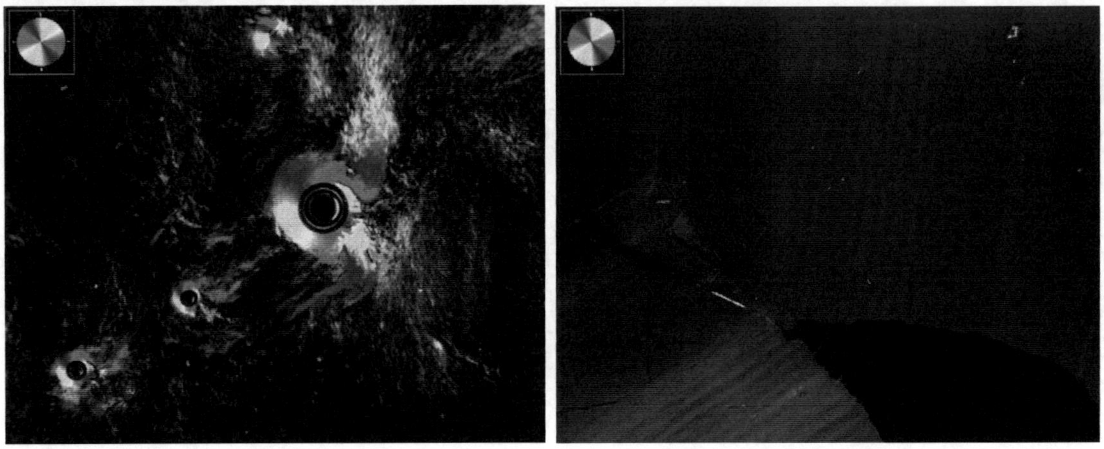

Figure 1: The polarized light LC-PolScope cross-polar processed images displaying retardance by brightness and azimuth of the slow axis by color (as shown by the circular legend) of (right) an air bubble trapped in GO suspension, triggers anchoring forces on the nematics phase, and (left) rGO film formed with the assistance of capillary forces at the meniscus of three phases interface.

# 379087 Award Submission: Molecular Barrier Functions of Graphene Oxide in Liquid-Liquid Systems

**Abstract:**

Graphene oxide (GO) is an effective stabilizer for a range of organic-in-water emulsions. The high stability of GO-based emulsions is due to in part to the formation of robust multilayers at the interface that effectively cover the droplet surface, allowing for full passivation of droplet surface patches left uncovered by packing inefficiencies in the first monolayer. In this work, we show that these GO interfacial films demonstrate unique molecular barrier properties at the liquid– liquid interface, a phenomena that is not observed when non-platelike stabilizers are used. These films are able to inhibit transport from the droplet interior and cause a significant suppression of dispersed-phase evaporation rates with potential applications in controlled release. They are also able to protect the dispersed phase from attack by reactive oxygen species, which typically react at diffusion-limited rates. The mechanism for protection is hypothesized to be a combination of hydroxyl radical scavenging and creation of tortuous pathlengths that hinder diffusion to the interior phase.

# 392239 Study on the Surface Energy of Supported Graphene By Contact Angle Measurements

**Abstract:**

Due to the atomic thinness of graphene, its integration into a device will always involve its interaction with at least one supporting substrate, making the surface energy of graphene critical to its fabrication and real-life applications. In the current work, the contact angle of graphene synthesized by chemical vapor deposition (CVD) was monitored temporally after synthesis using water, diiodomethane, ethylene glycol, and glycerol. The surface energy was then calculated based on the contact angle data by Fowkes, Owens-Wendt (extended Fowkes), and Neumann models. The surface energy of fresh CVD graphene grown on a copper substrate (G/Cu) immediately after synthesis was determined to be 62.2 $mJ/m^2$ (Fowkes), 53.0 $mJ/m^2$ (Owens-Wendt) and 63.8 $mJ/m^2$ (Neumann), which decreased to 45.6 $mJ/m^2$, 37.5 $mJ/m^2$ and 57.4 $mJ/m^2$, respectively, after 24 hours of air exposure. The ellipsometry and ATR-FTIR characterization indicates that the surface energy of G/Cu is affected by airborne hydrocarbon contamination. G/Cu exhibits the highest surface energy immediately after synthesis and the surface energy decreases gradually after airborne hydrocarbon contamination occurs. The intrinsically mild polarity of G/Cu surface has been discussed and could be attributed to high-energy surface defects, pi-hydrogen bonding, and partial wetting transparency.

# 392240 Strengthened Magnetic Epoxy Nanocomposites with Protruding Nanoparticles on the Graphene Nanosheets

**Abstract:**

Magnetic graphene (Gr) nanocomposites (Gr nanosheets coated with iron core iron oxide shell nanoparticles, named Gr/Fe@$Fe_2O_3$) have successfully served as nanofillers for obtaining magnetic epoxy resin polymer nanocomposites (PNCs) to be compared with the epoxy nanocomposites with pure graphene. Due to the high stiffness of Gr and the stress releasing function of the protruding Fe@$Fe_2O_3$ nanoparticles, the PNCs displayed improved mechanical properties and the highest tensile strength was observed in the PNCs with 1.0 wt% Gr/Fe@$Fe_2O_3$, which is 58% higher than that of the pure epoxy. In addition, the Gr/Fe@$Fe_2O_3$ was found to favor the char formation from the epoxy resin, and the morphology study of char residue indicated that Fe@$Fe_2O_3$ could assist the porous char layer formation in the PNCs with Gr/Fe@$Fe_2O_3$ which led to reduced flammability. The magnetic properties of the PNCs were also studied and due to the decreased interparticle dipolar interaction, the value of coercivity ($Hc$) was observed inversely proportional to the loading of Gr/Fe@$Fe_2O_3$ in the PNCs. A reduced viscosity was observed in the 1.0 wt% Gr–epoxy resin liquid nanosuspensions and the viscosity was increased with further increasing the Gr loading. The dynamic storage and loss modulii were studied together with the glass transition temperature ($T_g$) obtained from the peak of tand. The effects of nanofiller loading levels on electrical conductivity and dielectric property were systematically studied. Both nanofillers could increase the electrical conductivity of the epoxy matrix. And the increased real permittivity observed in the PNCs is attributed to the interfacial polarization.

# 389060 A Model for Nanoscale Friction on Few Layer Graphene

**Abstract:**


**A Model for Nanoscale Friction on Few Layer Graphene**

Sesha Hari Vemuri, Pil Seung Chung, Robert Smith and Myung S. Jhon

One of the most researched nanoscale materials is graphene and its thermal, mechanical, chemical, and electrical properties. The effect of the dimensional characteristics of graphitic materials on their tribological properties was elucidated in Lee et al.'s work on the frictional characteristics of atomically thin materials, with a particular emphasis on few layer graphene (FLG) where atomic force microscopy (AFM) were utilized to reveal the effect of the number of layers on the frictional response.

In the present work, we constructed a nano-friction model for the first time to compare our result with the experimental friction force on FLG materials for the entire range of the data available [1]. In contrast to the bulk material, the FLG exhibits a sloping frictional force profile over short AFM scanning distances. We develop a model to replicate the tilted friction phenomenon as a function of the number of layers. Our model compares well with the experimental friction profile of FLG and demonstrates the ability of our model to reveal the transition from 2-D to 3-D tribological characteristics.



[1] C. Lee, Q. Li, W. Kalb, X.K. Liu, H. Berger, R.W. Carpick, J. Hone, *Science* **328**, 76 (2010).

# 366591 Crumpling and Unfolding Behavior of Spray Dried Pristine Graphene Sheets

**Abstract:**

In the present work, we report crumpling of pristine graphene sheets using a scalable industrial spray drying process. Aqueous dispersions of pristine graphene sheets stabilized by pyrene derivatives were atomized in a spray dryer. The morphology transition from a 2D sheet to a 3D crumpled particle was investigated by collecting samples at various heights inside the spray dryer. During drying, a shell of graphene sheets form at the droplet surface. As the droplet shrinks, the capillary forces induce deformation of the shell surface. Initially, the pristine graphene shell folds inward at the edges; where the stiffness is less. Further increase in surface tension forces results in bending of the graphene sheets and buckling of the shell. Graphene oxide (GO) sheets which were spray dried using the same procedure showed a different morphology; these particle surfaces were highly wrinkled, and less buckling was observed. This difference in final morphology is attributed to differences in surface chemistry of pristine graphene and GO sheets. Spray drying parameters such as atomizer pressure, drying temperature and concentration of sheets in the dispersions were varied in order to observe their effect on the final morphology of crumpled particles. Also, other stabilizers such as polymers and surfactants were applied to assess the role of second component in the final morphology of crumpled pristine graphene sheets. Unfolding behavior of crumpled graphene and GO sheets was also studied by redispersing the spray-dried back into various solvents. While pristine graphene sheets immediately unfolded in the water due to the high affinity of pyrene derivatives for water, GO particles preserved their crumpled morphology in water. This can be explained by hydrogen bonds formed between GO sheets in crumpled particles which prevent the unfolding in the solvent.

# 378673 Quantitative Theory of Adsorptive Separation for the Electronic Sorting of Single-Walled Carbon Nanotubes

**Abstract:**

Recently, several important advances in techniques for the separation of single-walled carbon nanotubes (SWNTs) by chiral index have been developed. These new methods allow for the separation of SWNTs through selective adsorption and desorption of different (n,m) chiral indices to and from a specific hydrogel. Our group has previously developed a kinetic model for the chiral elution order of separation; however, the underlying mechanism that allows for this separation remains unknown. In this work, we develop a quantitative theory that provides the first mechanistic insights for the separation order and binding kinetics of each SWNT chirality (n,m) based on the surfactant-induced, linear charge density, which we find ranges from 0.41 e/nm for (7,3) SWNTs in 17 mM sodium dodecyl sulfate (SDS) to 3.32 e/nm for (6,5) SWNTs in 105 mM SDS. Adsorption onto the hydrogel support is balanced by short-distance hard-surface and long-distance electrostatic repulsive SWNT/substrate forces, the latter of which we postulate is strongly dependent on surfactant concentration and ultimately leads to gel-based single-chirality semiconducting SWNT separation. These molecular-scale properties are derived using bulk-phase, forward adsorption rate constants for each SWNT chirality in accordance with our previously published model. The theory developed here quantitatively describes the experimental elution profiles of 15 unique SWNT chiralities as a function of anionic surfactant concentration between 17 and 105 mM, as well as phenomenological observations of the impact of varying preparatory conditions such as extent of ultrasonication and ultracentrifugation. We find that SWNT elution order and separation efficiency are primarily driven by the morphological change of SDS surfactant wrapping on the surface of the nanotube, mediated by SWNT chirality and the ionic strength of the surrounding medium. This work provides a foundational understanding for high-purity, preparative-scale separation of as-produced SWNT mixtures into isolated, single-chirality fractions.

# 383262 Evaluation of Critical Parameters in the Separation of Single-Wall Carbon Nanotubes through Selective Adsorption Onto Hydrogels

**Abstract:**

The electronic type separation of single-wall carbon nanotubes (SWCNTs) by selective adsorption onto hydrogel surfaces has become a highly researched method. However, large variations in both SWCNT dispersion and separation protocols have been reported, making it difficult to elucidate whether changes to dispersion state or separation parameters are responsible for the observations. This study systematically evaluates the role that hydrogel type (dextran- or agarose-based) and SWCNT loading concentration has on separation quality, throughput, and reproducibility. For dextran-based gels, increased dextran concentration rather than cross-linker concentration improves retention and selectivity, suggesting that hydroxyl groups rather than linker molecules are the active adsorption sites. SWCNTs have much stronger interactions with agarose- than dextran-based gels. This stronger adsorption combined with the improved retention of SWCNTs at higher agarose concentration suggests that the double helices formed by agarose are important to the adsorption of nanotubes. The quality and selectivity of the separations are heavily influenced by the initial concentration of SWCNTs; the purity of the metallic fraction falls rapidly as the concentration increases from 20 to 70 mg/L in Sephacryl 200 HR columns. Reproducibility of the separation is greatly affected by the amount of cross-linking in the gel but re-stabilization of the hydrogel remains a critical concern for continued use. The results confirm that both surfactant orientation on the SWCNTs and hydroxyl group concentration on hydrogels are important parameters driving selective adsorption during separation. In general, the highest throughput separations are obtained with dextran-based gels while agarose-based gels produce higher purity of metallic SWCNT fractions (>98 %) and better reproducibility in consecutive separations. Using our understanding of the mechanism behind selective adsorption of SWCNTs onto hydrogel, we have developed high-fidelity separations based on a single elution profile in a single column.

# 374667 Two Phase Extraction Separation of Single-Wall Carbon Nanotubes

**Abstract:**

Aqueous two-phase extraction (ATPE) is a powerful technique for the separation of single semiconducting and metallic species of single-wall carbon nanotubes (SWCNTs). By controlling the relative affinity of the dispersed particles for each phase through co-surfactant concentrations, nanotubes as small as 0.6 nm and as large as 1.7 nm have been separated into single species populations. In this presentation we will first describe the general dextran – polyethylene glycol (PEG) aqueous two-phase system with bile salt surfactants, and then demonstrate the sequential isolation of single SWCNT species across a wide diameter range. Beyond the remarkable tunability that allows isolation of many single nanotube chiralities at high purity, the partitioning method can also be automated and used at large scale. Details of the separation, analysis of the underlying colloidal forces driving the partition phenomenon, and selected application examples of the separated materials will be presented.

# 371256 DNA-Controlled Carbon Nanotube Purification in Polymer Aqueous Two-Phase Systems

**Abstract:**

Sorting single-wall carbon nanotubes (SWCNTs) of different chiralities is both scientifically interesting and technologically important. Polymer aqueous two-phase (ATP) extraction has been demonstrated recently as a very effective way to achieve nanotube sorting. In this work, we report a systematic study of sequence-dependent partition of DNA-wrapped SWCNTs in several polymer ATP systems. This unique approach of combining DNA sequence and polymer ATP system enabled the purification of many single-chirality SWCNT species at high purity. In addition, the DNA-controlled partition of SWCNTs in ATP system suggests the sensitive dependence of hydration energy on the exact spatial distribution of hydrophilic groups which are determined by the DNA sequence and the SWCNT chirality (n,m).

# 392119 Separation Mechanism of Carbon Nanotubes Via Aqueous Two-Phase Method: Towards Simple Isolation

**Abstract:**

Single-wall carbon nanotubes (SWCNTs) are polydisperse in diameter and type; therefore multiple separation techniques have been developed to obtain selective tube distributions. In this work, we determined that the aqueous two-phase (ATP) separation mechanism is driven by the hydrophobicity of the surfactant, or combination of surfactants, covering the SWCNTs, which can be tuned by mixing different surfactants in well-defined ratios. The procedure is successfully applied to SWCNTs from different sources with various diameter (from 0.53nm to 1.2nm) and length (from 300nm to ~20um) distributions.

The development of non-precious transition metal catalysts is of great interest as next-generation technologies such as fuel cells and electrolyzers continue to grow in the alternative energy commercial market. Many commercial fuel cells, and in particular, acidic polymer electrolyte fuel cells, use platinum or other precious metals as the catalyst material. In addition, research that supports industries that have traditionally used costly, high-energy processes for chemical production is now considering alternative methods that would use less energy. For example, ammonia is currently produced using the Haber-Bosch process, where nitrogen gas is reduced to ammonia at high temperature and high pressure. For both energy conversion and chemical production applications, electrocatalysis under alkaline conditions avoids high energy conditions of elevated temperature and pressure, and alkaline electrocatalysis can be performed with non-precious metal catalysts. However, key challenges in the use of non-precious metal catalysts include material performance, as well as material stability and lifetime.

Here, we will present on a set of iron-nickel core-shell nanoparticles that have shown promise as electrocatalysts under alkaline conditions for both fuel oxidation and water electrolysis. The nanoparticle performance can be tuned for either fuel oxidation or oxygen evolution during water electrolysis through modification of the oxygen content and oxide morphology of the nanoparticles. Furthermore, we are now working on optimizing this metal combination for electrocatalytic production of ammonia from nitrogen. While the performance appears to be tailorable through optimization of the nanoparticle synthesis procedure, nanoparticle lifetime continues to be a challenge. Over a period of several months, the performance of the nanoparticles decreases, and an increase in the oxygen content and presence of oxides is observed. The nanoparticles appear to be polycrystalline with a nanocrystalline domain size of approximately 2-3 nm, and portions of the nanoparticles may be amorphous. The presence of nanocrystallinity and/or amorphous character suggests that the nanoparticles are thermodynamically metastable, and increased stability is likely necessary to retain catalytic performance over time. However, the crystalline/amorphous nature may also be the cause of the observed performance in electrocatalysis applications. An annealing study will be discussed as a possible method to stabilize the nanoparticle structure, as well as the effect of annealing on catalytic performance.

# 376598 Controlled Deposition of Iron Oxide Nanoparticles Using a Novel Gas-Expanded Liquid (GXL) Process to Generate Supported Fischer-Tropsch Catalysts

**Abstract:**

Fischer-Tropsch synthesis is a catalytic process used in industry to convert carbon monoxide and hydrogen to heavier hydrocarbons. Active metals like iron, cobalt, nickel and ruthenium are required as catalysts for this process. Supporting these metals on an oxidic support (silica/alumina) has many advantages including an influence in mass selectivity due to the pore structure of the support and stability of the support. Attrition or fracturing inside a fixed-bed or slurry reactor is also reduced. Nano-sized particles also offer several advantages due to their easy dispersability in organic solvents which makes homogeneous loading on supports easier. The usual methods of supported catalyst preparation (ex. incipient wetness impregnation, ion exchange and coprecipitation) rely on various solvent-based transport phenomenon including diffusion and capillary action to disperse the iron into the catalyst pores. Gas expanded liquids (GXLs) have various advantages over conventional solvents due to their pressure tunable transport properties. The diffusion of various molecules is usually enhanced in a GXL due to the lower density/viscosity of solvents at high pressure. We have developed a novel method to controllably deposit iron oxide nanoparticles on a support in the presence of a GXL. This method allows for a more uniform deposition of the metal particles into the pores of the support hence improving the performance of the catalyst in Fischer-Tropsch synthesis. In this study we have carried out preliminary experiments on the synthesis of these supported iron catalysts and tested the effect of various operating parameters on the performance of this catalyst under Fischer-Trosch synthesis conditions.

# 379001 Controlling Metal Nanoparticles in Fischer Tropsch Synthesis

**Abstract:**

## Controlling Metal Nanoparticles in Fischer Tropsch Synthesis


Jorge Gascon, Freek Kapteijn

The growing concerns about oil depletion have spurred worldwide interest in finding alternative feedstocks for important petrochemical commodities and fuels. Fischer-Tropsch synthesis (FTS) is a surface polymerization reaction where syngas (a mixture of CO and $H_2$), derived from natural gas, coal, or biomass, is catalytically converted to hydrocarbons over Co, Ru, or Fe catalysts.[1,2] This technology has long been considered as an alternative process to produce long chain paraffins *via* the so-called gas-to-liquids (GTL) process. Owing to their high activity and selectivity to long chain hydrocarbons, Co-based catalysts are the preferred choice. However, earth abundant and cheap Fe displays superior properties over Co-based catalysts, especially for syngas derived from coal (CTL) or biomass (BTL). As compared to Co, Fe is more resistant to poisoning by contaminants that may be present in the syngas feed such as sulfur and ammonia. In addition, Fe-based catalysts are active in the water-gas-shift reaction under typical FTS conditions, enabling the *in-situ* re-adjustment of the $H_2$/CO molar ratio for the conversion of hydrogen lean syngas originating from coal or biomass.[1] Furthermore, the product slate of Fe-based catalysts is more directed to short chain (unsaturated) hydrocarbons and short chain oxygenates, both among the most important chemical building blocks.

The main challenge in the design of stable FTS catalysts lies in overcoming the high deactivation rate due to sintering, carbon deposition and phase changes. During this lecture, we will highlight different synthesis strategies able to deliver very stable, active and selective FTS supported catalysts.

**References**

(1) Abelló, S.; Montané, D. ChemSusChem 2011, 4, 1538.

(2) Sun, B.; Xu, K.; Nguyen, L.; Qiao, M.; Tao, F. ChemCatChem 2012, 4, 1498.

# Coke-Resistant Core-Shell Catalysts for Dry Reforming of Methane

*Junshe Zhang, Fanxing Li**

Department of Chemical and Biomolecular Engineering
North Carolina State University, Raleigh, NC 27606
*fli5@ncsu.edu

Dry reforming of methane (DRM) represents a promising approach for carbon dioxide utilization and syngas generation. One of the key challenges that hinders the commercialization of DRM resides in catalyst performance. While supported noble metal catalysts are active and less susceptible to coke formation, they are highly expensive. Compared with noble metals, Ni-based catalysts are more commercially attractive. However, industrial application of Ni catalysts in dry reforming reactions is limited by its poor stability, caused by sintering of active metallic particles and carbon deposition on the metal surface.

Previous studies have indicated that coke formation is primarily catalyzed by step and edge sites. In addition, coke formation over small Ni particles is not energetically favored. To date, effective options to inhibit coke formation include poisoning coke-formation sites with sulfur, introducing alkali, alkaline earth or transition metal promoters, chemically or physically dispersing Ni in other metal oxides, confining Ni particles in mesoporous materials, and encapsulating catalyst particles in an inorganic shell. In the current study, we demonstrate that core-shell structured $NiO@SiO_2$ nanoparticles can be fabricated from NiO nanoparticles by the sol-gel method. The core-shell catalysts exhibit excellent catalytic performance toward dry reforming of methane with carbon dioxide at 850 $^oC$; carbon deposition is negligible in 39 h on stream.

# 390429 Reactions Catalyzed By Metal Particles Supported on Functionalized Carbon Nanotubes in Biphasic Systems

**Abstract:**

Carbon nanotubes in biphasic systems have recently gained attention as Pickering emulsion stabilizers in the areas of underground oil recovery and heterogeneous catalyst supports due to their ability to undergo functionalization and tune the surface chemistry while maintaining the nanotube's unique electronic and physical properties [1-4]. While a great deal of literature exists for nanoparticle stabilized Pickering emulsions, very little is known about the fundamental behavior of nanotubes as Pickering emulsion stabilizers. In this contribution, Pd nanoparticles are supported on multiwalled carbon nanotubes with varying degrees of defects and hydrophilic functional groups. A combination of detailed emulsion characterization is coupled with phase specific chemical reactions to quantify both the location of functionalized carbon nanotubes at an oil/water interface, as well as the role of diffusion across the interface for enhanced reaction control.

A combination of acoustic spectroscopy, optical microscopy, SEM and TEM images of thin cross sections of the Pickering emulsions is used to quantify the influence of various surface functional groups on the interfacial area stabilized by carbon nanotube emulsions, as well as their orientation at the oil/water interface. This information is coupled with chemical reactions of olefins and small oxygenates with exclusive solubility in either the oil or water phase to measure the number of catalytic particles in each phase, as well as the rate of diffusion across the oil/water interface. Finally, this information is used to prepare novel hybrid emulsions containing blends of carbon nanotubes with various functional groups to create a gradient in catalytic particles along the oil/water interface enabling the enhanced control of chemical reactions in biphasic systems.

References:

[1] Crossley, S., Faria, J., Shen, M., & Resasco, D. E. (2010). Solid nanoparticles that catalyze biofuel upgrade reactions at the water/oil interface. *Science*, *327*(5961), 68-72.

[2] Shen, M., & Resasco, D. E. (2009). Emulsions Stabilized by Carbon Nanotube− Silica Nanohybrids. *Langmuir*, *25*(18), 10843-10851.

[3]Drexler, S., Faria, J., Ruiz, M. P., Harwell, J. H., & Resasco, D. E. (2012). Amphiphilic nanohybrid catalysts for reactions at the water/oil interface in subsurface reservoirs. *Energy & Fuels*, *26*(4), 2231-2241.

[4]Shi, D., Faria Albanese, J. A., Pham, T. N., & Resasco, D. E. (2014). Enhanced Activity and Selectivity of Fischer-Tropsch Synthesis Catalysts in Water/Oil Emulsions. *ACS Catalysis*.

# 388691 Solid State Chemistry Mass Production of Platinum Group Metal Catalysts with Tailored Particle Morphology

**Abstract:**

Catalytic property of platinum group metal (PGM) nanoparticles can be altered significantly by the nature of facets exposed. For example, Pt (100) plane excels in activity/selectivity compared to other planes in many reactions, including ring-opening hydrogenation of pyrroles, benzene hydrogenation, methanol oxidation (MOR), and electro-oxidation of ammonia (AOR). Pt-Ni (111) surface can exhibit exceptionally high activity towards oxygen reduction reaction (ORR) for polymer electrolyte membrane fuel cells (PEMFCs). The findings have stimulated the exploration of new methods for preparing PGM catalysts with tailored particle morphology, because the usage of PGMs and thus the cost can be largely decreased. However, to date, there have been no feasible methods for cost-effective and mass production of the shaped PGM catalysts.

We realize mass production of PGM catalysts with tailored particle morphology by developing a robust, green, and low-cost solid state chemistry approach. A variety of shaped PGM catalysts, including cubic Pt/$SiO_2$, cubic Pt/C, tetrahedral Pd/C, octahedral Pt-Ni/C, and cubic Pt-Cu/C, have been demonstrated for preparation. The experiments suggest that the formation of shaped PGM nanoparticles is resultant of employing both CO and $H_2$ gases, wherein $H_2$ aids transportation and reduction of the metal precursors on support and CO is responsible for the particle morphology formation. The prepared octahedral $Pt_{1.5}Ni$/C catalyst exhibits high ORR activities of 3.99 mA/$cm^2$ Pt and 1.96 A/mg Pt at 0.90 V vs. RHE, which are about 20 and 10 times the values for commercial Pt/C specifically. The cubic Pt/C catalyst shows 1.44 mA/$cm^2$ at 0.6 V vs. RHE in AOR, which is five times that of 0.30 mA/$cm^2$ using commercial Pt/C.

# OPTIMIZATION OF GREEN SYNTHESIS OF GOLD NANOPARTICLES USING *DELONIX REGIA* LEAF EXTRACT AND EVALUATION OF THEIR CATALYTIC ACTIVITY

*Preeti Dauthal, Mausumi Mukhopadhyay**

S.V. National Institute of Technology, Surat, Gujarat, India
Tel.: +91 261 2201645; Fax: +91 261 2227334
*Email: mmu@ched.svnit.ac.in; mausumi_mukhopadhyay@yahoo.com

## Introduction

Noble metals such as of gold, palladium, platinum and silver with nano-size dimensions have received considerable attention due to their endless promises in the field of material science. Among these noble metal nanoparticles (NPs), particularly gold, nanoparticles (Au-NPs) plays a significant role in various catalytic reactions due to high surface area to volume ratio, energetic surface atoms and great surface energy. Biofabrication methods for the synthesis of these noble nanoparticles are not only a noble approach to synthesize benevolent nano-structures, but also a footstep to diminish the utilization of toxic chemical and generation of perilous by-products, which are hazardous to environment and public health. These methods are considered as benign, sustainable, cost-effective and ecologically sound. Biofabrication of nanoparticles using microorganisms is reported during recent studies (1, 2). However, use of the plant extracts as the potential resource for fabrication of NPs is gaining significant importance as unlike microbial fabrication these methods do not involve isolation and culture preparation techniques.

In a recent study, antioxidant potential of *Delonix regia* (*D. regia*) has been utilized for cost-effective and eco-benign fabrication of palladium nanoparticles (3). However, literature survey revealed that there is no report available for biofabrication of Au-NPs using *D. regia.* Thus, the present study is intended to fabricate Au-NPs by utilizing the metal reducing potential of *D. regia* plant extracts. Catalytic potential of fabricated nanoparticles is evaluated for reduction of anthropogenic pollutant 2-Nitroaniline (2-NA). Use of biofabricated nanoparticles as a catalyst in bioremediation reaction is a cost effective, sustainable and eco-friendly alternative (3-6).

## Results and discussion

The effects of the concentration of *D. regia* leaf extracts, reaction temperature, reaction time and pH were evaluated for biofabrication of Au-NPs. Optimization of above reaction conditions was carried out based on surface plasmon resonance (SPR) using UV-visible spectroscopy analysis. Further biofabrication of Au-NPs were carried out using following optimized conditions: 30 °C reaction temperature, 4 mL of $1 \times 10^{-3}$ M $HAuCl_4.3H_2O$, 0.5 mL of the leaf extract, pH 6 and 10 min reaction time.

## *Characterization of nanoparticles*

Dynamic Light Scattering (DLS), ζ potential, Transmission Electron Microscopy (TEM), X-ray diffraction (XRD), Energy DIsperse X-ray Spectroscopy (EDX) and Fourier Transform Infrared Spectroscopy (FTIR) techniques were adopted for evaluation of size, stability, morphology, crystal nature and purity of the biofabricated Au-NPs at optimized condition. UV-visible spectra represented SPR peak at 542 nm which suggested the reduction of gold ions ($Au^{3+}$) into zero-valent gold ($Au^0$). The results were rapid for the biofabrication of spherical Au-NPs of about 4-24 nm (by TEM) size within 10 min of reaction time. In the TEM micrograph, the edges of the nanoparticles were lighter compared to that of the centre, suggested capping and adherence of bio-organics of leaf extracts on the surface of Au-NPs (7). This observation was further supported by the FTIR, EDX and DLS analysis. Crystalline nature of nanoparticles was evident to a bright circular spot in the Selected Area Electron Diffraction (SAED) pattern and broad diffraction peaks in the XRD pattern (Figure 1).

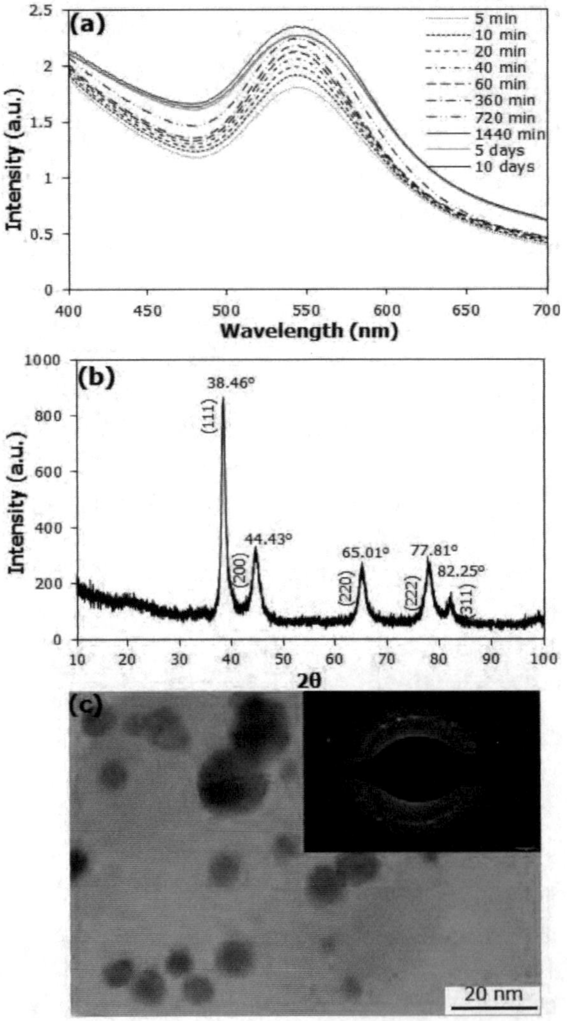


**Figure 1.** (a) UV-visible spectra (b) TEM micrograph (c) XRD spectra of biofabricated Au-NPs.

XRD pattern represented the five distinctive peaks present at 38.46° (111), 44.43° (200), 65.01° (220), 77.81° (311) and 82.25° (222) positions (JCPDS No. 04–0784), suggested the presence of crystalline gold at zero oxidation state. The lattice constant calculated from these patterns was 4.069 Å, confirmed the face centered cubic (fcc) geometry oriented along the (111) plane. The average particle size was calculated as 25.77 nm with the line width of the maximum intensity (111) diffraction peak by Debye Scherer equation. Thus, the XRD pattern proved strong evidence in favor of the TEM analysis for the presence of crystalline Au-NPs.

The appearance of optical absorption peak at 2.2 keV in EDX spectra further confirmed the presence of Au-NPs. High negative ζ potential (-15 mV) on the surface of biofabricated Au-NPs suggested long-term stability of colloidal Au-NPs. Particle size distribution of colloidal Au-NPs in DLS analysis was found in the range of 12-55 nm with an average particle size of about 39 nm with polydispersity index (PdI) 0.279, which appears to be higher as compare to TEM and XRD measurements. This was due to interference created by electric double layer on charged particles and overlapping NPs. Particles size observed in DLS is usually higher as the various forces of interaction in the solution like van der Waals forces of attraction is prominent (8). However, aggregation of NPs was avoided in the TEM imaging technique lead more precise observation (3).

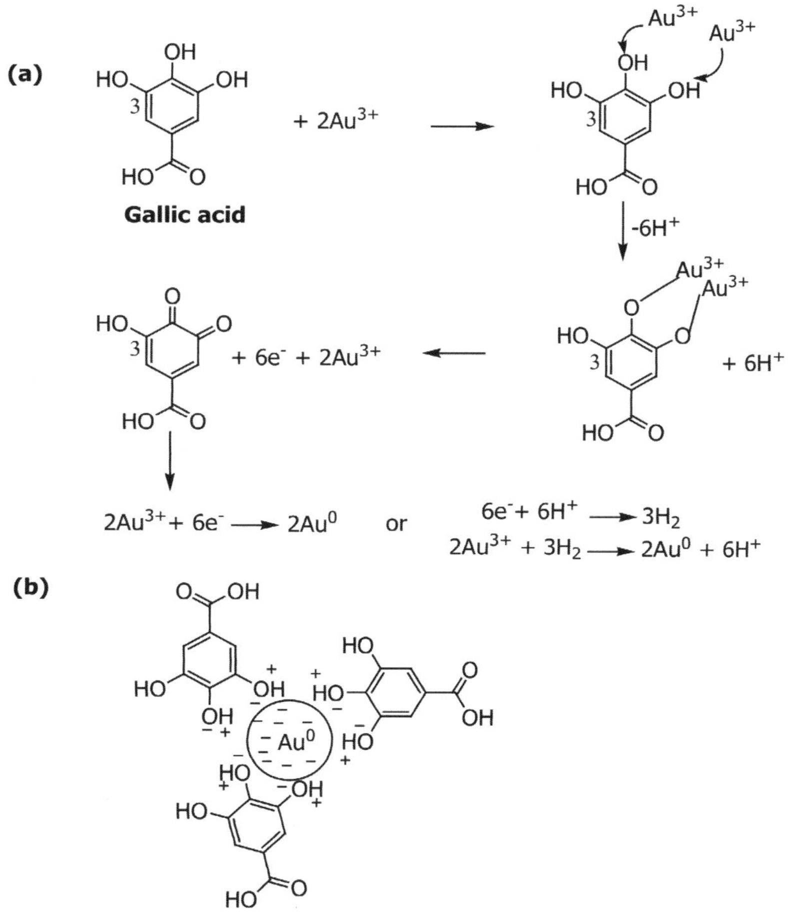


**Figure 2.** Probable reaction mechanism for (a) bio-reduction and (b) stabilization of Au-NPs using gallic acid.

FTIR analysis confirmed further the role of polar polyphenolic compounds of *D. regia* leaf extracts for biofabrication of Au-NPs and their chemical frameworks were also provided effective wrapping around the nanoparticles and robustness against agglomeration. Presence of water soluble polyphenolic compounds, including protocatechuic acid, gallic acid and hydroxybenzoic acid (phenolic acid) and catechins (flavonoids) are reported earlier in leaf extract of *D. regia* by HPLC analysis (3). Schematic representation of gallic acid mediated reduction of $Au^{3+}$ ion and formation of respective quinone are shown in Figure 2.

Long-term stability (upto 10 days) of the biofabricated Au-NPs was also confirmed by UV-visible and ζ potential analysis. No significant change was observed in the SPR position and intensity, suggested long-term stability of nanoparticles. The ζ potential was found to be -14.3 mV even after 10 days of the incubation period.

### *Catalytic activity of nanoparticles*

Catalytic activity of biofabricated Au-NPs was evaluated for the reduction of anthropogenic pollutant 2-NA. These nanoparticles were demonstrated to have a good catalytic activity for the reduction of 2-NA to 1, 2 Benzenediamine. Figure 3 represented reduction of 2-NA with $NaBH_4$ in aqueous environment at the surface of biofabricated Au-NPs as the catalyst.

The reduction of reaction followed pseudo first-order kinetics with respect to the substrates. It was also found that high concentration of Au-NPs and high temperature reduced the reaction time for catalytic reduction and increased the reduction rate. This observation suggested that catalytic reduction usually took place at the surface of catalyst (9). Catalytic reduction of 2-NA was found to increase by a factor of 1.6-1.9 for every 10 °C rise in temperature. The pseudo first-order rate constant was estimated to be $12.28 \times 10^{-2}$ $min^{-1}$ at room temperature (28 ± 2 °C) for this catalytic reduction reaction. The activation energy ($E_a$) for this catalytic reduction was calculated as 44.26 kJ $mol^{-1}$.

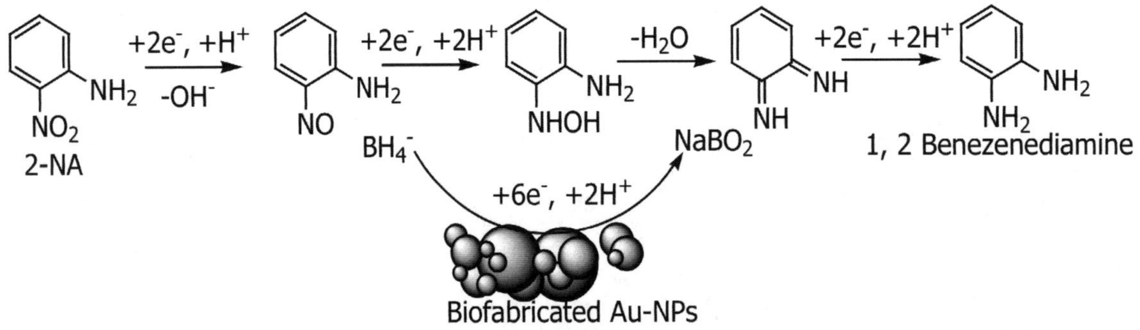


**Figure 3.** Schematic catalytic reduction of 2-NA to 1, 2 benzenediamine in the presence of $NaBH_4$ and biofabricated Au-NPs as the catalyst.

Au-NPs were recycled up to 10 cycles. It was observed that no significant change observed in the catalytic activity of the recycled catalysts up to 10 cycles. The reaction rate constant obtained at seventh cycle was found to be $10.57 \times 10^{-2}$ $min^{-1}$. However, catalytic activity of biofabricated Au-NPs decreased slightly after 7 cycles of reduction reaction. This was due to the agglomeration of NPs (10, 11) or not fully recovered during recycling process. Thus, biofabricated Au-NPs proved as a potent recyclable catalyst for the industrial applications. These results suggested the fabrication of stable Au-NPs with the available biological resource *D. regia*, which offers an inexpensive, sustainable and eco-friendly alternative for fabrication of catalytically active Au-NPs.

## References


1. Srivastava, Nishant and Mausumi Mukhopadhyay (2013), "Biosynthesis and Structural Characterization of Selenium Nanoparticles Mediated by *Zooglea ramigera*," Powder Technology, 244, pp. 26-29.
2. Srivastava, Nishant and Mausumi Mukhopadhyay (2014), "Biosynthesis of $SnO_2$ Nanoparticles Using Bacterium *Erwinia herbicola* and their Photocatalytic Activity for Degradation of Dyes," Industrial & Engineering Chemistry Research, 53, pp. 13971-13979.
3. Dauthal, Preeti and Mausumi Mukhopadhyay (2013), "Biosynthesis of Palladium Nanoparticles using *Delonix regia* Leaf Extract and its Catalytic Activity for Nitro-aromatics Hydrogenation," Industrial & Engineering Chemistry Research, 52, pp. 18131-18139.
4. Dauthal, Preeti and Mausumi Mukhopadhyay (2012), "*Prunus domestica* Fruit Extract-Mediated Synthesis of Gold Nanoparticles and Its Catalytic Activity for 4-Nitrophenol Reduction," Industrial & Engineering Chemistry Research, 51, pp. 13014-13020.
5. Sharma, Nilesh C., Shivendra V. Sahi, Sudip Nath, Jason G. Parsons, Jorge L. Gardea-Torresdey, Tarasankar Pal (2007), "Synthesis of Plant-Mediated Gold Nanoparticles and Catalytic Role of Biomatrix-Embedded Nanomaterials," Environmental Science & Technology, 41, pp. 5137-5142.
6. Dauthal, Preeti and Mausumi Mukhopadhyay (2014), "Biofabrication, Characterization, and Possible Bio-reduction Mechanism of Platinum Nanoparticles Mediated by Agro-industrial Waste and their Catalytic Activity," Journal of Industrial and Engineering Chemistry, DOI: 10.1016/j.jiec.2014.07.009.
7. Suresh, Anil K., Dale A. Pelletier, Wei Wang, Michael L. Broich, Ji-Won Moon, Baohua Gu, David P. Allison, David C. Joy, Tommy J. Phelps and Mitchel J. Doktycz (2011), "Biofabrication of Discrete Spherical Gold Nanoparticles using the Metal-Reducing Bacterium *Shewanella oneidensis*," Acta Biomaterialia, 7, pp. 2148-2152.
8. Dauthal, Preeti and Mausumi Mukhopadhyay (2013), "In-vitro Free Radical Scavenging Activity of Biosynthesized Gold and Silver Nanoparticles using *Prunus armeniaca* (Apricot) Fruit Extract," Journal of Nanoparticle Research, 15, pp. 1366-1376.
9. Murugadoss, Arumugam and Arun Chattopadhyay (2008), "Surface Area Controlled Differential Catalytic Activities of One-Dimensional Chain-like Arrays of Gold Nanoparticles, " The Journal of Physical Chemistry C, 112, pp. 11265-11271.
10. Kuroda, Kyoko, Tamao Ishida and Masatake Haruta (2009), "Reduction of 4-Nitrophenol to 4-Aminophenol over Au Nanoparticles Deposited on PMMA," Journal of Molecular Catalysis A: Chemical, 298, pp. 7-11.
11. Koga, Hirotaka and Takuya Kitaoka (2011), "One-Step Synthesis of Gold Nanocatalysts on a Microstructured Paper Matrix for the Reduction of 4-Nitrophenol," Chemical Engineering Journal, 168, pp. 420-425.

# 369839 Elastic Properties and Fracture Mechanics of Graphene Nanomeshes

**Abstract:**

Graphene nanomeshes (GNMs) are graphene nanostructures consisting of a periodic arrangement of nano-scale holes or pores in the graphene lattice with neck widths less than 10 nm, mimicking dense arrays of ordered nanoribbons. Establishing rigorous structure-property-function relationships in such patterned graphene nanostructures is of utmost importance for their optimal design toward enabling a broad range of technological applications.

In this presentation, we report the results of a systematic computational study on the elastic response and fracture dynamics of GNMs based on molecular-statics and molecular-dynamics simulations of uniaxial tensile deformation tests. Both the elastic properties and the dynamical response to mechanical loading are determined as a function of the nanomesh architecture, namely, the lattice arrangement of the pores, pore morphology, material density ($\rho$), and pore edge passivation. Scaling laws for the density dependence of the elastic modulus $M$, $M(\rho)$, are established. We find that, for circular unpassivated pores, $M$ scales with the square of $\rho$. Deviations from quadratic scaling are most strongly influenced by pore morphology and, to a lesser extent, by pore edge passivation and temperature. In addition, for the numerous GNM structures examined, stress-strain curves are generated, the corresponding fracture mechanisms are investigated, and the ultimate tensile strength, fracture strain, and toughness are determined as a function of the GNM architectural parameters. We find that GNM fracture is characterized by a brittle-to-ductile transition as the GNM density decreases and determine the critical density for the transition.

# 365004 Mechanical Behavior of Interlayer-Bonded Nanostructures Obtained from Bilayer Graphene

**Abstract:**

Based on first-principles density functional theory calculations, we have recently introduced a novel class of carbon nanostructures, formed due to interlayer covalent $sp^3$ C-C bonding in twisted bilayer graphene as a result of controlled chemical functionalization (hydrogenation or fluorination). Depending on the twist angle and local stacking of layers, these nanostructures are superlattices of diamond-like nanocrystals embedded within the bilayer. The electronic behavior of these $sp^2/sp^3$ hybrid configurations ranges from metallic to semiconducting with tunable electronic band gaps from a few meV to about 1.2 eV, to insulating with band gaps up to about 4 eV. This behavior makes this class of nanostructures very attractive for enabling technologies in nanoelectronics.

In this presentation, we report a comprehensive computational study of the mechanical behavior of these superlattices, based on molecular-dynamics simulations of tensile deformation and shear loading tests according to a reliable interatomic bond-order potential. We have found that the mechanical properties of these superstructures can be tuned precisely by controlling the fraction of $sp^3$-hybridized C-C bonds in the material through the extent of chemical functionalization. Their Young modulus and tensile strength decrease moderately compared to those of pristine bilayer graphene, but remain superior to those of most conventional engineering materials. On the contrary, their interlayer shear modulus increases strongly compared to that of pristine bilayer graphene and monotonically with the fraction of $sp^3$-hybridized C-C bonds. We have analyzed in detail the fracture mechanisms of the superstructures under tension as a function of the extent of interlayer bonding through chemical functionalization. In most cases, fracture is initiated at the interface between pristine graphene and nanodiamond domains in the superstructure and subsequently propagates across the superstructure leading to brittle failure. However, beyond a certain extent of interlayer bonding or a certain separation of nanodiamond domains within the unit cell of the superstructure, there is a transition to ductile failure with a structural response that is characterized by void formation and coalescence. In addition to the analysis of the fracture mechanisms, the predicted stress-strain curves are presented and discussed and the fracture toughness of all the nanostructures examined is determined.

# Encapsulation of Particle Ensembles in Graphene Nanosacks as a New Route to Multifunctional Materials

Yantao Chen, Fei Guo, Yang Qiu, Edward Walsh, Robert Hurt, et al.
*Brown University*

A major goal in the nanosynthesis field is the fabrication of hybrid nanostructures with multiple functions for use in smart materials, biomedical diagnostics, and theranostics. Hybrid nanostructures may be designed for combinations of photonic, magnetic, radiological, mechano-responsive, catalytic, and/or targeted delivery behaviors, and the designs typically require complex multistep chemical synthesis. Recent reports of a continuous, scalable aerosol process for graphene encapsulation of nanoparticles may provide a new route to such hybrids if different types of particles can be simultaneously wrapped.

Here we demonstrate a systematic way to create multifunctional hybrid materials through aerosol-phase graphene encapsulation of ensembles of simple uni-functional nanoparticles. We first develop a general theory of aerosol encapsulation based on pH-dependent colloidal interactions in drying microdroplets. It is shown that a wide range of cargo particles can be encapsulated, including some that are colloidally unstable alone, because of association with the highly-charged, stable graphene oxide sheets. High pH is shown to be a favorable operating regime for the aerosol encapsulation process because it leads to particle/sheet electrostatic repulsion, colloidal stability, and low proton activity that limits dissolution of metal and metal-oxide nanoparticles. The cargo-filled graphene nanosacks are also shown to be open structures that rapidly release soluble salt cargoes when reintroduced into water, but can be partially sealed by filler addition during synthesis to achieve slow release profiles of interest in controlled release or theranostic applications.

Finally we demonstrate an example multifunctional material by fabricating graphene/Au/$Fe_3O_4$ and graphene/$BaTiO_3$/$Fe_3O_4$ ternary particles as MRI/CT multimodal imaging probes. The graphene/Au/$Fe_3O_4$ probe is magnetically responsive and shows excellent MRI and CT contrast in CMC gel suspensions at low concentrations ranging from 0.05 - 2000 μg/ml based on tests using full-scale clinical scanners. Graphene nanosack encapsulation is shown to be a flexible approach for the fabrication of multimodal bioimaging probes and possible theranostic devices.

# 369872 Thermal Properties of Graphene Nanomeshes

**Abstract:**

Graphene has remarkable mechanical, electronic, and thermal properties, which can be tuned by properly modifying graphene structure and composition through patterning and chemical functionalization. Graphene nanomeshes (GNMs) are graphene nanostructures consisting of a periodic arrangement of nano-scale holes or pores in the graphene lattice with neck widths less than 10 nm, mimicking dense arrays of ordered nanoribbons. Establishing rigorous structure-property-function relationships in such patterned graphene nanostructures is significant for their optimal design toward enabling a broad range of technological applications.

In this presentation, we report the results of a systematic study based on molecular-dynamics (MD) simulations of the thermal properties of GNMs as a function of the nanomesh architecture determined by the lattice arrangement of the pores, pore morphology, material density ($\rho$), and pore edge passivation. Scaling laws for the density dependence of the thermal conductivity $k$, $k(r)$, are established. Specifically, we find that, for circular unpassivated pores, $1/k$ scales linearly with $1/r$. For elliptical pores with aspect ratio (semi-major : semi-minor ellipse axis) $f > 1$, the thermal conductivity of the GNMs becomes anisotropic; the anisotropy becomes stronger as the GNM density decreases. Passivating the nanomesh pore edges with H atoms increases the GNM thermal conductivity. Further analysis of the phonon mean free paths based on the geometry of the GNMs shows that phonon transport in GNMs is dominated by phonon-pore edge scattering. The results of Monte Carlo sampling of phonon mean free paths explain the scaling rules derived from the MD simulations.

Reduced graphene oxide and graphene oxide (GO) based materials have many unique properties that make them intriguing candidates for catalyst supports. These materials have high surface areas, high adsorption capacities, can transfer π-electron density to supported catalytic metal particles, can have high electron mobility which can facilitate electron transfer during catalytic reactions, and also can be chemically transformed through the large number of oxygenate groups on the surfaces. One of the challenges encountered when utilizing graphene oxide materials for catalyst supports is the regraphitization and reduction in surface area when the catalyst is dried to a powder. The reduced graphene oxide layers restack via the strong π-π interactions between layers. This leads to a reduction in access to active sites on the interior basal planes of the catalyst. A number of methods have been explored to reduce this restacking interaction, for example, by creating 3D structures such as graphene-organic-frameworks (GOFs), graphene aerogels, and graphene hydrogels.

We have explored using 3-D graphene/graphene oxide based framework to create catalytic "pockets" in which we can engineer a micro- or nano- environment that can influence catalytic performance. These 'pockets' are designed to create a partially nanoconfined or sterically hindered environment in which functional groups are placed near each other to work cooperatively and influence the catalytic active site or the reaction mechanism pathways. We made use of graphene oxides synthesized with nanosized in-plane defects or 'holes' that also have a high density of carboxylic acid groups and that allows for quicker diffusion of reactants and products to the interior areas of the 3-D structure. Functionalization of the oxygenate moieties on the surface of graphene oxide create a 'bridge' between two separate graphene oxide sheets. Utilization of selective chemistries such as 'click' chemistry via azide-alkyne cycloaddition enables us to distinguish non-bridging and bridging amide/amine groups. The introduction of catalytically relevant amine functional groups can then be accomplished by the utilization of diamine complexes, such as N-Boc-p-phenylenediamine or N-Boc-ethylenediamine, in which one of the amine functional groups is temporarily protected to prevent it from also becoming a bridging species. The protecting N-Boc group is then chemically or thermally removed which then result in a catalyst with a free primary amine can be located near residual surface carboxylic acid groups. We have also begun investigation into creating these catalytic pockets on the surface of traditional metal oxide (silica) supports. Preparing a catalyst that is only a few graphene oxide sheets thick could reduce diffusion limitations. This method can also be used to study the interaction between graphene oxide and metal oxide supports. If a catalytic nanoparticle is deposited on the metal oxide surface the thin graphene oxide layer could potentially act as a 'filter' to influence which reactants can approach the catalytic active site.The synthesis and characterization of 3-D structures prepared by this and similar methods will be described, including their structural, chemical and catalytic properties.

**Carbon black-derived graphene quantum dots as an efficient and a stable catalyst for $I_3^-/I^-$ redox couple**

Chun-Chieh Wang and Shih-Yuan Lu
Department of Chemical Engineering
National Tsing Hua University
Hsinchu 30013, Taiwan

A microwave-assisted oxidative cleavage process was developed to prepare graphene quantum dots (GQDs) from carbon black. The evolution in size and thickness of the resulting carbonaceous products was studied. In one hour, GQDs with a size of less than 10 nm and thickness less than 2 nm were obtained. The GQDs exhibited an onset absorption at 290 nm and photoluminescence emission at 410 nm. These GQDs were further composited with mesoporous carbon aerogels (CA) by a filtering process to form GQD decorated CA composites (GQD/CA). The GQD/CA composite was applied as a catalyst electrode for reduction conversion of $I_3^-$ to $I^-$, a critical electrolyte regeneration reaction in dye-sensitized solar cells (DSSC). Also investigated are Pt electrode, the expensive traditional counter electrode material for DSSCs, and plain CA electrode for comparison. Based on data derived from cyclic voltammograms and Tafel plots, the GQD/CA composite exhibited catalytic activities comparable to those of the Pt electrode and better than those of the plain CA electrode. The GQD/CA electrode however surpassed the Pt electrode in terms of long term stability. The reduction current dropped and the onset reduction potential increased significantly after 500 cycles for the Pt and plain CA electrodes, while the reduction current was maintained and the onset reduction potential increased only slightly for the GQD/CA electrode. The GQD/CA composite thus proves to be an inexpensive, an efficient, and a stable alternative for Pt as the counter electrode material for DSSCs.

# 387255 Strengthened Magnetic Epoxy Nanocomposites with Protruding Nanoparticles on the Graphene Nanosheets

**Abstract:**

Magnetic graphene (Gr) nanocomposites (Gr nanosheets coated with iron core iron oxide shell nanoparticles, named Gr/Fe@$Fe_2O_3$) have successfully served as nanofillers for obtaining magnetic epoxy resin polymer nanocomposites (PNCs) to be compared with the epoxy nanocomposites with pure graphene. Due to the high stiffness of Gr and the stress releasing function of the protruding Fe@$Fe_2O_3$ nanoparticles, the PNCs displayed improved mechanical properties and the highest tensile strength was observed in the PNCs with 1.0 wt% Gr/Fe@$Fe_2O_3$, which is 58% higher than that of the pure epoxy. In addition, the Gr/Fe@$Fe_2O_3$ was found to favor the char formation from the epoxy resin, and the morphology study of char residue indicated that Fe@$Fe_2O_3$ could assist the porous char layer formation in the PNCs with Gr/Fe@$Fe_2O_3$ which led to reduced flammability. The magnetic properties of the PNCs were also studied and due to the decreased interparticle dipolar interaction, the value of coercivity (*Hc*) was observed inversely proportional to the loading of Gr/Fe@$Fe_2O_3$ in the PNCs. A reduced viscosity was observed in the 1.0 wt% Gr–epoxy resin liquid nanosuspensions and the viscosity was increased with further increasing the Gr loading. The dynamic storage and loss modulii were studied together with the glass transition temperature ($T_g$) obtained from the peak of tand. The effects of nanofiller loading levels on electrical conductivity and dielectric property were systematically studied. Both nanofillers could increase the electrical conductivity of the epoxy matrix. And the increased real permittivity observed in the PNCs is attributed to the interfacial polarization.

# 388057 High Electrical Conductive Ethylene 1-Octene Copolymers Reinforced with Low Loading Carbon Nanotubes

**Abstract:**

In the paper, the high performance ethylene 1-octene copolymer (EOC) nanocomposites have been successfully fabricated by innovatively coating carbon nanotubes (CNTs) on the surface of gelated/swollen EOCs pellets. The effects of nanofiller loading levels and the processed temperature on the morphology, thermal stability, crystal structure, melt rheological behaviors, optical properties, electrical conductivity and dielectric properties were systematically studied. The increased thermal stability of EOC nanocomposites is observed with increasing the CNT loadings by thermogravimetric analysis (TGA). Moreover, compared with pure EOC, a stronger shear-thinning process was observed in the nanocomposites and was attributed to the orientation of nanofillers. The electrical conductivity ($\sigma$) was observed to depend on the processing temperature and loading of CNTs [1]. Meanwhile, the CNTs formed network structure more easily in the polymer matrix resulting in a higher $\sigma$ at lower processing temperature. The calculated band gap of the nanocomposites was observed to depend on the processing temperature and its variation was attributed to the strain on the CNTs in nanocomposites. Finally, the negative real permittivity was observed in the nanocomposites and the variation of real permittivity was attributed to the plasma frequency variation of CNTs in nanocomposites [2].

[1] J. Zhu, H. Gu, Z. Luo, N. Haldolaarachchige, D. P. Young, S. Wei and Z. Guo; Carbon Nanostructures Derived Polyaniline Metacomposites: Electrical, Dielectric and Giant Magnetoresistive Properties; Langmuir, 28(27), 10246–10255 (2012)

[2] H. Gu, J. Guo, Q. He, Y. Jiang, Y. Huang,* N. Haldolaarachchige, Z. Luo, D. P. Young, S. Wei and Z. Guo; Magnetoresistive Polyaniline/Multi-Walled Carbon Nanotubes Nanocomposites with Negative Permittivity; Nanoscale, 6, 181-189 (2014)

# Development and Characterization of β-Dicalcium Silicate Nanoparticles for Greener Cementitious Materials

Scarlett Widgeon,[1] Marianė Silva de Miranda,[2] Rahul Sangodkar,[1] Elizabeth Cisneros,[1] Flávio A. Rodrigues,[2] Bradley F. Chmelka[1]*

[1]Department of Chemical Engineering, University of California, Santa Barbara, CA 93106 USA
[2]Laboratório de Materiais e Superficies, Universidade de Mogi das Cruzes, SP 08780-911, Brazil

## Abstract

Cement manufacturing is responsible for approximately 5% of the world's carbon dioxide ($CO_2$) emissions, due to the release of $CO_2$ from limestone ($CaCO_3$) during high-temperature calcination to yield CaO. CaO is used as a source of calcium cations during the formation of anhydrous cement powders, which are comprised of calcium silicates and aluminates with typical particle sizes of several microns. The major components of anhydrous cement powders are *tri*calcium silicate ($Ca_3SiO_5$) and *di*calcium silicate ($Ca_2SiO_4$), which are responsible for the development of mechanical strength properties as cements hydrate. Though they are chemically similar, they hydrate at dramatically different rates: *tri*calcium silicate hydrates over days to weeks, compared to *di*calcium silicate which typically hydrates over months and years, due to its intrinsically slower hydration kinetics and large particle sizes. Increasing the overall hydration rate of *di*calcium silicate is considered to be an attractive option for replacing a portion of the *tri*calcium silicate to achieve similar mechanical strength properties with lower total calcium contents. Correspondingly, this could reduce the carbon foot print of the cement manufacturing process.

Recently, we have synthesized β-*di*calcium silicate by using a sol-gel process, rather than by the conventional solid-state reaction, to yield small (ca. 100 nm) $Ca_2SiO_4$ particles that hydrate much more rapidly: 85% complete within 30 days, compared to 20 days for conventionally prepared (micron sized) $Ca_3SiO_5$. By comparison, β-*di*calcium silicate synthesized using the conventional approach typically reach 85% hydration in about 180 days. The increased rate of hydration is thought to be due principally to the smaller particle sizes. Solid-state $^{29}$Si nuclear magnetic resonance (NMR) spectroscopy, X-ray diffraction (XRD), and scanning electron microscopy (SEM) provide new understanding of the molecular structures and compositions of β-dicalcium silicate before and after hydration. In particular, solid-state $^{29}$Si NMR spectra allow the mean lengths of hydrating silicate chains to be determined, as functions of particle size, preparation conditions, and extent of hydration. The analyses yield detailed molecular insights that are correlated with macroscopic properties, such as the viscosities and compression strength of hydrating slurries that help to understand the development of mechanical strength in cement materials.

# 383752 Multiscale Modeling for Nanoclearcoat Curing

**Abstract:**

Nanocomposite materials, if applied properly, can demonstrate significantly improved performance and/or new functionalities of the resulting coating layers, such as surface texture alteration control, self-healing, and toxic chemical/acid/corrosive agent repelling. Nanopaint is such a type of nanomaterials, which is being seriously evaluated in the coating industries. A key nanocoating development step is the curing process. As a nanoparticle-incorporated polymer material, four types of complex operational phenomena occur in a multistage curing process: heat transfer from the curing environment to the coating material, solvent and/or water evaporation, coating layer thickness change, and cross-linking reaction. Fundamental questions on nanocoating curing are as follows: (i) how much will the presence of nanoparticles of different sizes and volume fractionsaffect the solvent removal and curing performance, (ii) how much will the nanocoating curing will affect energy consumption, and (iii) how can we correlate the coating microstructure to the coating macroscopic quality?

In this paper, we introduce an integrated multiscale modeling approach to investigate the mechanisms of multistage heat and mass transfer, and chemical reaction kinetics at the presence of nanoparticles in the coating material during curing, which is a reactive drying process. The effect of the size of nanoparticle and the composition of nanoclearcoat on solvent evaporation, cross-linking reaction, process efficiency and product quality are studied at different time-length scales. It is demonstrated that the integrated model can provide a thorough analysis of drying/curing operations, guide the design of nanocoating materials, and form the basis for developing optimal operational strategies for improving coating quality and reducing defects. Model-based simulation also provides an opportunity to optimize the reactive drying process, where product quality and process efficiency goals can be achieved simultaneously.

# 390751 Continuous Sizing and Characterization of Suspension-Based Nanomaterials

**Abstract:**

Nanoparticle characterization is an area of intense interest owing to its increasing importance in areas such as photovoltaics, catalysis, and sensors. Most of these applications involve processing nanoparticles in suspension, and demand precise control of properties such as chemistry, size, morphology and/or crystalline structure to achieve desired functionality. Unfortunately, characterization methods have generally failed to keep up with this rapid pace of discovery, introducing a significant bottleneck between scientific discovery and commercialization of nanomaterial-based products. Here we introduce an approach that overcomes limitations of conventional analysis methods, enabling continuous online quantification and characterization of nanoparticle composition, size, and morphology, independent of agglomeration state. Our microfluidic approach enables continuous analysis by exploiting localized fluorescent complexation of nanoparticles in suspension triggered by a sharp chemical discontinuity, yielding a signature easily detectable over a broad concentration range. We evaluated our approach by co-injecting fluorescein dye into a microchannel in parallel with suspensions of ZnO and $TiO_2$ (anatase and rutile) nanoparticles. Interfacial fluorescence enhancement is observed in ZnO and anatase $TiO_2$, whereas quenching is observed in rutile $TiO_2$. In the case of ZnO, a pronounced concentration dependence can be clearly distinguished over 4 orders of magnitude in nanoparticle concentration. Interfacial fluorescence is also sensitively dependent on particle size regardless of the material's agglomeration state. Our approach is also highly sensitive to morphology, as demonstrated in the case of $TiO_2$ where fluorescence enhancement observed in the anatase form dramatically transforms to quenching in rutile samples. We also formulated a reaction-diffusion model applied it to verify experimental findings and capture the fundamental underlying transport and adsorption phenomena. The computational model enables the fluorescence signature to be correlated with particle size, and we have further refined the model to enable characterization of mixtures containing different particle species (e.g., anatase and rutile $TiO_2$). In addition to continuous environmental monitoring, our analysis format enables online characterization of solution-based nanomaterials routinely employed as additives and coatings owing to its inherent ability to function in a portable and automated manner.

# 381825 Novel Exposure and Toxicological Methods to Define the End-of-Life Environmental Health and Safety Implications of Nano-Enabled Products

**Abstract:**

Proliferation of Nano-enabled-Products (NEPs) has inevitably raised the urgent question of nano-release during their synthesis, integration, processing, assembly, usage and eventually recycling or disposal at the end of their life cycle (LC). Apparently, there is a need to study and understand in a systematic manner the release mechanisms and possible exposure routes across the LC of NEPs in particular during incineration. Here, we focus on the development of a novel Integrated Exposure Generation System which enables the assessment of possible environmental health and safety implications during the incineration scenario of nanocomposite materials. A specific target is the employment of the developed exposure platform for a variety of polymer nanocomposites that are currently in use in many industries and products such as automotive (engineering plastics for multiple components), electrical (plastics for switches, plugs), construction (insulation foams), packaging (extruded polymers, polystyrene), textile (polyamides, monofilaments). Finally, a detailed physicochemical, morphological, and toxicological characterization of the released aerosols for each nanocomposite is performed utilizing the developed exposure system. Biological responses and properties of released aerosol were linked to specific NEP properties and combustion parameters. Through such an understanding, safer-by-design polymer nanocomposites can be manufactured that retain the superior properties without exhibiting adverse effects to the environment and human health.

# 381263 Self-Assembly of Aqueous Surfactants on Chemical and Morphological Heterogeneous Surfaces

**Abstract:**

The widespread interest in the self-assembly of surfactants on solid surfaces is motivated by many technological applications, including the manufacture of nano-structured materials using bottom-up strategies. Although much is known about the adsorption of surfactants on homogeneous flat surfaces from experiments, theory, and/or simulations, limited information is available, in quantifiable terms, regarding the adsorption of surfactants on surfaces with chemical and/or morphological heterogeneity. In this study the adsorption of model surfactants on patterned substrates has been investigated using equilibrium dissipative particle dynamics (DPD) simulations. Surfactant adsorption is allowed on troughs (considered at least in part hydrophobic) surrounded by a surface that effectively repels the surfactants. Our results suggest that the trough depth, the lateral size, and the distribution of hydrophobic sites on its surface lead to rich surfactant behaviour. Some of the structures observed in our simulations are not possible on flat homogeneous, nor on flat chemically heterogeneous surfaces. Once verified experimentally, our predictions could be useful, for example, for the design of new nano-structured materials and coatings.

# 380372 Folding By Design

**Abstract:**

A form of self-assembly, "self-folding" presents an alternative approach to the creation of reconfigurable, responsive materials with applications ranging from robotics to drug design. However, the complexity of interactions present in biological and engineered systems that undergo folding makes it challenging to isolate the main factors controlling their assembly and dis-assembly. Here we use computer simulations of simple, minimalistic self-foldable structures and investigate their stochastic folding process. By dynamically accessing all the states that lead to, or inhibit, successful folding, we show that the mechanisms by which general stochastic systems can achieve their "native" structures can be identified and used to design rules for optimized folding propensity.

# 391681 Chiral Templating of Self-Assembling Nanostructures By Circularly Polarized Light

**Abstract:**

Materials possessing molecular and nanoscale chirality are well known to rotate the polarization direction of linearly and circularly polarized light (CPL) passing through them. Optical effects related to polarization rotation are relatively easy to observe and have multiple applications. The opposite effects, however, manifesting as the influence of circular polarization of incoming photons on molecular and especially nanometer organization of matter are less known and understood. We believe that it is essential to understand the chemical and physical processes being altered when in response to CPL. If chiral "templation" of organic or inorganic matter with CPL is possible, it can open new synthetic routs to chiral nanostructures, which currently attracting a lot of interest. Self-assembly of NPs is one of the mechanisms that could potentially be influenced by CPL due to chirality of individual NPs, potential amplification of circular polarization effects in them, and sensitivity of self-assembly processes to small changes in interparticle interactions. Following these rationales, here we show that circularly polarized light can drive the self-assembly of cadmium telluride nanoparticles (CdTe NPs) into nanoribbons controlling helical directions by transcription of chiral information from the light to NPs. Different helical directions of laser induce different light adsorption of CdTe NPs which lead to greater reactivity of a selective chirality. This simple method for chiral nanoribbons can open the door to understanding life's homochirality and chiroptical devices.

# 387199 A Multiresolution Approach to Optimally Control the Dynamic Directed Self-Assembly of Nanostructures


**Abstract:**

Self-assembled nanoscale structures are the basis of various technological advancements such as molecular circuits and factories [1]. With significant progress in self-assembly of periodic nanostructures (such as monolayers [2]), the focus is now shifting towards non-periodic structures. Control of various interaction force fields (electrostatic, Van der Waals, etc.) [3] between the nanoparticles and external actuators can result in the formation of nanostructures with desired geometry. The aim is to design the nanoparticles and the external actuators such that the desired structure can be self-assembled rapidly with high reliability and avoiding any kinetic trapping that an ill-designed energy landscape might cause.

Lakerveld et al. [4] showed that a master equation based dynamic model can be used to account for kinetic trapping explicitly. However, their model will quickly become intractable when designing larger systems. Therefore, the Adaptive Finite State Projection method [5] was introduced to deal with simulation of directed self-assembly in larger domains. The Adaptive Finite State Projection method identifies the important configurations and thus, simulates a manageable number of master equations.

In this contribution, control of the self-assembly process is achieved using a combination of the multiresolution approach [1] and the Adaptive Finite State Projection method. The proposed method also restricts the physical domain, which enhances the Adaptive Finite State Projection method and mitigates the increasing size of the model. In the multiresolution approach, a desired nanostructure is developed in stages by dividing the physical domain into various partitions, and ensuring that these partitions correspond to ergodic subdomains. The physical domain is divided at the lowest resolution initially, i.e. two parts, in the first stage, which is then refined, i.e. four parts, eight parts and so on, in the subsequent stages. Every stage of this multiresolution approach aims to isolate a desired number of nanoparticles in each part of the domain, which is based on the final desired structure. Since the domain under consideration is too large to be simulated in detail, some of the state space is made inaccessible using repulsive external actuators and the remaining accessible part is simulated using the Adaptive Finite State Projection method. The external actuators are positioned and varied in such a way that the desired number of nanoparticles is obtained in each part of the divided domain, with a high probability in the minimum possible time. The time varying strengths of the external actuators are obtained by solving an optimization problem that minimizes the final time. The optimizer uses the Adaptive Finite State Projection innovatively to simulate the system and calculate the sensitivities for optimization.

Once the isolation of nanoparticles in their respective parts of the domain is achieved, the particles are locked in those parts of the domain using external repulsive actuators and those parts are further refined in the next stage of the multiresolution approach. This process is repeated till the final desired structure is obtained. The proposed method is a novel mechanism to generate a nanostructure of desired geometry, which can be accurately simulated and effectively avoids kinetic traps during the process.


## References


[1] N. Stephanopoulos, E. O. P. Solis, and G. Stephanopoulos, "Nanoscale process systems engineering: Toward molecular factories, synthetic cells, and adaptive devices," *AIChE Journal*, vol. 51, no. 7, pp. 1858-1869, 2005.

[2] J. C. Love, L. A. Estroff, J. K. Kriebel, R. G. Nuzzo, and G. M. Whitesides, "Self-assembled monolayers of thiolates on metals as a form of nanotechnology," *Chemical Reviews*, vol. 105, no. 4, pp. 1103-1170, 2005.

[3] K. J. M. Bishop, C. E. Wilmer, S. Soh, and B. A. Grzybowski, "Nanoscale forces and their uses in self-assembly," *Small*, vol. 5, no. 14, pp. 1600-1630, 2009.

[4] R. Lakerveld, G. Stephanopoulos, and P. I. Barton, "A master-equation approach to simulate kinetic traps during directed self-assembly," *The Journal of Chemical Physics*, vol. 136, no. 18, 2012.

[5] S. Ramaswamy, R. Lakerveld, G. Stephanopoulos, and P. I. Barton, "Approximating the solution to the master equation to simulate directed self-assembly of nanostructures," *AIChE Annual Meeting*, poster session, 2012.

# 382609 Multiscale Modeling Approach to Determine the Role of Amphiphilic Building Block in the Stability of Paclitaxel Drug Delivery Nanocarriers

**Abstract:**

Polymeric micelles are a promising class of drug delivery carriers in field of cancer therapy due to their tunable physiochemical properties. Typically, drug delivery nanocarriers are prepared from self-assembled amphiphilic block copolymers that have a biocompatible hydrophilic shell and a hydrophobic core capable of encapsulating cancer drugs such as paclitaxel (PTX). Designing optimal nanocarriers, however, is challenging due to stringent design requirements such as stability, drug-loading capacity, size distribution, and target specificity. Here we adopted a multiscale computational approach in conjunction with experiments to evaluate the building blocks of highly tunable linear polyethylene glycol-b-dendritic oligo(cholic acid) copolymer called telodendrimer. The multiscale approach involved molecular dynamics simulations of the telodendrimers and paclitaxel in coarse grained simulations for over 0.1 μs to capture micelle formation and drug encapsulation followed by reverse mapping to atomistic representation for structural analysis. Among the possible telodendrimeric structures studied, the simulations showed that $PEG^{5k}CA_8$ telodendrimer forms stable monodisperse micelles with 27.7±5.5 nm diameter and highest drug loading capacity of 25% w/w, which is in excellent agreement with the experimental results (21.4±4.0 nm). This multiscale approach allows us to overcome the computational bottleneck of simulating large systems without sacrificing the relevant chemical interactions in the nanocarrier-drug micelle.

# 390474 Predicting Process Windows for Block Copolymer Directed Self-Assembly with Chemoepitaxial Guiding Underlayers with Pattern Density Multiplication

**Abstract:**

Directed self-assembly (DSA) of block copolymers (BCPs) using pattern density multiplication is a technique of producing patterns with small pitches utilizing guiding template patterns printed as larger feature sizes and pitches. A chemoepitaxial guiding underlayer is one such method for achieving this density multiplication. This nominally topographically flat guiding underlayer is composed of a pinning region, or stripe if referring to lamellae, which will chemically interact preferentially with one microphase of the BCP, as well as a second region that is often referred to as "neutral" to both phases of the BCP. Typically, the pinning stripe is the width of a single lamellae of the phase separated BCP, while the neutral stripe is some odd number of lamellae widths. In this work, detailed simulation studies have been performed to explore the effects that variables such as guiding stripe size or chemical composition of the neutral stripes have on the process window of DSA pitch sub-division patterning processes. The impact of such factors on both the overall alignment of the BCP film as well as the quality of the resulting lamellae (e.g. whether various compositions are more prone to induce footing in their respective lamellae or not) have been evaluated. Such process windows and lithographic parameters have been studied for different pitch sub-division conditions for ratios of pinning stripe width to neutral stripe width of 1:3 and 1:5. It is found that the center of the processing window with respect to alignment occurs at a composition of the "neutral stripe" such that it is slightly to somewhat strongly preferential to the type of polymer of opposite type to that attracted by the pinning stripe, and that this ideal "neutral stripe" composition becomes more neutral as the density multiplication increases. This ideal composition for alignment is near, though not necessarily the same as, the ideal composition for achieving the best lamellae quality, such as in the case of a 1:3 pattern where the best lamellae quality occurs at a higher composition than the best alignment.

# WORM-LIKE micelles of polymerizable surfactant as a template for polymerization

S. Rima[1], M. Lattuada[1]

[1] *Adolphe Merkle Institute, University of Fribourg, Fribourg, Switzerland*
*simonetta.rima@unifr.ch*

The use of self-assembled materials as templates in materials synthesis is a rapidly growing area. However, the subset of work that focuses on rodlike and wormlike self-assembled micelles for polymerization of organic solution is under-utilized given the potential of these systems for synthesizing anisotropic materials. The use of particles with anisotropic shape is of significant interest since it allows fabrication of structures with special symmetries and degree of packing, as well as with anisotropic properties [1]. Rod-like polymer particles could have interesting properties and could find many practical applications; however, few methods for the production of such particles are available. In this work we introduced and investigate a method for synthesis of a new class of polymeric nanorods and nanofibrils based on emulsion polymerization using wormlike polymerizable micelles as template. Wormlike micelles, are elongated and semiflexible aggregates of surfactant molecules [2]. The equilibrium and dynamics of the micellar structures, however, are determined by a delicate balance of intermolecular forces that can be easily disrupted. In the current study it has been observed that addition of oils or polymers typically reduces the average micelle length and results in a dramatic decrease of the solution viscosity on the microstructure and loss of the highly elongated and entangled micelles. Solubilization and polymerization of styrene in different Cetyl trimethyl ammonium Tosylate (CTAT) wormlike solutions at various concentrations, temperatures and initiator systems have been performed, and have been showed to preferentially form spherical particles. This is explained as follows: during the polymerization process, the viscosity is reduced by several orders of magnitude to a water-like value, and no viscoelasticity can be observed either visually or rheometrically. These data suggest a transition from rodlike to spherical micellar aggregates. In order to overcome the problem, a polymerizable surfactant, cetyltrimethylammonium 4-vinylbenzoate (CTVB), containing a polymerizable counterion, has been prepared and investigated with the purpose of "locking in" the micellar structure over the oil polymerization so that the template is less sensitive to environmental changes. Elongated particles and long polymer fibrils have been formed depending on cross linker concentration used, partially preserving the structure of the parent micelle upon polymerization. Therefore, the presented approach represents a possible method for the production of anisotropic particles, with the purpose of studying their assembly for the preparation of nano- and microstructured material

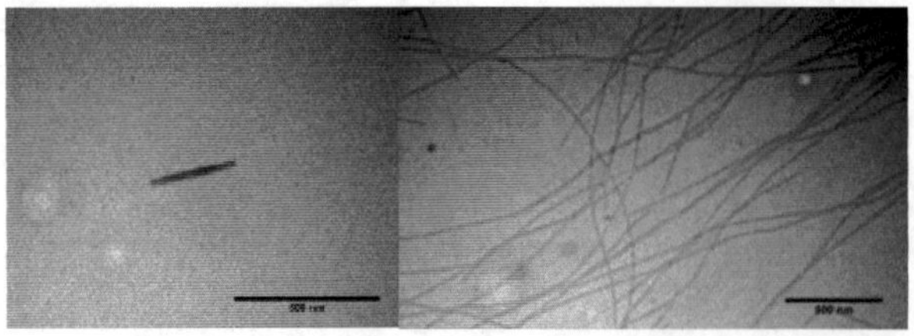

Figure 1 – TEM images of p(STY/DVB) fibrils and pSTY elongated nanoparticles.

[1] Ward, I.M., Structure and properties of oriented polymers. 1997: Springer.

[2] Dreiss, C.A., Wormlike micelles: where do we stand? Recent developments, linear rheology and scattering techniques. Soft Matter, 2007. 3(8): p. 956-970.

**Simultaneous Synthesis and Assembly of Noble Metal Nanoclusters with Variable Micellar Templates**

Yao Zhou[1] and Hua Chun Zeng*
NUS Graduate School for Integrative Sciences and Engineering and Department of Chemical and Biomolecular Engineering, Faculty of Engineering, National University of Singapore, 10 Kent Ridge Crescent, Singapore 119260
KEYWORDS: nanoclusters, synthesis, self-assembly, micelles, noble metals
[1]Email□a0103385@nus.edu.sg *Email: chezhc@nus.edu.sg

**ABSTRACT:** In the present work, one-phase simultaneous synthesis and assembly of Au, Pt and Pd nanoclusters (NCs) into giant assemblages with well-defined geometric shapes is achieved with cetyltrimethylammonium bromide (CTAB) used as surfactant and a moderate reductant. Shapes, hierarchical complexity and packing density of the metal NCs assemblages vary mainly depending on the metal species and CTAB concentration. As shown in Figure 1, in general, "solid nanorod" assemblages are formed from Au NCs, giant vesicles and/or dandelions-like spheres for the Pt case and rhombic/hexagonal structures for the Pd case, and owing to the dynamic nature, they demonstrate various structural reforming behaviors. Such NCs assemblages are found to inherit the shapes directly from the giant micelles of $CTA^+$-metal halide complex in which the metal halides distribute themselves homogeneously as the counterions of $CTA^+$. The self-assembly behaviors of $CTA^+$-metal halide complex are explained at molecular and mesoscopic scale. The metal precursors, which serve as counterions of the positively charged surfactant heads, screen the electrostatic repulsive forces among the surfactants as well as the micelles, which therefore provides the driving force for formation of these soft templates. Meanwhile, the formation of the small NCs is well explained from the perspectives of nucleation and growth kinetics. Primary catalytic study indicates that the as-prepared NCs have high catalytic activity. The as-formed bicontinuous inorganic/organic hybrid assemblages not only provide the potential to construct active components-preloaded nanoreactors but also visualize the versatile micelles interactions.

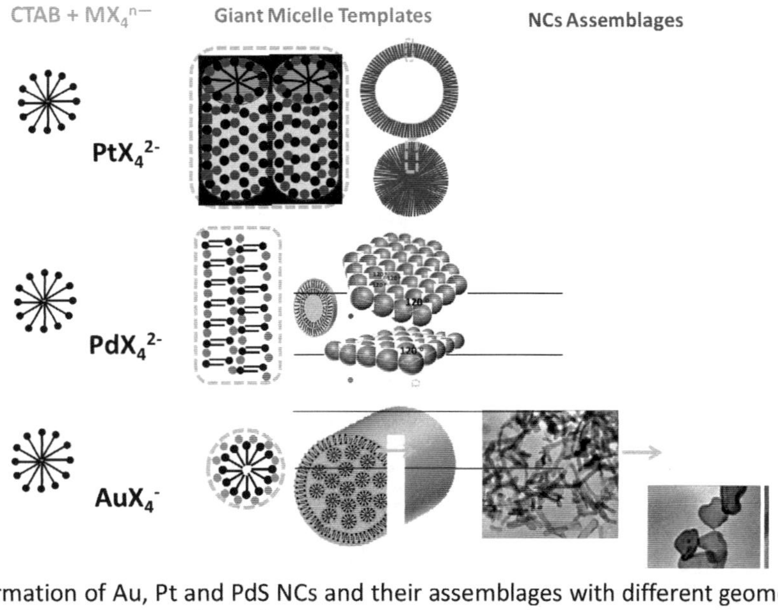


Figure 1. Formation of Au, Pt and PdS NCs and their assemblages with different geometric shapes with the assistance of micelles.

# 362900 Nanocrystals and Nanostructures for Study of Catalysis

**Abstract:**

This presentation focuses on fundamental understanding of catalytic processes and development of highly active (or selective) and durable catalysts for the reactions of industrial importance, especially for energy conversion and storage. Here I present a research system that connects fundamental investigation on well-defined extended surfaces (e.g. single crystal surfaces), extrapolation onto nanocrystals with highly controlled shape and size, exploration of interfacial interaction using novel nanocrystal superlattices as platform, and finally design of high performance catalysts in which all the possible beneficial properties from complex functional structures are implemented.

# Scalable process control of highly luminescent and environmentally friendly ZnS quantum dot synthesis by anaerobic bacteria in aqueous solutions

Gyoung Gug Jang[a], Ilia N. Ivanov[b], Harry M. Meyer III[c], Michelle Kidder[d], David E. Graham[a], Ji-Won Moon[a]*

[a] *Biosciences Division, Oak Ridge National Laboratory (ORNL), Oak Ridge, TN 37831, USA*
[b] *Center for Nanophase Materials Sciences Division, ORNL, Oak Ridge, TN, 37831*
[c] *Materials Science and Technology Division, ORNL, Oak Ridge, TN 37831*
[d] *Chemical Science Division, ORNL, Oak Ridge, TN 37831*

## Abstract

Highly luminescent, green ZnS quantum dots (QD) were synthesized by a biological process with the *in-situ* addition of oleic acid across scales, from 10 to 24,000 mL. This biological process, called NanoFermentation, uses thermophilic bacteria to catalyze the reduction of metal ions or sulfur species, producing nanomaterials on the outside of the cell surface in aqueous solutions. The NanoFermentation technology is a single-step process, which requires a relatively low temperature, inexpensive oxidized chemicals and environmentally friendly aqueous media without organic solvents. Optical, crystalline and chemical structural characterizations including spectrofluorometry, XRD, XPS, FTIR and UV-vis absorption indicated that oleic acid capping exhibited significant size-confinement effects to produce zero-dimensional quantum dots with 1.3 nm of average crystal size, resulting in ~22-fold higher luminescence than abiotic, L-cysteine capped ZnS. We found several process factors critical for QD biosynthesis to reproduce desirable properties across scales, reduce synthesis time and produce high yields: 1) bacterial growth and the development of $HS^-$ ion, 2) incubation time after Zn and capping agent dosing, 3) composition of bacterial growth medium, and 4) aqueous solubility of capping molecules. Each factor was evaluated to optimize biological ZnS QD production with a reproducible optical quality. Finally, the scale-up conditions of the process were evaluated using a 24,000-mL reactor to produce 10's of grams of ZnS QD product.

# 388679 Real-Time Analysis of the Cellular Mechanisms Associated with Carbon Nanotubes Exposure

**Abstract:**

The unique intrinsic properties of carbon nanotubes (CNTs) make them potential candidates for a wide range of biomedical applications spanning bio-imaging, disease targeting, and delivery of genes and drugs. Therefore, in-depth analysis of their toxicity and fate inside biological systems is crucial prior to their application. Several reports have shown that CNTs toxicity can be attributed to their length, surface chemistry, aggregation or metal impurities to name a few. However, due to the various types of CNTs and the different surface functionalizing techniques, there are many discrepancies in literature and limited understanding of their toxicological and pharmacological profiles. In this research, we employed an electric cell impedance sensing device to provide a comprehensive analysis of the cellular behavior of human lung cells post exposure to CNTs with different physical and chemical properties. This non-invasive approach employs arrays with gold electrodes as immobilization platforms to provide real-time measurements of cellular adhesion, cell-cell interactions as well as changes in cellular morphology. Such measurements were further supplemented with microscopy and viability assays to derive structure-function relations associated with the cytotoxic and apoptotic events induced by cellular exposure to CNTs. Our real-time results provided new insights into the underlying cellular mechanisms associated with CNTs exposure and promise to extend such cellular-based studies to advance the biomedical applications of CNTs and other nanomaterials.

# 388917 Thermal Management in MEMS/Nems

**Abstract:**


### Thermal Management in MEMS/NEMS

Sesha Hari Vemuri, Yu Liu, Pil Seung Chung, and Myung S. Jhon

Micro/Nanoscale electromechanical systems play a an important role in integrating transistor-like nanoelectronics with mechanical actuators, pumps, or motors, and thereby form physical, biological, and chemical sensors This places a range of stringent operating requirements in the inside the nanoscale length dimensions, where heat transfer during operation on the order of nanosecond becomes critical.

To shed light on this issue, in this study we develop a mesoscale level model based on kinetic theory called lattice Boltzmann method (LBM) to understand the nanoscale heat transfer in the MEMS/NEMS. In this study, we examine the effects of main energy carriers at nansocale (electrons & phonons) via novel LBM based methodology. We investigate the overall heat transfer contribution of the coupled heat carriers and the electron-phonon coupling issues.

# 389049 Gold Nanoparticle Interactions with Model Biological Membranes

**Abstract:**

The unique surface area properties of nanoparticles (NPs) make them attractive for a variety of applications and industries that did not exist before, such as drug delivery applications and consumer products. However, those properties may alter their cytotoxicity posing a potential health threat with increased use ultimately leading to NP exposure and disposal within our environment. There is little information on how NPs will interact once in contact with environmental habitats and us, making it important to understand these interactions to ensure their safety.

A Quartz Crystal Microbalance with Dissipation (QCM-D) was used to create a supported lipid bilayer (SLB) formed from L-α-phosphatidylcholine (PC) on quartz crystal substrate. This SLB is representative of a typical biological membrane. This instrument allows for nano-scale readings in real time, making it ideal for our goal. The QCM-D measures frequency and dissipation changes at five overtones (3, 5, 7, 9, 11). Frequency is inversely proportional to mass changes, and dissipation results lend an understanding of surface rigidity. The overtones measured by the QCM-D give a depth perspective to our results. The 3$^{rd}$ overtone is indicative of the fluid furthest from the crystal and the membrane's surface, while the 11$^{th}$ overtone is indicative of the fluid closest to the crystal and the furthest depth of the membrane. The diameters of gold NPs studied were 2, 5, 10, and 40 nm gold NPs at a constant concentration of $7.14*10^{10}$ particles/mL. Gold NPs were either diluted in deionized water or in poly(methacrylic acid) (PMA). NPs suspended in water were used to investigate how they would interact in a sterile lab condition, while PMA was used to understand how gold NP interactions may be altered in the presence of organic coatings. NPs will ultimately come in contact with natural organic material in the environment during disposal, having the ability to alter the NP surface potential and hydrophobicity.

When gold NPs were suspended in water, all four sizes exhibited similar results, with some variation in the 10 nm NPs. All overtones resulted in similar magnitudes, suggesting even insertion throughout the depth of the bilayer. When gold NPs were suspended in PMA, a size dependence was clearly observed, with the largest interaction occurring with 40 nm gold NPs, which exhibited significant mass removal. Our results show gold NPs in water all have a similar trend. However, once gold NPs are in contact with PMA, the mechanism for their interaction changes. The smaller size gold NPs interact with the SLB by adding small amounts of mass to the surface. This is a very different mechanism than the 10 nm and 40 nm gold NPs suspended in PMA, where mass is removed. The 10 nm gold NPs contribute to a relatively small amount of mass or lipid removal in comparison with the 40 nm gold NPs. With a SLB mass of 350.4 ng, the 40 nm gold NPs removes approximately 1/3$^{rd}$ of the bilayer.

# 389467 Preparation of High Performance Graphene Fibers By Wet-Spinning Method

**Abstract:**

In this study, we use lamellar graphite to prepare GO, and then integrate the large GO sheets into macroscopic fibers by wet-spinning method. We use amines alcohol solution as a new kind coagulation liquid. In the process, meters-long graphene fibers were continuously spun from GO suspensions followed by thermal reduction. By varying wet-spinning conditions, a series of graphene fibers were prepared, then, the performances of the fibers were investigated, through all of these works we find the optimal conditions to fabricating high mechanical strength graphene fibers. The as-prepared graphene fibers's mechanical strength mechanical strength can be achieved 200 MPa, and they possessed with good flexibility, the electronic conductivity reached 38 S/cm.

Accelerated generation of free radicals by iron oxide nanoparticles in the presence of an alternating magnetic field

Robert J. Wydra, Kimberly W. Anderson, Thomas D. Dziubla, J. Zach Hilt

Iron oxide-based magnetic nanoparticles are studied for a wide range of applications due to their unique physical properties. One property of particular interest is the ability to be heated remotely by an alternating magnetic field (AMF). The nanoparticles absorb the energy from the AMF and convert it into heat through two relaxations: Brownian relaxation (physical rotation of the particles) and Neel relaxation (rotation of the magnetic moment). In recent work by others, magnetic nanoparticles were shown to kill cancer cells without a measurable temperature rise, which has generated great interest in the potential of intracellular hyperthermia. In related work, it has been demonstrated that when the solution temperature does not rise, the magnetic nanoparticle surface still undergoes a significant increase in temperature when exposed to an AMF. At this time, it is unknown as to if this explains the toxicity of intracellular hyperthermia or if other phenomena contributing. At the molecular level, it has been observed that a significant contributing factor to the iron oxide toxicity is the generation of reactive oxygen species (ROS) through the Fenton and Haber-Weiss reactions catalyzed at the nanoparticle surface. Fenton chemistry itself has been shown to be temperature dependent with an increase in activity up to 40 °C – beyond this temperature the hydrogen peroxide degrades into oxygen and water limiting the reaction. To date, no one has studied the effects of an AMF has on the surface mediated production of ROS, and we hypothesize that it could be a key contributor to the toxicity that has been demonstrated as intracellular hyperthermia.

The rate of ROS production from the surface of iron oxide magnetic nanoparticles was studied at different temperatures and differing AMF conditions using a methylene blue decolorization assay. Briefly, iron oxide nanoparticles catalyze the decomposition of hydrogen peroxide into radicals which degrade methylene blue and can be measured colorimetrically. Recently, we have successfully demonstrated an increase in ROS generation from a suspension of magnetic nanoparticles when exposed to an AMF. The increase in ROS generation compared to the Arrhenius prediction was both time and concentration dependent; in which we observed a decrease in ROS enhancement with increased time of exposure and concentration. We postulate that the decrease is due to agglomeration in the presence of the field. As the nanoparticles agglomerate, there is a decrease in surface area per mass limiting the reaction rate. Preliminary intracellular hyperthermia studies were performed on colon cancer cell lines. Cells with internalized nanoparticles were exposed to an AMF and the oxidative damage to the cells was measured using a DCF assay. Functionalized nanoparticles displayed a significant increase in ROS generation with field exposure compared to uncoated nanoparticles.

# 390844 Mechanism of Nanoparticle Assembly in Polymer Thin Films from Quantitative SEM Image Analysis

**Abstract:**

Nanoparticle self-assembly is being widely explored as a powerful, scalable, and inexpensive method for fabricating nanostructures. Recently, we demonstrated the self-assembly of shaped metal nanoparticles into chains of nanoparticles within a polymer thin-film for plasmonic applications [1]. The mechanism of assembly in such systems is still poorly understood, although such knowledge is essential for controlling the structures that are formed. We have developed an automated image analysis tool to extract detailed information about the assembly morphology from SEM images. This information in turn allows us to make quantitative comparisons with theory and simulations to predict the mechanism of assembly. I will present the results of our study, where we have found that the assembly process is well described by the diffusion-limited cluster-cluster aggregation mechanism. I will also describe our image analysis tool, scaling analyses, and general approach, which should prove helpful to other researchers in the field.

[1] B. Gao, G. Arya, and A. R. Tao, *Nature Nanotechnology* 2012, **7**, 433-436.

# Catalyst-Free Growth of Ultra-High Density ZnO Nanowires for Dye-Sensitized Solar Cells

**Cheng Xu*, Yang Zhao, Luping Li, and Kirk J. Ziegler**

Department of Chemical Engineering, University of Florida, 1006 Center Drive, Gainesville, FL 32611, US

*Presenting Author (Ph: 352-392-3122; Email: chengotopia@ufl.edu)

**Abstract:**

The excellent optical and electrical properties of ZnO such as direct bandgap, high electron mobility, large exciton binding energy, and excellent transparency makes it an ideal choice for the use of field transistors, piezoelectrics, photovoltaics, and sensor devices. Since device performance is often directly proportional to the active surface area, fabricating vertically aligned nanowire arrays with high density and aspect ratio is important to maximize their efficiencies beyond the current limit. The traditional approach of growing ZnO nanowires with excellent crystallinity include catalyst based vapor-liquid-solid (VLS) growth and seeded hydrothermal growth. However, the density of nanowires grown by both methods is limited since the density is controlled by the metal catalyst for VLS growth and ZnO texture seeds for the hydrothermal growth. As a result, the density of ZnO nanowires can only reach $10^9$ $cm^{-2}$ for VLS growth and $10^8$ $cm^{-2}$ for hydrothermal growth. Furthermore, both methods struggle to obtain nanoparticle catalysts or seeds with small diameter and to maintain nanowires with small diameter during growth due to high surface energy associated with the substrate. Although templated approaches provide an excellent way of demonstrating nanowire density control, it is difficult to form and remove the back electrode and integrate these nanowires into useful devices.

In this study, we focused on not only using the anodized alumina oxide (AAO) as a template to control nanowire density but also providing a method of transferring the template onto a substrate of choice that could be integrated into device fabrication. AAO provides maximum density ($10^{10}$-$10^{12}$ $cm^{-2}$) with its closely-packed hexagonal pore arrangement and, by carefully tuning the experimental conditions, the pore diameter and interpore spacing can be precisely controlled. First, to demonstrate the transfer process, ultra-thin AAO templates were fabricated on aluminum using 2-step anodization. After dissolving aluminum, AAO was transferred to a foreign substrate. To study the mechanism of catalyst-free vapor-solid (VS) growth on AAO, the growth condition of temperature, flow rate, oxygen partial pressure, amount of source, and distance between source and substrate had to be optimized to control the initial nucleation of ZnO nanoparticles in the AAO pores. With the optimized condition, ZnO nanoparticles were deposited in AAO pores with identical density of AAO, and then ZnO nanowires were grown on the nanoparticles. To study the effect of experimental parameters on the final ZnO nanowire morphology, SEM snapshots were taken at different time intervals and XRD was done for different growth parameters. As a result, nanowires were fabricated on a substrate of choice with densities of $10^{11}$ $cm^{-2}$. This technique provides a new route of fabricating large-scale, cost-effective optoelectronic devices.

*Keywords: photovoltaics, ZnO nanowires, alumina template, catalyst-free nanowire growth*

# 391854 Steps Towards Achieving Uniform Electrodeposition in Nanoporous Aluminum Oxide Templates

**Abstract:**

Nanowires have garnered much interest in recent years for their potential in increasing performance of devices such as solar cells, thermoelectrics, batteries, and piezoelectrics. Many methods exist for the fabrication of nanowires, but one of the most promising methods is the use of nanoporous aluminum oxide templates due to its extremely high density and self-ordering through a hexagonal close packed structure. However, fabricating uniform arrays of nanowires in contact with the conductive aluminum substrate still remains a challenge. The problems are most evident when a greater degree of control over the electrodeposition into the aluminum oxide templates is needed for applications such as creating catalyst nanoparticles.
A vital part of the template process is removal of the barrier layer between the pore and the underlying substrate. This is needed to establish direct contact from the pores to the substrate. The objective is to remove the barrier without sacrificing the pore walls. Several methods exist for barrier layer thinning, including stepping down the anodization voltage or ramping down the current. However, the thinning method will influence the final pore opening, which creates the contact between the pore and aluminum substrate. In this presentation, different barrier layer thinning techniques will be discussed and their effect on the pore opening process examined. Both step-down voltage and ramped current methods were used to understand the barrier layer thinning process. These results were correlated to the amount of pore opening achieved, measured through a combination of chronoamperometry and electrochemical impedance spectroscopy. Ellipsometry was used to characterize the barrier layer thinning and pore opening processes of the aluminum oxide templates. Finally, nanowires were grown within these templates to study the effects of different barrier layer thinning and pore opening techniques on nanowire growth and uniformity. This research provides a comprehensive approach towards fabricating uniform, high-density nanowire arrays.

# 358346 Self-Assembly of Amphiphilic Peptide $(AF)_6H_5K_{15}$: Roles of Hydrophobic and Hydrophilic Residues

**Abstract:**

A molecular dynamics simulation study is reported to investigate the roles of hydrophilic and hydrophobic residues in the self-assembly of $(AF)_6H_5K_{15}$ peptide derivatives. The peptide, as well as water and counterions, are represented by the MARTINI coarse-grained model. The assembly is observed to follow a three-step process: formation of small clusters, large clusters and micelles. With increasing length of hydrophilic Lys residues in $(AF)_6H_5K_n$ (n = 10, 15, 20 and 25), assembly capability is found to be reduced with the formation of smaller micelles or the presence of individual peptide chains. Upon replacing Ala by more hydrophobic Phe in $A_mF_nH_5K_{15}$ (m + n = 12), larger micelles are formed. With increasing length of hydrophobic Phe residues in $F_nH_5K_{15}$ (n = 4, 8, 12 and 16), micelle size increases and the morphology shifts from spherical to fiber-like. The simulation study provides mechanistic insight into the crucial roles of hydrophilicity and hydrophobicity in the assembly of $(AF)_6H_5K_{15}$ derivatives; it reveals that assembly capability is reduced by increasing hydrophilicity, whereas increasing hydrophobicity leads to morphology transition.

# 362495 Self Assembly in of Soft Particle Fluids, and Hard-Soft Particle Mixtures

**Abstract:**

Soft, penetrable particles exhibit highly anomalous clustering behavior. We present a study of these systems using integral equation theory, and predict a striking contact to overlap transition. In the contact regime, the structure is that of a 'normal' fluid, while the overlap regime is characterized by entropy enhancement through cluster formation. These clusters are formed due to multiple overlaps of purely repulsive particles. At very large volume fractions, a cluster-crystal phase is predicted. Both transitions are a function of the density of soft particles, their stiffness, and their penetrability. While cluster size is independent of volume fraction, cluster density grows linearly with it. Anomalous structure factors are predicted in both the overlap, and the cluster-crystal regimes. At very low temperatures, we predict ordered structures similar to charged colloidal crystals, having large lattice constants. These lattice constants vary with the stiffness of the inter-particle potential. We also present some initial results of hard and soft particle mixtures. In these systems, the cluster-crystal phase is largely destroyed. However, under certain conditions, the hard particles are encapsulated by soft particles. This encapsulation is a function of the particle densities, and hence is highly tunable.

Institute of Thermodynamics and
Thermal Process Engineering
Prof. Dr.-Ing. J. Gross

University of Stuttgart
Germany

itt

# Different ways of looking at the force between nano particles

**Alexander Lange, Fabian Danecker, Gernot Bauer, Joachim Gross**
Institute of Thermodynamics and Thermal Process Engineering, University of Stuttgart, Germany

## Introduction

Nano particles – consisting of hundreds of atoms – are the building blocks of new materials with applications in various fields of modern engineering. Much about them, especially about their interaction with other nano particles, is known from experiments and molecular simulations. Due to their complex structure and interaction, a good theoretical foundation is often missing.

In this work we study capped gold nano particles and their effective interaction potential obtained from molecular dynamics (MD) simulations (Fig. 1). Computations of this kind are expensive; about $10^3$ molecular groups ($SH_2$, $CH_2$, $CH_3$, forming the thiol ligands of our nano particles) are assumed to interact with each other via Lennard-Jones potentials. By exploiting theoretical arguments from thermodynamics and statistical physics we try to reduce the number of necessary simulations.

**Figure 1** Potential of mean force (PMF) over the distance $R$ of two nano particles $Au_{147}(SC_8)_{58}$ – obtained from MD simulations performed with GROMACS [1]. The PMF $\phi(R)$ is computed by integration of the mean force $F_m=((F_1-F_2)\cdot e_x)/2$.

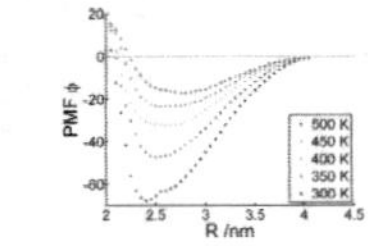


$$\phi(R) = \int_R^\infty F_m(S)\,dS$$

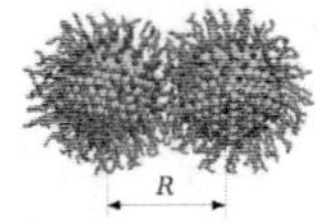


## Methods

For different temperatures $T$, we run MD simulations computing the potential of mean force (PMF), which is given by the Helmholtz free energy $A$ of two nano particles at distances $R$ and $\infty$ [2],

$$\phi(R) = \frac{A(R) - A(\infty)}{kT}. \qquad (1)$$

### High temperature expansion of the Helmholtz free energy at low order

The perturbative description of Zwanzig [3] offers an expansion in the inverse temperature $\beta$,

$$\beta A = \sum_{n \geq 0} a_n \beta^n, \quad \text{where} \quad \beta = \frac{1}{kT} \qquad (2)$$

with $k$ being the Boltzmann constant. The coefficients $a_n$ are determined by the molecular interaction; in practice only a finite number of them is relevant. They can be specified by theoretical methods and/or molecular simulations. Even if not indicated, these coefficients depend on $\beta$, but for the considered temperature range ($300\mathrm{K} \leq T \leq 500\mathrm{K}$) the dependence is unimportant (cf. Results). For dispersive molecular interactions (e.g., Lennard-Jones interactions, as being the case for the molecular groups of our nano particles), first order expansion is sufficient

$$\beta A = a_0 + a_1\beta + \mathcal{O}(\beta^2). \qquad (3)$$

Without knowledge about the precise interaction, simulations for two (or more) temperatures are required to estimate the two coefficients (Fig. 2) [1]. For polar interactions, second order expansion is needed, etc., and the number of necessary simulations increases.

**Figure 2** PMF curves $\phi(R)$ for five temperatures (A). The ten pairs of PMF curves are used to calculate coefficient curves, $a_0(R)$ (B) and $a_1(R)$ (C): the pointwise average (red dots) and the area between upper and lower envelopes (red).

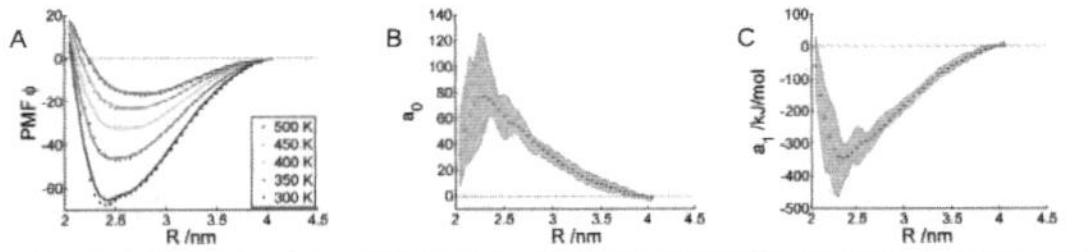


### Physical particle density – obtained via principle component analysis

To reduce the number of simulations, we calculate one of the coefficients using models of density functional theory (cf. Results). Applying this methodology, we must determine the density of the relevant interaction centers. Naive sampling may not account for the momenta that are exchanged when coupled to a thermostat or the symmetry that is broken when running a simulation. Therefore we propose a method that corrects for all the mechanical constraints imposed on the nano particles during the simulation. For each time step, we re-adjust the center of mass and the tensor of inertia (Fig. 3) by applying principal component analysis (PCA).

**Figure 3** For each time step of the simulation, the coordinate system/ lattice is chosen via PCA (A), leading to one small and two very similar large EWs. Panal B shows the EVs of the large EWs; they form a circle, which justifies the use of cylindrical symmetry and a corresponding lattice.

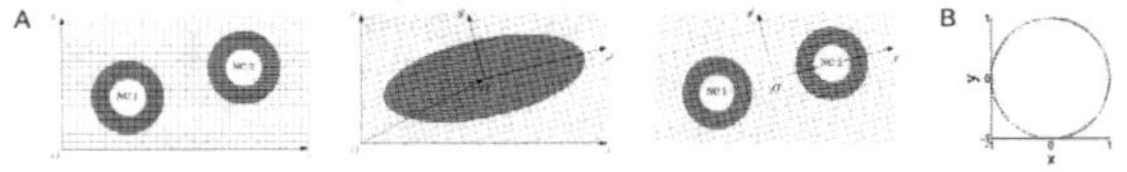


### The free energy within thermodynamics and statistical physics

According to the first law of thermodynamics,

$$A = TS + U, \qquad (4)$$

the constant $a_0=S/k$ in (2) represents the entropy and the remainder of the expansion – the internal energy. The partition function of stat physics $Q_N=(\Lambda^{3N}N!)^{-1}\int\exp(-\beta V_N)\mathrm{d}r^N$ defines the free energy,

$$A = -\beta^{-1}\ln Q_N, \qquad (5)$$

in terms of the molecular interactions $V_N=\sum_{i,j\leq N}v(|r_i-r_j|)$. Thermodynamic perturbation theory (TPT), which models the transition from a reference fluid (well-known, with pair potential $v_0$) to a target fluid (unknown, with pair potential $v$), provides us with formulas for the coefficients $a_n(R)$.

## Acknowledgments

Support from DFG through Collaborative Research Center 716, Project A.7 is gratefully acknowledged.

## Results

Using different thermodynamic models, we calculate one of the coefficients, $a_0$ or $a_1$, and compute the PMF via MD simulation for some reference temperature $T_0$. The other coefficient is then determined by (3) for any temperature $T$. The simplest model for the entropic contribution (cf. (4)) is an ideal gas:

### Ligand segments as ideal gas – w/ & w/o a physical particle density

$$a_0^{\mathrm{IG}} = \int \rho\,(\ln \Lambda\rho - 1)\,dr^3, \qquad (6)$$

where $\rho$ is the particle density and $\Lambda$ de Broglie wavelength, which cancels when computing the PMF. Following a naive, straightforward sampling procedure with a fixed lattice in space, the resulting density leads to an entropic contribution that is too large (10–20%, cf. Fig. 4A) in order to consistently explain the simulation data for different temperatures. Applying our physical density – obtained through PCA – we end up with an entropic contribution that is significantly smaller than the one obtained from naive sampling and may thus explain the temperature dependence correctly.

A much more advanced model (cf. Fig. 4CD) is given by Wertheim's hard sphere chains [4]:

### Ligands as hard sphere chains – applying a group contribution method

$$a_0^{\mathrm{HC}} = \int \bar\rho \left( \frac{4\eta - 3\eta^2}{(1-\eta)^2} - (1 - 1/m)\ln\frac{1-\eta/2}{(1-\eta)^3} \right) dr^3, \qquad (7)$$

where $\eta=\pi\sigma^3\bar\rho/6$ is the packing fraction, $\sigma$ the hard sphere diameter, and $m$ the fractal # of ligand segments. According to [5], a weighted particle density $\bar\rho$ must be utilized; therefore the method also works w/o employing the PCA-corrected density [6]. The disadvantage of this method is the extra parameters ($\sigma$, $m$), which are determined here by a homo-segmented group contribution method [7].

**Figure 4** Coefficients $a_0(R)$ (AC) and resulting PMFs (BD). Panels AB show the ideal gas – Panels CD the hard sphere chains. For all distances $R$ of the nano particles (in A), the coefficients based on a naive particle density (green) are larger than the coefficients based on the PCA-corrected particle density (blue). Right to the minimum of the PMFs, the (blue) coefficient curves approximate the averaged coefficients (red dots, cf. Fig. 2) very well. Left to the minimum, however, the deviation between the coefficients becomes large. Error fluctuations (<5%, light color) in the coefficient curves $a_0(R)$ were smoothed out. The fit of the PMF curves is similar – with better results towards larger distances $R$. For both methods, the reference temperature $T_0$=350K was used.

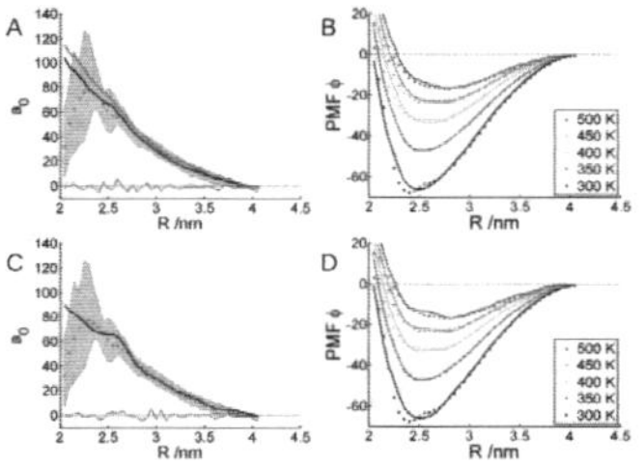


Finally, we compute the PMF via the interaction term in first order perturbation theory (Fig. 5):

### Lennard-Jones interaction between ligand segments at first order TPT

$$a_1^{\mathrm{LJ}} = \frac{1}{2}\iint \rho(r_1)\rho(r_2)g(|r_1-r_2|)v_1(|r_1-r_2|)dr_1^3 dr_2^3, \qquad (8)$$

where $v_1$ is the perturbation in Barker-Henderson's theory [8] and $g$ the pair distribution.

**Figure 5** Coefficients $a_1(R)$ (A, smoothed) and PMFs (B, $T_0$=350K). Right to the minimum of the PMFs, the coefficients (blue curve) approximate the averaged coefficients (red dots, cf. Fig. 2) well. Left to the minimum, the deviation between the coefficients becomes large. We also calculated $a_1$ via WCA [9], but the splitting of $v$ into repulsive and attractive parts did not improve the result.

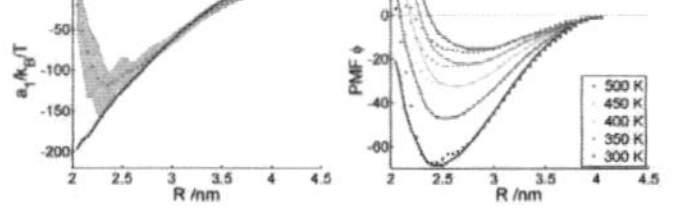


## Discussion

All three approaches lead to PMF curves of satisfactory quality at higher particle distances, although they differ considerably with respect to the number of input parameters and the computational effort (cf. Table 1). The ideal gas method returns the best PMFs and requires no extra parameters, yet one must correct the simulated particle distribution via PCA. As MD simulations usually involve constraints on parts of the molecular structure, such a correction will be applicable to many problems.

**Table 1** Comparison.

| Method | PRO | CON | Error | cf. Fig | Computation time |
|---|---|---|---|---|---|
| Introduced in [1] | Best fit | # simulations double | 55 | 2A | > 24 |
| Ideal gas | Excellent fit | PCA correction | 92 | 4A | 2 |
| Hard sphere chains | No PCA correction | Extra parameters | 127 | 4B | 1 |
| Lennard-Jones fluid | Close to simulation | Error & comp. time | 279 | 5B | 12 |

### The origin of the non-monotonic behavior of $a_0(R)$ – a depletion force?

Change of monotony of the coefficient curves at distance $\sigma$ between the gold core surfaces (Fig. 6) suggests that there is a depletion force caused by the ligand segments, which seem to act like a solvent. When including the number of particles the magnitude of the force is of expected size [10].

**Figure 6** Coefficients (blue curves) and its errors (light blue areas) are obtained by the ideal gas formula (6). The averaged coefficients (red dots, cf. Fig. 2), which represent the best prediction, strongly deviate from our computed coefficients at small distances – more than the error. Therefore, it seems reasonable that these averaged coefficients show the true behavior.

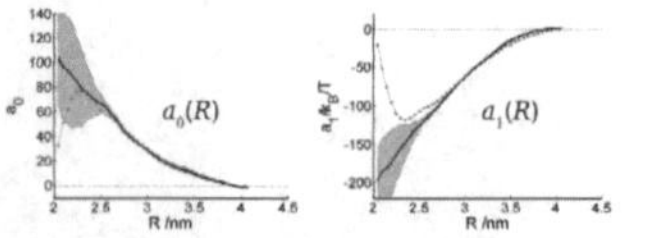


## References

[1] Bauer G, Lange A, Gribova N, Holm C, Gross J. Mol Sim (2014)
[2] Jarzynski C. Phys Rev Lett 78:2690 (1997)
[3] Zwanzig RW. J Chem Phys 22:1420 (1954)
[4] Wertheim MS. J Chem Phys 87:7323 (1987)
[5] Tarazona P. Mol Phys 52:81 (1984)
[6] Lo Verso F, Yelash L, Egorov SA, Binder K. J Chem Phys 135: 214902 (2011)
[7] Sauer E, Stavrou ME, Gross J. Ind Eng Chem Res 53:14854 (2014)
[8] Barker JA, Henderson D. J Chem Phys 47:4714 (1967)
[9] Weeks JD, Chandler D, Andersen HC. J Chem Phys 54:5237 (1971)
[10] Goetzelmann B, Evans R, Dietrich S. Phys Rev E 57:6785 (1998)

Dr. A. Lange, ITT, Uni Stuttgart, Pfaffenwaldring 9, 70569 Stuttgart, Germany
**lange@itt.uni-stuttgart.de**

# Preparation and Characterization of Curcurmin Loaded Polymeric Nanoparticles by Rapid Precipitation Method

Maulick Chopra, Sonal Mazumder*

*Department of Chemical Engineering, BITS Pilani, Pilani, Rajasthan – 333031, India.*
*maulick.chopra@pilani.bits-pilani.ac.in*
* Corresponding Author's E-mail: *sonal.mazumder@pilani.bits-pilani.ac.in;* **Tel: +91-1596-515725**

---

**Abstract:** Nanoparticles have provided a promising approach to obtain desirable bioavailability owing to their high surface to volume ratio. Curcurmin, is a low molecular weight natural polyphenol present in turmeric (Curcuma longa) with low intrinsic toxicity but a wide range of pharmacological activities, which include antitumor, antioxidant, anti-amyloid and anti inflammatory properties. Curcurmin is a potent inhibitor of NFB, a transcription factor implicated in the pathogenesis of several malignancies and also inhibits the productions of various cytokines including tumor necrosis factor and interleukin – 1, but the problem associated with curcurmin its poor water solubility and poor bioavailability poses considerable challenge in effective drug administration, often requiring higher doses at times. In the present study, nanoparticles made up of Curcurmin and polymer was produced by a rapid precipitation process for the first time for the treatment of colon cancer. The particles were processed by rotavap followed by freeze drying. Particle diameters as measured by dynamic light scattering were dependent on the synthesis method used. The particles prepared by the mixer showed well defined particle size ~ 130-500 nm before and after freeze drying respectively with lower polydispersity indices. Fourier-transform infrared spectroscopy showed chemical stability and intactness of entrapped drug in the nanoparticles. Differential scanning calorimetry showed that the drug was in amorphous state in the polymer matrix. Curcurmin drug loading was around 6 wt%. The release studies showed increase in solution concentration of drug from the nanoparticles compared to the pure drug.

***Keywords:*** Polymer Drug Nanoparticles; Curcurmin; Colon Cancer; Multi Inlet Vortex Mixer (MIVM)

# Rheological Study of Order-to-Disorder Transitions (ODTs) in Block-Copolymer/Small-Molecule-Additive Complexes

Rohit Kothari[1], H. Henning Winter[2], James J. Watkins[1]
[1]Polymer Science and Engineering
[2]Chemical Engineering
University of Massachusetts Amherst

Acknowledgment : Center for Hierarchical Manufacturing (CMMI-1025020)

## Microphase Separation in Block Copolymers

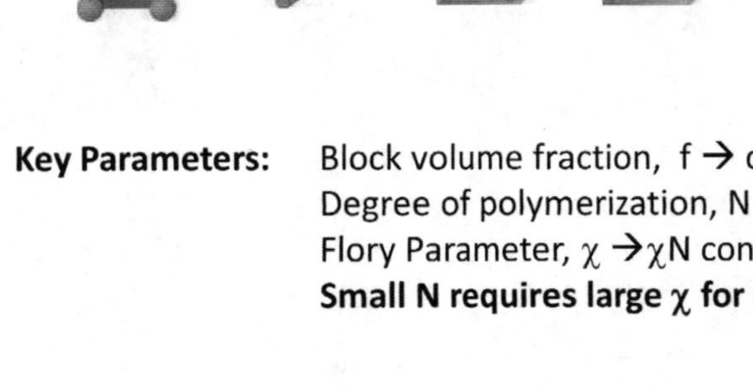


**A-B diblock copolymer Phase Diagram**

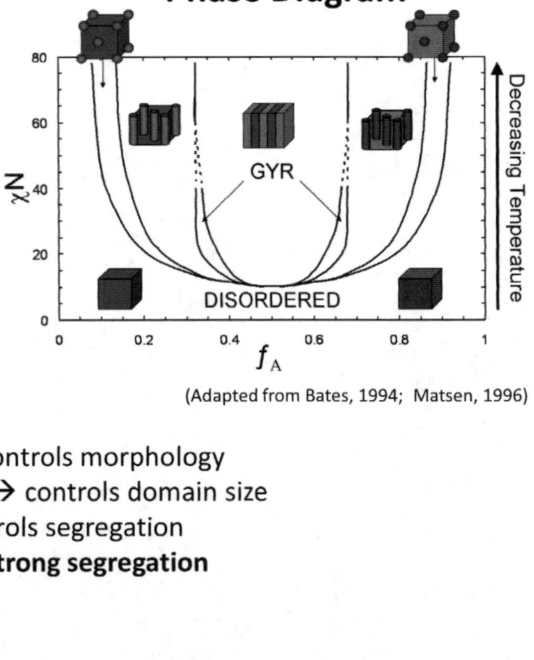


(Adapted from Bates, 1994; Matsen, 1996)

**Key Parameters:** Block volume fraction, f → controls morphology
Degree of polymerization, N → controls domain size
Flory Parameter, χ → χN controls segregation
**Small N requires large χ for strong segregation**

# Block Copolymer with Small Domain Size: Uses

- High-density magnetic recording devices
- Quantum confinements for light emission
- Nanowires
- Filters with nanometer pore

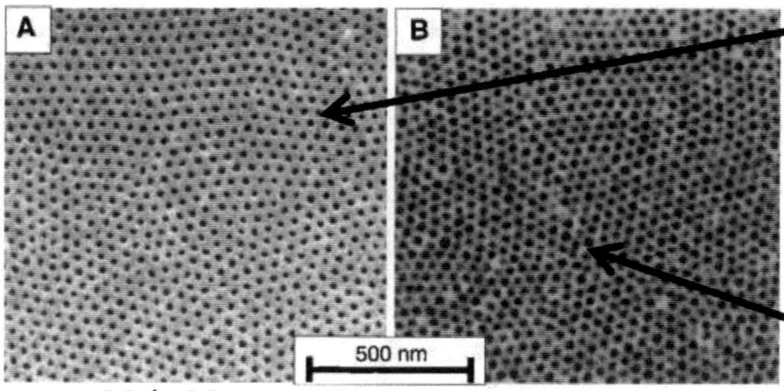


PS-b-PB
68k 12k

Distance between holes : 30 nm

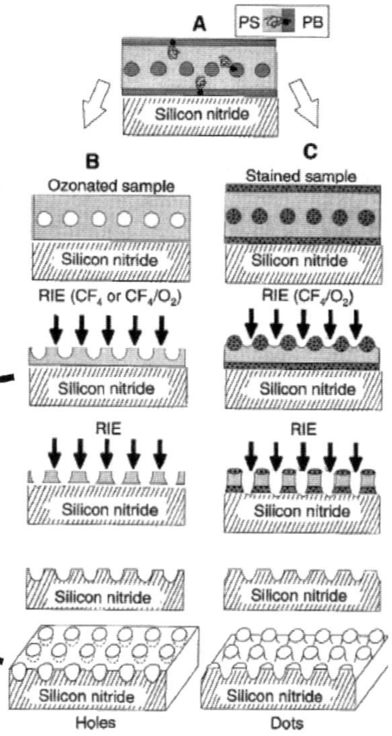


***Science* 1997**, 276, 1401-1404

# Additive Driven Assembly of Pluronic

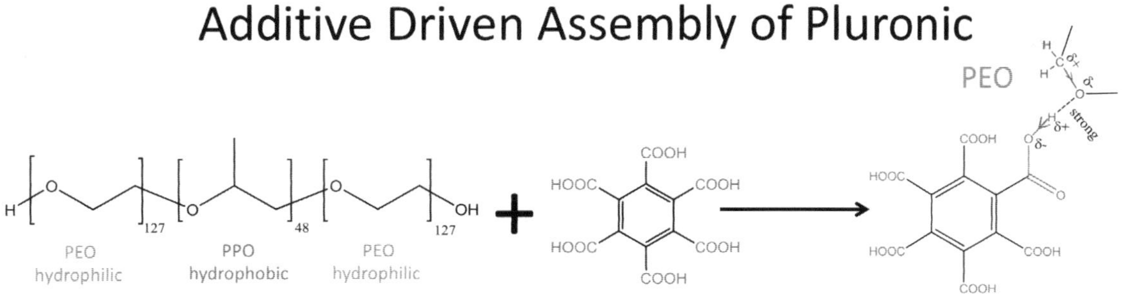


Pluronic® F108 (80 wt% PEO, $M_w$ ~ 14,000)

BHCA

H-bonding between BHCA and PEO

$\chi_{BHCA-PEO/PPO} >> \chi_{PEO-PPO}$

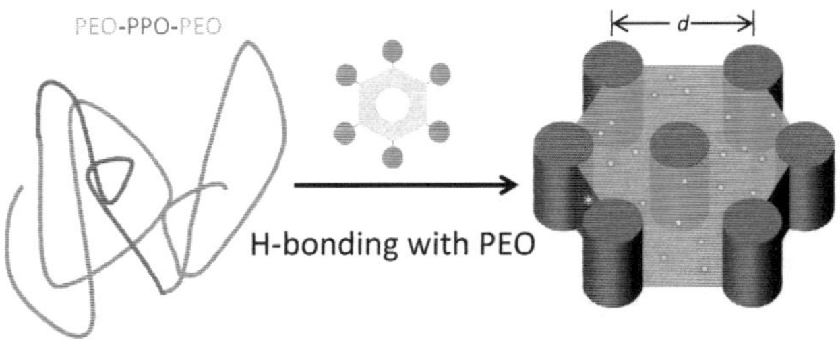


$d$ = 15 nm

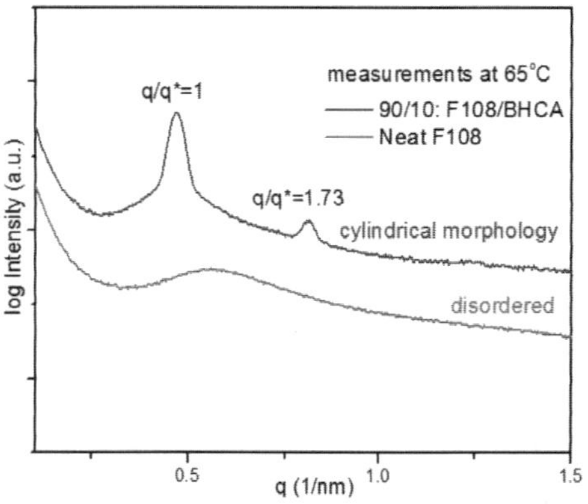

# Order-to-Disorder Transition Temperature($T_{ODT}$)

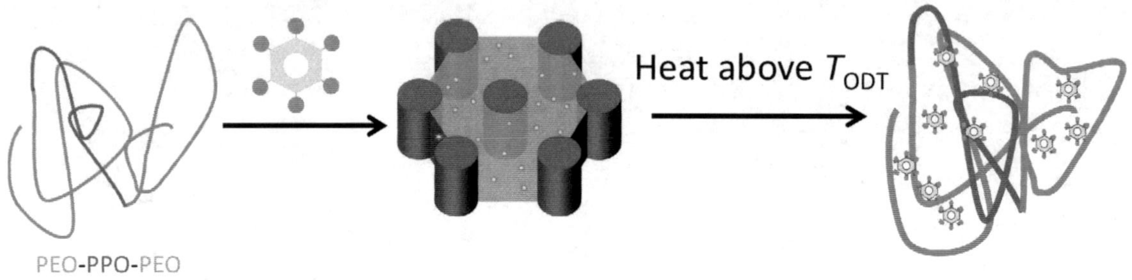


To explore dependence of $T_{ODT}$ on additive

- Amount of additive
- Number of functional groups on the additive
- Type of functional group on the additive

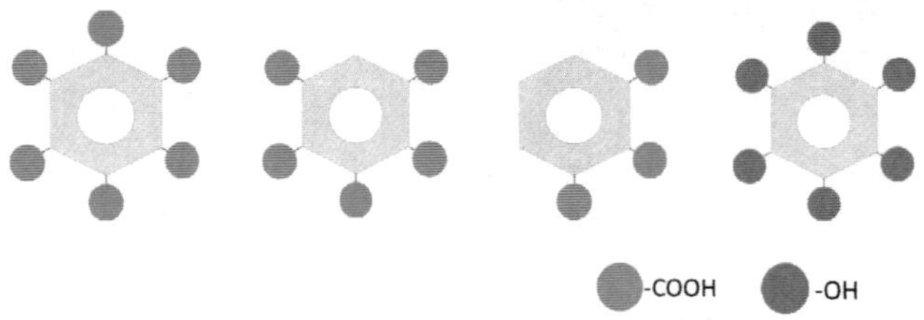


# Determination of $T_{ODT}$ using Rheology

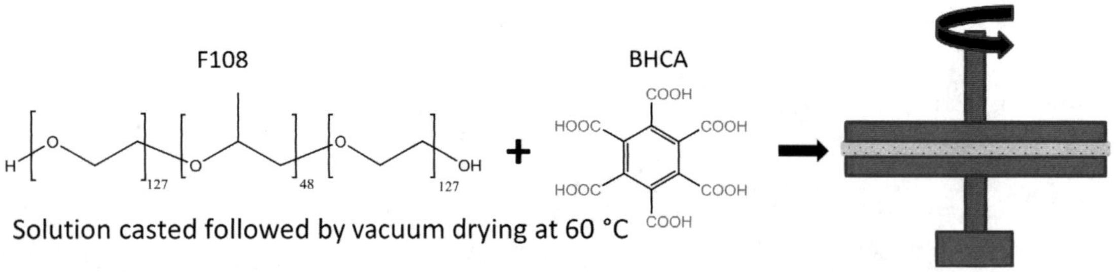


**90/10 F108/BHCA**

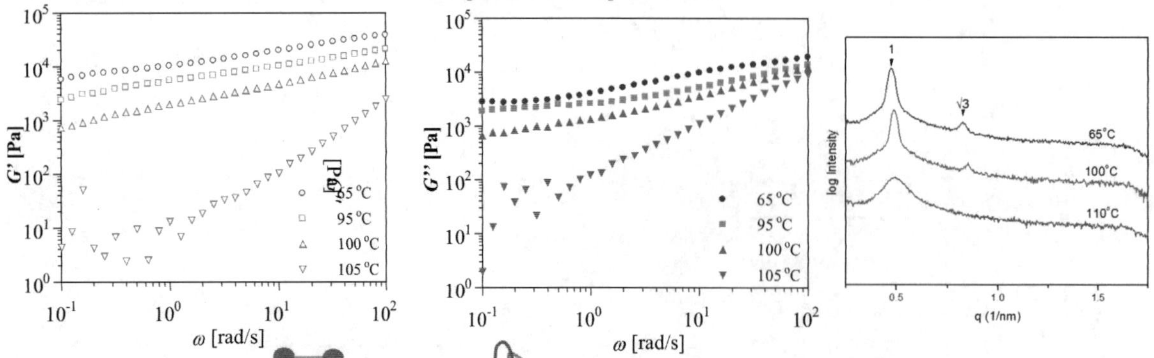


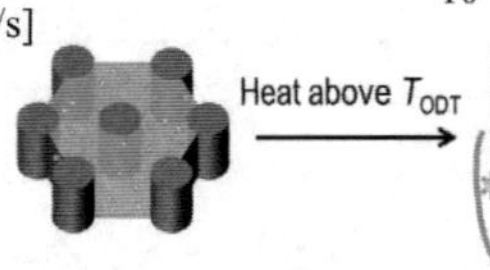


**$T_{ODT}$~100-105°C**

# Determination of $T_{ODT}$ using Rheology

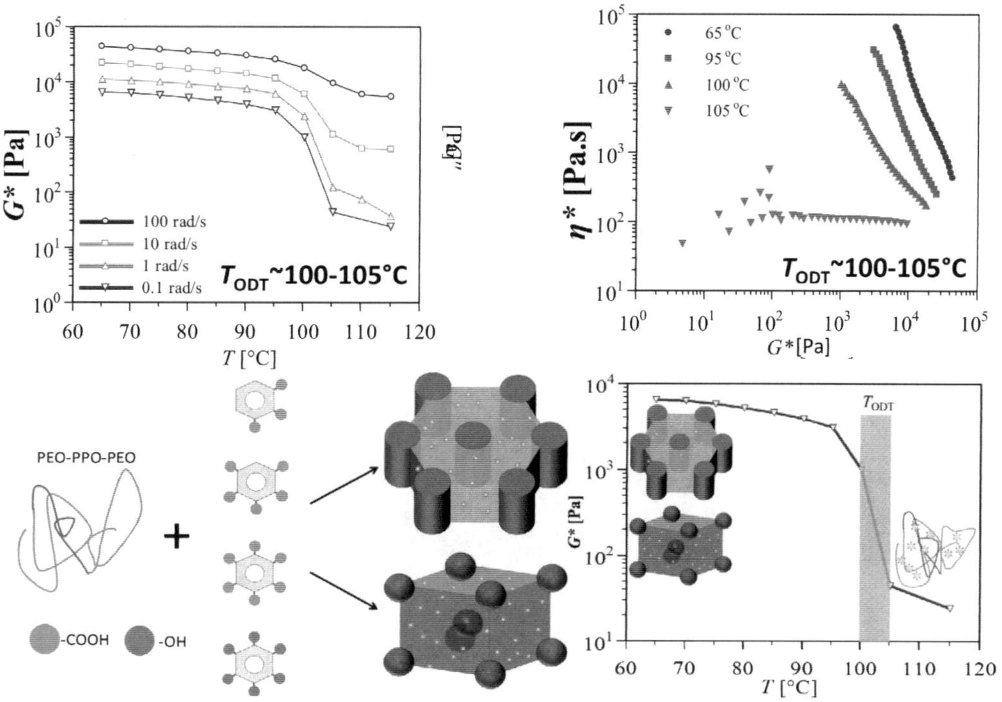


R. Kothari, et al., ***Macromolecules*** **2014**, **DOI:** 10.1021/ma501816d

# $T_{ODT}$ Dependence on Additive Loading

Additive:

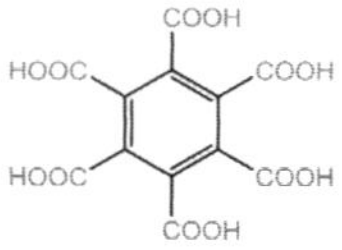


BHCA

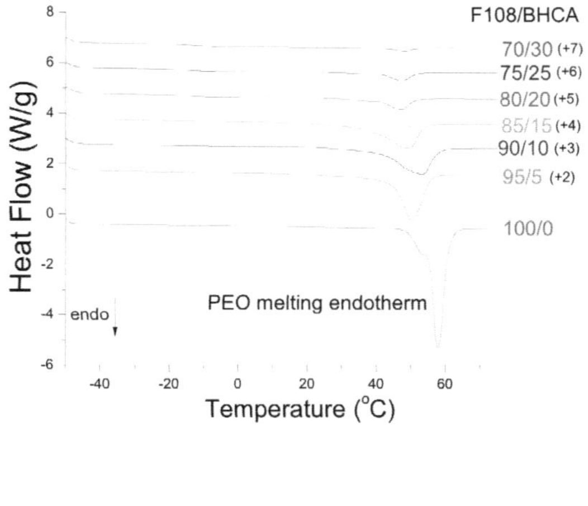


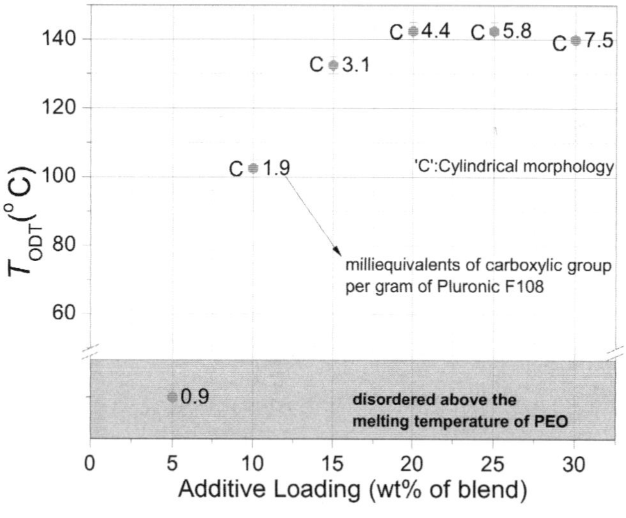


R. Kothari, et al., ***Macromolecules*** **2014**, **DOI:** 10.1021/ma501816d

## Crystallization of Additive at High Loading

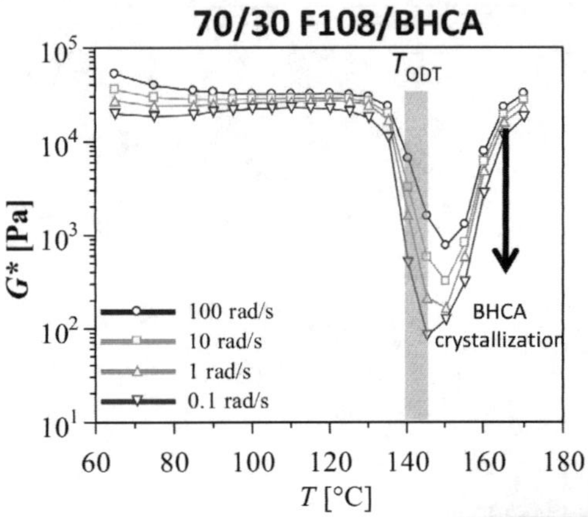


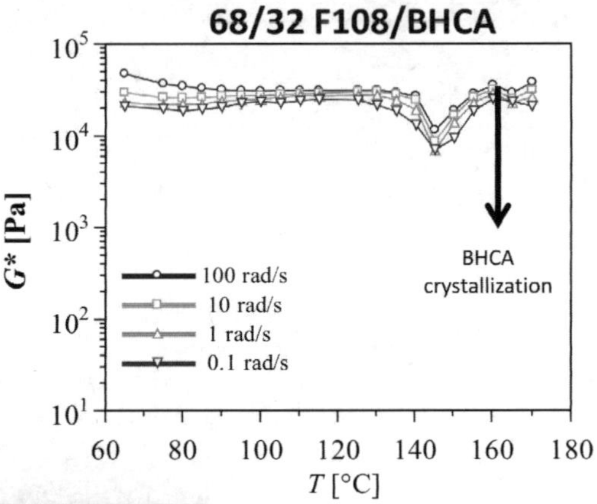


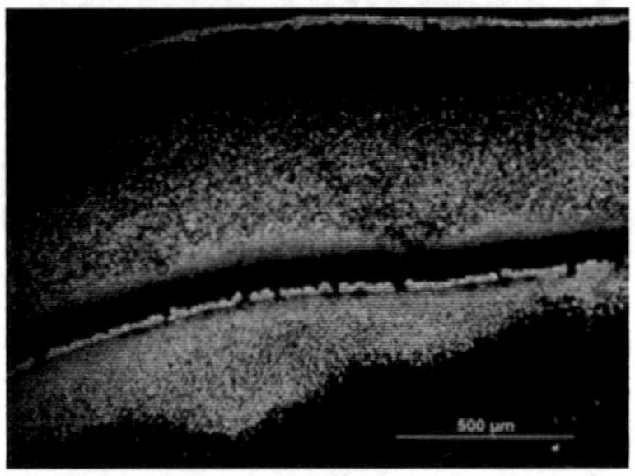


BHCA crystals under cross-polarizer

## Disorder-to-Order Transition at low Temperature: High Additive Loading Complexes

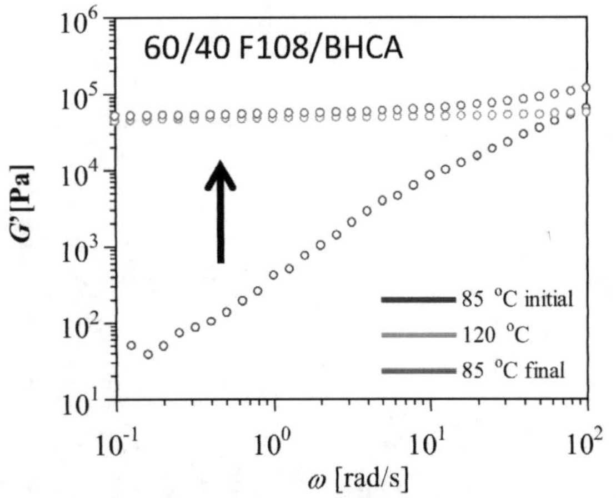


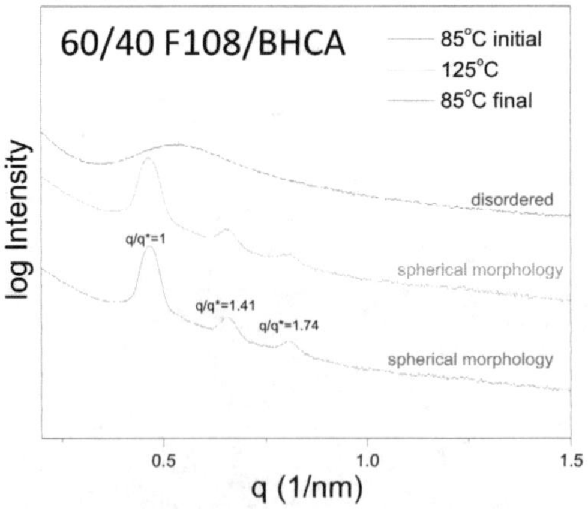


- Sudden increase in storage modulus above 115°C
- Irreversible transition

Disorder-to-Order transition occur upon heating, resulting in increase of modulus in low frequency region

# $T_{ODT}$ Dependence on The Number of Bonding Sites

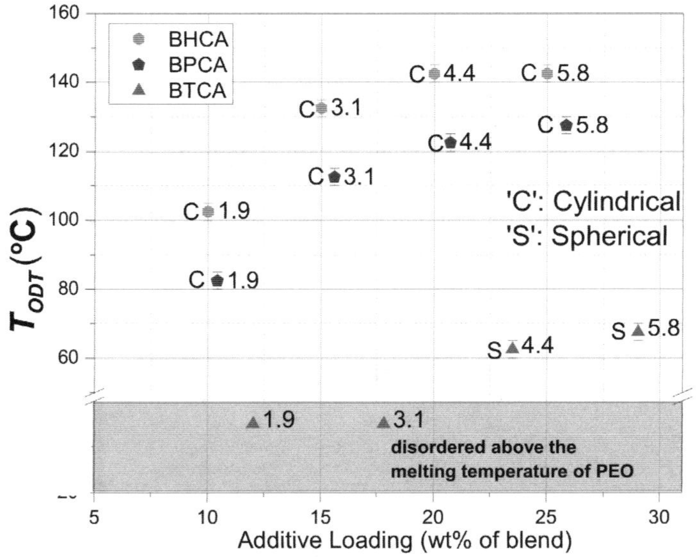


Volume added to incorporate **30** functional groups

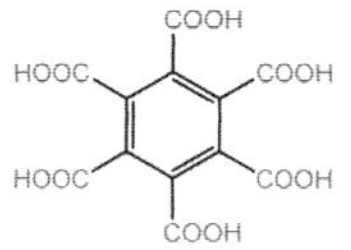


BHCA:5x164Å$^3$= 820 Å$^3$
Strongly Segregated:
Lower volume fraction
Low pKa (pKa$^1$=1.40)#

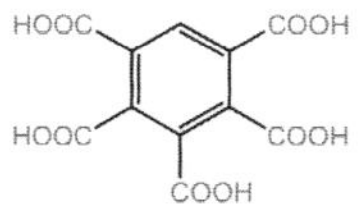


BPCA: 6x152Å$^3$= 912 Å$^3$
Volume added w.r.t. BHCA
912/820 = **1.1**

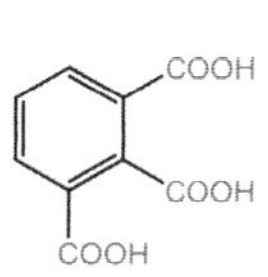


BTCA: 10x135Å$^3$= 1350 Å$^3$
Volume added w.r.t. BHCA
350/820 = **1.6**
Weakly Segregated:
Higher volume fraction
High pKa (pKa$^1$=2.88)*

**Precipitate out**

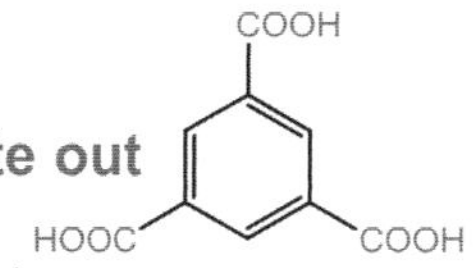


**Thermochimica Acta, Vol.414, Issue2, 263-270*
\# Brown, H.C., et al., in Baude, E.A. and Nachod, F.C., *Determination of Organic Structures by Physical Methods*, Academic Press, New York, 1955.

# $T_{ODT}$ Dependence on The Type of Functional Group

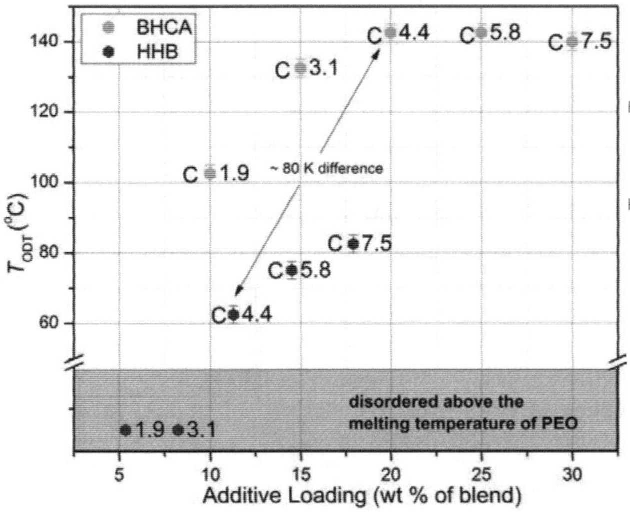


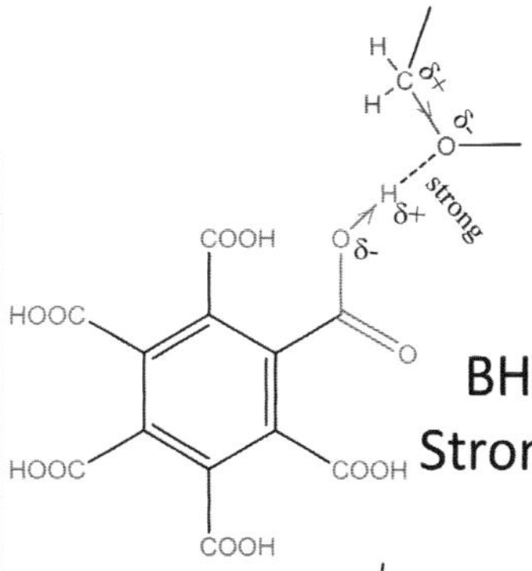


BHCA : 164 Å$^3$
Strong H-bonding

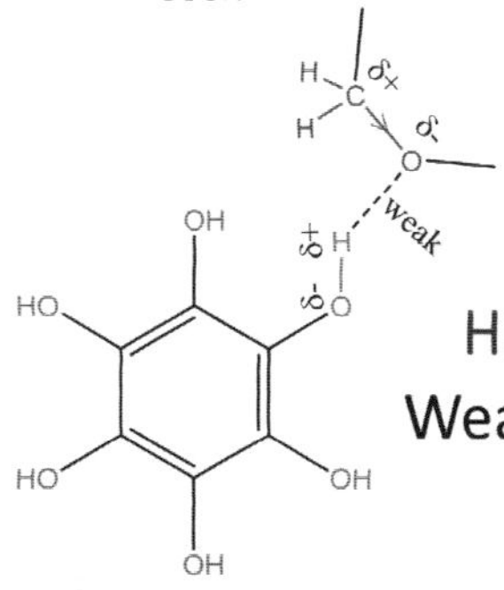


HHB : 88 Å$^3$
Weak H-bonding

R. Kothari, et al., ***Macromolecules* 2014**, **DOI:** 10.1021/ma501816d

# Disorder-to-order (DOT) transition dynamics

85/15: F108 (14.6K)/BHCA
Lattice parameter ~ 15nm
$T_{ODT}$: 100-105°C

styrene-isoprene-styrene (56K)
Lattice parameter ~ 28nm
$T_{ODT}$: 130-135°C

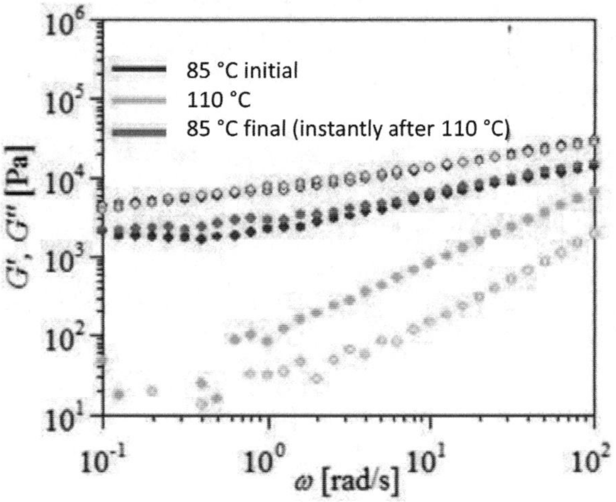


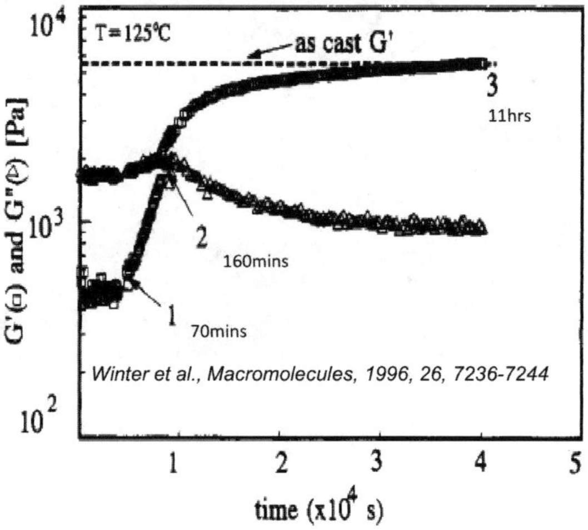


ODT: fast, DOT: fast

ODT: fast, DOT: slow

Pluronic® F108 /BHCA can be coated in disordered state (low viscosity) and quickly converted into ordered nanoscale template by quenching

# Summary

- Disordered Pluronic® F108 block copolymer can be converted into ordered nanoscale material by blending with hydrogen bond donating small molecule additives
- As additive loading increase $T_{ODT}$ initially increases followed by level-off
- Additive crystallized at high loading and high temperature
- DOT occur at very high additive loading
- Segregation strength decreases on reducing number of bonding sites on molecule
- Carboxyl functional additive induce much stronger segregation in Pluronic® compared to hydroxyl functional additive
- DOT in these Pluronic®/additive complexes is fast and comparable to ODT dynamics

# Analysis of Lipid Transfer Rates of Phospholipid Nanodiscs (Bicelles) Using Time-Resolved Differential Scanning Calorimetry and Small Angle Neutron Scattering

*Yan Xia*[1*], *Kamil Charubin*[1], *Ying Liu*[1], *Ming Li*[1], *Frederick A. Herberle*[2], *Drew Marquardt*[2], *John Katsaras*[2], *Mu-Ping Nieh*[1,3,4*]

[1]*Department of Chemical & Biomolecular Engineering, University of Connecticut, Storrs, CT, USA*

[2]*Neutron Sciences Directorate, Oak Ridge National Laboratory, Oak Ridge, TN, USA*

[3]*Polymer Program, Institute of Materials Science, University of Connecticut, Storrs, CT, USA*

[4]*Department of Biomedical Engineering, University of Connecticut, Storrs, CT, USA*

## Abstract

Phospholipid-based nanodiscs are versatile biological model membranes. Understanding the association-dissociation behavior of the lipids can facilitate their biomedical or biochemical applications. In this study, the lipid transfer rates of phospholipid nanodiscs (also known as "bicelles") made of zwitterionic long chain1,2-dihexadecanoyl-*sn*-3-phosphocholine (DPPC), charged long chain 1,2-dihexahecanoyl-*sn*-glycero-3-phospho-(1'-rac-glycerol) (DPPG), and zwitterionic short chain1,2-dihexanoyl-*sn*-glycero-3-phosphocholine (DHPC), were investigated in parallel using time-resolved differential scanning calorimetry (DSC) and small angle neutron scattering (SANS) at 20 ºC, 25ºC and 30 ºC. The kinetics of lipid transfer among nanodiscs can be expressed by a simple Arrhenius equation yielding an activation energy $E_a$ of 122 KJ/mol and 159 KJ/mol, through DSC and SANS methods, respectively. Both suggest an enthalpically unfavorable ($\Delta H > 0$) but entropically favorable ($\Delta S > 0$) lipid transfer process. The lipid transfer rate increases with increased compositions of short chain lipid DHPC. Moreover, the half-life of lipid transfer of liposomes made of DPPC and DPPG from our previous studies were found to be more than 10-fold of that of nanodisc.

# Mesoporous magnetic nano-assemblies for thermo-chemotherapy

Sunil Kumar[a], Amita Daverey[b c], and Dhirendra Bahadur[b]

*[a]Department of Chemical Engineering [b]Department of Metallurgical Engineering and Materials Science, Indian Institute of Technology Mumbai-400076, India. [c]Department of Chemical and Bimolecular Engineering, University of Nebraska, Lincoln, NE-68588.*

A PEG-diacid based polyol method was used for the synthesis of size tunable mesoporous MgFe2O4 magnetic nanoassemblies (MNAs). The PEG-diacid coated MNAs exhibit a significant specific surface area of 92 $m^2/g$ and saturation magnetization of 57emu/g (Mg+Fe). These MNAs exhibit a very good colloidal stability in PBS (pH 7.4) with non-appreciable cytotoxicity in mouse fibroblast (L929) and cervical cancer (HeLa) cells. We have demonstrated the potential application of MNAs as an integrated nanosystem for drug delivery and magnetic hyperthermia (MHT) through *in vitro* studies. The anti-cancerous drug doxorubicin (DOX) has been loaded with 80% efficiency due to highly negative surface charge and mesoporous nature of MNAs. It is observed that 65-70% of HeLa cells undergo apoptosis through DNA fragmentation after 24 h of incubation with DOX loaded MNAs. MHT alone induces 40-45% cells death, whereas synergistic effect of combination of DOX and MHT leads to about 90% cells death. Our results show that MHT significantly increase the therapeutic efficacy of DOX to induce more apoptosis in cancer cells. Hence, a combination of MHT with chemotherapy makes MNAs as a powerful multimodal system for synergistic cancer therapy.

***KEYWORDS:*** Mesoporous magnetic nanoassemblies (MNAs), HeLa, magnetic hyperthermia treatment (MHT), doxorubicin (DOX).

**Structure and Transport Property of Sodium Dodecyl Surfate Micelle in Water Phase: Molecular Dynamics Simulation Approach**

Byeong Jae Chun and Seung Soon Jang

Computational NanoBio Technology Laboratory, School of Materials Science and Engineering, Georgia Institute of Technology, 771 Ferst Drive NW, Atlanta, GA 30332-0245

We investigate the structure and transport property of micelle consists of sodium dodecyl surfate (SDS) in water phase using full atomistic molecular dynamics (MD) simulation. For this, first, SDS is geometry-optimized using B3LYP and 6-31G** and then Mulliken population analysis is performed for atomic charges. Second, a spherical self-assembled micelle is prepared by arranging SDS molecules to have a centro-symmetric inner structure and subsequently is located in water phase. By performing MD simulation, the structure and stability of micelle are characterized by 1) radius of gyration; 2) density distribution; 3) Connolly surface area; 4) distribution of sodium atoms; 5) formation energy. The simulated structure and energy will be compared to experimental ones in order to validate our simulation protocols as well as force field. Once the equilibrated SDS micelle is validated, the transport of various small atomic and molecular species through micelle is analyzed using potential of mean force (PMF) to characterize the permeability of SDS micelle in terms of free energy barrier. We believe this study provides a molecular-level understanding of micelle for its applications towards molecular carriers and molecular reactors.

# 383513 Linking the Electronic Character of Single-Wall Carbon Nanotubes to Their Biological Impacts Using a Model Aquatic Organism

**Abstract:**

Over the last decade, extensive research has been completed on the potential threat single-walled carbon nanotubes (SWCNTs) cause to living organisms upon release to aquatic systems. Most studies have focused on the effect of SWCNT length, diameter, and metallic impurities in traditional dose response assays. However, very little work has been done attempting to elucidate the biological response of test organisms based on the electronic structure or SWCNTs, i.e., metallic (m-) or semiconducting (s-) species, if any. A commonly used method to produce well separated fractions of these two SWCNT types is column-based, through interaction with agarose gels. Our previous work describing the driving forces at play during separation, and the source of selectivity of the process has allowed us to create high purity fractions of both m-, and s-SWCNTs appropriate for use in eco-toxicity. Within, we describe our methods for the separation, purification, and stabilization type separated SWCNTs to be used in traditional dose response studies. Additionally, we report significant differences to the biological response of the test organism *P. subcapitata* when exposed to these isolated fractions of m- and s-SWCNTs. Importantly, our results suggest that the observed toxicity of each species is influenced by a variety of factors including electronic type, surfactant concentration, and synthesis method. Finally, we propose that hydrogel separation of SWCNTs has potential for use as a simple bench scale tool to predict SWCNT toxicity in natural systems.

# 383533 Rheology of Nanodiamond Particles Aligned in DC Electric Fields

**Abstract:**

Carbon nanoparticles including carbon multiwall nanotubes and single wall nanotubes, fullerenes, and graphite are semiconducting forms of carbon, but unlike these other forms of carbon, nanodiamond is unique because it is a form of carbon that is highly insulating both electrically and thermally. The use of nanosize carbon solids as additives for changing the rheology of fluids and the rheological behavior of fluids with the application of electric fields has previously been shown. The ER effect for fluids containing micron-sized diamond is currently used in mechanical polishing operations. The rheological properties of diamond (<10nm) in silicone oil (PDMS) were analyzed using steady shear and oscillatory shear measurements. The apparent viscosity of the diamond/PDMS mixtures increased by an order of magnitude when the suspension is electrified at lower shear rates. The effect of electric field and concentration on diamond rheology was mainly targeted. The rate of shear thinning for the mixtures is high when the applied electric field is high.

# Structure, Dynamics and Rheology of Surfactant Micelle and Micelle-Nanoparticle Solutions from Molecular Dynamics Simulations

A. Sambasivam, S. Dhakal, and R. Sureshkumar
Department of Biomedical and Chemical Engineering
Department of Physics
Syracuse University, Syracuse, New York

Coarse grained molecular dynamics (MD) simulations [1-2], utilizing the LAMMPS MD simulation package, are used to study the equilibrium structure, shear-induced structure changes and rheology of cationic cylindrical micelle solutions. Systems that consist of up to a million coarse grained particles are simulated in the presence of explicit solvent and salt interactions. By varying the salt and surfactant concentrations, a phase diagram that consists of dilute, semi-dilute and entangled regimes is obtained. We also evaluate the micelle contours and orientation correlations from which micelle lengths, persistence length and shape anisotropy are calculated as functions of salt concentration. Addition of nanoparticles (NPs) to cationic wormlike micelles results in the formation of electrostatically stabilized micelle-NP junctions. The molecular mechanism and the kinetics of such junction formation will be elucidated. The effect of salt concentration and NP volume fraction on the structure and rheology of micelle-NP networks will be discussed. At sufficiently high salt concentrations, it is observed that branching in micelles results in a decrease in zero-shear viscosity of the solution. Non-equilibrium MD simulations are performed for micelle and micelle-NP solutions, showing shear thickening due to shear-induced structure (SIS) formation and shear thinning behavior depending on the salt and surfactant concentration. A shear-induced isotropic to nematic transition was observed for Weissenberg number $Wi$ O(1), where $Wi$ is defined as the product of shear rate and the intrinsic structure relaxation time of the system. For $Wi > 1$, micelles align in the flow direction. Mean square displacement of NPs in such solutions suggests that particles could undergo transient trapping due to entanglement constraints arising from the micelle structure.

1. A.V. Sangwai & R. Sureshkumar, Langmuir, 27, 6628 (2011)
2. A.V. Sangwai & R. Sureshkumar, Langmuir, 28, 1127 (2012)

# 390407 Equilibrium and Non-Equilibrium Self-Assembly of Nanostructured Materials

**Abstract:**

Self-assembly of nanoscopic components into higher-order architectures defines the forefront of fundamental nanoscience research and is important for the development of new materials with potential applications in optoelectronics, high-density data storage, catalysis, and biological sensing. In my talk, I will discuss how the peculiar nature of electrostatic and photoinduced dipole-dipole forces acting between nanoscale components can mediate their self-assembly into various superstructures and materials. I will show how the interactions underlying self-assembly can be studied and understood in quantitative detail, and how they can be tailored to synthesize unusual higher-order architectures: ionic-like crystals of nanoparticles, crystalline aggregates that can be assembled and disassembled by light, as well as extremely durable and yet very flexible metallic structures. Since these materials display a range of novel optical, electrical and mechanical properties, the discussion of experimental results will be accompanied by theoretical analyses combining elements of thermodynamics, statistical mechanics, electrodynamics and elasticity.

**References:**

**1.** A.M. Kalsin, M. Fialkowski, M. Paszewski, S.K. Smoukov, K. J.M. Bishop & B.A. Grzybowski Electrostatic self-assembly of binary nanoparticle crystals with a diamond lattice, *Science,* **312**, *420* (2006).

2. R. Klajn, K.J.M. Bishop, M. Fialkowski, M. Paszewski, C.J. Campbell, T.P. Gray & B.A. Grzybowski Plastic and moldable metals by self-assembly of sticky nanoparticle aggregates, *Science* 316, 261-264 (2007).

**3.** H.Nakanishi, K.J.M. Bishop, B.Kowalczyk, E.A. Weiss, A. Nitzan, K.V. Tretiakov, M.M. Apodaca, R. Klajn, J.F. Stoddart & B.A. Grzybowski Photoconductance and inverse photoconductance in films of functionalized metal nanoparticles *Nature* **460**, 371-375 (2009).

# Synthesis FeCo Nanoparticles By $NaBH_4$ Reduction in Oleylamine

Xia Xu[a], Jihoon Park[b], Yang-Ki Hong[b], and Alan Lane[a]
[a]*Department of Chemical & Biological Engineering, The University of Alabama*
[b]*Department of Electrical & Computer Engineering, The University of Alabama*

14AIChE
Annual Meeting, Atlanta, GA

## Background

**FeCo Nanoparticles Application**

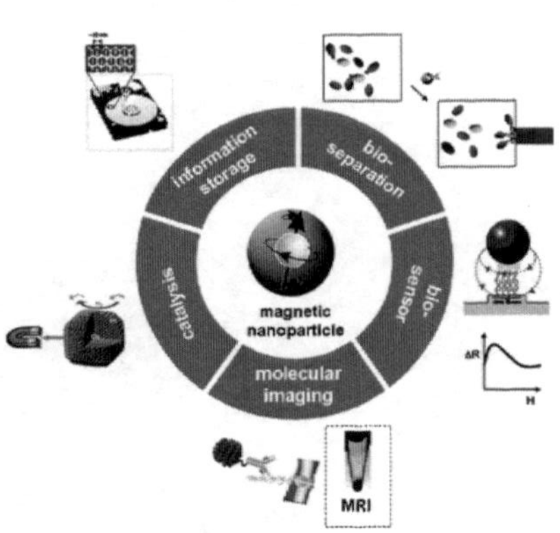


**FeCo Nanoparticles Chemical Synthesis**

- Method: $NaBH_4$ reduction in aqueous solution at room temperature.
  Drawback: amorous products; $Fe_xB$ and $Co_xB$ in products; products agglomeration.
  Glavee, G. N. et al. *Inorganic Chemistry*, **34**, 28-35 (1995).
- Method: Thermal decomposision of $Fe(CO)_5$ and $Co_2(CO)_8$.
  Drawback: Toxic precursors; environmental hazardous.
  Wang, C. et al. *Nano Research*, **2**, 380-385 (2009).
- Method: Reductive salt-matrix annealing of Fe-Co-O.
  Drawback: Great difficulty to separate prodcuts.
  Poudyal, N. et al. *Nanotechnology*, **24**, 345605 (2013).

## Experiments

**Chemical Synthesis**

$Fe(acac)_3$ | $Co(acac)_2$ | $NaBH_4$ | Oleylamine → Heating at 300 °C → Washing → Products

**Notes**

$Fe(acac)_3$:Ion(III) acetylacetonate.
$Co(acac)_2$: Cobalt(II) acetylacetonate.
(Organic precursors used to get rid of hydrate in the reaction).

Oleylamine: Solvent and surfactant.

$NaBH_4$: Reducing agent.

High temperature (300 °C): to crystalize the products.

## Acknowledge

This work was supported in part by the U.S. Department of Energy ARPA-E REACT Program under Award Number DE-AR0000189. This work utilized TEM and XRD owned and maintained by the Central Analytical Facility, which is supported by the University of Alabama.

THE UNIVERSITY OF ALABAMA

## Results

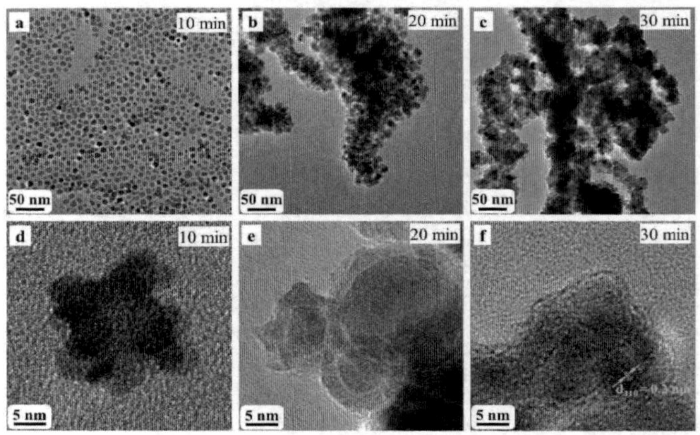


TEM images of as-synthesized FeCo nanoparticles: (a) and (d): 300 °C/ 10 min; (b) and (e): 300 °C/ 20 min; (e) and (f): 300 °C/ 30 min.

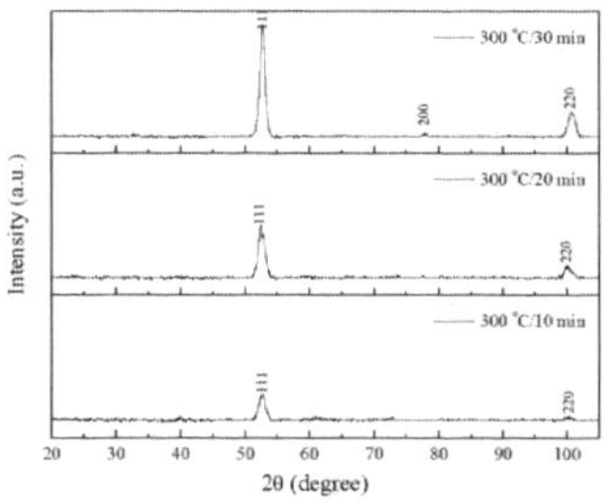


XRD patterns of as-synthesized FeCo nanoparticles.

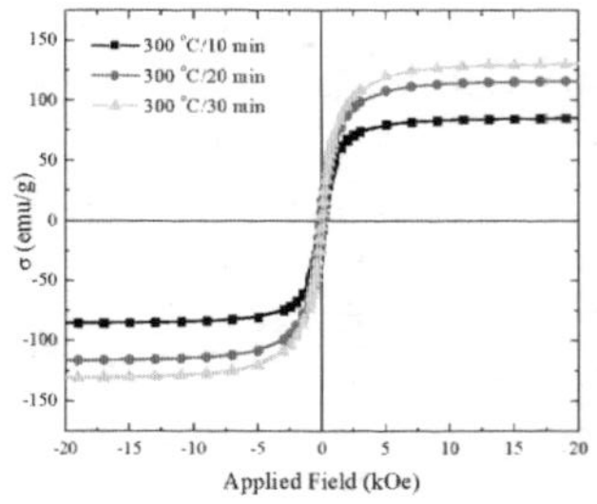


Magnetic hysteresis loops of as -synthesized FeCo nanoparticles.

## Conclusions

- FeCo nanoparticles were synthesized by $NaBH_4$ Reduction in Oleylamine at 300 °C.
- Longer reaction time leads to a larger particle size.
- Compared with previous methods, there are several advantages for this reported method: monodispersed products; better crystallization; no side products (such as $Fe_xB$ and $Co_xB$ ); environmental friendly.

The designer peptide RADA16-I (RADARADARADARADA) can form nanofiber matrices with important biomedical applications[1,5,6]. Peptide self-assembly into nanofibers occurs at physiological conditions after peptide is dissolved in an aqueous solvent (e.g., pH 5 phosphate buffer). Once fully self-assembled, the peptide/solvent solution takes the form of a hydrogel. The peptide concentrations necessary for matrix formation depend on pH and buffer concentration, but in practical applications the gel will be more than 99% $H_2O$ (less than 10 mg/mL RADA16-I). The resulting RADA16-I hydrogel has been used extensively in 3-dimensional cell culture[3], as well as animal models which have used the gel as a scaffold to promote the repair of neuronal tissue following a traumatic brain injury[2,4]. These studies have shown no detectable immune response due to RADA16-I.

Using high resolution magic angle spinning (HRMAS) nuclear magnetic resonance (NMR) spectroscopy, we have observed a signal from water-soluble RADA16-I peptide within a nanofiber system. The immediate results of these experiments have yielded a proton ($^1H$) spectrum showing several mobile species in solution. We consider this result significant, as it is the first time these mobile species have been observed. Mobile peptide may play a role in nanofiber self-healing, which is the spontaneous re-growth of nanofibers fragmented by mechanical shear forces[6]. Sonication has been used in the past in order to fragment the nanofiber matrix and allow for observation of this reassembly process[1]. However, the $^1H$ spectrum we have acquired does not appear to be significantly affected by sonication. This suggests the components observed in the $^1H$ spectrum may not be key components in the structure of the nanofiber matrix. Up to this point, no one has succeeded in identifying the self-assembly and self-healing mechanisms of RADA16-I nanofibers. We believe the possible identification of mobile species in the RADA16-I nanofiber matrix will shed new light on the processes involved and eventually allow for the development of an accurate mechanistic model of a RADA16-I nanofiber system.


1. Arosio, P.; Owczarz, M.; Wu, H.; Butte, A.; Morbidelli, M. End-to-End Self-Assembly of RADA 16-I Nanofibrils in Aqueous Solutions. *Biophys. J.* **2012**, 102, 1617-1626.

2. Ellis-Behnke, R. G.; Liang, Y. X.; You, S. W.; Tay, D. K. C.; Zhang, S. G.; So, K. F.; Schneider, G. E. Nano Neuro Knitting: Peptide Nanofiber Scaffold for Brain Repair and Axon Regeneration with Functional Return of Vision. *P. Natl. Acad. Sci. U. S. A.* **2006**, 103, 5054-5059.

3. Gelain, F.; Bottai, D.; Vescovi, A.; Zhang, S. Designer Self-Assembling Peptide Nanofiber Scaffolds for Adult Mouse Neural Stem Cell 3-Dimensional Cultures. *Plos One* **2006**, 1, e119.

4. Guo, J.; et al. Self-Assembling Peptide Nanofiber Scaffold Promotes the Reconstruction of Acutely Injured Brain. *Nanomedicine* **2009**, 5, 345-346-351.

5. Yang, Y.; Khoe, U.; Xiumei, W.; Horii, A.; Yokoi, H.; Zhang, S. Designer Self-Assembling Peptide Nanomaterials. *Nano Today* **2009**, 4, 193-194-210.

6. Yokoi, H.; Kinoshita, T.; Zhang, S. G. Dynamic Reassembly of Peptide RADA16 Nanofiber Scaffold. *P. Natl. Acad. Sci. USA* **2005**, 102, 8414-8419.

# THE IMPACTS OF Ph AND TRYPTOPHAN CONCENTRATION ON CONTROLLABLE SYNTHESIS OF GOLD NANOPARTICLES

*Leilei Dai, School of Chemical Engineering and Technology, Tianjin University, Tianjin, P. R. China*
*Fang Cai, Wuhan Petrochemical Engineering Co.,Ltd, Wuhan, P. R. China*
*Hao Lv, School of Chemical Engineering and Technology, Tianjin University, Tianjin, P. R. China*
*Jinsheng Sun*, School of Chemical Engineering and Technology, Tianjin University, Tianjin, P. R. China*
*Jie He, School of Chemical Engineering and Technology, Tianjin University, Tianjin, P. R. China*
*Xijia Cao, School of Chemical Engineering and Technology, Tianjin University, Tianjin, P. R. China*
**Corresponding author. Tel. /fax: +86 22 27891755*
*E-mail addresses: jssun2006@vip.163,com*

## Abstract:

This article describes an easy and efficient method to synthetize gold nanoparticles (GNPs) using tryptophan as reducing/stabling agent in aqueous solution. The important reaction parameters including pH and tryptophan concentration were investigated to examine their influences on the reaction kinetics and the size/shape of the resultant GNPs. Spectrophotometric, kinetic, and transmission electron microscopic (TEM) analyses for the harvested gold GNPs were carried out. Experimental researches find that strong acid or alkaline conditions ( pH$\leqslant$1.5 or pH $\geqslant$9.5) are unfavorable for the reaction process. The appropriate reaction pH falls between 2.5 and 6.0, with 4.5 the optimal pH condition for the formation of gold nanospheres with relatively high monodispersity. Additionally, tryptophan concentration also has effects on size and shape of the formed nanoparticles to a certain extent. With the increase of tryptophan concentration, the reaction rate speeds up and thus results in smaller and higher spherical nanoparticles.

**Keywords**: Nanoparticles; Synthesis; Kinetics

## Introduction

Over the past decade, nanomaterials have been the subject of enormous interest due to their potential applications in industrial, biomedical and electronic applications. [1-3] Gold nanoparticles or GNPs, in particular, are of strong interest, mainly due to their high chemical and thermal stability, non-cytotoxicity and biocompatibility.[4,5] They are found wide spreading applications in electronics, [6] medical diagnostics [7] and catalysis [8] because of their unique optical and physicochemical properties. [9]

It is well known that the size, shape and surface morphology play pivotal roles in controlling the physical, chemical, optical and electronic properties of these nanoscopic materials. [10.11] Thus an important aspect of nanotechnology concerns about the development of experimental processes for the synthesis of nanoparticles of different chemical compositions, sizes, shapes and controlled dispersity. Conventional synthetic methods of GNPs have involved a number of physical and chemical methods. As synthesis methods with abundant theoretical and experimental conclusions, chemical syntheses have already achieved considerable successes and developments towards perfect. While most of these methods developed so far always take additional reagents for reducing, capping and stabilizing in the GNPs synthesis, which inevitably increase the complexity of the synthesis and violate the recent presented principle for greener nano-synthesis. [12.13]

As a result, quite a few researchers turn their interests to find the meaningful and green synthesis approach to nanoparticles. Synthesis metal nanoparticles with biomolecules have recently attracted more and more attentions in the rapidly developing area of nanobiotechnology. Amino acids as the composition unit of protein, are considered as suitable and ideal agents in the bionic green synthesis of GNPs, as protective layers and for their assembly, due to the presence of different functional groups, such as–SH, –COOH and $–NH_2$, with affinity to gold. Additionally, differences in molecular structures among the amino acids, such as the type of functional group in the side chain and the number of electron-donating groups per molecule, could result in differences in reaction kinetics and hence different product size and shape. [14.15] Meanwhile, nanoaprticles of various sizes and shapes can be obtained by regulating the reaction conditions, such as pH, reducing or capping agent. For example, Huang et al. [16] prepared various sized and shaped Au–Ag nanocomposites in amino acid solutions by regulating the reaction pH. Lingtao Yu et al. [17] studied the direct growth of gold nanocrystals by peptide sequence, Asn-Pro-Ser-Ser-Leu-Phe-Arg-Tyr-Leu-Pro-Ser-Asp, finding that the pH value has a significant effect on the growth rate of Ag nanocrystals and hence resulted in different shapes. Yen Nee Tan et al. [18] studied the synthesis of gold nanostructures by aspartic acid (Asp) and investigated the effect of Asp and gold precursor concentration on the size and shape of nanoparticles. Additionally, Abu Rafey et al. [19] investigated the growth of Ag-nanoparticles using aspartic acid (Asp) in aqueous solutions. The silver nanoparticles are spherical, uniform particle size, and strongly depend on the Asp concentration.

The above mentioned encourages this present work to explore the synthesis of GNPs using tryptophan. The possibility to manipulate the size/shape of GNPs by altering key reaction conditions such as pH and tryptophan concentration was investigated. In addition, UV–visible spectrometer was used to monitor the reaction process. Moreover, the particles size and morphology is monitored by the transmission electron microscopy.

## Materials and Method

### *Materials*

Hydrogen tetrachloroauric trihydrate ($HAuCl_4 \cdot 3H_2O$, AR) and tryptophan ($C_{11}H_{12}N_2O_2$, AR) were purchased from Guangfu Reagent, China. Ultrapure Millipore water (18MΩ) was used as the solvent.

### *Synthesis and characterization of GNPs*

The GNPs were synthesized under different reaction pH and tryptophan concentrations. To investigated the influence of pH, 50mL of $5\times10^{-4}$ mol $L^{-1}$ aqueous solution of $HAuCl_4 \cdot 3H_2O$ was reduced by 50mL of $5\times10^{-4}$ mol $L^{-1}$ aqueous solution tryptophan at physiological temperature (37℃) with pH from 1.5 to 12. As for tryptophan concentrations, 50mL of $5\times10^{-4}$ mol $L^{-1}$ $HAuCl_4 \cdot 3H_2O$ aqueous solution was added to 50mL of tryptophan aqueous solution with concentrations varied as $2\times10^{-4}$ mol $L^{-1}$, $4\times10^{-4}$ mol $L^{-1}$, $6\times10^{-4}$ mol $L^{-1}$, $8\times10^{-4}$ mol $L^{-1}$, $10\times10^{-4}$ mol $L^{-1}$, respectively. Besides, UV-visible absorbance spectra were recorded on a Cary100 spectrophotometer with a scan rate of 600nm/min to monitor the optical properties of the resultant GNPs. Transmission electron microscopy (TEM) was performed using a JEM-2100F transmission electron microscope at an accelerating voltage of 200 kV to survey size and morphology of the formed nanoparticles.

## Results and Discussion

### *Effect of pH*

Tryptophan is a neutral, polar amino acid with an indole on the side chain (Fig. 1). Its pKa values are 2.46 (for the carboxylic group), 9.41(for the amine group). [20] Thus the isoelectric pH (the average of the two pKa values, which correspond to equilibria involving the neutral zwitterions) is 5.94. Fig. 1 shows the protonation equilibria of tryptophan, and symbols $AH_3^+$, $AH_2$ and $AH^-$ represent corresponding three tryptophan species in aqueous solutions. The curves giving the concentrations (mole fractions) of the different protonated species of tryptophan versus pH are shown in Fig. 2. The neutral zwitterionic state of tryptophan is stable in the pH range from about 3 to 9, the positive cationic state with pH $<$ 1, and the negative anionic state with pH $>$11. According to this dissociative diagram, the experimental pH points were selected as follows: 1.5, 2.5, 3.4 (initial mixed value), 4.5, 6.0, 9.5 and 12.

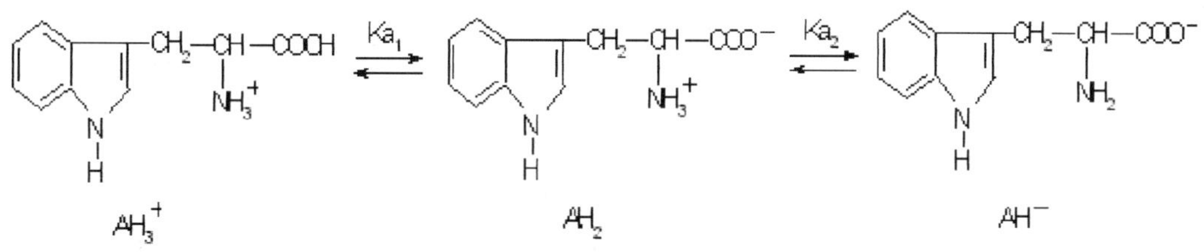


**Fig. 1.** The protonation equilibria of tryptophan (three species AH3+, AH2, AH-) in aqueous solutions.

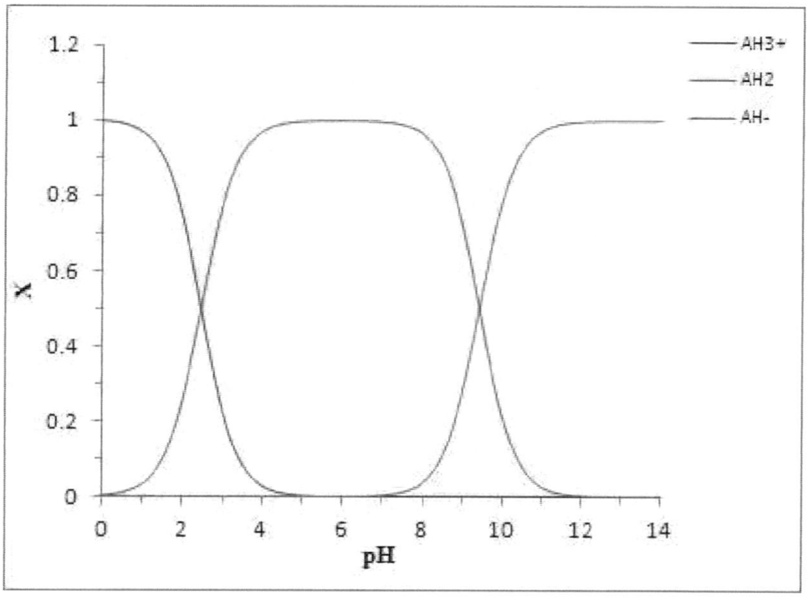


**Fig. 2.** Mole fractions(X) of tryptophan species: $[AH_3^+]$ (— ), $[AH_2]$ (— ), and $[AH^-]$ (— ) as a function of pH of aqueous solutions

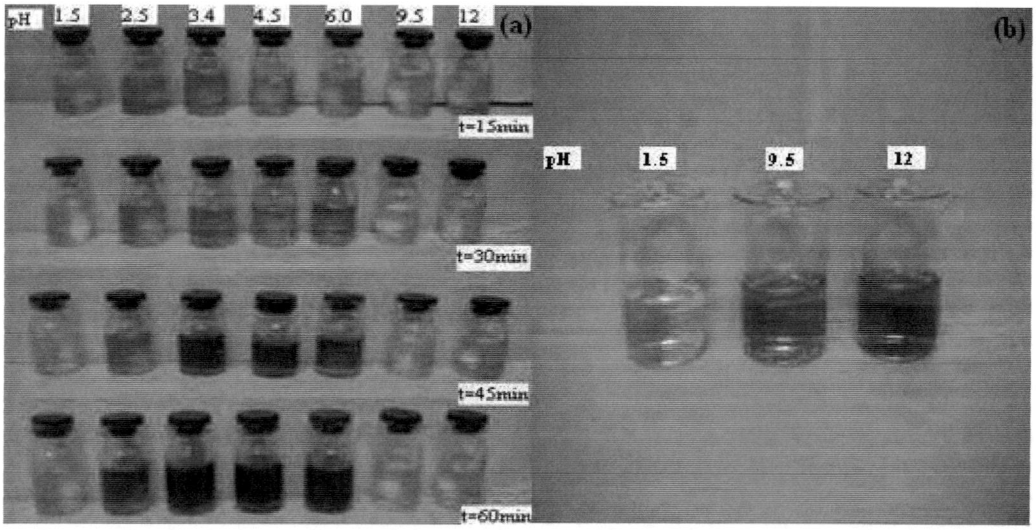


**Fig. 3.** Solution colors of the gold colloidal solution: **(a)** pH values (1.5, 2.5, 3.4, 6.0,

9.5, and 12) of time within 1h; **(b)** pH values (1.5, 9.5 and 12) of time after 32h

Fig. 3a shows the color changes of the gold colloidal solution at different pH values with reaction time ranging from 15 to 60 min. There were obviously color changes from wine red to atropurpureus within pH from 2.5 to 6, which indicated the formation of GNPs, while almost no color changes happened for pH values of 1.5, 9.5 and 12. In order to further determine whether the reaction happens or not, the reaction time was extended. The solution colors after reaction of 32h at pH values of 1.5, 9.5 and 12 were shown in Fig. 3b, which were faint yellow, purple, claret-red, respectively. Obviously, in the above three pH state points, the reaction was quite slow or didn't proceed at all, especially for pH 1.5 almost with no changes. It is believed that the lower pH leads to higher concentrations of protons, which compete with gold ions for reduction sites. Therefore, hinder the process of the reaction. It is preliminary explained that the appropriate reaction pH is between 2.5 and 6.0. Additionally, from the perspective of color changes with the growth of time, the reaction speed order is as follow: pH 2.5 < 6.0 < 4.5 < 3.4, which was also be confirmed by the UV–vis spectra recorded from the colloidal gold solutions as a function of time shown in Fig. 4. From Fig. 4a-c, it is obviously that the reaction proceeded tardily for pH 2.5, while relatively fast for pH 4.5 and 6. It is also inferred that tryptophan species $AH_2$ in aqueous solutions played an important role in the reaction. Additionally, as time goes on, the adsorption peaks are all bathochromic-shift, which may be due to the size increasing of the formed GNPs or some aggregation in the course of reduction. [21] The TEM images of the formed GNPs and corresponding size distribution histograms at different pH values are shown in Fig. 5 and Fig. 6, respectively. It can be seen that: pH=2.5, the particles appear like clusters of pine needles, which are seldom reported to the best of our knowledge; pH=4.5, almost all spherical particles with sizes concentrated at ~44 nm; pH=6, most of the particles with edges and comers, which can be attributed to their anisotropic growth, with sizes of ~45 nm. Hence pH=4.5 is believed to be an optimal pH condition to formed relatively high monodispersity and sphericity gold nanoparticles. Fig. 7 shows UV-visible adsorption spectra of the gold colloidal solution after reaction of 32h at pH 9.5 and 12. The peaks are relatively sharp and appear at wavelengths of about 537nm and 528nm, respectively, which indicated small size and distribution of the resultant nanoparticles. From Fig. 8 and Fig. 9, it is obviously that the particles are quite small with size between 4~16nm. Additionally, it is found that the sizes of GNPs under pH 12 are slightly smaller than that under pH 9.5, which is consistent with above UV-vis adsorption spectra analysis. It is reported that nanoparticles with small sizes are often required in the field of catalysis due to their higher activities. [22] For example, sub-5nm Au-Ag alloy nanoparticles showed good activity in catalysis of CO reduction at low temperature. [23] Therefore, the above formed pony-sized GNPs are believed to well exert their application.

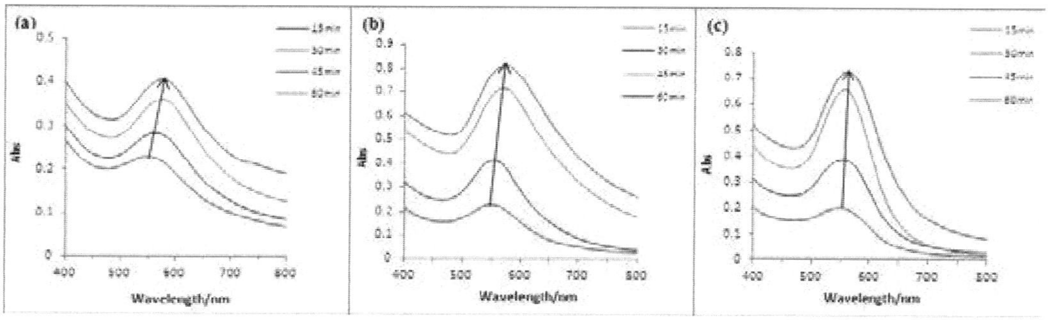


**Fig. 4.** Time-resolved UV-vis adsorption spectra of the gold colloidal solution synthesized at different pH values: **(a)** pH=2.5; **(b)** pH=4.5; **(c)** pH=6.0

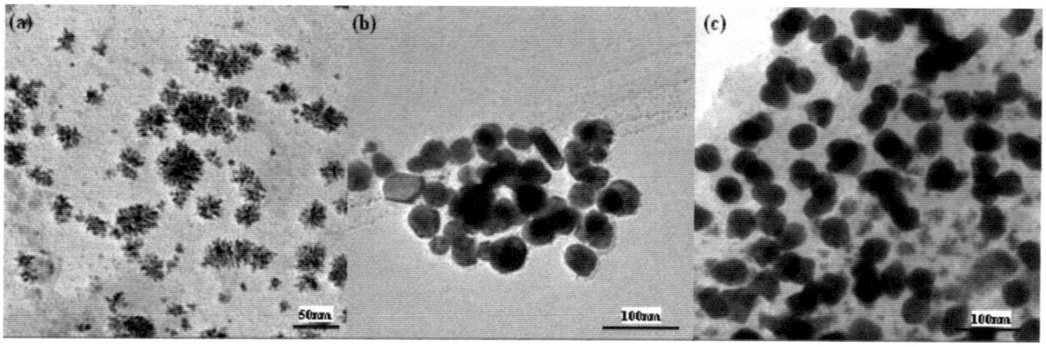


**Fig. 5.** Representative TEM images of GNPs synthesized at: **(a)** pH=2.5; **(b)** pH=4.5; **(c)** pH=6.0

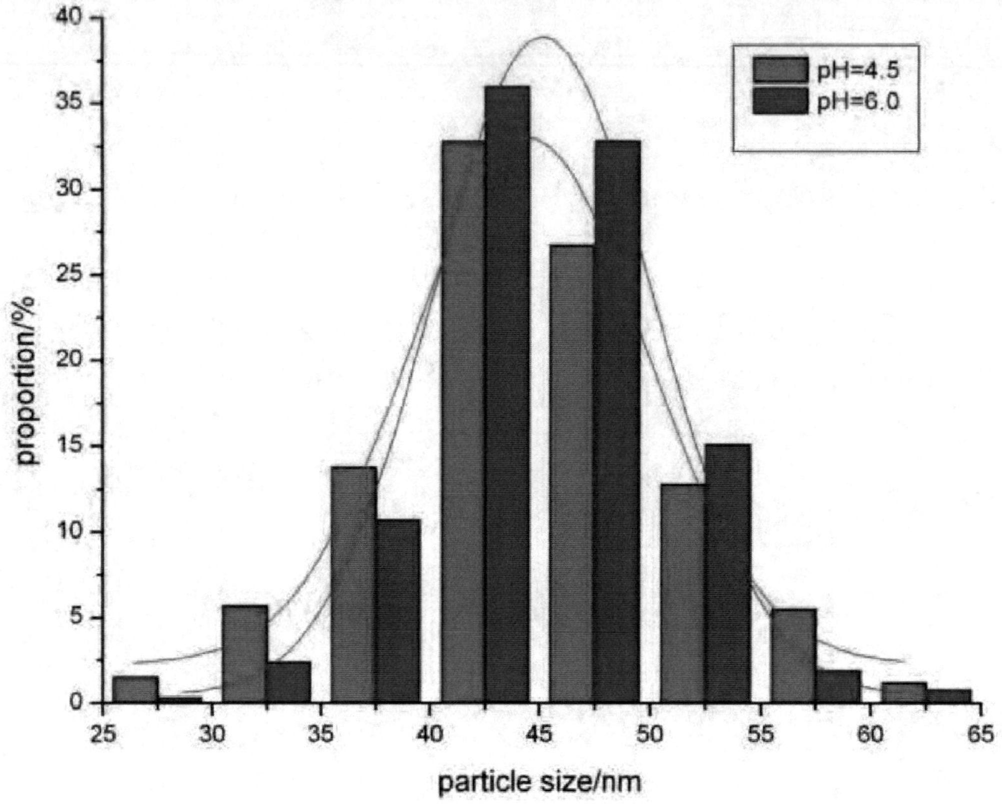


**Fig. 6.** Size distributions of GNPs synthesized at pH 4.5 and 6.0

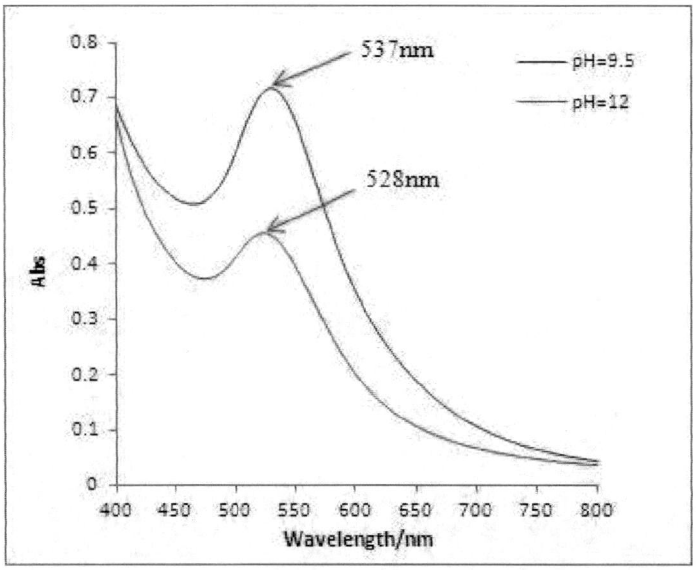


**Fig. 7.** UV-vis adsorption spectra of the gold colloidal solution at pH=9.5 and pH=12 after reaction of 32h

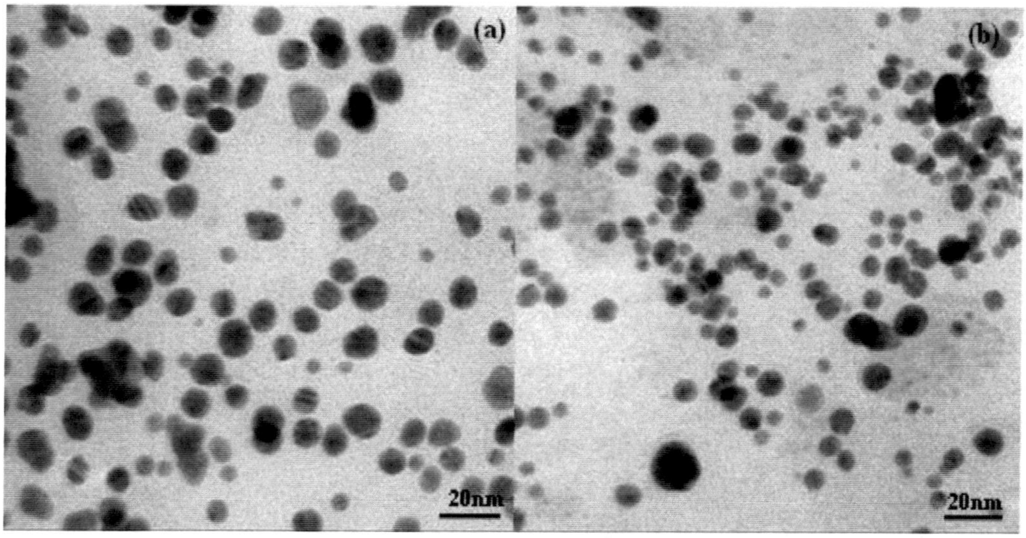


**Fig. 8.** Representative TEM images of GNPs synthesized at: **(a)** pH=9.5; **(b)** pH=12

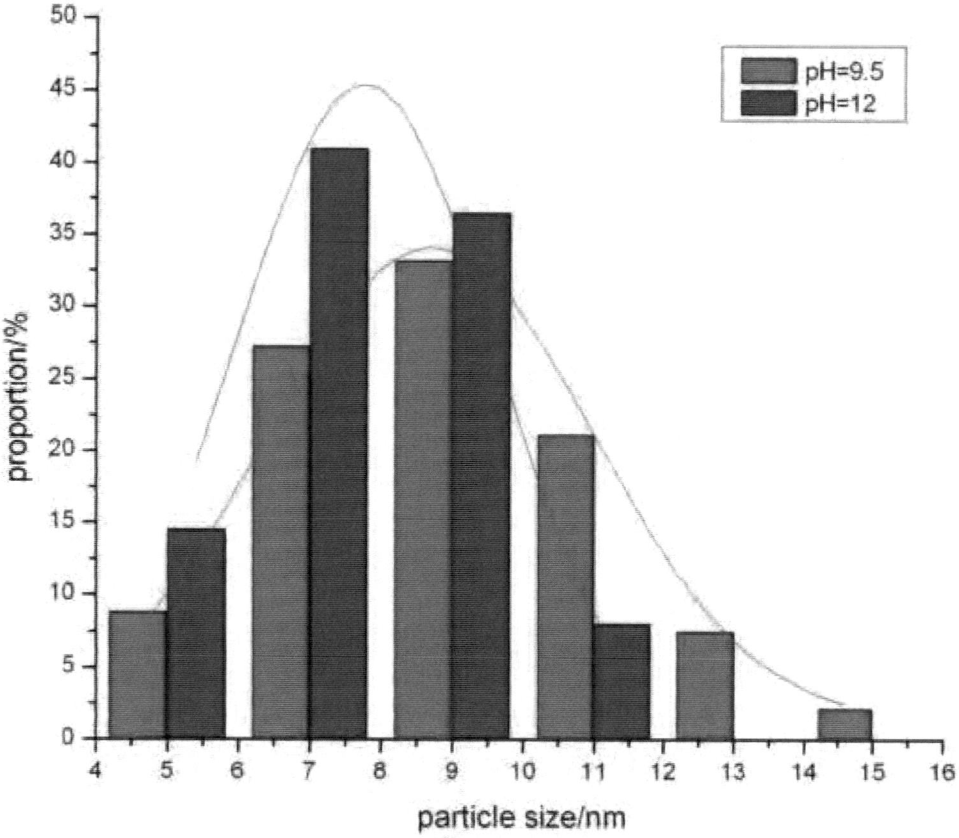


**Fig. 9.** Size distributions of GNPs synthesized at pH 9.5 and 12

***Effect of tryptophan concentration***

The concentration of reducing agent tryptophan is another considered factor in the synthesis of GNPs. The UV-vis adsorption spectra of gold aqueous colloidal solution synthesized at different concentration of tryptophan as a function of time is presented in Fig. 10. An increasing in the SPR intensity with time was observed, and with the increasing of tryptophan concentration the adsorption peaks become more and more strong and sharp, which indicate a fast formation of nanoparticles. Additionally, the adsorption peaks appear bathochromic shift or hypsochromic shift of varying degree, which indicates size increasing of the formed GNPs or some aggregation, or a change in the particle crystallinity or dissociation of large particles into smaller ones. [21,24] The corresponding time course plots of adsorption intensity at typical gold surface plasmon resonance frequency are shown in Fig. 11. As can be seen from the curves, when the concentration of tryptophan is $2\times10^{-4}$ mol $L^{-1}$, the reaction proceed considerably slows down. With the increase of tryptophan concentration, a sharp growth of reaction rate arises. However, when tryptophan concentration raises above $8\times10^{-4}$ mol $L^{-1}$, the reaction quickly achieved a nearly completion within 30min, and in the follow-up 10min the adsorption peaks practically unchanged, which indicated the completely reduction of Au(III). Therefore, the optimal concentration of tryptophan for the rapid and complete reduction of Au(III) would be $8\times10^{-4}$ mol $L^{-1}$. The TEM images and corresponding size distribution of GNPs synthesized at different concentration of tryptophan are shown in Fig. 12 and Fig. 13, respectively. It is obviously that: [tryptophan]=$2\times10^{-4}$ mol $L^{-1}$, only some tiny unshaped crystal nucleus; [tryptophan]= $4\times10^{-4}$ mol $L^{-1}$, particles with edges and comers with a wide size range; [tryptophan]= $6\times10^{-4}$ mol $L^{-1}$, well dispersion spherical particles with size concentrated at 45nm; [tryptophan]= $8\times10^{-4}$ mol $L^{-1}$, particles with decreased size ~42nm and relatively higher spherical and dispersity; [tryptophan]= $10\times10^{-4}$ mol $L^{-1}$, particles with further reduced size ~38nm. These observations can be explained as: when the reducing agent concentration is too low, it first leads to the retardatory and incomplete reaction, and then results in tiny unshaped crystal nuclei; with the increase of tryptophan concentration, the reaction proceeds relatively quick which produced atypical particles with edges and comers; when [tryptophan]$\geqslant6\times10^{-4}$ mol $L^{-1}$, the reaction proceeds so quickly and thoroughly that more and more smaller particles with high sphericity form. It was also reported that the average particle size of gold nanoparticles prepared with higher concentration of 3-thiopheneacetic acid was much smaller and the particles were isotropic in shape. [25]

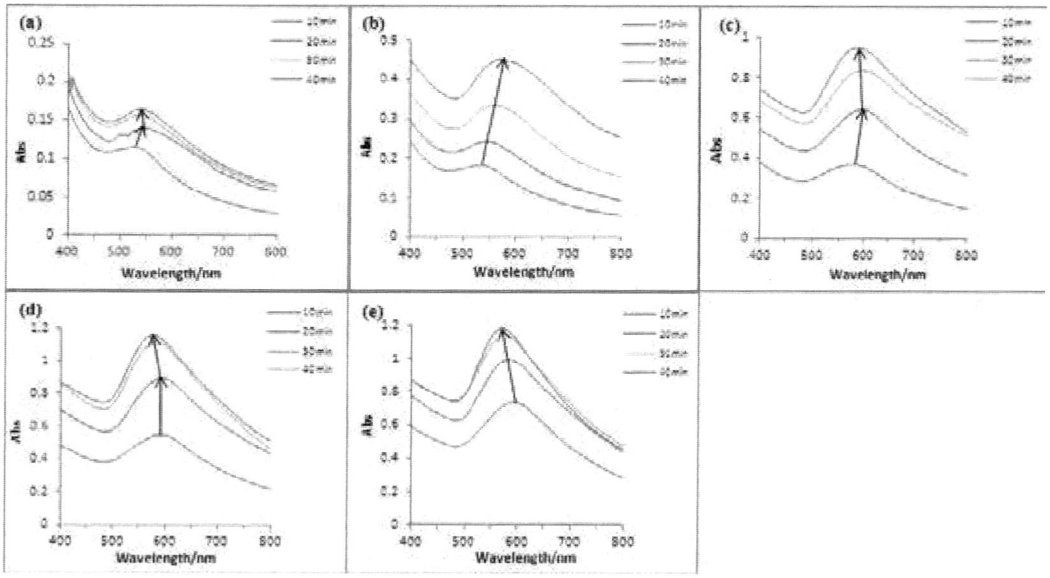


**Fig. 10.** Time-resolved UV-vis adsorption spectra of the gold colloidal solution synthesized at different concentration of tryptophan: **(a)** $2\times10^{-4}$ M; **(b)** $4\times10^{-4}$ M; **(c)** $6\times10^{-4}$ M; **(d)** $8\times10^{-4}$ M; **(e)** $10\times10^{-4}$ M

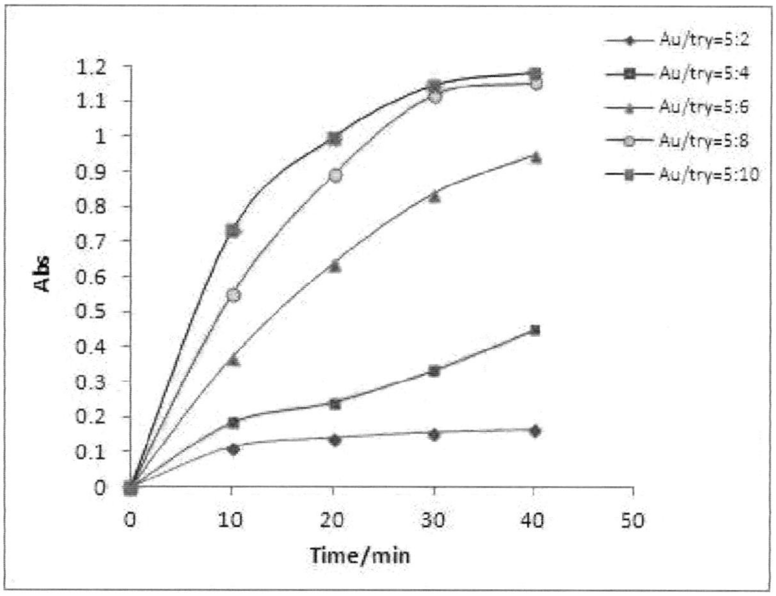


**Fig. 11.** Time course plots of adsorption intensity at typical gold surface plasmon resonance frequency

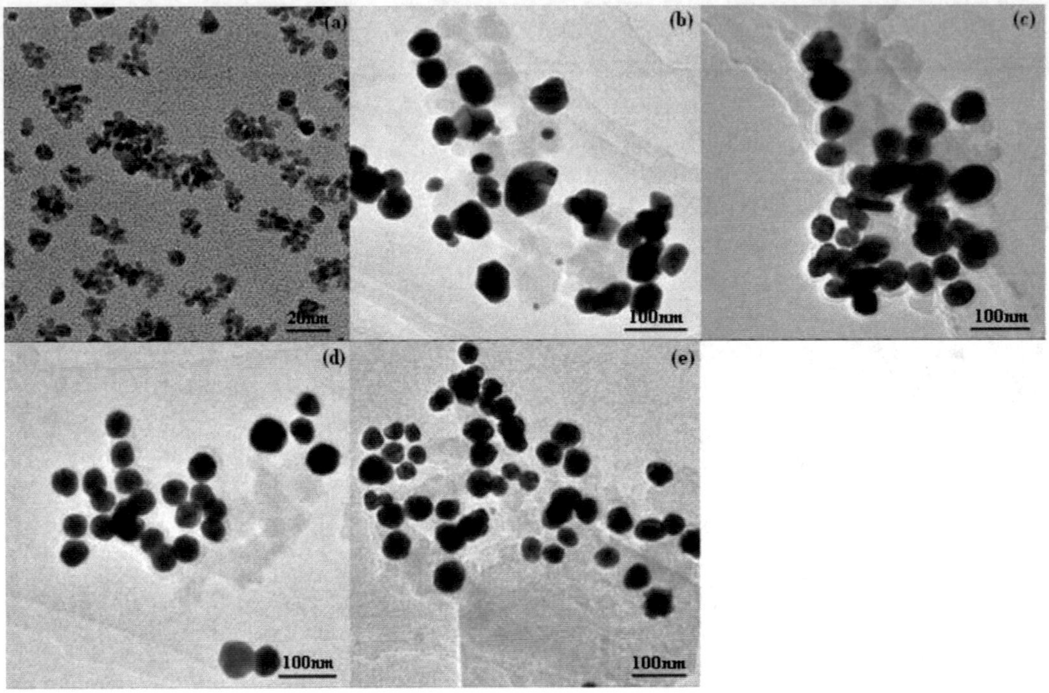


**Fig. 12.** Representative TEM images of GNPs synthesized at different concentration of tryptophan: **(a)** $2\times10^{-4}$ M; **(b)** $4\times10^{-4}$ M; **(c)** $6\times10^{-4}$ M; **(d)** $8\times10^{-4}$ M; **(e)** $10\times10^{-4}$ M

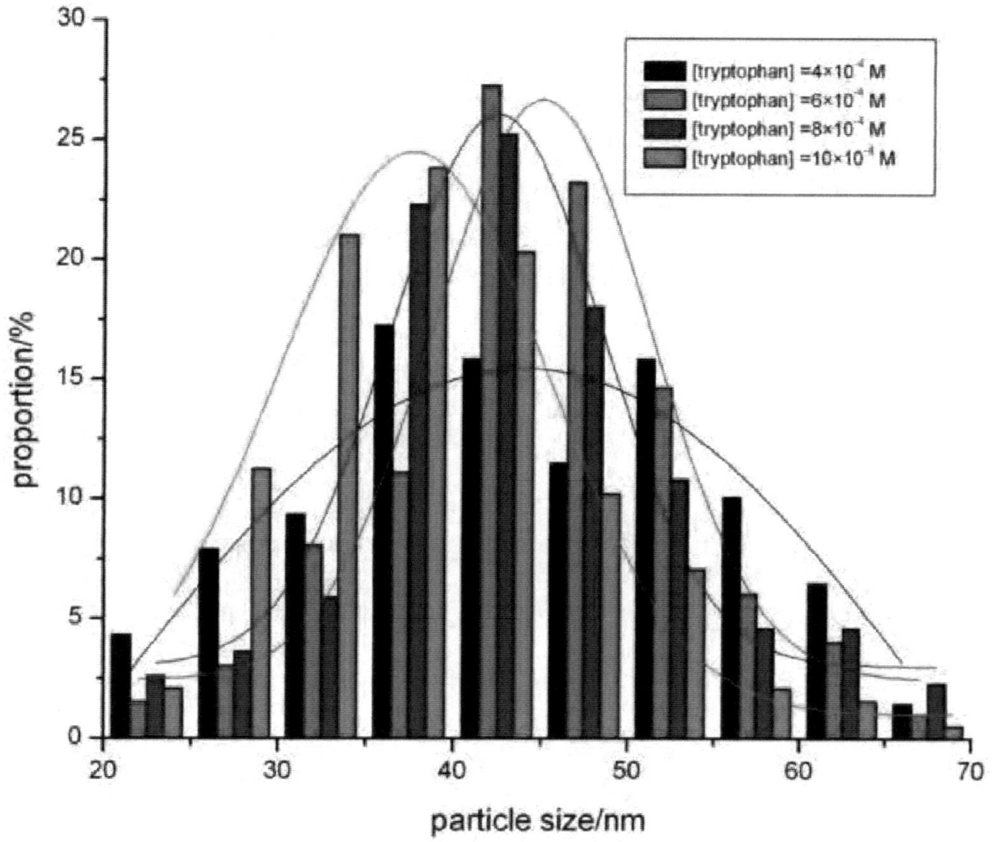


**Fig. 13.** Size distributions of GNPs synthesized at different concentration of tryptophan

## Conclusions

In this article, we successively demonstrated the synthesis of GNPs by the reduction of chloroaurate ions using tryptophan as reducing agent. The size and shape of the nanoparticles are affected not only by reaction pH but also tryptophan concentration. Strong acid or alkaline conditions( pH $\leqslant$ 1.5 or $\geqslant$ 9.5) that is unfavorable for the reaction, especially for pH 1.5. The relatively optimal pH point state for the formation of gold nanospheres with relatively high monodispersity is 4.5. Besides, at pH 2.5, pine needles liked GNPs were generated, which are seldom reported to the best of our knowledge. The tryptophan concentration can impact on the synthesis by increasing the reduction rate. The higher the concentration, the more rapidly and thoroughly the reaction, and thus result in smaller and high sphericity nanoparticles. And the optimal concentration of tryptophan for the rapid and complete reduction of Au(III) is $8\times10^{-4}$ mol $L^{-1}$. These investigations have a certain reference value in the subsequent synthesis to obtain well size/shaped GNPs.

## Acknowledgements

The authors greatly acknowledge the financial support from the National Natural Science Foundation, P.R. China (Project No. 21076155). And the program of Introduction of Talents of Discipline to Universities (B06006) is also greatly acknowledged.

***References***

1. C.W. Corti, R.J. Holliday and D.T. Thompson (2002), "Developing new industrial applications for gold: Gold nanotechnology," *Gold Bulletin* , pp. 111-117.
2. H.W. Liao, C.L. Nehl and J.H. Hafner (2006), "Biomedical applications of plasmon resonant metal nanoparticles," *Nanomedicine* , pp. 201-208.
3. X.L. Luo, A. Morrin, A.J. Killard and M.R. Smyth (2006), "Application of Nanoparticles in Electrochemical Sensors and Biosensors," *Electroanalysis*, pp. 319-326.
4. T. Jennings and G. Strouse (2007), "Past, present, and future of gold nanoparticles," *Adv Exp Med Biol*, pp. 34-47.
5. R. Shukla,V. Bansal, M. Chaudhary, A. Basu, R. R. Bhonde and M. Sastry (2005), "Biocompatibility of Gold Nanoparticles and Their Endocytotic Fate Inside the Cellular Compartment: A Microscopic Overview," *Langmuir*, pp. 10644-10654.
6. Y.L. Wu, Y. N. Li, P. Liu, S. Gardner and B.S. Ong (2006), "Studies of gold nanoparticles as precursors to printed conductive features for thin-film transistors," *Chem Mater*, pp. 4627–4632.
7. X.H. Huang, P.K. Jain, I.H. EI-Sayed and M.A. EI-Sayed (2007), "Gold nanoparticles: interesting optical properties and recent applications in cancer diagnostic and therapy," *Nanomedicine*, pp. 681–693.
8. A.S.K. Hashmi and G.J. Hutchings (2006), "Gold catalysis," *Angew Chemie–Int Edn*, pp. 7896–7936.
9. M.C. Daniel and D. Astruc (2004), "Gold nanoparticles: assembly, supramolecular chemistry, quantum-size-related properties, and applications toward biology, catalysis, and nanotechnology," *Chem Rev* , pp. 293–346.
10. A.P. Alivisatos (1996), "Semiconductor clusters, nanocrystals, and quantum dots," *Science*, pp. 933-937.
11. P.V. Kamat (2002), "Photophysical, photochemical and photocatalytic aspects of metal nanoparticles," *J Phys Chem B*, pp. 7729–7744.
12. J. Zeng, Y.Q. Zheng, M. Rycenga, J. Tao, Z.Y. Li, Q. Zhang, Y.M. Zhu and Y.N.Xia (2010), "Controlling the Shapes of Silver Nanocrystals with Different Capping Agents," *J. Am. Chem. Soc.*, pp. 8552–8553.
13. J. A. Dahl, B. L. S. Maddux and J.E. Hutchison (2007), "Toward Greener

Nanosynthesis," *Chem. Rev*, pp. 2228–2269.

14. Y. Zhou, W. Chen, H. Itoh, K. Naka, Q. Ni, H. Yamane and Y. Chujo (2001), "Preparation of a novel core–shell nanostructured gold colloid–silk fibroin bioconjugate by the protein in situ redox technique at room temperature," *Chem. Commun*, pp. 2518-2519.
15. Y.N. Tan, J.Y. Lee and D.I.C. Wang (2008), "Aspartic acid synthesis of crystalline gold nanoplates, nanoribbons, and nanowires in aqueous solution," *J Phys Chem C*, pp. 5463-5470.
16. Y. Huang, Y. Lin and H. Chang (2006), "Growth of various Au–Ag nanocomposites from gold seeds in amino acid solutions," *Nanotechnology*, pp. 4885-4894.
17. L.T. Yu, I.A. Banerjee and H. Matsui (2003), "Direct growth of shape-controlled nanocrystals on nanotubes via biological recognition," *J. Am. Chem. Soc*, pp. 14837-14840.
18. Y.N. Tan, J.Y. Lee and D.I.C. Wang (2008), "Aspartic Acid Synthesis of Crystalline Gold Nanoplates, Nanoribbons, and Nanowires in Aqueous Solutions," *J. Phys. Chem. C*, pp. 5463-5470.
19. A. Rafey, K. B. L. Shrivastavaa, S.A. Iqbal and Z. Khan (2011), "Growth of Ag-nanoparticles using aspartic acid in aqueous solutions," *Journal of Colloid and Interface Science*, pp. 190–195.
20. D.R. Lide (2003), "Handbook of Chemistry and Physics," 84th ed, , CRC Press, Boca Raton.
21. L. Polavarapu and Q.H. Xu (2008), "A single-step synthesis of gold nanochains using an amino acid as a capping agent and characterization of their optical properties," *Nanotechnology*, pp. 075601-075606.
22. F. Raimondi, G. G. Scherer, R. Kötz and A. Wokaun (2005), "Nanoparticles in Energy Technology: Examples from Electrochemistry and Catalysis," *Angew. Chem., Int. Ed*, pp. 2190–2209.
23. C.-W. Yen, M.-L. Lin, A. Wang, S.-A. Chen, J.-M. Chen and C.-Y. Mou (2009), "CO Oxidation Catalyzed by Au–Ag Bimetallic Nanoparticles Supported in Mesoporous Silica," *J. Phys. Chem. C*, pp. 17831–17839.
24. R. Sardar and J.S. Shumaker-Parry (2011), "Spectroscopic and microscopic investigation of gold nanoparticle formation: ligand and temperature effects on rate and particle size," *J Am Chem Soc*, pp. 8179–8190.
25. H.Z. Huang and X.R. Yang (2005), "One-step, shape control synthesis of gold nanoparticles stabilized by 3-thiopheneacetic acid," *Colloids and Surfaces A: Physicochem. Eng. Aspects*, pp. 11-17.

# 371160 Purification of DNA-Wrapped Carbon Nanotubes in Polymer Aqueous Two-Phase Systems

**Abstract:**

Specific short single-stranded DNA sequences have been identified to show recognition abilities for specific chiralities of SWCNTs from their synthetic mixture. In addition, polymer aqueous two-phase (ATP) extraction has been demonstrated recently as an efficient and robust method for the purification of SWCNTs dispersed by both surfactants and DNA. Specifically, DNA-SWCNT hybrids showed sequence-dependent partition in ATP systems, which enabled isolation of many small-diameter SWCNT species at high purity. In this work, we report a systematic study of designing DNA sequence patterns within 40-mers by mutation of previously identified recognition sequences for purifying specific species of SWCNTs with improved yield and purity.

# 376993 A Biofabrication Approach for Controlled Synthesis of Silver Nanoparticles with High Catalytic and Antibacterial Activities

**Abstract:**

Simple, facile and size-controllable synthesis of uniform Ag nanoparticles with tobacco mosaic virus (TMV) as a biomediator was developed in the absence of external reducing agents. Ag nanoparticles with average diameter of 2, 4 and 9 nm were obtained by simply tuning the ratio of TMV/$Ag(NH_3)_2^+$. Autocatalytic growth followed by coalescence was observed from Ag formation in the presence of TMV. The as-prepared TMV-mediated Ag nanoparticles show substantially higher catalytic and antibacterial activities than previous results. For the 4-nitrophenol hydrogenation reaction, the rate constants per surface area for 2 and 9 nm Ag nanoparticles were determined to be 0.64 and 1.2 L $m^{-2}$ $s^{-1}$ respectively. Both Kirby–Bauer disk diffusion test and tube culture results demonstrate high antibacterial activity of TMV-mediated Ag particles against *Escherichia coli*, with minimal inhibition concentration (MIC) of 2.3 and 2.5 ppm for 2 and 9 nm Ag nanoparticles respectively. We expect that our biomediated Ag synthesis approach can be readily extended to other biomaterials and metal nanoparticle systems.

# 377176 Low Temperature Catalysts for Direct Remediation of H2S Malodor in Air

**Abstract:**

**Low temperature catalysts for direct remediation of $H_2S$ malodor in air**

**Gabriel Kei Bo Cheung[1], Hao Chen[1,3], Wei Han[1], King Lun Yeung[1,2,*]**

*[1]Department of Chemical and Biomolecular Engineering, [2]Division of Environment, [3]Nano Science and Technology Program, The Hong Kong University of Science and Technology, Clear Water Bay, Kowloon, Hong Kong, P.R. China*

*Corresponding author: kekyeung@ust.hk

Division: 22 Nanoscale Science and Engineering Forum

Subdivision: 22001 Poster Session: Nanoscale Science and Engineering

Format: Poster

Abstract:

Malodor causes discomfort, and long term exposure can have serious health consequences. $H_2S$ is one of the major malodor pollutants generated from anaerobic decomposition of sulfur-containing organic matter, and often found in household and industrial wastes, especially along the sewer line.

Existing solutions to $H_2S$ including Claus process, which uses catalyst to convert $H_2S$ into sulfur at high temperature (about 500 K), and adsorbents, is energy-consuming and not sustainable. This work presents a new route to convert $H_2S$ under room temperature and humid conditions, which suits for the applications near the malodor source such as sewer.

In this study, the vanadia-titania catalyst ($V_2O_5/TiO_2$) is successfully developed to treat $H_2S$ at room temperature. The above catalyst was prepared by using rotary evaporator to evenly distribute ammonium vanadate on UV100 powder. The obtained catalysts are tested under different concentration ranging from 3.33 to 33.3ppm with a flow of 210cm$^3$/min at room temperature. A conversion of 35% can be maintained for 25mg catalyst with 33.3ppm $H_2S$, 210cm$^3$/min. The fresh and used catalysts were characterized by XRD, micro-Raman and X-ray photoelectron spectroscopy in order to identify the active sites and confirm the formation of elemental sulfur on the catalyst. The catalytic reaction under humid condition was also conducted, and the result shows that the catalytic performance of $V_2O_5/TiO_2$ catalyst is not sensitive to humidity of reaction system.

# 377389 Block Copolymer Templating of Hydrothermally-Grown ZnO Nanorod Arrays with Controlled Density for Photovoltaic Devices

**Abstract:**

The ability to control the diameter and density of ZnO nanowires in hydrothermally-prepared forests is critical for customization of ZnO arrays for targeted applications. Accomplishing this in a facile and scalable manner remains a significant challenge. Self-assembly of block copolymers is well-suited to ordering nanoscopic domains over macroscopic areas, and block copolymer micelles in particular offer advantages of regular, long-range periodicities when cast in a monolayer without any post-processing. We present a fully solution-based method for high-fidelity templated synthesis of ZnO nanorod arrays directed by a confluent monolayer film of self-assembled micelles of a block copolymer (BCP), poly(styrene-*b*-4-vinyl pyridine) (PS-b-P4VP). Growth of the ZnO nanorods proceeds from an underlying seed layer by diffusion of Zn salts through the micelles. The location and diameter of the resulting nanorods are controlled by the location and size of the P4VP micelle cores with growth in intervening spaces inhibited by the PS corona. Using simple geometrical relationships, density and inter-particle spacing of nanorods can be rationally designed based on block copolymer molecular weight and volume fraction. To demonstrate the significant benefits of this ability, we prepare ZnO:P3HT ordered bulk heterojunction solar cells using templated arrays and show improved PV efficiency proportional to our ability to match charge-transport limiting length scales.

# 377741 Designing Composite Polymer Electrolyte Interfaces for Stable Electrodes

**Abstract:**

Polymer electrolytes are promising materials for high energy density rechargeable batteries. However, typical polymer electrolytes are not electrochemically stable at the charging voltage of advanced positive electrode materials. Decomposition of polymer electrolyte would occur when the cathodic potential is above 4V. The initial goal of this project is to understand polymer electrolyte oxidation and design a polymer electrolyte/cathode interfacial layer to prevent such oxidation. This cathode and interfacial layer comprises inorganic metallic nanoparticles and organic protecting ligands. The interfacial layer must be permeable to ions and impermeable to electrons so that further oxidation of polymer electrolyte can be prevented

Currently, thin gold nanoparticle (Au NP) films are being studied. Au NPs have been chosen for their large surface-to-volume ratio to provide sufficient contact area and signal. In addition, gold is inert and hence non-reactive with most polymer electrolytes and lithium salts. Finally, it is facile to modify the surface. In our study, poly (ethylene oxide) (PEO) will serve as the polymer and lithium bis(trifluoromethanesulfonyl) imide salt (LiTFSI) will be the salt. Characterization of structure and conductivity of Au NP thin films with different length ligands will be presented. Electrochemical stability of polymer electrolyte as a function of ligand length will also be reported.

# 377859 Experimental Characterization and Multiphysics Modeling of Surface Tethered, Environmentally Responsive Poly(acrylic acid) Brushes on Thin Film Au and Cellulose Substrates

**Abstract:**

In this study, surface-tethered poly(acrylic acid) (PAA) brushes grown from gold and cellulose substrates were created and characterized for surface morphology, chemical composition and adsorptive behavior. PAA brushes are of interest in the design of environmentally responsive ion exchange (IEX) materials; brush morphology and adsorptive properties can be controlled in large part by environmental variables such as salt concentration, ionic strength, and pH. PAA brushes have been utilized in biosensing applications as a three dimensional precursor layer with high loading capacity for subsequent biomolecular functionalization. In bioseparations, grafted polymer brushes can be used for protein separation and purification applications. This study aimed to characterize PAA brushes tethered to two types of substrates, a thin film of gold (for biosensing applications) and regenerated cellulose (for bioseparations applications). Scanning electron microscopy (SEM) and Fourier Transform Infared Spectroscopy (FTIR) were used to characterize sample morphology and chemical composition of the grafted materials. Surface Plasmon Resonance (SPR) was used to characterize layer thickness and adsorption of common proteins for the PAA brushes attached to a gold substrate while batch adsorption studies were used study adsorption on the modified cellulose materials. Using results from these characterization studies, multiphysics models were developed using Comsol Multiphysics to probe the adsorptive behavior of lysozyme (PI~11) and bovine serum albumin (BSA) (PI~4.8) to these complex tethered polymer materials for a variety of environmental conditions and brush geometries. Initial models were based on experimental investigations; hence the brush geometry, protein fluid samples, environmental parameters, and operating conditions corresponded with experimental studies. Further models were developed to investigate how different flow conditions, brush densities, grafted layer thicknesses, and brush conformations affected protein binding and elution, with the overall goal of improving adsorptive behavior of these materials.

# TRIBOELECTRIC NANOGENERATOR FROM CARBON NANOTUBES

ABSTRACT

The increasing permeation of technology through every facet of human existence has given rise to the development of mobile technology and the associated need for lightweight energy supply systems. A lightweight energy harvesting device based on the concept of triboelectric charging has been assembled to harvest energy from everyday repetitive motion. The mechanical, structural and electrical properties of carbon nanotubes make them prime candidates for use as electrodes in fabricating a Triboelectric Nanogenerator (TENG). TENG cells based on aligned carbon nanotube forests and mica electrodes were fabricated and it was shown how triboelectricity could be generated from this device and used to power an external circuit. We report the maximum output voltage and current density of the cell and we demonstrate methods of serial connection (in series or parallel) of these cells to a circuit. Also, we present an insight into the position of carbon nanotubes (and other forms of graphitic carbon) on the triboelectric series and study the triboelectric generation capabilities of other electrodes (besides mica) on the opposite side of the series to carbon nanotubes.

# 381340 Fabrication of Graphene Nanofiber from Graphene Oxide By Ice Template

**Abstract:**

We reported the fabrication of highly ratio surfance area graphene nanofibers (GNFs) by ice template method from two-dimensional graphene oxide to one or one point five dimensional graphene nanofibers. GNFs exhibited good structural stability, high surface area, and electrical conductivity twice that of carbon nanotube aggregates. It's used as electrode materials in supercapacitor, exhibited high ratio capacitance and energy density in a wide range of high power density and excellent cycling stability, higher than carbon nanotubes.

# 381460 Preparation of Large Area Transparent Conductive Graphene Film Utilizing Langmuir-Blodgett Method


**Abstract:**

Transparent conductors have been used in a wide variety of photoelectronic and photovoltaic devices, such as flat displays, solar cells, optical communication devices, and solid-state lighting. Graphene, a two-dimensional monolayer of $sp^2$-bonded carbon atoms, has attracted significant interests recently because of the unique transport properties.

Many technical approaches were reported. However, LB assembly method is a proffered technique that can realize controllable deposition of graphene oxide (GO) in a layer-by-layer manner. The thickness of GO films can be accurately controlled upon repeated deposition, leading to optimized optical and electrical properties of the final products.

We have successfully demonstrated Langmuir-Blodgett assembly of GO, By dispersion graphene or graphene oxide on organic solvent /water, and dropping on water, which the surface pressure and pulling speed form assembly, the film density and the formation of wrinkling can be tailored to produce different topological features. We fabricate flexible transparent graphene film on PET as substrate and preparation of large area graphene film (GFs) with the new technological process. The transparent conductors made from reduced graphene oxide sheets show a high conductivity with sheet resistance of ~650 Ω/sq and transmittance of 90% (2 or 4 layers). The large-scale using graphene transparent conducting film for potential application in solar cells, OLED, are on the way.

# 383477 Effectiveness and Toxicity of Magnetic Micelles for Use As Cancer Therapy Agents

**Abstract:**

To investigate the effectiveness of magnetic micelles for cancer treatment, it is essential to evaluate their toxicity and therapeutic efficacy. These magnetic micelles can also be used to trigger chemotherapy drug release upon application of external magnetic field. In this study, the effects of hyperthermia (at 43 $^{o}$C and 46 $^{o}$C) were tested on Chinese hamster ovary (CHO) cells to observe the change in cell viability. The effectiveness of hyperthermia using a water bath was compared to magnetic hyperthermia using iron oxide magnetic nanoparticles (MNPs) placed in a high frequency magnetic field to elucidate any enhanced toxicity caused by magnetic heating. Finally, the toxicity of poly(caprolactone-b-ethylene glycol) micelles and MNPs was evaluated by studying CHO cell viability, when exposed to these materials over a period of 48 hrs. It was observed that hyperthermia using 46 $^{o}$C led to complete cell death, whereas hyperthermia at 43 $^{o}$C partially killed the cells in a period of 48 hours. There was no significant difference in cell viability while using water bath hyperthermia and magnetic hyperthermia. In the range of concentrations tested the MNPs and polymers displayed no statistically significant toxicity to CHO cells in a period of 48 hours, indicating that these materials have potential to be used in therapeutic formulations. These results have laid the foundation for testing magnetically-triggered drug release using polymer micelles that can deliver cancer therapy in a localized area.

# MODEL FOR THE SYNTHESIS OF AND MACROPHAGE RESPONSES TO NANOSCALE HYDROXYAPATITE

GARIMA BHARDWAJ[1], ARTHUR A.GONZALES III[1] _THOMAS J. WEBSTER[1]

1DEPARTMENT OF CHEMICAL ENGINEERING, NORTHEASTERN UNIVERSITY, BOSTON, MA-02115

**Introduction:** To monitor the immunological response of implants, the host response to materials must be determined prior to implantation. This response is highly dominated by macrophages which are the primary cells governing the immune response. Macrophages not only fuse to become multinucleated foreign body giant cells, but they also activate T lymphocytes by expressing co-stimulatory molecules (e.g., CD86 and CD80) and surface antigens (e.g., MHC II). T lymphocyte activation and giant cell formation have been observed on macrophages adherent to biomaterials and in retrieved adjacent tissues to failed implants. They are major effectors in the defense against bacterial infection and respond to interactions with gram-negative and gram-positive bacteria with a marked enhancement of their functional activities. Thus, the assessment and control of macrophage responses to various surface nano-modifications on titanium implants is of vital importance. This current study aims to modify the surfaces of titanium by coating it with nanoscale hydroxyapatite, in a range of sizes, using electrophoretic deposition with both AC and DC current. Macrophages were then seeded onto this surface and their behavior observed concerning how to further improve the modification process.

**Materials and Methods:** Nanoscale hydroxyapatite was synthesized using a wet chemical synthesis process in an array of different sizes in the nanometer scale. By varying the parameters like pH, reaction time and temperature in the synthesis process and analyzing experimental data, a MATLAB model predicting particle size for optimized synthesis was established using the BBD theory of statistics.TEM was used for particle size charecterization. It was then coated onto a titanium mesh purchased from Alpha Aesar (Catalog no.7440-32-6) by electrophoretic deposition with AC and DC current. The mesh was used both as the working and counter electrode. SEM was used for characterizing the coating. Macrophages were purchased from ATCC (RAW 264.7 (ATCC® TIB-71™)) and were cultured using EMEM (ATCC® 30-2003™), 10% FBS (ATCC® SCRR-30-2020™) and a 1% penicillin-streptomycin solution (ATCC® 30-2300™). Cell adhesion and proliferation was observed using an MTT assay for 1, 3, 5 and 7 days. Levels of TNF-α, IL-1, IL-6 and nitrite released by the macrophages were studied. Bacterial assays were conducted using *Staphylococcus aureus* (ATCC® 29740™) and *Pseudomonas aeruginosa* (ATCC® 39324™) strains of bacteria. Experiments were conducted in triplicate and repeated at least three separate times each.

**Results and Discussion:** The particle size was predicted using the following equation:

**R= a*A + b*B + c*C + d*A^2 + e*B^2 + f*C^2 + g*A*B + h*A*C + i*B*C + j**

Where R= response (hydroxyapatite particle size) and A =pH, B=temperature, and C =reaction time.

The change in the surface topography of the material affected the adhesion and proliferation of the macrophages leading to reduced activation of macrophages with increased nanometer roughness. The level of TNF-α, IL-1, IL-6 and nitrite released decreased with increasing hydrophilicity of the surface but increased overall with time. Bacterial activity decreased with smaller particle size of hydroxyapatite.

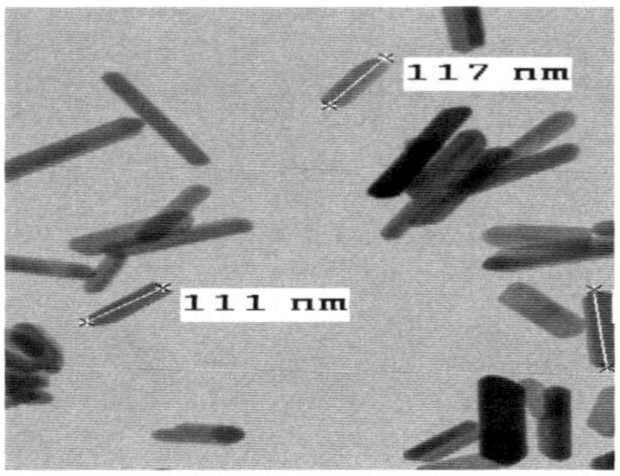


Figure 1: TEM image of the 110 nm HA powder. Scale bar- 100 nm

Figure 2: SEM image of the 110 nm HA powder coated onto the titanium mesh using EPD. Scale bar- 1μm

**Conclusions:** The model helps us determine the particular parameters needed to synthesize a particular particle size.Controlling the size of the nanophase hydroxyapatite, the coating procedure and its parameters can help influence the fate of macrophage attachment, activation and functionality at the implant surface by reducing levels of TNF-α, IL-1, IL-6 and nitrite released with increased nanoscale topography.

**Acknowledgements:** The authors would like to thank Northeastern University for funding.

**References:** 1.) Y. Chun et al, Control of macrophage responses on hydrophobic and hydrophilic carbon nanostructures, Carbon 49 (2011) 2092–2103. 2.) D. Khang et al, Reduced responses of macrophages on nanometer surface features of altered alumina crystalline phases, Acta Biomaterialia 6 (2010) 3864–3872.

# 383827 The Role of Structure and Surface Composition on the Contact Angle and Wetting Characteristics of Carbon Nanofiber Coatings

**Abstract:**

In recent years, superhydrophobic surfaces have attracted considerable interest because of their potential applications with great significance in various domains such as microfluidics, paints, and anticorrosive agents. The role of roughness and composition on the wetting characteristics of a series of carbon nanofiber based coatings were studied in order to evaluate its superhydrophobic properties. In this study, idealized surfaces were created from a smooth stainless steel and aluminium plates and two other stainless steel plates with different textured surfaces. All surfaces were coated with carbon nanofiber alcohol solutions in order to generate coatings of variable compositions using mixtures of isopropanol, water and a commercial carbon nanofibre. Nanofiber coatings of various concentration were created and studied in this work. A general trend of increasing hydrophobicity was observed for coated surfaces as compared to the bare substrate. Individual contact angles were dependent on the nature of the underlying substrate, relative surface pattern, and roughness. Apparent surface contact angles on these substrates exhibited behavior that followed both the Wenzel and Cassie models. Overall wetting properties were dependent upon composition and micro scale roughness of the coatings.

Key words: Contact Angle, Wetting, Wenzel and Cassie model, Coatings, Carbon Nanofibre, Superhydrophobicity

# 383885 The Effects of Synthesis Method on the Physical and Chemical Properties of Iron Oxide Nanoparticles

**Abstract:**

The method by which dextran coated iron oxide nanoparticles are synthesized greatly affects the physical and chemical properties of the nanoparticles. To study this, dextran coated iron oxide nanoparticles were synthesized via four different synthesis methods, which varied the timing of ammonium hydroxide and dextran addition to the iron salt solution in the formation of iron oxide nanoparticles. The four synthesis methods were two-step, semi-two-step, simultaneous semi-two-step, and one-step reactions. The one-step reaction was then synthesized using different amounts of dextran. The nanoparticles were characterized for size, stability, coating percent, crystallinity/crystal size, and AMF heating properties. In the two-step synthesis, the iron oxide nanoparticles were formed and washed prior to coating with dextran. The semi-two-step first formed the iron oxide nanoparticles via addition of ammonium hydroxide prior to dextran coating but without a washing step in between. For the simultaneous semi-two-step method, the dextran and ammonium hydroxide solutions were injected simultaneously into the iron salt solution, and finally, the one-step reaction formed the iron oxide nanoparticles in the presence of dextran.

The physical and chemical properties of the dextran coated iron oxide nanoparticles were characterized for each synthesis method. The variability between the methods as well as the variability within each synthesis method was evaluated. The two-step method resulted in nanoparticles with high crystallinity and good heating properties. However, the stability in PBS over time was lower than the other synthesis methods due to the larger size and less dextran coating on the surface of the nanoparticles. The other extreme, the one-step method, resulted in highly stable nanoparticles due to a smaller hydrodynamic diameter, but these particles lacked the heating properties of the other synthesis methods. This is most likely due to the high viscosity of the dextran solution present during the nanoparticle formation. When the viscosity of the solution was reduced by decreasing the dextran amount to 1 gram, the heating properties dramatically increased, but the size and stability of the nanoparticles was compromised. The other two synthesis methods resulted in the physical and chemical properties of the iron oxide nanoparticles being somewhere in between the two-step and one-step methods. The semi-two step synthesis was found to have the most batch-to-batch variability due to the separate injections of ammonium hydroxide and dextran into the reaction vessel. The simultaneous semi-two-step had reduced batch-to-batch variability as well as adequate heating properties, size, and stability. Overall, this study demonstrated how the synthesis method of dextran coated iron oxide nanoparticles directly affects the chemical and physical properties of the nanoparticles. Additionally, each method led to a different amount of variability in the properties of the nanoparticles.

# Porous Carbon Microsphere-Encapsulated ZnO-CoO Nanoparticles Anode for High-Performance Lithium-Ion Batteries

Dr. Liu Lianjun, Zhao Cunyu, Zhao Huilei, Zhang Qianyi, Prof. Li Ying*

University of Wisconsin-Milwaukee, 3200 N. Cramer St. Milwaukee, WI, 53211 USA
E-mail: (liying@uwm.edu)

Cobalt monoxide (CoO) has a high specific capacity of 714 mAh $g^{-1}$ as lithium-ion battery (LIB) anodes, but its toxicity, high cost and rapid capacity fading due to poor conductivity and volume expansion presents a significant challenge for practical applications. Here we report a novel architecture of hierarchical porous carbon microspheres (CS)-encapsulated ZnO-CoO nanoparticles that serve as an advanced anode for high-performance LIB. Porous CS is fabricated by a facile aerosol spray pyrolysis method, and ZnO-CoO particles are in situ infiltrated in the CS by a simple hydrothermal or wet-impregnation procedure. The developed hybrid provides several advantages: (1) partially replacing CoO with ZnO to offer a low-cost and eco-friendly candidate anode, (2) a continuous and large surface area carbon network for improved electrical conductivity and uniform dispersion of ZnO-CoO nanoparticles, and (3) porous structure for good electrolyte diffusion and fast $Li^{+}$-ions transport and to buffer the large volume expansion of metal oxides. As a result, this new ZnO-CoO/CS nanocomposite demonstrates a higher reversible capacity (1250 mAh $g^{-1}$ at a current density of 100 mA $g^{-1}$ after 125 cycles), more excellent cycling stability, higher Coulombic efficiency (99%), and better rate capability than a conventional non-porous ZnO-CoO/carbon black material. Our confinement-in-3D porous carbon approach should shed light on the design and synthesis of other metal oxides electrodes for energy storage.

**Keywords**: ZnO-CoO nanoparticles, porous carbon microsphere, confinement, lithium-ion batteries, reversible capacity

# Characterization of Complex Solution-Nanotube Interfaces using AUC and SANS

Stephanie Lam and Jeffrey A. Fagan

*Polymers and Complex Fluids Group, Materials Science and Engineering Division, National Institute of Standards and Technology, Gaithersburg, MD*

Single-walled carbon nanotubes (SWCNTs) have many special properties including: high tensile strength, high thermal conductivity, as well as unique optical properties. However, their use in specialty applications (i.e. nanoelectronics and sensors) is often limited by their purity. Over the past decade, numerous methods have emerged for the separation of carbon nanotubes - one of the most scalable being Aqueous Two-Phase Extraction (ATPE). ATPE was recently demonstrated to have the capability of recovering (6,6), (5,5,), (7,4), (7,7), (6,5), (6,4), (7,5), (7,6), and (9,4) nanotubes at high purity.[1] The ATPE process employs colloidal interactions between bile surfactants, sodium dodecyl sulfate (SDS), and polymers (PEG and dextran) with different chirality SWCNTs to favor partitioning of (n,m) SWCNTs into one of the two aqueous phases. Thus, it is important to elucidate these interactions in order to better understand the driving forces for SWCNT separation in ATPE. To do this, we propose utilizing analytical ultracentrifugation (AUC) in conjunction with small angle neutron scattering (SANS) to resolve the density and spatial distribution of adsorbed bile surfactants and SDS atop (n,m) SWCNTs. Additionally, using AUC, we will also determine the size of the hydration shell around surfactant/co-surfactant/(n,m) SWCNT complexes. Such results would not only enhance our understanding of the ATPE process but also on a broader scale, contribute to knowledge of interactions between carbons of different hybridizations and various types of amphiphilic molecules and polymers.

## References

1. Fagan, J. A. et al. *Adv. Mater.* 2014, 26, 2800. doi: 10.1002/adma.201304873

**Scalable Synthesis Methods to Produce Janus Nanoparticles**

**Author: Alan Hanley**
**Co-Author/Advisor: Dr. Allan David**

Since research into the idea of Janus nanoparticles began in the chemical engineering community, a number of synthesis methods have been considered and tested for efficacy and yield. While the ability to create Janus nanoparticles is now a possibility, a synthesis method that can be scaled up to industrial levels remains an unanswered question. Using functionalized nanoparticles, we have developed a new method to synthesize Janus nanoparticles that has the potential to be scalable. Janus nanoparticles, because of their two distinct and different hemispheres, will self assemble depending on what they are suspended in, what type of electric current is run through them, and a number of other controllable conditions. Janus nanoparticles have a large number of potential uses considering the nature of their design, such as waterproof coatings, antimicrobial coatings, developing ultra thin monitors, and advancements in drug delivery. Once an effective method is discovered and verified, and broad new area of research in the chemical engineering community will be available. The possible combinations of compounds that can be bonded to these particles is extensive, and as such will provide many new areas of study in the realm of nanomaterials.

# Evaluating the potential of dynamic magnetic susceptometry for Ricin B detection

Lorena Maldonado-Camargo[1], Srinand Sreevatsan[2], and Carlos Rinaldi [1,2*]
[1]*Department of Chemical Engineering, University of Florida.*
[2]*College of Veterinary Medicine, University of Minnesota*
[3]*J Crayton Pruitt .Family Department of Biomedical Engineering, University of Florida.*
*Corresponding author E-mail: carlos.rinaldi@bme.ufl.edu

Ricin, a fatal toxin extracted from castor bean plant, could potentially be used as a biocontaminant in the food and water supply. Ricin is among the most lethal naturally occurring toxins and is easily accessible, inexpensive, and easy to prepare. Furthermore, ricin is highly thermostable and remains active at acidic and alkaline pH. Ricin is composed of an A chain (32 kDa), which is a cytotoxic ribosome-inactivating protein, and a B chain (32 kDa), which is catalytically inactive but serves to mediate transport of the A-B protein complex from the cell surface. Current systems used to detect Ricin contamination are time consuming, non-specific, and have limited detection range. Here we evaluate the potential of dynamic magnetic susceptometry for the detection of Ricin using magnetic nanoparticles conjugated to a single-stranded DNA aptamer (SSRA1) with high selectivity for the Ricin B chain. Dynamic magnetic susceptometry consists of monitoring the response of thermally blocked magnetic nanoparticles, relaxing through the so-called Brownian mechanism of particle rotation, to low amplitude alternating magnetic fields as a function of applied field frequency. Changes in nanoparticle hydrodynamic diameter due to binding to a target analyte can be detected through shifts in the dynamic magnetic susceptibility spectra of the nanoparticles. This technique is potentially interesting for the detection of Ricin because it requires relatively small sample volumes (~100 μL) and can be applied to optically complex and opaque samples, such as complex food mixtures. Thermally blocked magnetic nanoparticles were coated with poly(ethylene glycol) to minimize non-specific interactions and functionalized with SSRA1 or a non-targeting scrambled aptamer. The effect of particle size, controlled through the molecular weight of the PEG coating, on the ability to detect Ricin B binding to the nanoparticles was studied. The potential advantages and limitations of dynamic magnetic susceptometry for the detection of protein analytes such as Ricin will be discussed.

Keywords: Magnetic nanoparticles, Ricin, biosensor, aptamers

# 386679 Magnetic Nanoparticles for Non-Invasive Quantification of Prostate Cancer Aggressiveness

**Abstract:**

Prostate Cancer (PCa) is the most diagnosed non-skin malignancy and the second leading cause of death from cancer among men in the United States. Therapies for PCa, unfortunately, remain ineffective. In order to achieve more effective therapy, early and accurate detection of prostate cancer is necessary. Current Screening technologies for PCa, including the prostate-specific antigen (PSA) test, are limited by false-negatives and false-positives. Additionally, even when PCa is successfully detected, these screens provide no indication of the aggressiveness of the disease. Improved molecular imaging techniques, that are both quantitative and sensitive, and that can non-invasively detect processes deep within the human body are required to monitor changes which may be predictors of treatment outcomes. We have developed multimodal and bioprocess-sensitive "smart" magnetic nanoparticles (MNP) which respond to overexpressed protease activity in the tumor microenvironment – and thus to tumor aggressiveness – and which can be monitored and quantified non-invasively over the entire tumor volume. Cancer cells overexpressing legumain protease have been shown to have an invasive and metastatic phenotype *in vivo*. Legumain specifically degrades peptides containing an AANL-peptide sequence. We conjugated a fluorescence marker via the legumain cleavable AANL-peptide linker (LCP). This fluorescent-MNP displayed sensitivity to legumain activity; slowly releasing the fluorescent marker in low activity microenvironments and more rapidly in the presence of high legumain activity. Release of Dy747 resulted in a decrease in the fluorescence signal in tumors, due to wash-out into the blood flow, at a rate proportional to the local proteolytic activity. Quantitative MRI analysis indicated the amount of agent delivered to a particular location, providing the baseline to correct for tumor heterogeneity. Taken together, the fluorescent/MRI signal ratio is a function of the local proteolytic activity.

# 387381 Simulations of Phenol-Capped Alkanethiol-Coated Gold Nanoparticles in Organic Solvents


**Abstract:**

Introduction: Gold nanoparticles (AuNP's) represent a sub-micrometer form of colloidal gold. Nanoparticles composed of gold were first used in association with the preparation of colored glass. AuNP's have since been researched in areas such as of materials science and biomedicine. However, fundamental interactions between coated AuNP's and solvents are not completely understood. Molecular dynamics simulations were conducted to gain insights into the influence of (1) the thickness of an alkanethiol AuNP coating and (2) different organic solvents on the behavior of AuNP's.

Methods: Phenol-capped alkanethiol-coated gold nanoparticles were composed of a 1.5-nm-diameter gold core coated with 60 phenol-capped alkanethiol ligands. All simulations were conducted with the Assisted Model Building with Energy Refinement (AMBER) 12 software package using the General Amber Force Field (GAFF) parameters. Simulations were conducted at 300 K for 40 ns using a timstep of 2 fs. Data analysis was used to assess the effective size of the NPs (radius of gyration), the distance from the ligand end groups to the gold surface atoms, and the radial distribution of the phenol end groups.

Results and Discussion: The size and polarity of the solvent molecules influenced the conformations of the alkanethiol ligands and nanoparticle-solvent interactions. Smaller, polar solvents (acetonitrile and methanol) led to a greater distribution of phenol groups positioned further away from the gold core (i.e., more accessible to the solvent molecules), the largest radius of gyration, and the greatest distance from the phenol groups to the gold surface atoms. In contrast, the non-polar solvents (chloroform and toluene) led to a greater distribution of phenol groups positioned closer to the gold core and less accessible to the solvent, the smallest radius of gyration, and the smallest distance from the phenol groups to the gold surface atoms. The larger, polar solvents (acetic acid, acetone, and dimethyl sulfoxide) afforded results that were between the smaller, polar solvents and the non-polar solvents. These results suggest that the polar ligand end groups (phenols) had the most favorable interactions with the smaller, polar solvents; followed by the larger, polar solvents; and then by the non-polar solvents. Interestingly, the results were more sensitive to the sizes of the solvent molecules than to the dipole moments of the solvent molecules. Regarding ligand length, it was found that the aforementioned effects of the solvents were more prominent with longer alkanethiol chains.


References:


Louis, C. & O. Pluchery, O. (2012). Gold Nanoparticles for Physics, Chemistry and Biology. London: Imperial College Press.

# 387449 Catalytic Conversion of Methane to Methanol on Metal Phthalocyanine Functionalized Graphene

**Abstract:**

Functionalized Graphene involves the addition of functional groups to a Graphene support plate—these functional groups aid in the catalysis of oxidative reactions from hydrocarbons to alcohols. In this study, we have used first principles density functional theory (DFT) to study the structural, energetics, and kinetic properties of catalytic oxidation reaction of methane to methanol conversion on Iron-phthalocyanine (FePc) and Copper-phthalocyanine (CuPc) functionalized graphene. Our preliminary calculations show that $CH_4$ is adsorbed on both oxidized FeCuPc/graphene and CuPc/graphene surfaces, a favorable first step towards the catalytic reaction. The energetics of intermediate and final products of this reaction are also much more thermodynamically favorable steps. Our calculations will provide overall possible reaction mechanism and also the activation of the strong C–H bond of methane leading to its homolytic cleavage, which is one of the important rate limiting steps to methanol conversion reaction. A functional group of either FePc or CuPc will be added to the graphene surface to aid the catalysis reaction. Finally, we will present the role of the central metal, ligand and several single bridging atoms for tunable catalytic properties in case of methane to methanol conversion reaction.

**Acknowledgement:**

DFT calculation work is supported from NSF TeraGrid (XSEDE) resources under allocation numbers [TGSEE090006] and [TG-DMR130127]. Use of the Center for Nanoscale Materials is supported by the US Department of Energy, Office of Science, Office of Basic Energy Sciences, under Contract No. DE-AC02-06CH11357.

Introduction: Sorafenib is a multikinase inhibitor that was approved by thc FDA in 2007 for the treatment of hepatocellular carcinoma (HCC). Since then, it has quickly become the standard of care for patients with advanced stage disease. However, patient tolerance is an issue due to the side effects resulting from systemic delivery. To avoid systemic delivery of chemotherapies for HCC, a procedure called transarterial chemoembolization (TACE) has been developed to selectively place a catheter in tumor feeding arteries to locally deliver the chemotherapy. This is then followed by or combined with embolic agents to contain the chemotherapy and to further starve the tumor from nutrients and oxygen in the blood. Thus, poly(lactide-*co*-glycolide) (PLG) microspheres have been developed that can encapsulate sorafenib for local delivery to liver tumors via TACE as well as co-encapsulating iron oxide nanoparticles to enable MRI-monitoring of the delivery. Specifically, the purpose of this study aims to use MRI to correlate to microsphere concentrations at the tumor to determine estimated sorafenib doses delivered.

Methods: PLG sorafenib iron oxide microspheres were fabricated via a double emulsion/solvent evaporation method. The microspheres were then characterized for microsphere size, sorafenib and iron oxide loading and release, as well as MRI contrast effect properties *in vitro* (7T MRI, Bruker Clinscan, Billerica, MA, USA). From these *in vitro* MRI studies, the microsphere concentrations were plotted to correlate against changes in MRI signal changes. Afterwards, liver tumors were inoculated in 4 New Zealand White Rabbits in which a VX2 tumor fragment was injected through a biopsy needle under ultrasound guidance. The tumor was then allowed to grow to roughly 1 cm in diameter after 2-3 week after which the rabbits underwent TACE procedures and microsphere infusions (50 mg). The rabbits were also scanned pre-and post procedurally at the same 7T MRI Bruker Clinscan to compare tumor signals before and after microsphere infusion with $T_2$ and $T_2$* map protocols. Afterwards, the tumor $T_2$ and $T_2$* values were extracted from the MRI images via MATLAB analyses to determine changes in the $T_2$ and $T_2$*. These changes were then used to estimate the microsphere concentrations at the tumor using the correlations previously obtained following the *in vitro* MRI studies. The rabbits were then survived for 24 hours before sacrificing and harvesting the tissue for Prussian blue staining for iron.

Results: The resulting microspheres were of an average diameter of 13 microns. Loading and release studies indicated that the weight percentage of sorafenib and iron oxide in the microspheres were 8.8% and 0.89%, respectively and that 21% and 28% of the loaded sorafenib and iron oxide released in 72 hours, respectively. *In vitro* MRI studies showed that microsphere concentration was correlated to $1/T_2$* via the equation: **($1/T_2$*)** = 0.0205(**Microsphere Concentration**) + 0.1262. For subsequent *in vivo* studies, MRI indicated that the microspheres primarily deposited in the tumor periphery and the estimated microsphere concentrations were on average 45 mg/mL in those regions. After the rabbits were sacrificed, Prussian blue staining confirmed proper tumor targeting and that primary microsphere deposition was at the tumor periphery.

Conclusion: The developed PLG sorafenib iron oxide microspheres allow for targeted delivery of sorafenib to liver tumors with the advantage of MRI monitoring and proper estimations of the delivered dose. Further studies are evaluating biomarkers for tumor responses and possible correlations of the tumor responses to the microsphere doses that were effectively delivered.

# *Controlling Ferrofluids to Cross an in vitro Blood-Brain Barrier Model*

Di Shi[1], Linlin Sun[2], Gujie Mi[1], Soumya Bhattacharya[3], Suprabha Nayar[3], and Thomas J Webster[1*]
[1]Department of Chemical Engineeging, Northeastern University, Boston, USA.
[2]Department of Bioengineering, Northeastern University, Boston, USA.
[3]Materials Science and Technology Division, CSIR-National Metallurgical Laboratory, Jamshedpur, India.
*th.webster@neu.edu

## I. Introduction

Among several intriguing nanoparticles, superparamagnetic iron oxide nanoparticles (SPIONs) have been extensively studied because of their ability to be controlled by magnetic fields. For instance, magnetic fields and nanomaterials have been identified to locate and target nucleic acids within the last decade. Moreover, SPIONs have also been discovered as an excellent contrast agent which can potentially increase image contrast and improve MRI sensitivity [1]. Under normal conditions in the body and outside of the brain, the accumulation of magnetic nanoparticles can be removed by several detoxification and antioxidant mechanisms. But, if the SPIONs cross the blood brain barrier and have not been efficiently cleared, the accumulation of those iron oxide nanoparticles may cause neurodegeneration and be harmful to normal brain function in the long term [2].

As there are concerns about both the delivery of drugs across the blood-brain barrier and the accumulation of SPIONs in the body, this study established an *in vitro* blood-brain barrier model to test the ability of a new type of magnetic material (ferrofluids) to pass through the blood-brain barrier. In this specific case, since primary cultures of brain microvessel endothelial cells can be easily contaminated by other neurovascular cells and lose their function to tighten the blood-brain barrier; we used an immortalized cell line, bEnd.3 cells, in our blood–brain barrier model. We tested several variations of magnetic nanoparticles in an effort to both increase blood-brain barrier passage (for neural drug delivery applications) and decrease blood-brain barrier passage (to minimize toxicity).

## II. Materials and Methods

### *A. Material Characterization*

All the five samples, which were GGB (ferrofluids synthesized using 0.001% glycine, 0.001% glutamic acid and 0.01% BSA), GGC (0.001% glycine, 0.001% glutamic acid and 0.01% collagen), GGP (0.001% glycine, 0.001% glutamic acid and 0.01% PVA), BPC (0.001% BSA, 0.01% PEG and 0.01% collagen) and CPB (0.01% Collagen, 0.01% PVA and 0.001% BSA) [3], were characterized by zeta potential to determine their charge and dynamic light scattering for hydrodynamic diameter. TEM was also used to assess the iron oxide inner core diameter.

### *B. Confirmation of the Blood-Brain Barrier Model [4]*

For the *in vitro* blood-brain barrier model, murine brain endothelioma (b.End3) cells were used and were cultured in DMEM with 10% FBS and 1% P/S. After being diluted down to $3 \times 10^5$ cells per mL, cells in the previous cell culture section were pipetted onto membrane inserts combined with 24-well plates. Then, the well plates with inserts were incubated daily changing medium in the outer well for two days or until cell confluency was reached. Following the desired confluency, cell culture medium was replaced with 1:1 DMEM/Ham's F12 with 1% P/S for the next 96 hours.

To confirm the blood-brain barrier model, all the inserts were transferred to wells that contained 600 μL of HBSS before adding FITC-Dextran. Positive controls were set up by adding 100 μL of 10 μg/mL FITC-Dextran solution with 600 μL of HBSS and negative controls were set up by adding 700 μL of HBSS only. After 2 hours of incubation, 100 μL were taken from each well and transferred to a blank 96-well plate with a clear bottom. A fluorescent plate reader was excited at 490/20 nm and relative fluorescence was measured at 528/20 nm.

### *C. Experimental Samples Tested Through the Blood-Brain Barrier Model*

The previously described *in vitro* blood-brain barrier model was then used to test the permeability of GGB, GGC, GGP, BPC and CPB. For this, nanoparticles were diluted 1:19 with HBSS and then inserts were exposed to the nanoparticles for 2 hours. After 2 hours, a 100 μL solution was taken from each well and a iron assay kit was used to determine the iron concentration that passed through the model. Each experiment was conducted in triplicate and repeated at least three times.

## III. Results and Discussion

### *A. Material Characterization*

Dynamic light scattering revealed that the hydrodynamic diameters of the samples ranged from 20nm to 200nm. For zeta potential, the lowest absolute zeta value was 29mV for BPC and the highest one was 69mV for GGB. TEM images show that the iron core was about 5-10nm for all of the samples. In addition, according to the paper published by our co-author, all the five samples share the core-shell structure, in which iron oxide acts as the core and the three coatings on the outside were bound to each other by non-covalent, secondary bonds [5].

## B. *Confirmation of the Blood-Brain Barrier Model*

The value representing the permeability rate of the experimental blood-brain barrier model was in accordance with values from the literature. As it has been introduced by Bennett et al., the decline of permeability with increasing exposure times to serum-free medium indicated that the model was successfully established (Fig. 1) [6]. Confocal fluorescent images also proved that tight junctions were successfully established after 96 hours.

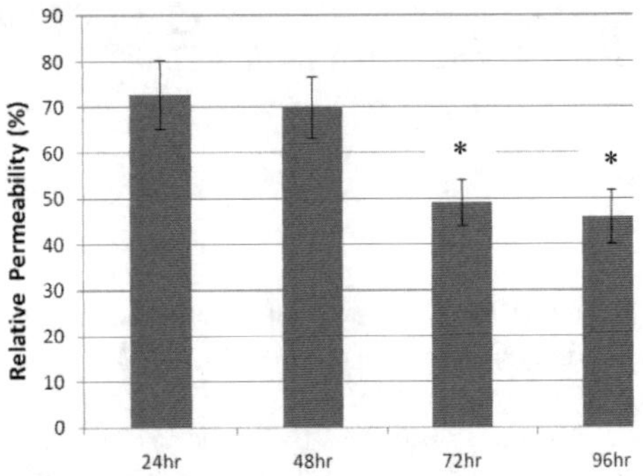


Fig. 1. Permeability of FITC-Dextran across the blood–brain barrier model confirming the establishment of an accurate blood–brain barrier. Data are shown as the mean ± SD; N = 3; *P < 0.01 compared with 24hr.

## C. *Experimental Samples Tested Through the Blood-Brain Barrier Model*

Results showed that the highest permeability was obtained from CPB (3.36 ng/μL) and the lowest permeability was obtained from GGB (2.19 ng/μL) (Fig. 2). Also, ferrofluids (FF) synthesized using a combination of collagen and PVA generally had higher permeability than those synthesized using glycine and glutamic acid. A regression study was run between blood–brain barrier FF permeability and FF size, and between FF permeability and FF zeta potential. Results support the theory that nanoparticles with a large diameter and low zeta potential (e.g., GGB for192 nm and -68.95mV) have difficulty passing through the blood-brain barrier. These results suggest that for nanoparticles that need to be delivered through the blood-brain barrier (i.e., for treating neurological diseases), FF should be coated with collagen while, FF should be coated with glycine and glutamic acid to keep the nanoparticles from penetrating the blood-brain barrier (i.e., for whole body MRI). Moreover, in order to determine the cytotoxicity that FFs have

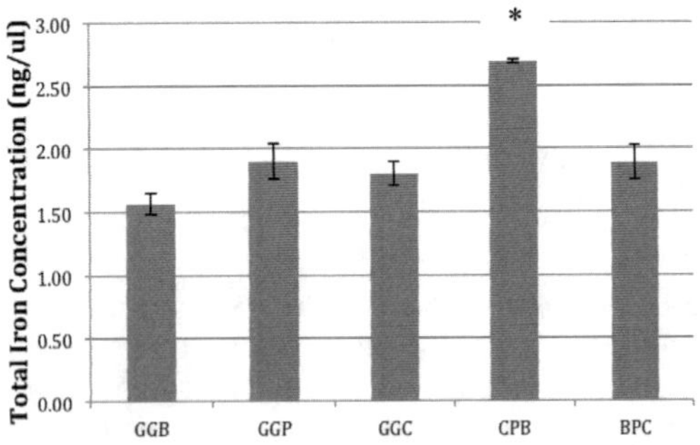


Fig. 2. Blood–brain barrier permeability by sample. Data are shown as the mean ± SD; N = 3; *P < 0.01 compared with all other samples.

to cells, a biocompatibility experiment was set up and carried out using an MTS assay. Results indicated that all the five samples have greater biocompatibility with cells when diluted 1:19 with HBSS.

# IV. Conclusions

An *in vitro* model of a blood-brain barrier was established using b.End3 cells. As the permeability decreased with increasing exposure to serum-free medium, the model was confirmed by comparing the permeability trend of FITC-Dextran in serum-free medium with previous research. With the successfully established model, the permeability of five magnetic nanoparticles ferrofluids was examined. The present results suggest a possibility to manipulate magnetic nanoparticle penetration across the blood–brain barrier by controlling bioactive coatings. Such data lay the foundation for the modification of ferrofluids to be either coated with collagen to pass through blood-brain barrier, or to be coated with glycine and glutamic acid to avoid penetration.

# Acknowledgment


The authors would like to thank Northeastern University for funding this research. One of the authors (S.B.) would like to thank Council of Scientific and Industrial Research (CSIR), Govt. of India for providing financial support in the form of a Senior Research Fellowship.

## References


[1] Christian P, Olivier Z, Olga M. Magnetically enhanced nucleic acid delivery. Ten years of magnetofection—Progress and prospects, Advanced Drug Delivery Reviews. 2011; 63: 1300-1331.

[2] G.D. Iannetti, Richard G. Wise, BOLD functional MRI in disease and pharmacological studies: room for improvement?, Magnetic Resonance Imaging, 2007; 25: 978-988.

[3] Nayar S, Sinha A, Pramanick AK. A biomimetic process for the synthesis of aqueous ferrofluids for biomedical applications. Application number 0672DEL2010.

[4] Dan H, Lubna S, Thomas JW. Comparison of ferrofluid and powder iron oxide nanoparticle permeability across the blood-brain barrier. Int J. of Nanomedicine. 2012; 8:703-710

[5] S. Bhattacharya, et al. Protein-Polymer Functionalized Aqueous Ferrofluids Showing High T 2 Relaxivity. J. Biomed. Nanotechnol. 2013; Vol. 9: 1–9.

[6] Bennett J, et al. Blood–brain barrier disruption and enhanced vascular permeability in the multiple sclerosis model EAE. J Neuroimmunol. 2010; 229:180–191.

# 382428 Development of Peptide-Conjugated Iron Oxide Nanoparticles for Hyperthermia Treatment of Cancer

**Abstract:**

Hyperthermia, the heating of tissue between 42 and 45°C, has been shown to enhance the effects of radiation and chemotherapeutics, but current methods of hyperthermia often result in severe side effects due to lack of localization and overheating of tissue. Magnetically mediated hyperthermia provides the opportunity for localized heating, however, this method is currently limited by the lack of particle penetration into tumor tissue. In this research, a nanoparticle system composed of an iron oxide core with a crosslinked dextran coating functionalized with the tumor homing peptide, CREKA, has been developed to overcome this limitation by homing to and penetrating into tumor tissue. The iron oxide core allows for particle heating upon exposure to an alternating magnetic field while the dextran coating stabilizes the particles in suspension and decreases the cytotoxicity. The overall goal of this study was to optimize CREKA-conjugated iron oxide nanoparticles for enhanced tumor homing and penetration for effective hyperthermia treatment applications. This specific particle system can enhance particle accumulation at the tumor site, providing a high enough concentration to induce hyperthermia conditions. The particles were characterized for size, stability, biocompatibility, and heating capabilities. The particles were stable in PBS and media over at least twelve hours, have a hydrated diameter of 50 nm, and can generate a significant amount of heat to raise solution temperatures well into the hyperthermia range. The cytotoxicity of the particles was analyzed through studies on A549 lung cancer cells for low particle concentrations with high exposure time and results determined that the particles have low cytotoxicity over both time frames and concentrations. Fibrinogen clots were used to determine the binding affinity of CREKA conjugated iron oxide nanoparticles compared to non-targeting nanoparticles. The binding and uptake of CREKA conjugated iron oxide nanoparticles into multicellular spheroids and monolayer cell cultures was also evaluated. This particle system is novel in that it incorporates the heating properties of iron oxide with the tumor homing properties of CREKA, to enhance particle homing to tumor sites.

**adolphe merkle institute**
excellence in pure and applied nanoscience

UNIVERSITY OF FRIBOURG SWITZERLAND

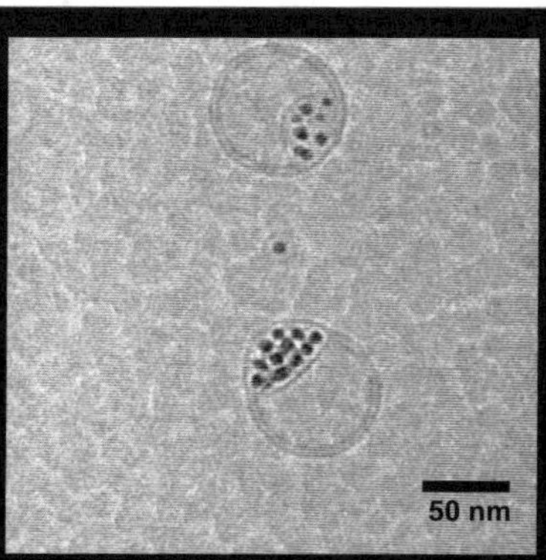


# Preparation, Characterization and Properties of Janus Magnetic Liposomes

**Cécile Bonnaud, Christophe A. Monnier, Davide Demurtas, Dimitri Vanhecke, Corinne Jud, Xavier Montet, Ruud Hovius, Marco Lattuada, Barbara Rothen-Rutishauser and Alke Petri-Fink**

Adolphe Merkle Institute
University of Fribourg, Switzerland

## Concept

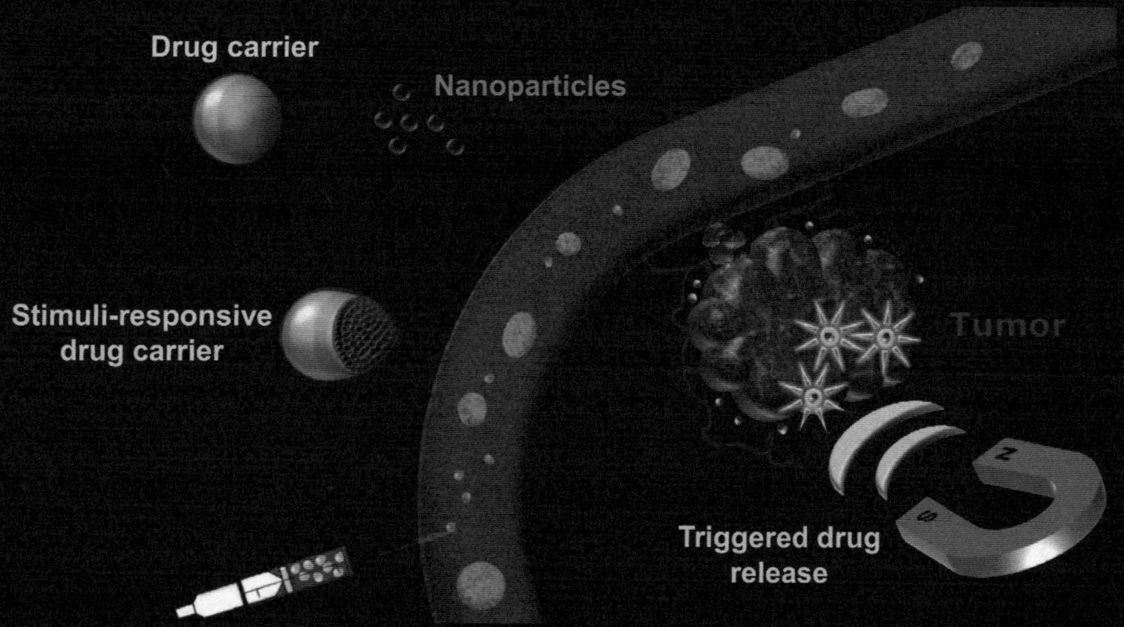

## Material 1: Liposomes

**Artificially constructed vesicles consisting of a phospholipid bilayer**

Biocompatible, nonimmunogenic drug carriers

On the market (eg. Doxil, DaunoXome)

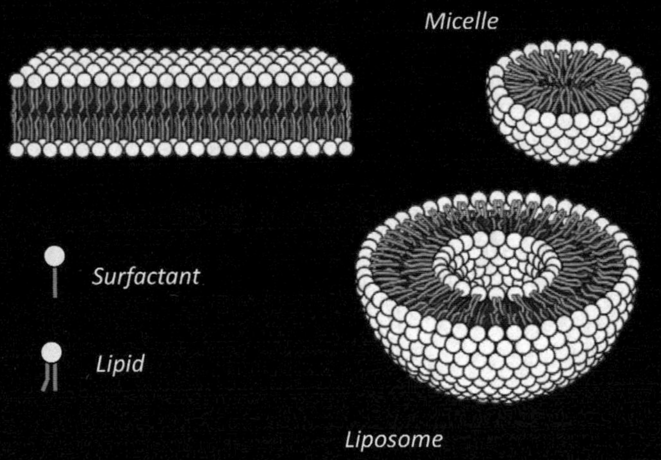


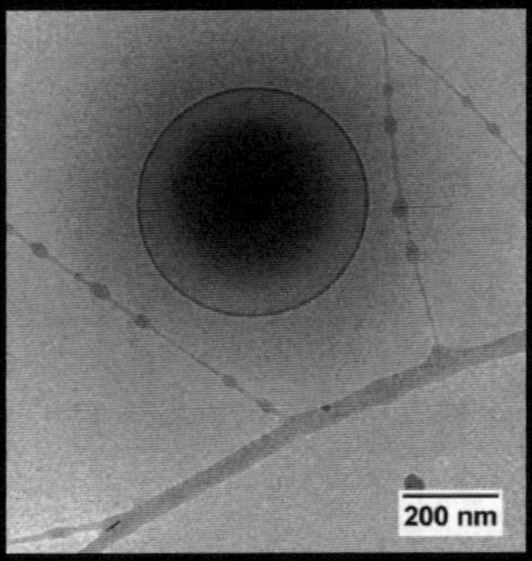


## Material 1: Liposomes

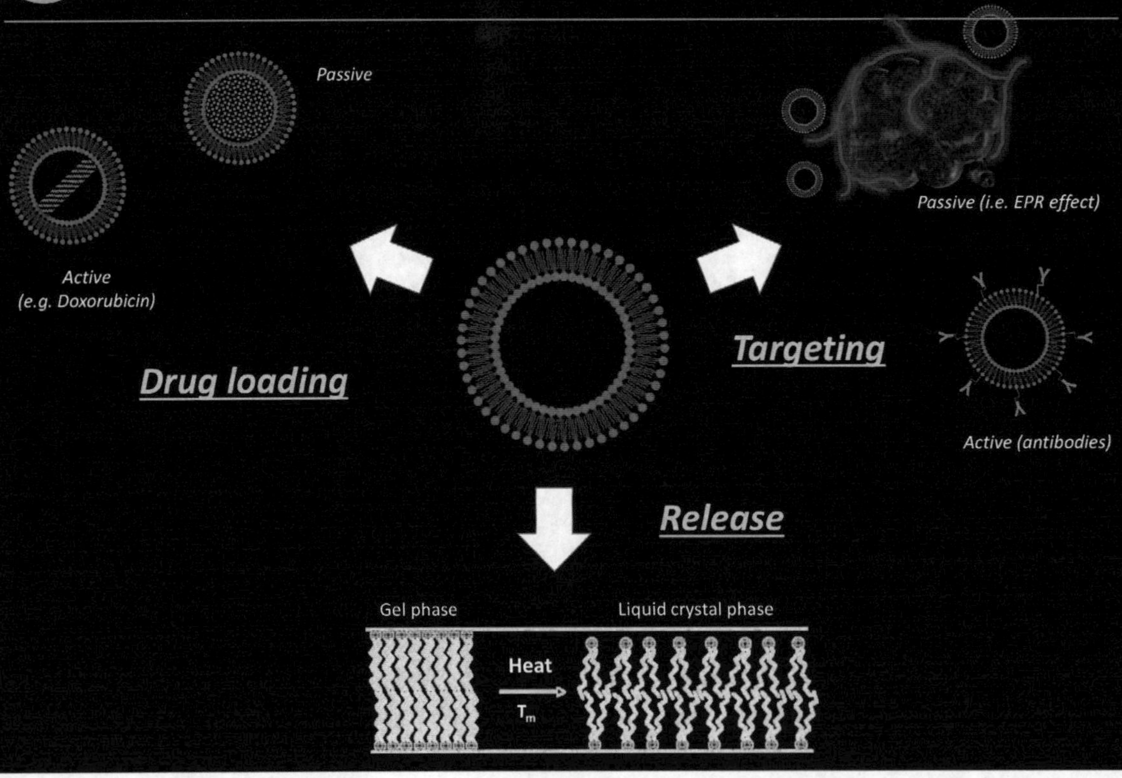

## Material 2: Active Triggers

**SuperParamagnetic Iron Oxide Nanoparticles (SPIONs)**

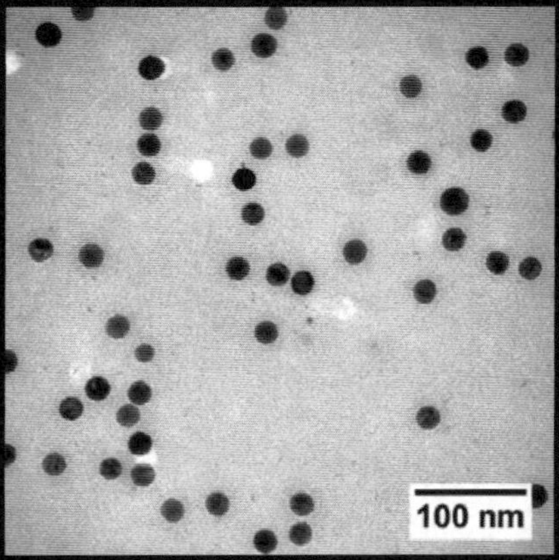


- Superparamagnetic properties between around d = 4 – 25 nm
- Heat dissipation when exposed to an alternating magnetic field
- Presently used in magnetic resonance imaging and hyperthermia

**Gold nanorods (AuNRs)**

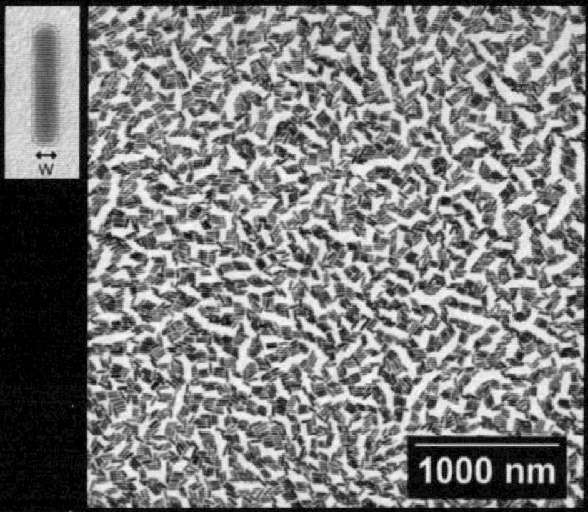


- Sensible near-infrared light
- Presently used in photodynamic therapy

## Particles + Lipsomes: Idea not new

### Existing work

Cryo-TEM

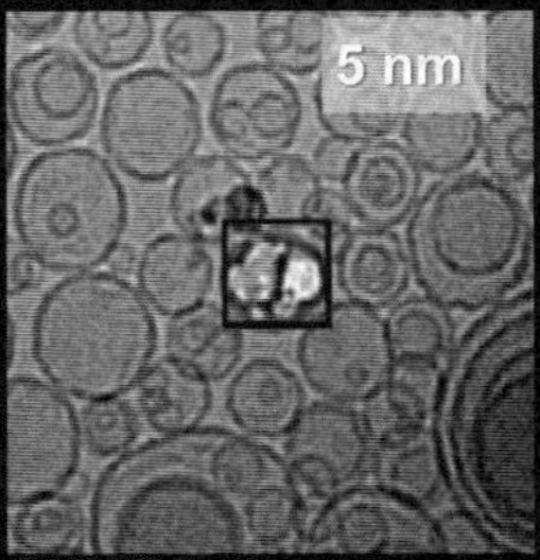


2010, ACS Nano

Cryo-TEM

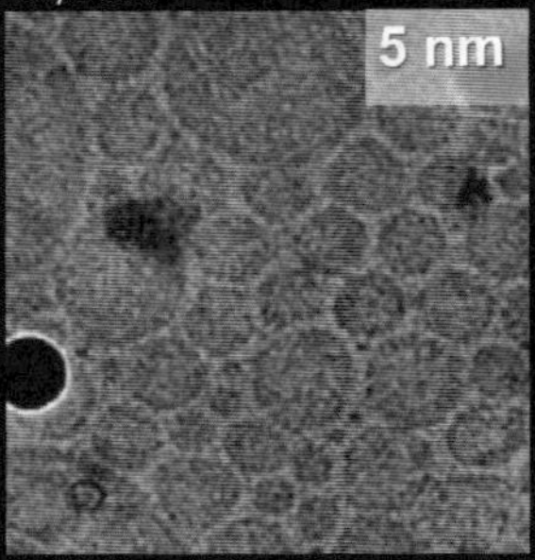


2011, Nano Letters

TEM

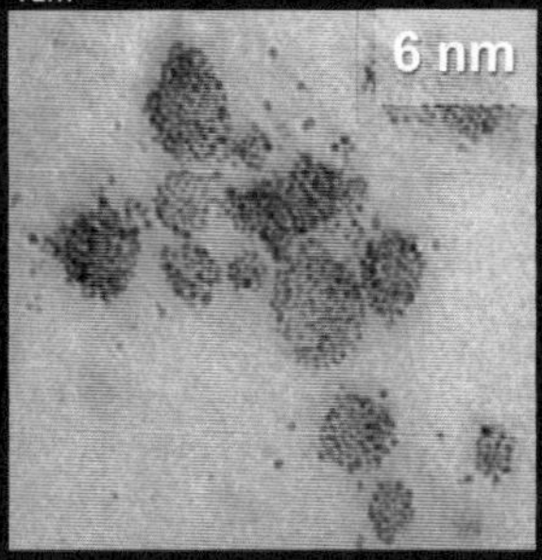


2013, Colloids and Surfaces : Biointerfaces

**Incorporate directly the colloid into the membrane**

- Control the self-assembly of phospholipids together with hydrophobic SPIONs
- Prove the incorporation of hydrophobic SPION into the bilayer
- Test size limitations
- Assess the release upon alternating magnetic field (AMF) exposure

## Particles + Lipsomes: Idea not new

**But…**

- Inclusive structures above 6.5 nm in diameter between the lipid bilayer are not possible (according to the literature)
- Small SPIONs (d < 6.5 nm) are not effective at clinically relevant field strengths and frequencies (i.e. specific absorption rate, SAR, is too low)

**Options:**

I. Increase frequency

II. Increase field strength

III. Increase particle diameter

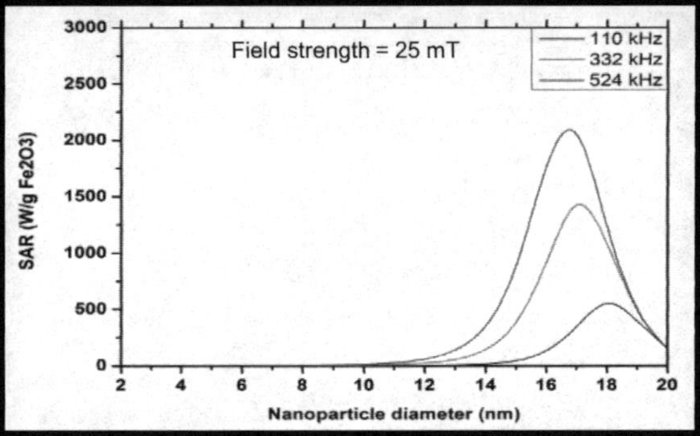


## Technical boundary conditions

Magforce Company (www.magforce.de)

- NanoTherm® therapy as an adjuvant therapy in combination with standard approaches
- Local treatment of solid tumors
- The method is based on the principle of introducing magnetic nanoparticles directly into a tumor and then heating them in an alternating magnetic field

Glioblastoma Multiforme: Clinical Phase I and II: finished

Alternating magnetic field: 100kHz; ≈ 19mT

NanoActivator®

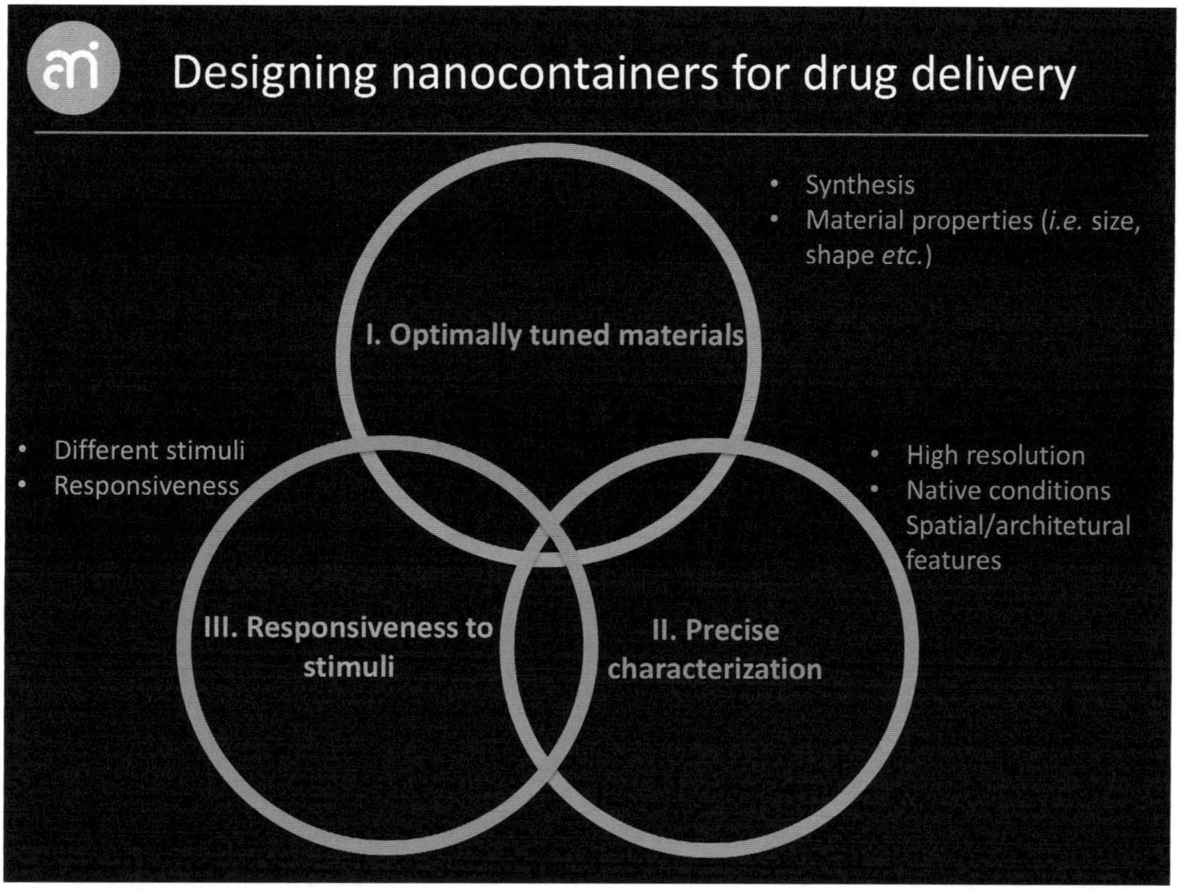
Designing nanocontainers for drug delivery
Synthesis
Material properties (i.e. size, shape etc.)
I. Optimally tuned materials
Different stimuli
Responsiveness
High resolution
Native conditions
Spatial/architetural features
III. Responsiveness to stimuli
II. Precise characterization

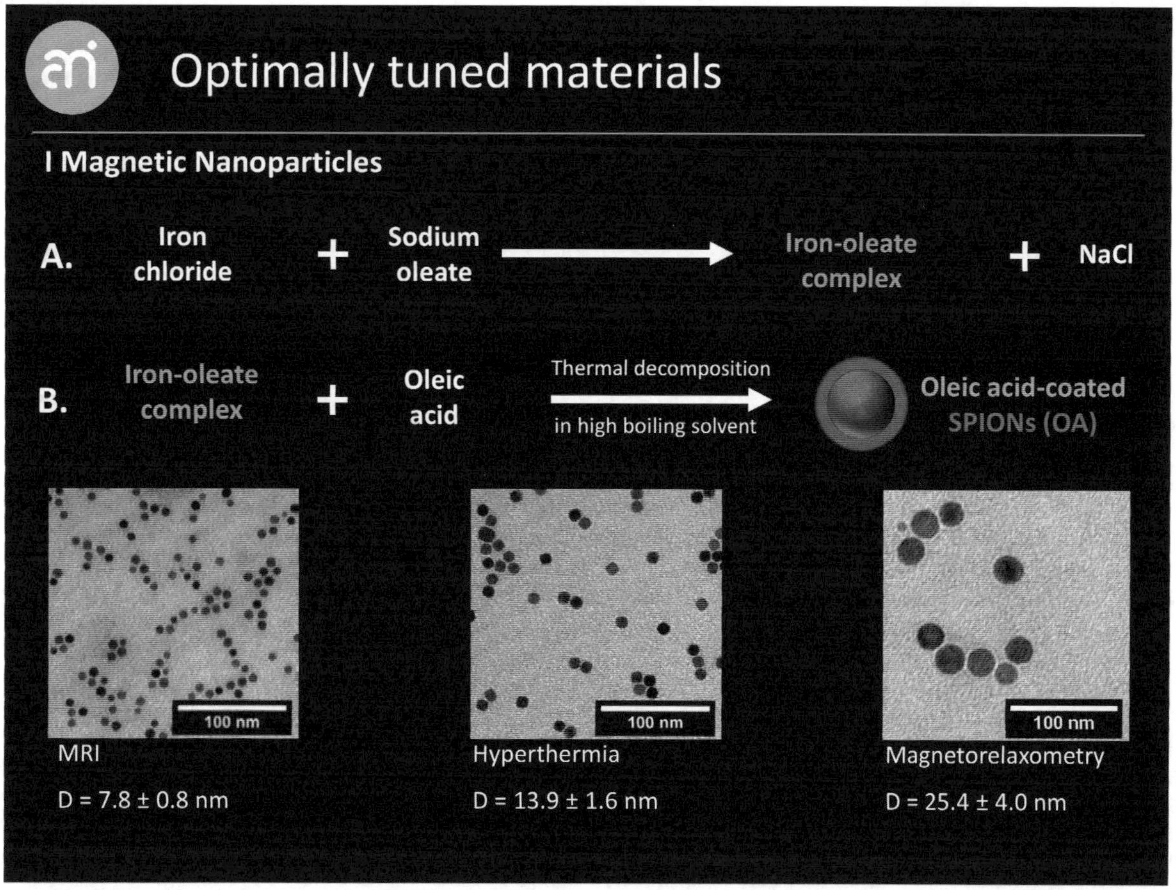
Optimally tuned materials
I Magnetic Nanoparticles
A.
Iron chloride
+
Sodium oleate
Iron-oleate complex
+
NaCl
B.
Iron-oleate complex
+
Oleic acid
Thermal decomposition
in high boiling solvent
Oleic acid-coated SPIONs (OA)
100 nm
MRI
D = 7.8 ± 0.8 nm
100 nm
Hyperthermia
D = 13.9 ± 1.6 nm
100 nm
Magnetorelaxometry
D = 25.4 ± 4.0 nm

# Optimally tuned materials

## II Liposomes & SPIONs

### Lipid film rehydration

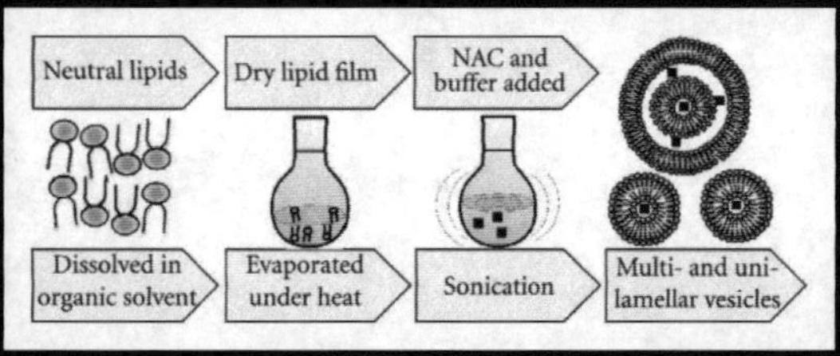


openeni.nlm.nih.gov

Lipid composition (Tm ≈ 42 °C)

- DPPC
- DMPE
- DPPE-PEG-2000
- DPPG
- Cholesterol

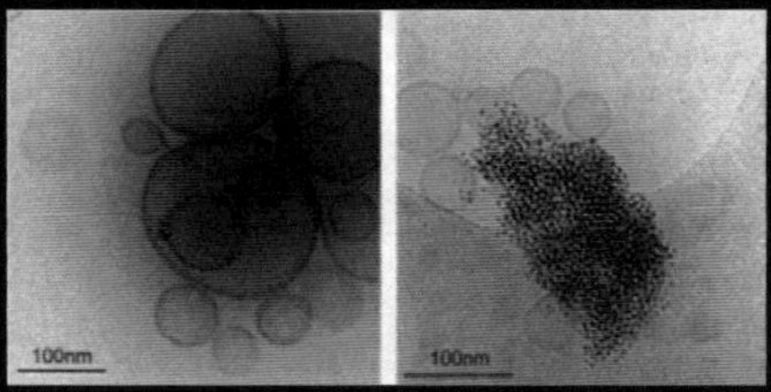


**Strong particle interaction within the film**

- Lateral aggregation
- Oleic acid steric stabilization not strong enough under these conditions

1,2-dipalmitoyl-sn-glycero-3-phosphocholine (DPPC), 1,2-dimyristoyl-sn-glycero-3-phosphoethanolamine (DMPE), 1,2-dipalmitoyl-sn-glycero-3-phospho-(1'-rac-glycerol) (DPPG)

# Optimally tuned materials

## II Liposomes and SPIONs

### I. Mixed micelle preparation

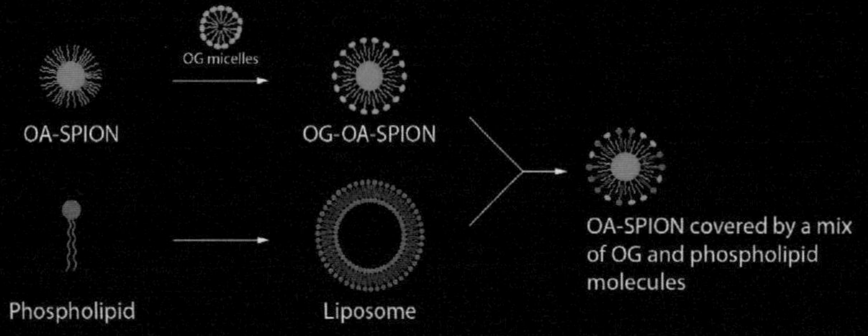


### II. J-MLs formation

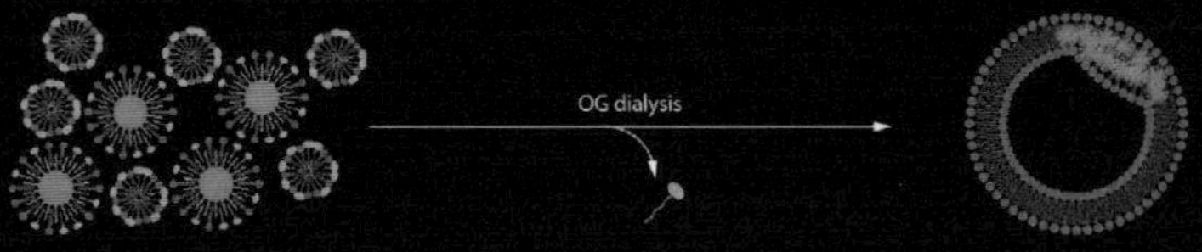


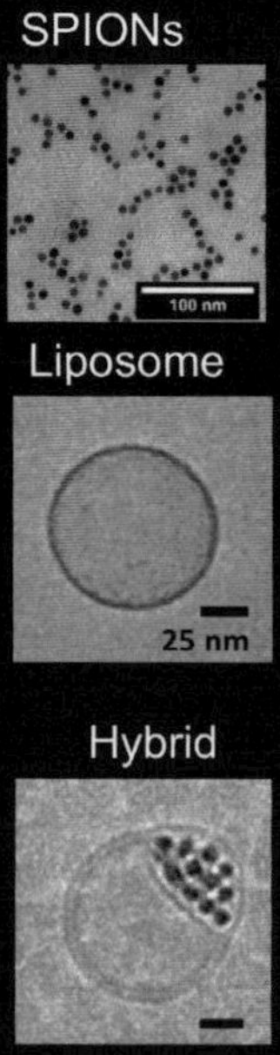


OA: Oleic acid; OG: $\beta$-Octylglucoside

# Optimally tuned materials

After dialysis of surfactant Octyl glycoside

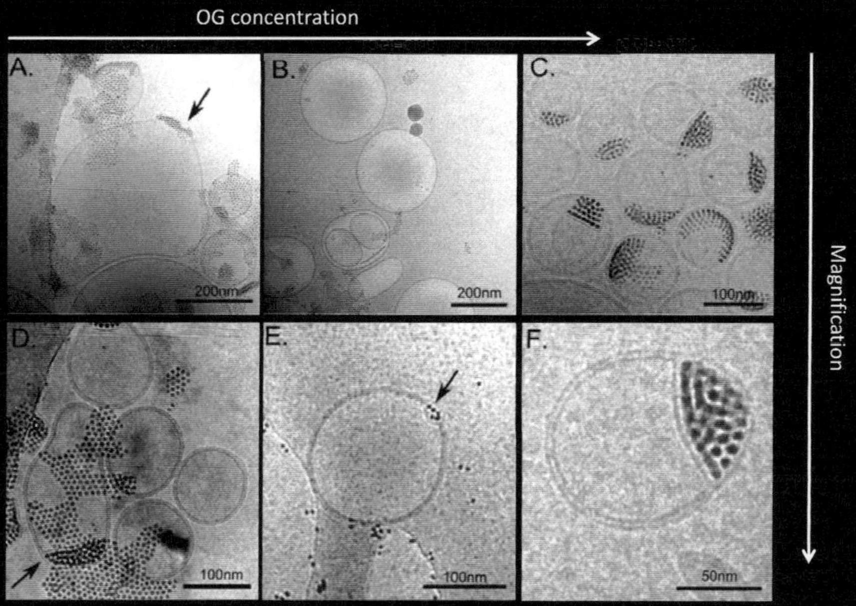


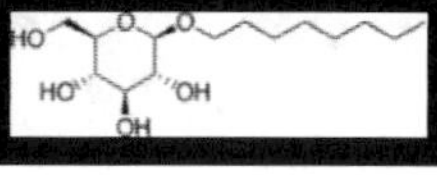


# Optimally tuned materials

**Janus magnetic liposomes**

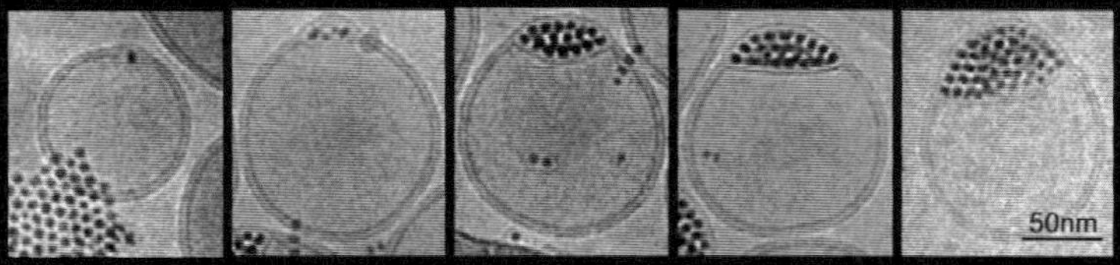


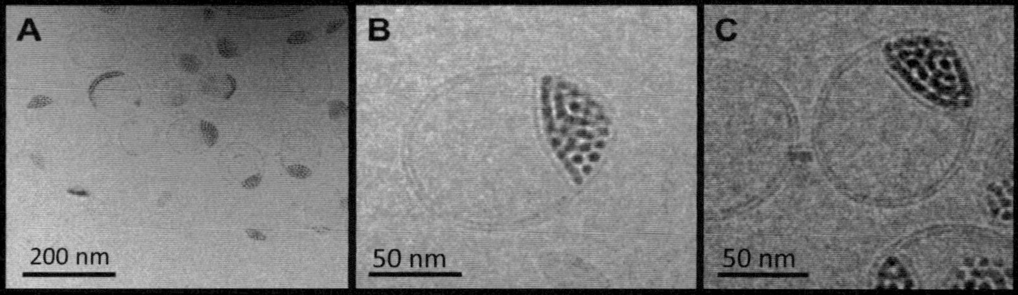


Bonnaud, Monnier *et al.*, ACS Nano 8/4 (2014) 3451.

# Precise characterization

## Incorporated?

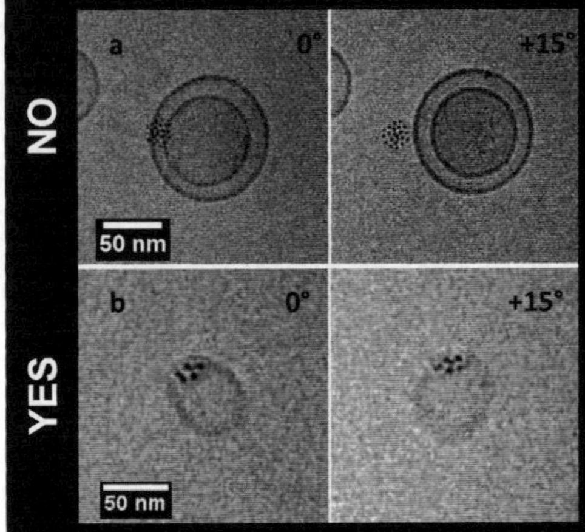


Cryo-TEM

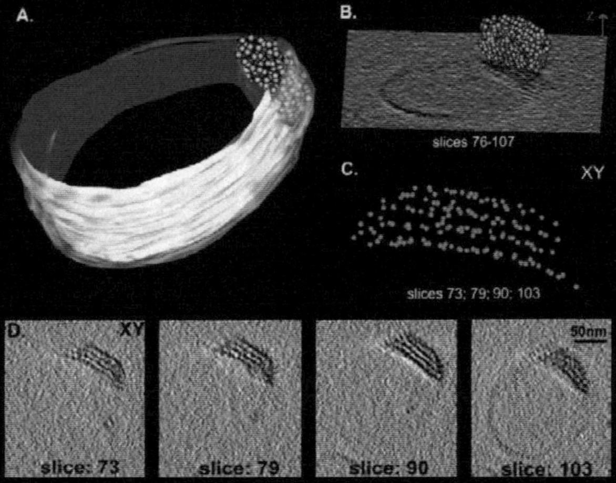


Cryo-EM tomography

Bonnaud *et al.*, IEEE Transactions on Magnetics 49/1 (2013) 166-171

# Optimally tuned materials

- Nanoparticles form flexible clusters before vesicle formation
- Larger quantities and volumes are incorporated

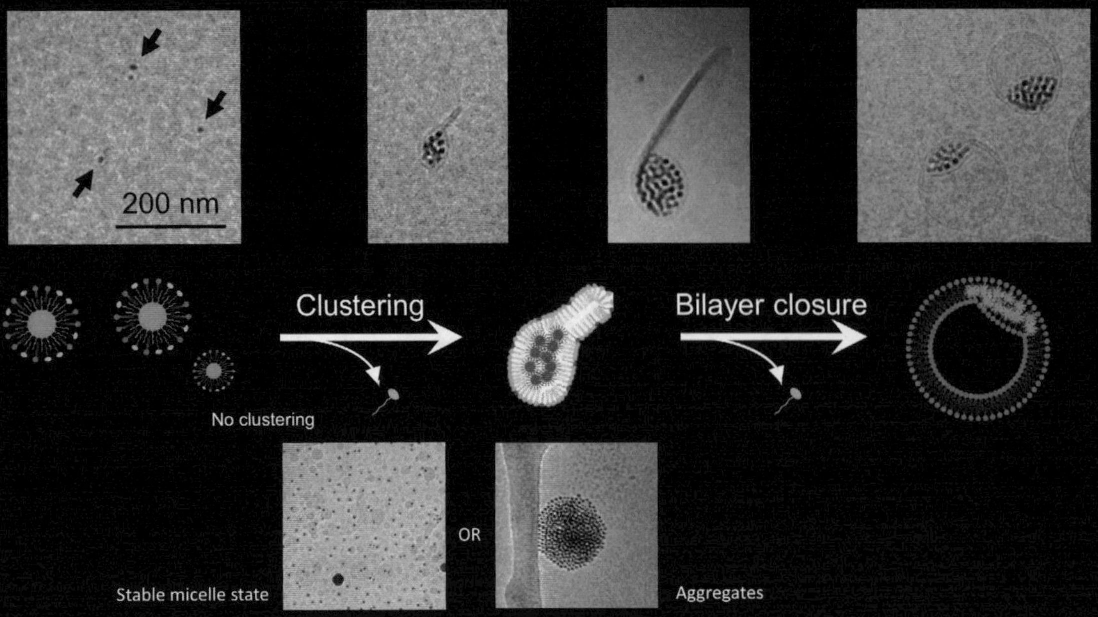

## Open questions

- Why do particles form these clusters inside the lipid membrane?
- Why particles are not dispersed individually inside the lipid bilayer?
- How is this energetically possible?

## Energetics of inclusions into a lipid membrane

- In the work by Wi et al. (J Phys. Cond. Matt., 2008), the authors analyzed the effect of a spherical inclusion in a lipid bilayer
- They assumed that the lipid bilayer is deformed because of the inclusion. **The shape of the lipid bilayer is given.**
- The presence of an inclusion has two effects:
  - It bends the membrane
  - It stretches the membrane

## Energetics of inclusions into a lipid membrane II

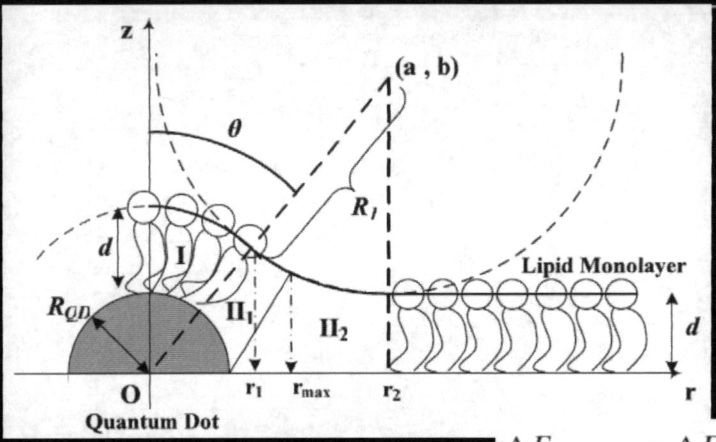


Wi et al. (J Phys. Cond. Matt., 2008)

$$\Delta E_{\text{def,QLC}} = \Delta E_{\text{bend}} + \Delta E_{\text{stretch}}$$

$$\Delta E_{\text{bend}} = \frac{\kappa}{2}\int_{\text{I}}\left(\frac{2}{R_{\text{QD}}+d} - \frac{1}{R_0}\right)^2 \mathrm{d}A + \frac{\kappa}{2}\int_{\text{II}}\left(-\frac{1}{R_1} + \frac{1}{R_2} - \frac{1}{R_0}\right)^2 \mathrm{d}A$$

$$\Delta E_{\text{stretch}} = \frac{K}{2}\int_{\text{II}}\left(\frac{u}{d}\right)^2 \mathrm{d}A.$$

## Energetics of inclusions into a lipid membrane II

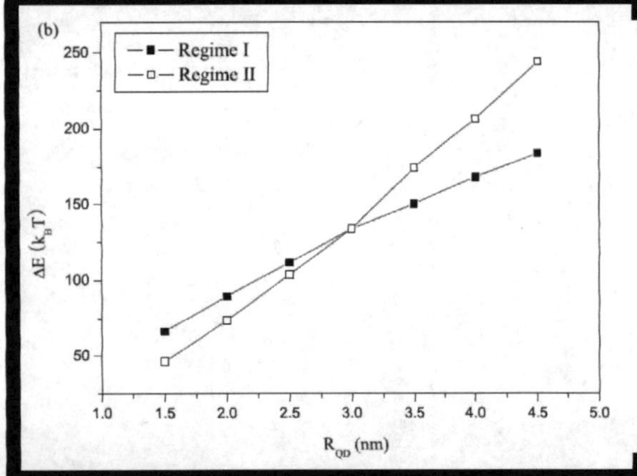


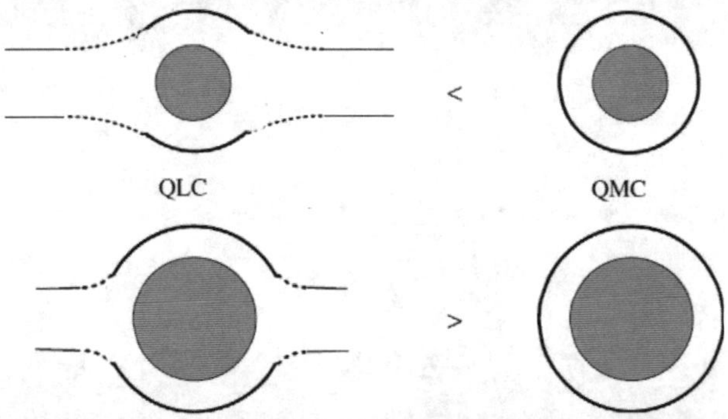


The model predicts that inclusions smaller than a certain size are incorporated in the lipid membrane, while larger inclusions are not

## What happens in the case of Janus magnetic liposomes?

- In the case of liposomes, the model was extended to accommodate non-spherical inclusions
- The shape of the inclusion has been assumed to be a spherical cap
- SPIONs in the inclusion can rearrange into a shape that minimizes the energy, keeping the volume constant

## Results

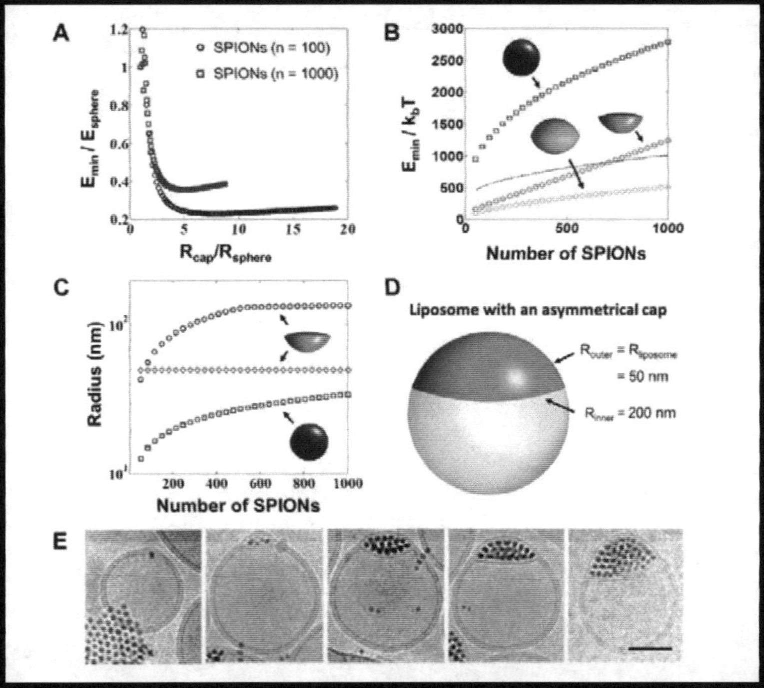

## Responsiveness to stimuli

### Can larger particles fit in the bilayer?

- By using clusters, present size limitations can be circumvented

6 nm 8 nm 12.5 nm 17.5 nm

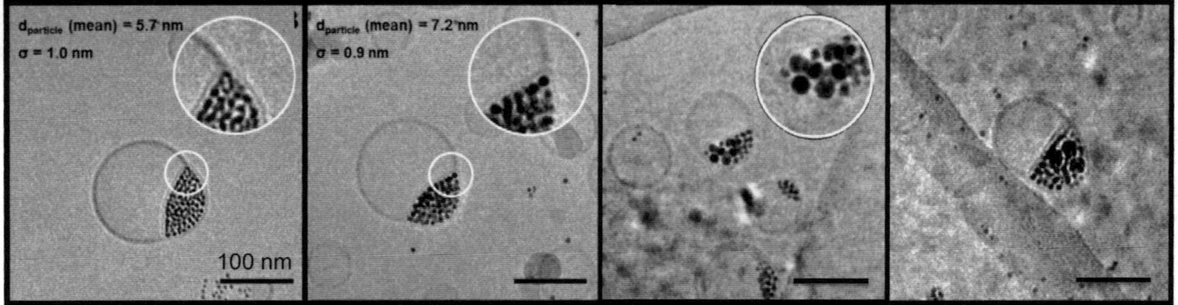


## Responsiveness to stimuli

### Triggered release

- Methylene blue (MB) as a hydrophilic model drug

Unloaded vs. loaded Before exposure After exposure

AMF exposure: f = 110 kHz, field strength = 17 mT, t = 30min

- Highly concentrated intraliposomal MB is released by an alternating magnetic field

# Summary

## Key points

- Janus magnetic liposomes are a new class of nanoparticle-vesicle hybrids which could qualify for theranostic application
- The applied materials are biocompatible/degradable and are already found in medical settings
- Their effectiveness can be optimized (i.e. with clusters of various NP sizes) to cover a broader field of application (MRI, drug release, hyperthermia, thermography etc.)

## Outlook

- Next steps will include further investigation on release kinetics and feasibility tests


# Acknowledgement

# 383429 In Situ Evaluation of Nanoparticle–Protein Interactions By Dynamic Magnetic Susceptibility Measurements


**Abstract:**

A technique to study protein-nanoparticle interactions *in situ* is proposed. The technique consists of magnetic measurements of the rotational diffusion of thermally blocked cobalt ferrite nanoparticles in protein solutions. To illustrate the technique we studied the effect of degree of carboxylic acid substitution in carboxymethyl dextran coated magnetic nanoparticles on their interaction with model proteins such as albumin (BSA), lysozyme (LYZ), immonuglobulin G (IgG), fibrinogen (FIBR), apo-transferrin (TRANS), and histone (HIS) in a wide range of protein concentrations. Experiments indicated that interactions between negatively charged particles and negatively charged proteins BSA, IgG, FIBR, and TRANS were negligible. On the other hand, positively charged proteins LYZ and HIS seemed to readily absorb on the particle surfaces, as evidence by an increase in size and eventual aggregation of the particles. Onset of this effect seemed to happen at lower concentration of HIS, compared to LYZ. The technique can be easily applied to the other particle surface coatings and to particles in complex protein mixtures such as serum, allowing systematic *in situ* studies of protein-nanoparticle interactions.

**Synthesis of a polymer-magnetic particle platform for "tailored" multimodal materials for imaging and treatment**

Roland Stone[1], Bin Qi[1], Terri Bruce[2], Tamara McNealy[3], Tzuen-Rong Jeremy Tzeng[3], O. Thompson Mefford[1]
[1]*Department of Materials Science and Engineering, Clemson University, Clemson, SC USA*
[2]*Light Imaging Facility, 190 Collings Street, Clemson University, Clemson, SC, USA*
[3]*Department of Biological Sciences, 132 Long Hall, Clemson University, Clemson, SC, USA*

00Magnetic nanoparticles have been studied for many years for use in biomedicine, not only for their high surface area, but also because of its unique magnetic properties. They can magnetically interact with their environment, be guided to a specific location, and manipulated to release energy in the form of heat. To ensure that these magnetic nanoparticles survive in the circulatory system, they must be modified with materials to make them colloidally stable in water and shield them from the body's immune response to foreign objects.[1]

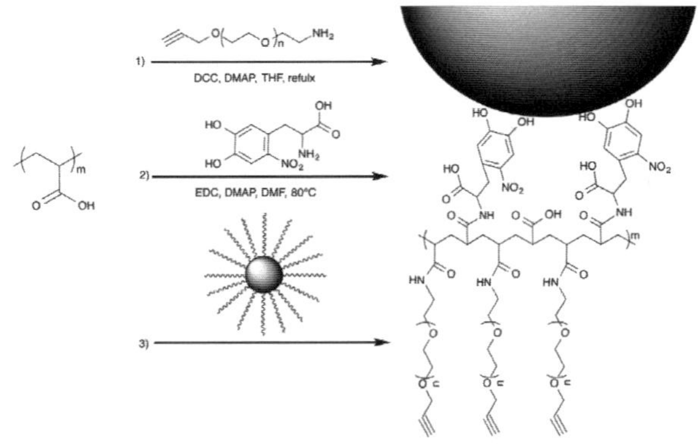


**Figure 0: Reaction schematic to produce colloidal stable platform for multimodal theranostic magnetic nanoparticle**

This presentation will describe the synthesis of a multi-anchored universal ligand for iron oxide nanoparticles, with improved stability in biological environments, while also providing a platform for additional functionality. The particles reported in this talk are coated with a poly(acrylic acid) oligomer that has been modified with a heterobifunctioal polyethylene oxide (PEO) with an terminal end capable of 'click' chemistry. In addition, the oligomer was modified with nitroDOPA to provide strong binding to the surface and used in a mulitdentate approach provides biocompatibility and enhanced stability in fetal bovine serum and phosphate buffer saline.[2,3] These colloidally stable biocompatible polymer-particles complexes were then be modified with a near-infrared dyes (e.g. Cy5) and utilized in characterizing the integration of magnetic nanoparticles in biofilms *(Legionella pneumophila)*. The particles were then added to the biofilm, and their position was then observed using a super resolution ground state depletion microscope. In addition to imaging, we have also utilized the same platform for the targeting of different strains of bacteria through 'clicking' on species-specific moieties. The modified particles adhere to the targeted bacteria strains and agglomerate. Through the application of an alternating field, magnetic energy can be transformed to promote cellular death, resulting in a multi-log reduction in bacteria population.

What will be presented represents the initial findings of the research opportunities available with this new platform for diagnostic and therapeutic applications. These universal magnetic nanoparticles can be modified with different fluorescent dyes imaging biofilms, carbohydrates for targeting bacteria, and other moieties for multifunctional diagnostic probes to show the versatility of this design.

1. Stone, R., Willi, T., Rosen, Y., Mefford, O. T. & Alexis, F. Targeted magnetic hyperthermia. *Therapeutic Delivery* **2,** 815–838 (2011).
2. Saville, S. L., Stone, R. C., Qi, B. & Mefford, O. T. Investigation of the stability of magnetite nanoparticles functionalized with catechol based ligands in biological media. *J. Mater. Chem.* **22,** 24909 (2012).
3. Roland Stone, Bin Qi, David Trebatoski, Ragini Jetty, Iurii Bandera, Stephen Foulger, and Olin Thompson Mefford, A Versatile Stable Platform for Multifunctional Applications: Synthesis of NitroDOPA-PEO-Alkyne Scaffolding for Iron Oxide Nanoparticles, *Journal of Materials Chemistry B* (2014) *In Press.*

Mefford@clemson.edu
Meffordresearch.com

# 364284 Nanostructured Size-Selective Catalysts Prepared By Molecular Layer Deposition

**Abstract:**

Heterogeneous catalysts are widely used. However, heterogeneous catalysts cannot selectively convert specific molecules in the reactant mixture to catalyze only desired reactions. Size-selective catalysts with a metal core and porous oxide shell have a promising structure that can increase the reaction selectivity through reactant molecular discrimination. This paper describes a general method to create porous metal oxide films with well-defined porous structures and precisely controlled thickness down to several angstroms from dense organic/inorganic hybrid metal alkoxide films grown by molecular layer deposition (MLD). A novel nanostructured catalyst with an ultra-thin porous shell obtained from the thermal decomposition of an aluminum alkoxide film deposited by MLD for size-selective reactions was developed. The molecular sieving capability of the porous metal oxide films was verified by examining the liquid-phase hydrogenation of n-hexene versus cis-cyclooctene. The success of making these materials by MLD opens up a new method for preparing size-selective catalysts.

# Platinum Nanoparticles Supported on Multiwalled Carbon Nanotubes Encapsulated by Cerium Oxide: A Highly Reactive and Stable Catalytic System

Dang-guo Cheng
College of Chemical and Biological Engineering,
Zhejiang University, Hangzhou 310027, China

# Content

- Background
- Catalyst preparation
- Characterizations and Performance
- Summary
- Acknowledgement

# Background

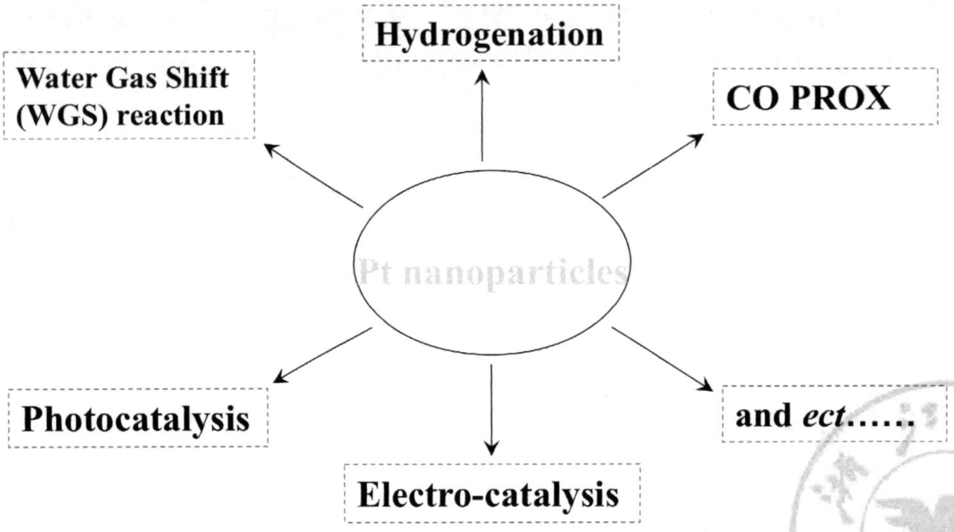


# Background

## Supported Pt nanoparticles catalyst

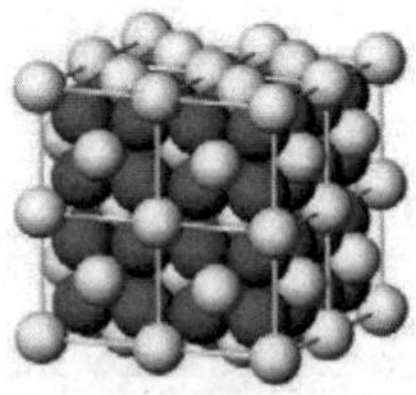

- **Metal oxide: $CeO_2$, $TiO_2$, $Al_2O_3$**

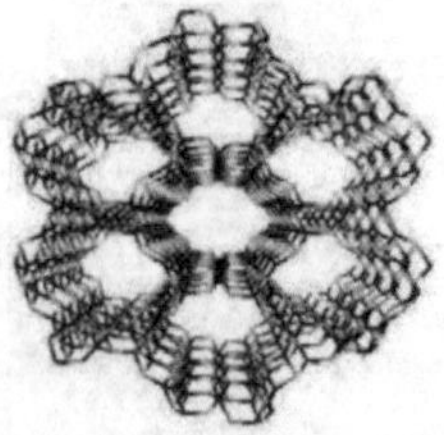

- **Zeolites: HZSM-5, HY, SBA-15**
- **Carbon material: Nanofiber, Nanotube**

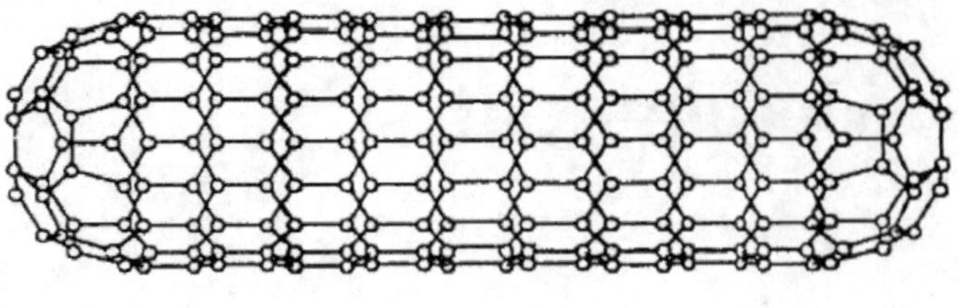

# Background

ZHEJIANG UNIVERSITY

- Aggregation
- Sintering
- Leaching

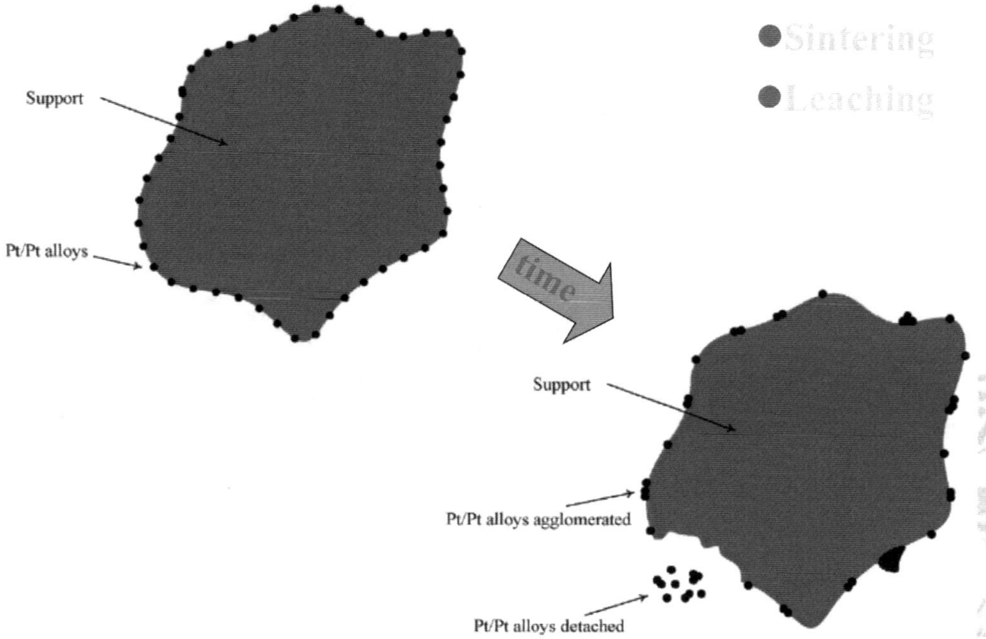


# Background

ZHEJIANG UNIVERSITY

Strategies to prevent the aggregation, sintering, and leaching

- **Exploitation of the strong metal–support interaction**
- **Formation of multi-metallic alloys**
- **Addition of co-catalysts**

Alternative method: Encapsulation

- **To encapsulate NPs in a thermally stable inorganic shell layer to prevents their mobility under reaction condition**

# Catalyst preparation

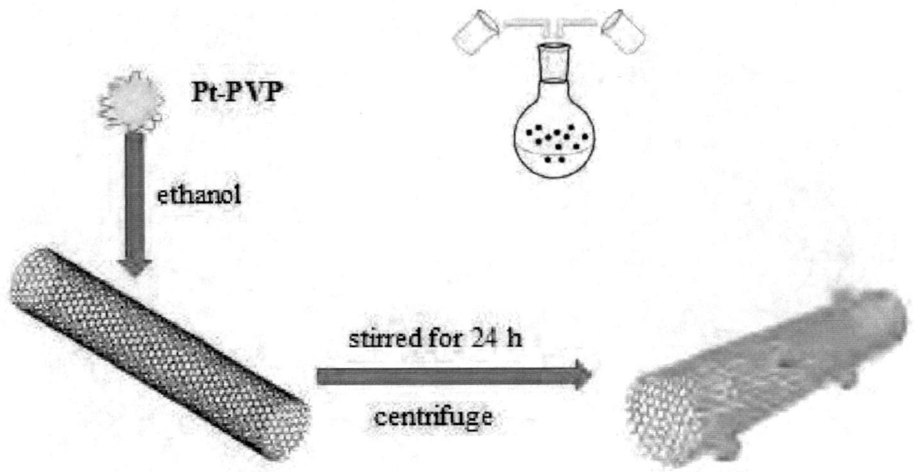


# Catalyst preparation

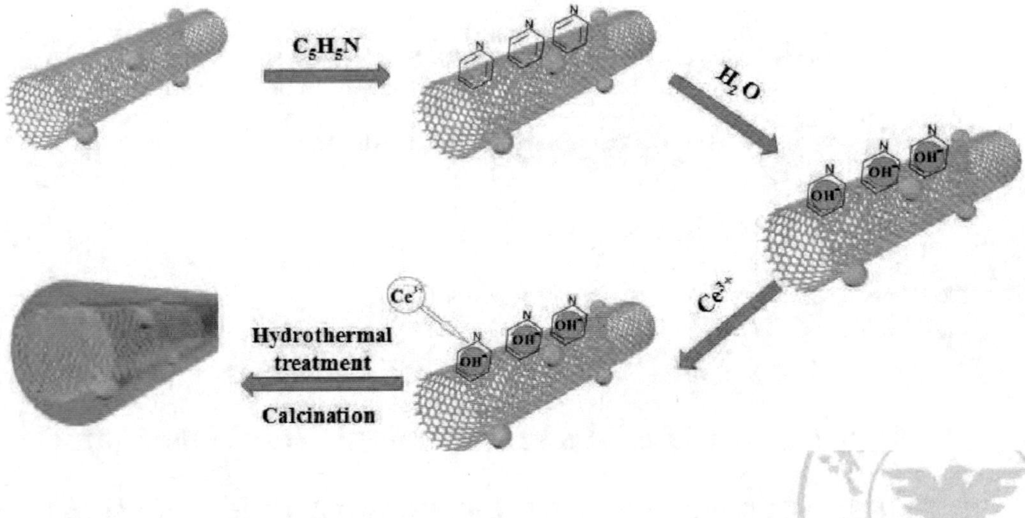

# Characterizations and Performance

Pt-PVP, Polyol reduction method

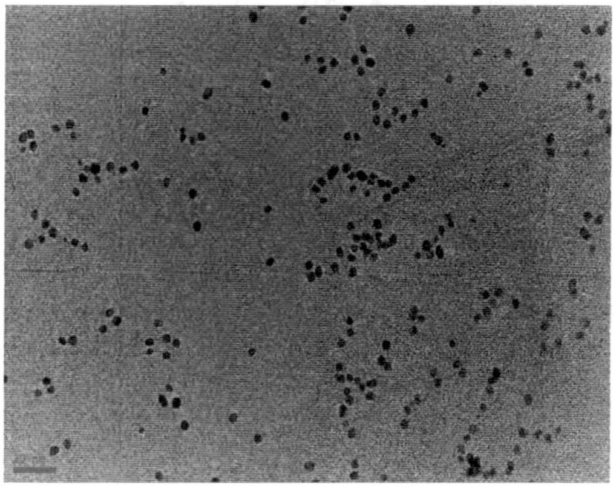

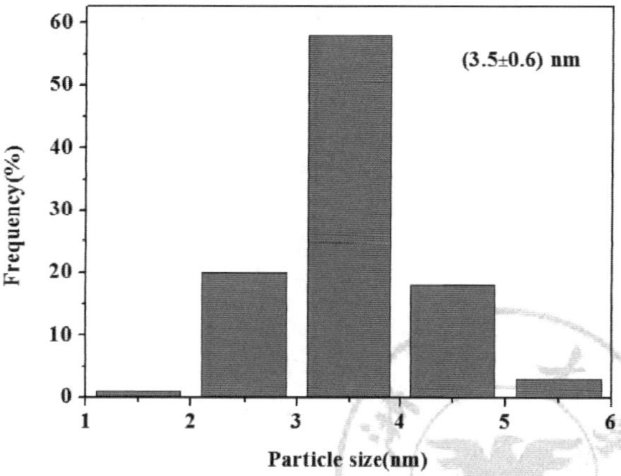


# Characterizations and Performance

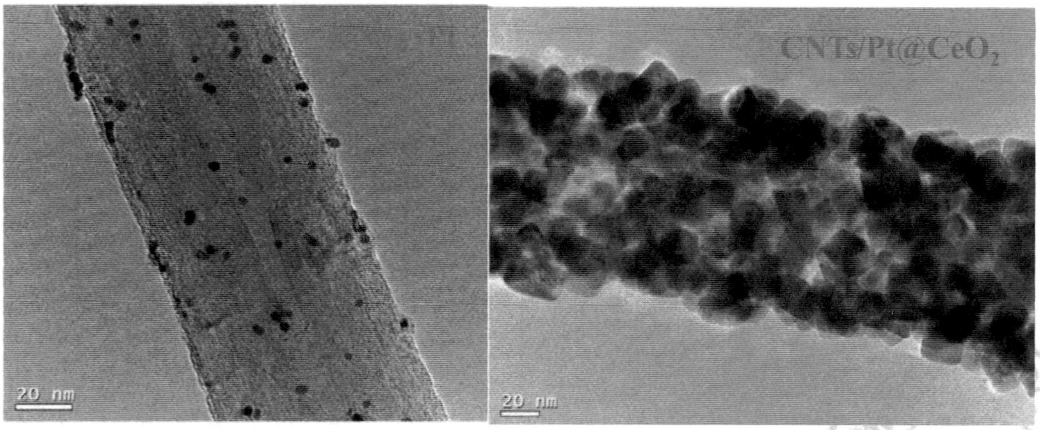

# Characterizations and Performance

浙江大学 ZHEJIANG UNIVERSITY

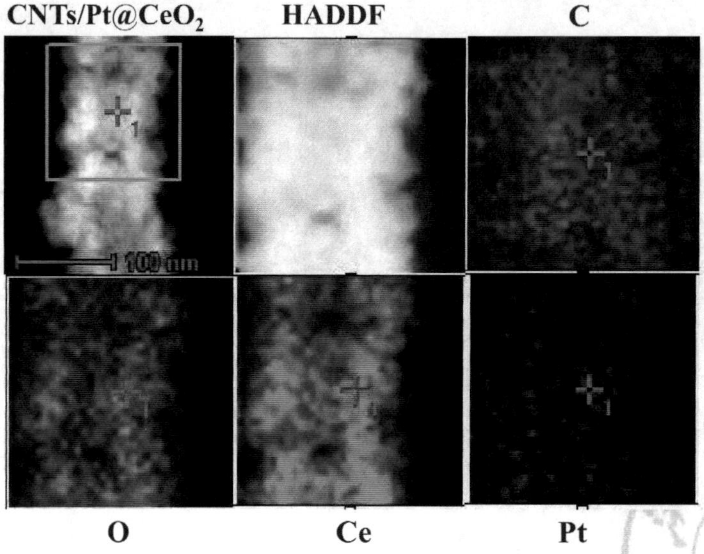


# Characterizations and Performance

浙江大学 ZHEJIANG UNIVERSITY

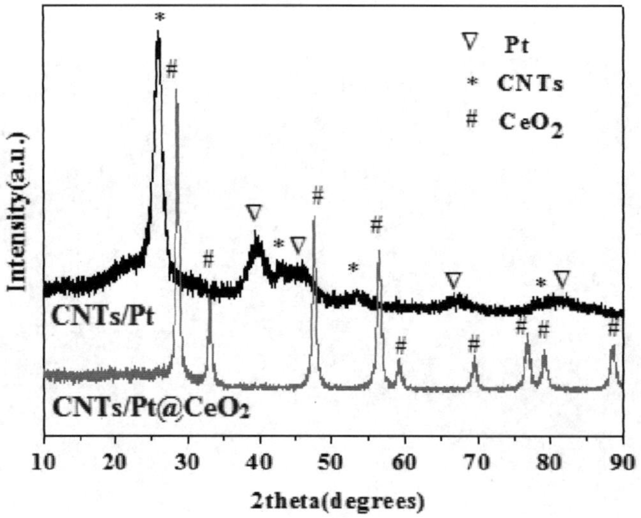

## Characterizations and Performance

浙江大學 ZHEJIANG UNIVERSITY

| Sample | $S_{BET}$($m^2g^{-1}$) | Pore Volume($cm^3g^{-1}$) | Pore Diameter (nm) |
|---|---|---|---|
| CNTs | 110 | 0.38 | 3.2 |
| CNTs/Pt | 72 | 0.24 | 3.4 |
| CNTs/Pt@$CeO_2$ | 47 | 0.17 | 2.8/3.8 |

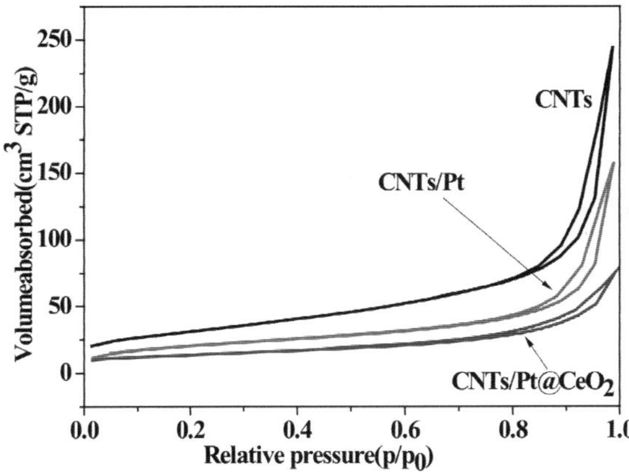


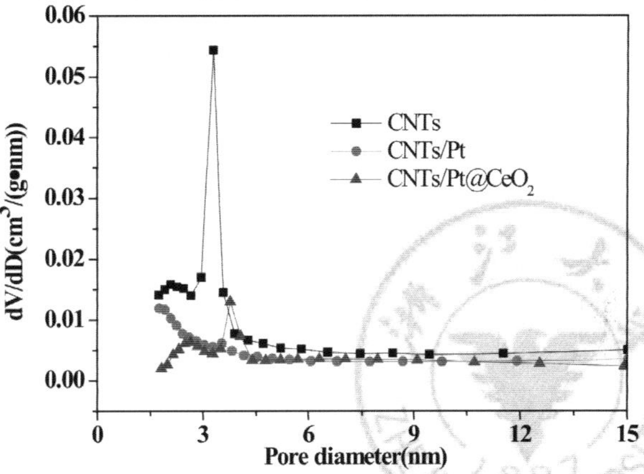



## Characterizations and Performance

浙江大學 ZHEJIANG UNIVERSITY


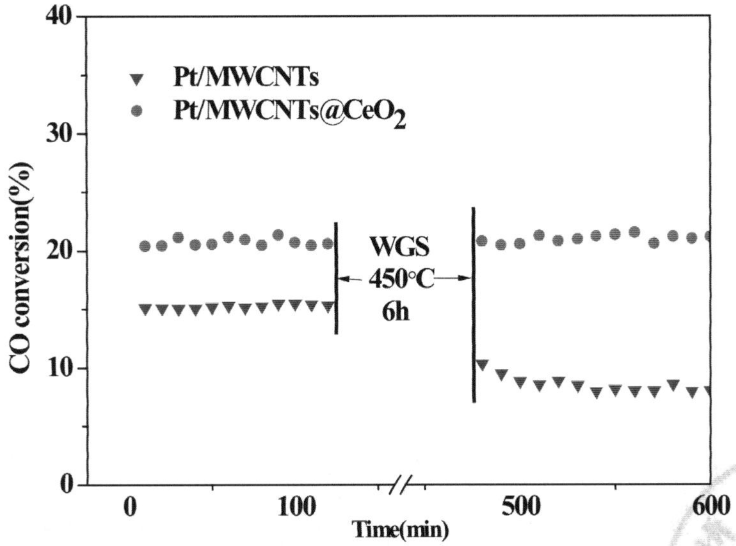


**Reaction conditions: 250°C; 0.05 g catalyst; 80 ml/min: 1.0 vol % CO, 3.0 vol % $H_2O$, diluted in Ar**

# Summary

浙江大學 ZHEJIANG UNIVERSITY

- CNTs/Pt@$CeO_2$ catalysts have been synthesized successfully by the Pyridine-thermal synthesis methods.
- Compared with CNTs/Pt, both the activity and stability of CNTs/Pt@$CeO_2$ catalysts have been enhanced in WGS reaction.
- Further studied need to confirm the existence of the interfacial effect and nanometer effect in CNTs/Pt@$CeO_2$ catalysts

## Acknowledgment

- Financial supports from National Natural Science Foundation of China and Zhejiang Provincial Natural Science Foundation
- The experimental work by Mr. Chao Wan and valuable discussions with Prof. [illegible] Chen

# 371281 Using Active Site Confinement and Organic Ligand Approaches to Control Catalysis

**Abstract:**

The local environment of a catalyst's active site plays a critical role in determining the catalyst's activity and selectivity. This effect is seen most profoundly in enzymatic systems where the environment of the active site is controlled through the chemical entities surrounding the active site and their precise geometric locations.

Inspired by the GroEL/GroES system in *E. coli*, examples with a metal complex/cluster that is immobilized inside of mesoporous materials, particularly SBA-15, will be discussed. These nano-confined active sites demonstrate an increased degree of activity and stability which provide valuable insight into structure-function relationships in catalysis. [1] [2]

Additionally, in different gold cluster systems, we explore approaches to controlling the environment that surround these sites as well as their ultimate effects on catalysis. The use of organic ligands, such as calixarenes, is investigated to confine and to control catalytically active noble metal sites on metal oxide supports and in homogeneous solution, using an approach that is conceptually similar to that used by enzymes to control reactivity through organic side chains acting as ligands. Gold catalysts have been synthesized with calixarene ligands bound to the metal nanoparticle, confining the gold cluster, yet these clusters still demonstrate reactant accessibility to the metal nanoparticle surface as measured by chemisorption probe experiments. [3] [4] Catalytic activity has been demonstrated for these systems using different reaction systems such as reduction reactions on these gold catalysts. Results provide a unique perspective into the structure-activity relationships present in gold cluster catalysis.

References:

[1] F. Marras, J. Wang, M.-O. Coppens, J. N. H. Reek. Chem. Comm. **2010**, *46*, 6587.

[2] L.-C. Sang, A. Vinu, and M.-O. Coppens, *Phys. Chem. Chem. Phys,* **2011**, *13*, 6689.

[3] N. de Silva, J.-M. Ha, A. Solovyov, M. M. Nigra, I. Ogino, S. W. Yeh, K. A. Durkin, A. Katz. *Nature Chem.* **2010**, *2*, 1062.

[4] M. M. Nigra, A. J. Yeh, A. Okrut, A. G. DiPasquale, S. W. Yeh, A. Solovyov, A. Katz. *Dalton Trans.* **2013**, *42*, 12762.

# Sinter-resistant metal nanoparticles encapsulated by zeolite nanoshell

Shilpi Saxena[a], Rupesh Singh[a], Raj Ganesh S. Pala[a] and Sri Sivakumar[a,b]

shilpi@iitk.ac.in, rupeshs@iitk.ac.in, rpala@iitk.ac.in, srisiva@iitk.ac.in

[a]Department of Chemical Engineering, Indian Institute of Technology Kanpur, (UP), India.
[b]Material Science Programme, DST Thematic Unit of Excellence on Soft Nanofabrication, Centre for Environmental Science & Engineering, Indian Institute of Technology Kanpur, (UP), India.

*KEYWORDS: nanoshell, sinter-resistant, gold nanoparticles, ZSM-5 and MCM-22.*

Nanocatalysts have distinct activity and selectivity, which can be achieved by controlling their size, shape[1], surface composition, and electronic structure. In particular, gold nanoparticle possesses unique catalytic properties when they are supported on metal oxides/carbon and are utilized for various reactions such as CO oxidation[2], acetylene hydrochlorination[3], hydrogenation[4], selective oxidation[5] and epoxidation[6]. However, they may suffer from sintering and subsequent loss in activity due to harsh reactions conditions (e.g. high temperature, pressure, solvents etc.). This can be circumvented by encapsulating the gold nanoparticles with different shell materials such as silica, mesoporous carbon, and polymer shell. Although such approaches decrease sintering, the protective shell does not augment the product selectivity of the composite "nanoparticle core"-"protective shell" catalysts. Towards this goal, the present work explores and demonstrates the feasibility of structured encapsulants with well-defined pores to suppress sintering of encapsulated nanoparticles. We propose a general hydrothermal based synthetic strategy to assemble bifunctional catalysts in which a single gold nanoparticle is encapulated by a variety of zeolites-nanoshell (e.g. MCM-22 and ZSM-5). Gold nanoparticles were synthesized by reduction of chloroauric acid by sodium citrate[7]. These nanoparticles were coated with silica (denoted as Au@SiO2) followed by transformation into Au@MCM-22 and Au@ZSM-5 by modified hydrothermal synthesis, by varying amount of Si/Al. The zeolite nanoshells are of uniform shell thickness, morphology, possess high thermal stability and play the dual role of stabilizing the nanoparticles against sintering and their distinct pore structure can facilitate shape selective catalysis.

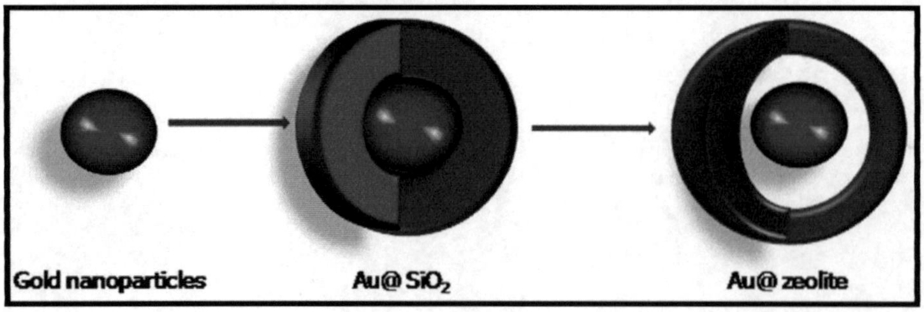


**Schematic representation of synthesis of gold nanoparticle encapsulated by zeolite-shell.**

The prepared catalyst was characterized using TEM (Transmission Electron Microscopy), SEM (Scanning Electron Microscopy), UV (Ultraviolet–visible spectroscopy), and XRD (X-ray diffraction pattern). TEM images of zeolite-coated gold nanoparticles confirm that gold nanoparticles are present inside the zeolite. XRD spectrum of MCM-22 and ZSM-5 matches with the respective JCPDS XRD datas. While, this general approach can be validated by many reactions, we investigate the selective oxidation of cyclohexane[8] as its product KA-oil (a mixture of ketone and alcohol) is an important intermediate in the production of nylon 6 and nylon-6,6. These gold nanoparticle encapsulated by zeolite nanoshell catalysts showed increased activity/selectivity for cyclohexane oxidation reaction at high temperature and pressure.

Following the same scheme other nanoparticles like Palladium, Platinum, Nickel, and Cobalt will also be coated with different zeolites to cater towards various reactions performed at harsh conditions.

**References:**


1. *Oliver-Meseguer, J., et al., Small Gold Clusters Formed in Solution Give Reaction Turnover Numbers of 107 at Room Temperature. Science, 2012. 338(6113): p. 1452-1455.*
2. *Haruta, M., et al., Low-Temperature Oxidation of CO over Gold Supported on TiO2, Î±-Fe2O3, and Co3O4. Journal of Catalysis, 1993. 144(1): p. 175-192.*
3. *Nkosi, B., et al., Hydrochlorination of acetylene using carbon-supported gold catalysts: A study of catalyst reactivation. J. Catal., 1991. 128(2): p. 378-386.*
4. *Mitsudome, T. and K. Kaneda, Gold nanoparticle catalysts for selective hydrogenations. Green Chemistry. 15(10): p. 2636-2654.*
5. *Turner, M., et al., Selective oxidation with dioxygen by gold nanoparticle catalysts derived from 55-atom clusters. Nature, 2008. 454(7207): p. 981-983.*
6. *Gajan, D., et al., Gold Nanoparticles Supported on Passivated Silica: Access to an Efficient Aerobic Epoxidation Catalyst and the Intrinsic Oxidation Activity of Gold. Journal of the American Chemical Society, 2009. 131(41): p. 14667-14669.*
7. *Ojea-JimeÌ□nez, I., et al., Small Gold Nanoparticles Synthesized with Sodium Citrate and Heavy Water: Insights into the Reaction Mechanism. The Journal of Physical Chemistry C. 114(4): p. 1800-1804.*
8. *Hereijgers, B.P.C. and B.M. Weckhuysen, Aerobic oxidation of cyclohexane by gold-based catalysts: New mechanistic insight by thorough product analysis. Journal of Catalysis. 270(1): p. 16-25.*

Controlling metal nanoparticle dispersion with 'nanocavity' oxide supports

Controlling growth of metal nanoparticle on oxide support during synthesis is one of the defining challenges of materials synthesis for catalysis. Photodeposition[1] and strong electric absorption (SEA)[2] methods are reported to generate highly dispersed Ag nanoparticles on TiO2 support and Pt nanoparticles on SiO2 support. But none of these methods can intrinsically prevent metal nanoparticles sintering in harsh reaction environment. Previously, researchers have reported using a second oxide overcoat to cover the supported metal nanoparticles and successfully enhancing their thermal stability[3]. However the complete over-coating approach can lead to loss of metal nanoparticle surface area, which eventually decreases catalytic activity of catalyst. Here we present new synthesis strategies in combining these ideas to template and stabilize supported metal nanoparticles inside oxide nanocavity materials. Following method from our previous report[4], oxide nanocavity materials are synthesized by grafting a sacrificial template molecule on a carrier particle (e.g. $SiO_2$ or $TiO_2$), then partially overcoating the surface with a second oxide using a self-limiting technique such as atomic layer deposition or sol-gel under reactant-limiting conditions. Finally, the templates are removed to create nanocavities. $SiO_2$ deposited over templates grafted $TiO_2$ would be for example, named as $SiO_2$-number of $SiO_2$ coating cycles-$TiO_2$. Recent work in our group shows these nanocavities on the oxide surface have characteristic dimension of ~1nm[4]. The as-synthesized nanocavities are then backfilled with metal nanoparticles by selective photoreduction or SEA. In the photodeposition case, Ag nanoparticles are selectively deposited onto exposed $TiO_2$ domains on $TiO_2$ support containing nanocavities by photoreduction with a sacrificial reductant. Results show that $SiO_2$-8cycles-$TiO_2$ has better dispersion of Ag nanoparticles and controlled particle size distribution (2.9 ± 0.6 nm by Transmission Electron Microscope (TEM)) compared to Ag nanoparticles on $TiO_2$ surface without nanocavities (4.7 ± 2.7 nm by TEM). Likewise, Ag nanoparticles on $SiO_2$-8cycles-$TiO_2$ show better sintering resistance compared to Ag photodeposited onto unmodified $TiO_2$. The Extended X-ray Absorption Fine Structure (EXAFS) results confirm Ag particle size increases from 7 to 8.9 nm for unmodified $TiO_2$ after sintered at 673 K for 3 hours, while Ag particle size decreases from 4.9 to 3.7 nm on $SiO_2$-8cycles-$TiO_2$.

In a second route, SEA is used to selective deposit $Pt(NH_3)_4^{2+}$ onto the exposed $SiO_2$ domains of $Al_2O_3$-10cycles-$SiO_2$ and deposit $PtCl_6^{-}$ onto the exposed $TiO_2$ domains of SiO2-8cycles-$TiO_2$. The former takes place at pH~9 and the latter at PH~3. For $Pt(NH_3)_4^{2+}$ deposition, preliminary results show when the nanocavity surface density is changed from 0.08 to 0.24 nanocavity/$nm^2$, Pt deposition increases from 0.05 to 0.1 complexes/$nm^2$. With this, we are looking to convert a workhorse technique for catalyst synthesis into tool for precise synthesis of robust, tunable, supported metal nanoparticles.

[1] Chan, S. C., & Barteau, M. A. (2005). Preparation of highly uniform Ag/TiO2 and Au/TiO2 supported nanoparticle catalysts by photodeposition. Langmuir, 21(12), 5588-5595.

[2] Miller, J. T., Schreier, M., Kropf, A. J., & Regalbuto, J. R. (2004). A fundamental study of platinum tetraammine impregnation of silica: 2. The effect of method of preparation, loading, and calcination temperature on (reduced) particle size. Journal of Catalysis, 225(1), 203-212.

[3] Lu, J., Fu, B., Kung, M. C., Xiao, G., Elam, J. W., Kung, H. H., & Stair, P. C. (2012). Coking-and sintering-resistant palladium catalysts achieved through atomic layer deposition. Science, 335(6073), 1205-1208.

[4] Canlas, C. P., Lu, J., Ray, N. A., Grosso-Giordano, N. A., Lee, S., Elam, J. W., ... & Notestein, J. M. (2012). Shape-selective sieving layers on an oxide catalyst surface. Nature chemistry, 4(12), 1030-1036.

# Single-Walled Copper Silicate Nanotubes Doped with 3*d* Transition Metals for $CO_2$ Hydrogenation

Yuan Sheng† and Hua Chun Zeng*, †, ‡

†Department of Chemical and Biomolecular Engineering, National University of Singapore, 10 Kent Ridge Crescent, Singapore 119260. ‡Institute of Materials Research and Engineering (IMRE), 3 Research Link, Singapore 117602

* Corresponding author, email: chezhc@nus.edu.sg

Transition metal silicate nanotubes have great potential in heterogeneous catalysis due to their large surface area and high metal loading. To achieve a fully accessible surface and versatile catalytic properties, a single-walled structure and tunable composition are highly desirable. However, copper silicate has so far been the only transition metal silicate to form single-walled nanotubes. As shown in Figure 1, in this work we developed a general method to synthesize a series of new single-walled silicate nanotubes (with an average outer diameter and wall thickness of 6 nm and 1 nm respectively) containing other 3*d* transition metal elements (M = Mn, Fe, Co, Ni and Zn) by wet-chemical doping of pre-synthesized copper silicate nanotubes; up to 80% of Cu can be replaced with these metals (M) through complex-assisted ion exchange, while the original tubular structure remains intact. These metal silicate nanotubes can serve as solid precursors to transform into Cu-based nanocatalysts under a reducing environment. Using $CO_2$ hydrogenation as a model reaction, enhanced activity and methanol selectivity were demonstrated with the doped catalysts at 240°C, 30bar. A specific activity calculated based on total mass of metals which is superior to an industrial $Cu/ZnO/Al_2O_3$ catalyst was also achieved.

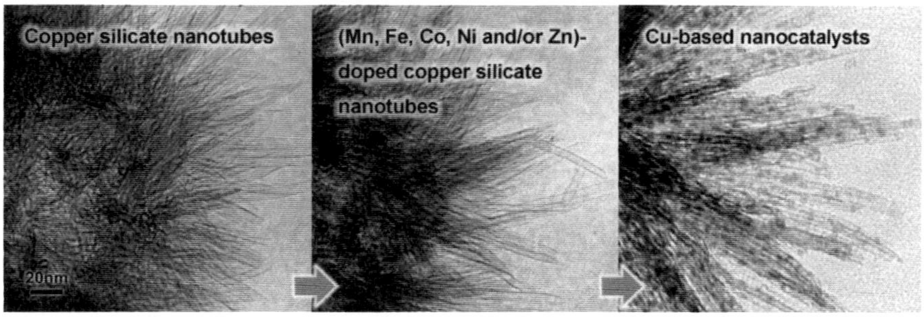


Figure 1. TEM images showing the typical morphologies of pristine and doped copper silicate nanotubes and active catalysts derived from them. The background is colored for visual effects. The scale is the same for the three images.

Driven by increasing costs of waste disposal and raw materials, catalysis research in the 21st century is shifting from a focus on activity to one on selectivity.[1] While breakthroughs in catalyst design research have provided great insights into how the rates of various steps in a reaction mechanism can be tuned by altering the catalyst composition,[2] tuning catalysts to discriminate between competing reactions that are chemically similar remains to be difficult. An example of such a reaction is the oxidation of an alkane to a 1-alcohol with high purity, i.e., with little production of 2- and higher alcohols, polyols, etc., which is important in the pharmaceuticals and specialty chemicals industries. In this contribution, we discuss our work on designing catalysts with unprecedented regioselectivity to produce 1-butanol from butane, accomplished by encapsulating a metal nanoparticle catalyst within a metal-organic framework (MOF).[3] MOFs are porous, crystalline solids built from metal or metal oxide "nodes" connected by organic "linker" compounds. The encapsulation provides a way to discriminate against reagents in a reaction environment and can be engineered to maximize selectivity. For example, the MOF pore diameters can be adjusted to only allow certain sized molecules to access the catalyst; likewise, the pore polarities can be configured to draw in polar molecules and push out non-polar molecules, or vice versa. Additionally, the pore networks can be designed to force reagents to interact with the catalyst in specific geometries (Figure 1). MOF-encapsulated catalysts have much to offer in terms of maximizing catalytic selectivity; however, little is known about the MOF/catalyst interface at the molecular level.

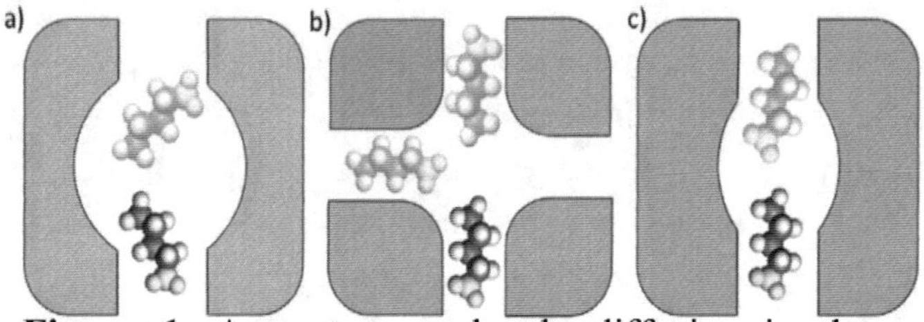


**Figure 1.** A pentane molecule diffusing in three different MOF pore networks. The desired reaction is pentane oxidation to 1-pentanol, i.e., oxidation of only the yellow carbon. All three scenarios only allow a terminal carbon to access the catalyst, thus preventing production of 2-pentanol, but only scenario c prevents the pentane from "turning around" in the pore, thus preventing over-oxidation to 1,5-pentanediol.

In this work, we use density functional theory calculations to investigate butane oxidation to 1-butanol,

$$2CH_3CH_2CH_2CH_3 + O_2 \rightarrow 2CH_2OHCH_2CH_2CH_3$$

We use "featureless" rings of He atoms to simulate the steric restrictions imposed by the MOF pore (Figure 2). Using climbing image-nudged elastic band calculations, we provide insight on 1) the optimal pore size that is required to activate butane but prevent oxidation of the secondary carbons, and 2) the catalyst composition that is required to activate butane, dissociate oxygen, and promote O-C bond formation, without over-dehydrogenating or over-oxidizing. Our results will feed future synthesis and design experiments to be performed by our collaborators.

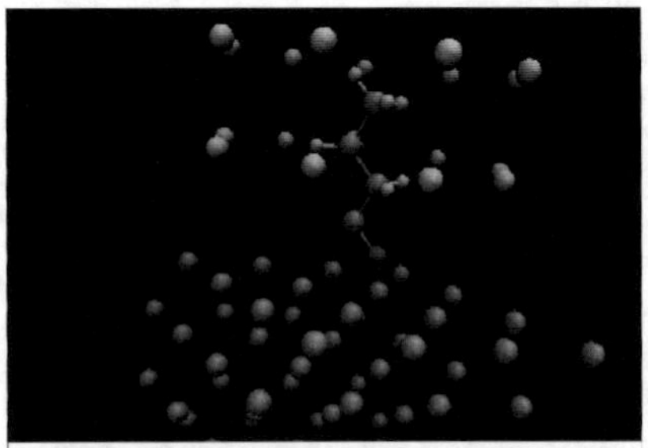

**Figure 2.** A terminally oxidized butane molecule adsorbed on a transition metal catalyst encapsulated by helium rings (pink=transition metal, red=O, blue=C, white=He, grey=H).

[1] Somorjai, G. A. and Rioux, R. M. "High technology catalysts toward 100% selectivity: Fabrication, characterization and reaction studies." *Catalysis Today*, 100, 2005, p. 201-205.

[2] Grabow, L. C. "Computational Catalysis Screening." In: Computational Catalysis, A. Asthagiri and M. J. Janik, Eds. RSC Publishing: Cambridge, UK, 2005.
and reaction studies. Catalysis Today, 100, 2005, p. 201-205.

[3] G. Lu, S. Li, Z. Guo, O. K. Farha, B. G. Hauser, X. Qi, Y. Wang, X. Wang, S. Han, X. Liu, J. S DuChene, H. Zhang, Q. Zhang, X. Chen, J. Ma, S. Chye, J. Loo, D. Wei, Y. Yang, J. T. Hupp, and F. Huo, "Imparting functionality to a metal–organic framework material by controlled nanoparticle encapsulation." *Nature chemistry*, 4, 2012, p. 310-316.

# 378606 Semiconductor Nanowire Fabric

**Abstract:**

Solution-based synthetic routes have been developed to make significant quantities of nanowires of various semiconductors, including Si, Ge, InP and $CuInSe_2$. The nanowires are threadlike crystals with nanometer-scale widths and lengths that can extend for millimeters. Their remarkably high surface area-to-volume ratios and nanoscale dimensions give rise to range of unique material properties that might be useful in a variety of new technologies, including printable electronics, solar cells, batteries and sensors with unprecedented functionality. We have been developing and exploring the use of these nanowires to make a new class of semiconductor "fabrics." The fabrics are essentially tangled, non-woven mats of nanowires. Of course, underlying these efforts is the ability to generate sufficient quantities of nanowires, which must be long enough, and free of particulates, to be extensively intertwined into mechanically stable layers. The nanowires can also be combined with intervening polymer to make composites. Since the nanowires are semiconducting and mechanically flexible, but also have very high strength-to-weight ratio (far exceeding steel for example) because of their lack of extended defects, they have the potential to provide an unprecedented range of new properties in a fabric-like material. This presentation will explain the nanowire synthesis, the approach to forming the fabric, and a summary of their properties. For example, non-woven fabric of Si nanowires can be formed with extremely high optical densities; Si nanowires can be combined with polymer and elastomeric hosts to create new semiconductor-based materials with unique combinations of optical, electronic and mechanical properties; and Si and Ge nanowires in lithium ion batteries can increase the storage capacity of the anode by more than five times its current capacity.

# 392287 Piezotronics and Piezo-Phototronics in Semiconductor Nanowires for Active Electronics and Optoelectronics


**Abstract:**

New technologies for developing electronics and optoelectronics with tunable/adaptive functionalities and performance are critical to emerging applications in wearable technology, communication, pervasive computing, human-machine interfacing and biomedical diagnostics, in which the active and adaptive interactions between devices and stimuli from the ambient or host (e. g. the human body) are essential. Mechanical stimuli are ubiquitous and abundant in the environment for interacting with or controlling these electronics and optoelectronics. This is, however, not facile to implement using the state-of-the-art technology using semiconductor materials such as silicon, which rely on electrically-gated modulation of charge carriers by externally-applied voltage to perform intelligence-bearing operations. Piezoelectric effect has been widely utilized for electromechanical sensing, actuating and energy harvesting, which produces polarization charges in response to mechanical deformation in materials lacking inversion symmetry or with polarization domains. Conventional piezoelectric materials such as PZT and PVDF are insulators and hence not feasible for constructing functional electronics/optoelectronics. The effect of mechanically-induced polarization on electronic transport and optoelectronic processes of charge carriers in piezoelectric materials has consequently been long overlooked.

Semiconductor materials such as ZnO, GaN and CdS also possess piezoelectric properties but are not as extensively utilized in piezoelectric sensors and actuators due to their relatively small piezoelectric coefficients. The coupling among piezoelectric polarization, semiconductor properties and optical excitation in piezoelectric semiconductor nanowires has resulted in both novel fundamental phenomenon and unprecedented device applications, leading to the increasing research interests in the emerging fields of *piezotronics* and *piezo-phototronics*. For wurtzite-structured semiconductor nanowires which have non-central symmetry, the piezoelectric polarization charges can effectively tune/control the charge transport across interface/junction (piezotronic effect) and modulate the optoelectronic processes, such as generation, separation, diffusion and recombination, of charge carriers (piezo-phototronic effect). Utilizing piezotronic effect, novel nanoelectronic devices such as force/pressure controlled sensors, logic units and memories have been demonstrated. Large-scale vertical nanowire transistor array has also been developed based on piezotronic effect for tactile imaging. Based on the principle of piezo-phototronic effect, optoelectronic devices such as solar cells, photon detectors and light emitting diodes fabricated using wurtzite nanowires show enhanced and tunable performance, which is not readily available in existing technologies. Large-scale array of piezoelectric nanowire light-emitting diode has also been demonstrated for mapping and imaging the distribution of applied strain through the output of light-emission.

The objective of this talk is to introduce the fundamentals of piezotronics and piezo-phototronics using ZnO nanowire as the material platform, and to give an updated progress of their applications in energy science and sensing. These studies provide the novel approach for modulating device characteristics by tuning the junction/contact properties, which has been unavailable in conventional technologies without modifying the interface structure or chemistry, and may open up opportunities in enabling technology advances in sensing, adaptive human-electronics interfacing, robotics, biomedical therapy, prosthetics, optical MEMS, efficient solid-state lighting and more.

*In situ* X-ray Absorption Spectroscopy (XAS) has been used to study the reduction of palladium during the hydrothermal synthesis, thereby serving as an entryway into fundamentally understanding the relationship between the coating properties and the underlying reduction and growth processes governing mineralization on biotemplates. Additionally, the electrical conductivity of the synthesized nanorods was found to be comparable with those of electrochemically grown palladium nanowires, thus demonstrating the potential of these materials in biotemplated nanoelectronic devices.Genetically engineered Tobacco Mosaic Virus (TMV1Cys) has been applied as an expedient template for facile and controlled nanorod synthesis at mild operating conditions. In recent years, the discovery of the hydrothermal synthetic scheme, performed in the absence of a reducing agent, has allowed the production of higher quality nanorods with more uniform coating, narrower size distribution and easily controllable sizes. In this study, the mechanism for reduction and growth in this process and the potential opportunity it offers to control nanorod properties during applications has been explored. Reduction and grain growth were analyzed by monitoring the evolution of absorption spectra over the course of the reaction. Valuable information about the nature and kinetics of the process is also acquired. The resulting organic-inorganic hybrid nanorods were subsequently electrically characterized through a focused ion beam (FIB) assisted nano-gap approach. Differences in nanorod conductivities after different growth cycles were studied. Results acquired indicate that the reduction of palladium on the biotemplate is autocatalytic and first order in palladium. Also, the resulting electrical conductivity of double coated palladium nanorods compare reasonably with other grown palladium nanowires. In summary, insight into the hydrothermal synthesis is attained via XAS, the nanorods acquired are comparable with those produced from more complex synthesis and a simple electrical characterization method has laid a foundation for design and characterization of nanorods with specific electrical properties for nanoelectronics applications.

**Kinetic Control of Catalyst-free Growth of Ultra-high Density ZnO Nanowires**

**Cheng Xu*, Yang Zhao, Luping Li, and Kirk J. Ziegler**

Department of Chemical Engineering, University of Florida, 1006 Center Drive, Gainesville, FL 32611, US

*Presenting Author (Ph: 352-392-3122; Email: chengotopia@ufl.edu)

**Abstract:**

The excellent optical and electrical properties of ZnO such as direct bandgap, high electron mobility, large exciton binding energy, and excellent transparency makes it an ideal choice for the use of field transistors, piezoelectrics, photovoltaics, and sensor devices. Since device performance is often directly proportional to the active surface area, fabricating vertically aligned nanowire arrays with high density and aspect ratio is important to maximize their efficiencies beyond the current limit. The traditional approach of growing ZnO nanowires with excellent crystallinity include catalyst based vapor-liquid-solid (VLS) growth and seeded hydrothermal growth. However, the density of nanowires grown by both methods is limited since the density is controlled by the metal catalyst for VLS growth and ZnO texture seeds for the hydrothermal growth. As a result, the density of ZnO nanowires can only reach $10^9$ cm$^{-2}$ for VLS growth and $10^8$ cm$^{-2}$ for hydrothermal growth. Furthermore, both methods struggle to obtain nanoparticle catalysts or seeds with small diameter and to maintain nanowires with small diameter during growth due to high surface energy associated with the substrate. Although templated approaches provide an excellent way of demonstrating nanowire density control, it is difficult to form and remove the back electrode and integrate these nanowires into useful devices.

In this study, we focused on not only using the anodized alumina oxide (AAO) as a template to control nanowire density but also providing a method of transferring the template onto a substrate of choice that could be integrated into device fabrication. AAO provides maximum density ($10^{10}$-$10^{12}$ cm$^{-2}$) with its closely-packed hexagonal pore arrangement and, by carefully tuning the experimental conditions, the pore diameter and interpore spacing can be precisely controlled. First, to demonstrate the transfer process, ultra-thin AAO templates were fabricated on aluminum using 2-step anodization. After dissolving aluminum, AAO was transferred to a foreign substrate. To study the mechanism of catalyst-free vapor-solid (VS) growth on AAO, the growth condition of temperature, flow rate, oxygen partial pressure, amount of source, and distance between source and substrate had to be optimized to control the initial nucleation of ZnO nanoparticles in the AAO pores. With the optimized condition, ZnO nanoparticles were deposited in AAO pores with identical density of AAO, and then ZnO nanowires were grown on the nanoparticles. To study the effect of experimental parameters on the final ZnO nanowire morphology, SEM snapshots were taken at different time intervals and XRD was done for different growth parameters. As a result, nanowires were fabricated on a substrate of choice with densities of $10^{11}$ cm$^{-2}$. This technique provides a new route of fabricating large-scale, cost-effective optoelectronic devices.

*Keywords: photovoltaics, ZnO nanowires, alumina template, catalyst-free nanowire growth*

# 377320 Interplay Between Defect Propagation and Surface Hydrogen in Si Nanowire Kinking Superstructures

**Abstract:**

Semiconductor nanowires containing kinking superstructures – multiple, user-programmable changes to growth orientation – offer tantalizing opportunities to fabricate nanoelectronic bioprobes, generate metamaterials with chiraloptical response, and manipulate thermal conductivity. However, the synthesis of kinking superstructures with long-range structural coherence remains exceedingly challenging. Here, we combine high-resolution electron microscopy with operando infrared spectroscopy to show why this is the case for Si nanowires and, in doing so, reveal the subtle interplay between defect propagation and surface chemistry during <211> → <111> and <211> → <211> kinking. Our experiments show that adsorbed hydrogen atoms are responsible for selecting <211> oriented growth and indicate that a twin boundary imparts structural coherence. The twin boundary, only continuous at <211> → <211> kinks, reduces the symmetry of the trijunction and limits the number of degenerate directions available to the nanowire. These findings constitute a general approach for rationally engineering kinking superstructures and also provide important insight into the role of surface chemical bonding during vapor-liquid-solid synthesis. [Research supported by ORNL's Shared Research Equipment (ShaRE) User Program, which is sponsored by the Office of Basic Energy Sciences, the U.S. Department of Energy.]

# 383499 Controlled Chirality in Nanoscale Semi-Conducting CdTe Nanowires

**Abstract:**

Chirality is ubiquitous in nature – from the conformation of proteins within our cells to the photonic response seen in butterfly wings – and plays an important role in the functionality of a given molecule, building block, or particle assembly. Understanding how to harness and control the chirality of a nanoparticle assembly paves the way for devices such as chiral beam splitters and filters, as well as other photonic devices. Toward this end, the Kotov group has recently demonstrated the ability to dictate the handedness (left or right) of semi-conducting CdTe nanowire assemblies by capping the nanoparticle precursors with a chiral amino acid stabilizer, cysteine.[1] Using Monte Carlo and Molecular Dynamics computer simulations, we present a model for the assembly of these particles based upon the inter-particle forces present in the system. We first show that chirality can emerge spontaneously, but without predictability, in simple hard particle systems. We next refine this model based on experimental parameters and show what mechanisms can be introduced to control this chirality and control the size and pitch of the nanowire.

[1] "**Self-Assembly of Chiral Nanoparticles into Chiral Helices.**" Yunlong Zhou, Ryan Marson, Greg Van Anders, Peter Ercius, Kai Sun, Bongjun Yeom, Sharon Glotzer, Nicholas A. Kotov. *(preprint)*

# 376491 Solid-Liquid-Vapor Anisotropic Etching of Semiconductor Nanowires

**Abstract:**

Semiconductor nanowires synthesized by bottom-up techniques are promising materials for next generation electronic, photonic, and energy conversion devices because one can add multiple functionalities along the nanowire length. To achieve this, however, temperature and gas phase composition must be modulated multiple times during growth. These changes frequently result in graded composition profiles that are undesirable in many applications. Here, we demonstrate anisotropic *in situ* etching of semiconductor nanowires, a process that permits the removal of these transition regions. Ge nanowires are first synthesized via the vapor-liquid-solid technique with a eutectic Au-Ge catalyst at 370 $^{o}$C and a $GeH_4$ partial pressure of 0.27 Torr. An intentional diameter expansion near the nanowire base serves as an internal reference point to accurately measure etch rate. The introduction of 1,3 butanedione (BD) at a partial pressure of 0.0035 Torr initiates etching selectively from the eutectic catalyst. We find that the atomic etch rate is constant regardless of sidewall taper. Furthermore, we study this etching behavior as a function of temperature, BD partial pressure, and nanowire diameter. Our data indicate a solid-liquid-vapor etching mechanism, where Ge atoms initially dissolve from the nanowire into the eutectic catalyst through the solid-liquid interface prior to reaction with and removal by BD at the liquid-vapor interface. This etching technique, which effectively reverses vapor-liquid-solid growth, opens new opportunities to precisely control nanowire structure.

# 389881 Characterization and Effects of Barrier Layer Thinning and Pore Opening in Porous Aluminum Oxide for Nanowire Growth

**Abstract:**

Nanowires have garnered much interest in recent years for their potential in increasing performance of devices such as solar cells, thermoelectrics, batteries, and piezoelectrics. Many methods exist for the fabrication of nanowires, but one of the most promising methods is the use of nanoporous aluminum oxide templates due to its extremely high density and self-ordering through a hexagonal close packed structure. However, fabricating uniform arrays of nanowires in contact with the conductive aluminum substrate still remains a challenge. The problems are most evident when a greater degree of control over the electrodeposition into the aluminum oxide templates is needed for applications such as creating catalyst nanoparticles.

A vital part of the template process is removal of the barrier layer between the pore and the underlying substrate. This is needed to establish direct contact from the pores to the substrate. The objective is to remove the barrier without sacrificing the pore walls. Several methods exist for barrier layer thinning, including stepping down the anodization voltage or ramping down the current. However, the thinning method will influence the final pore opening, which creates the contact between the pore and aluminum substrate. In this presentation, different barrier layer thinning techniques will be discussed and their effect on the pore opening process examined. Both step-down voltage and ramped current methods were used to understand the barrier layer thinning process. These results were correlated to the amount of pore opening achieved, measured through a combination of chronoamperometry and electrochemical impedance spectroscopy. Ellipsometry was used to characterize the barrier layer thinning and pore opening processes of the aluminum oxide templates. Finally, nanowires were grown within these templates to study the effects of different barrier layer thinning and pore opening techniques on nanowire growth and uniformity. This research provides a comprehensive approach towards fabricating uniform, high-density nanowire arrays.

# 384053 Optimizing Heat Generation in Magnetic Iron Oxide Nanoparticles for Use in Cancer Hyperthermia

**Abstract:**

Using magnetic fluid hyperthermia (MFH) for cancer treatment is a viable option to achieve local heat generation in tumors, and avoid severe side effects of conventional cancer treatments like radiation and chemotherapy. However to achieve maximum therapeutic efficiency with a minimal patient exposure to magnetic nanoparticles (MNPs), high heat generation using the least possible concentration of iron oxide MNPs is essential. In this study, it was determined that heat generation in MNPs depends on magnetic field strength and frequency, size of MNPs, and viscosity of the medium surrounding these MNPs. The heat generation in these MNPs was standardized using specific absorption rate (SAR), which was used to find the set of parameters that gave the highest SAR values. It was found that the heating rates and SAR increased with corresponding increase in magnetic field strength (15.1 – 47.7 kA/m) and frequency (123 kHz – 430 kHz) in the range tested. Also, SAR and heating rates increased with increase in MNP size in the range of 8 – 18 nm. Increasing the viscosity of the medium decreased the heating rates and SAR values. After analyzing the SAR values it was concluded that maximum heating rate could be obtained for the 18 nm MNPs at the highest field strength and frequency in media with lower viscosity.

# 388481 Magnetic Nanoparticles for Measuring Prostate Cancer Aggressiveness

**Abstract:**

Prostate Cancer (PCa) is the most diagnosed non-skin malignancy and the second leading cause of death from cancer among men in the United States. Therapies for PCa, unfortunately, remain ineffective. In order to achieve more effective therapy, early and accurate detection of prostate cancer is necessary. Current Screening technologies for PCa, including the prostate-specific antigen (PSA) test, are limited by false-negatives and false-positives. Additionally, even when PCa is successfully detected, these screens provide no indication of the aggressiveness of the disease. Improved molecular imaging techniques, that are both quantitative and sensitive, and that can non-invasively detect processes deep within the human body are required to monitor changes which may be predictors of treatment outcomes. We have developed multimodal and bioprocess-sensitive "smart" magnetic nanoparticles (MNP) which respond to overexpressed protease activity in the tumor microenvironment – and thus to tumor aggressiveness – and which can be monitored and quantified non-invasively over the entire tumor volume. Cancer cells overexpressing legumain protease have been shown to have an invasive and metastatic phenotype *in vivo*. Legumain specifically degrades peptides containing an AANL-peptide sequence. We conjugated a fluorescence marker via the legumain cleavable AANL-peptide linker (LCP). This fluorescent-MNP displayed sensitivity to legumain activity; slowly releasing the fluorescent marker in low activity microenvironments and more rapidly in the presence of high legumain activity. Release of Dy747 resulted in a decrease in the fluorescence signal in tumors, due to wash-out into the blood flow, at a rate proportional to the local proteolytic activity. Quantitative MRI analysis indicated the amount of agent delivered to a particular location, providing the baseline to correct for tumor heterogeneity. Taken together, the fluorescent/MRI signal ratio is a function of the local proteolytic activity.

# 388136 Controlled, Wireless Heating of Iron Oxide Nanoparticle Composites

**Abstract:**

Magnetic induction heating of superparamagnetic iron oxide nanoparticles (SPIONs) has been proposed to wirelessly power polymer composite films for *in situ*thermal sterilization of implanted medical device infections. Wide variations in device application demand coatings with a wide variety of mechanical and chemical properties. For example, orthopedic implant coatings must withstand large mechanical stresses while heart valves require significantly more elasticity. To determine the effect of the composite matrix and fabrication processes on the coating's ability to deliver wireless heating, we investigated the synthesis, characterization, and performance of SPIONs in hydrophobic (polystyrene (PS)) and hydrophilic (poly(vinyl alcohol) (PVA)) polymer films and their application to controlled surface heating in an alternating magnetic field (AMF). Magnetic composites were characterized for particle size, agglomeration, and dispersion (scanning electron microscopy); crystal structure and oxide state (x-ray diffraction); and magnetic susceptibility to an AMF (specific absorption heating rate (SAR) measurements).

SPION suspensions were synthesized in-house via an aqueous co-precipitation method and were shown to exhibit good dispersion and particle size (10-20nm) as commonly reported in literature. Aqueous suspensions were further prepared for hydrophobic applications with retained particle size and dispersion from the original hydrophilic colloid. PVA and PS films were fabricated via solvent casting to produce uniformly-thick, dry films with minimum particle loadings of 10dry-wt% iron oxide. PVA sols in water were crosslinked to produce stable, hydrated films up to 80°C. Normalized for variations in the magnitude of the strength of the AMF, a film's heat flux was characterized for uniformity via infrared thermal imaging. A film's intrinsic power output was analyzed via SAR measurements using a five-turn, 50mm ID coil and a 5kW AMF generator at 350kHz. Films were fabricated for various maximum SAR values by controlling the particle concentration and film thickness. Due to the low pH requirement of the PVA crosslinking mechanism, iron oxide particles exhibited noticeable oxidation at concentrations above 10dry-wt%, thereby decreasing the intrinsic SAR value. Ultimately, PS and PVA films were produced with similar SAR values but drastically different mechanical properties and utility.

For SPION films with known SAR, temperature control at the composite's surface was implemented with PID control of the magnetic field power output and fiber-optic temperature probe measurements; transient heating times en route to a steady state surface temperature are a function of the heat-sink condition and controller tuning parameters. Maximum surface temperatures on both PVA and PS composites targeted an 80°C setpoint for a conduction-only heat-sink.

# 386148 Shape-Controlled Iron Oxide Nanoparticles Fpr Bioimaging

**Abstract:**

Shape control is one of the most exciting aspects of the nanoparticle synthesis, which allows manipulating the chemical and physical properties of nanomaterials. In this talk, the shape control of iron oxide nanoparticles using a modified "heat-up" method will be presented. In particular, we will discuss the formation mechanisms of various shaped iron oxide nanoparticles and their shape-dependent magnetic properties. For example, ultrathin iron oxide nanowires were formed by selectively decomposing certain ligands of the Fe-oleate complex precursor. In contrast, iron oxide nanoworms, nanoplates, and nanoflowers can be achieved if a weakly binding ligand is added during synthesis to tailor the nucleation and growth processes. The shape of the iron oxide nanoparticles directly affects their magnetic properties, such as ferromagnetic nanoworms, superparamagnetic nanocubes, paramagnetic/superparamagnetic mixed nanoplates and ultrathin nanowires, subsequently define their application as magnetic resonance imaging (MRI) contrast agents .

# Preparation and Characterization of Magnetic Gene Transfection Agents Consisting of Polyethyleneimine Coated Iron Oxide Nanoparticles

**Melissa Cruz-Acuña**[1], Lorena Maldonado-Camargo[2], Jon Dobson[1, 3], and Carlos Rinaldi[1, 2]

[1]Department of Biomedical Engineering, University of Florida, Gainesville, FL;
[2]Department of Chemical Engineering, University of Florida, Gainesville, FL;
[3]Department of Material Science and Engineering, University of Florida, Gainesville, FL.

The biocompatibility of iron oxide magnetic nanoparticles (MNPs) and the ability to manipulate them using magnetic fields is of great interest in a variety of applications, including non-viral gene delivery. Monodisperse, superparamagnetic iron oxide magnetic nanoparticles were synthesized by the thermal decomposition method and coated with polyethylenimine (PEI) of different molecular weights after a simple aqueous phase transfer procedure. The stability, size, and charge of the grafted nanoparticles were studied in water, phosphate buffered saline (PBS), and cell culture media. As expected, coating molecular weight influenced colloidal stability. The colloidal stability of the particles changed from electrostatically-mediated for a graft molecular weight of 800 Da to sterically-mediated at molecular weights of 2000 Da and above. The ability of the particles to form DNA polyplexes was evaluated, as well as their effects on cell viability and ability to transfect cells with and without stimulation using an array of oscillating permanent magnets, to identify optimal coating molecular weight for use as magnetic field responsive gene transfection agents.

# BIOCOMPATIBLE IRON OXIDE NANOPARTICLES OBTAINED VIA THE NON-AQUEOUS SYNTHESIS IN HYDROPHILIC SOLVENTS FOR HIGH GRADIENT MAGNETIC PROTEIN PURIFICATIONS

Magnetic nanoparticles have a broad field of applications including biomedicine, magnetocaloric refrigeration, magnetic sealing and magnetic storage. For all applications it is of crucial importance that the characteristics such as the physical properties, morphology, composition and surface structure of the particles are well defined and controllable.

The non-aqueous sol-gel method is a bottom-up synthesis which allows synthesizing particles with the desired properties. Particles with tunable size and high crystallinity can be obtained employing mild conditions and using environmental friendly high-boiling solvents. Whereas the use of hydrophobic solvents such as benzyl alcohol is well established, we show that by using hydrophilic solvents like triethylen glycol (TEG), water-compatible nanoparticles with high stability against agglomeration can be directly achieved.

We characterized the physical and chemical properties of the particles such as their size and morphology but also magnetic properties throughout the synthesis process in order to gain knowledge about the parameters which have to be tuned to obtain particles with the optimum properties for applications such as high gradient magnetic protein purification.

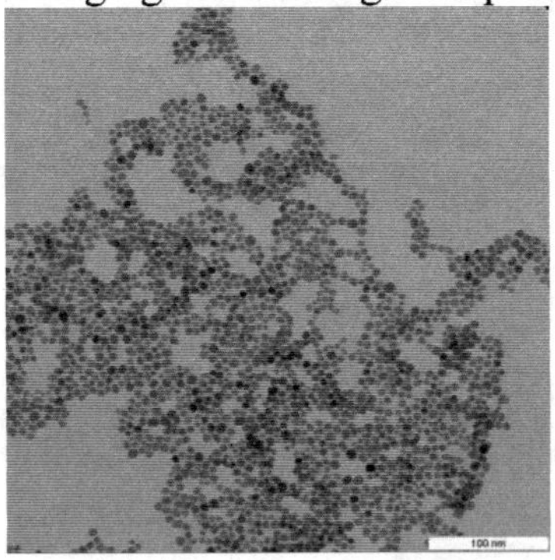


Figure 1 Iron oxide nanoparticles after 24h of synthesis in TEG.

We found that depending on the solvent used for the synthesis, the surface of the nanoparticles strongly differs, and a further functionalization can be applied to adjust and optimize the surface chemistry. The physical properties can be accustomed by terminating the synthesis after certain periods, thus being able to tune e.g. the anisotropy energy and blocking temperature of the particles.

In a second step the acquired particles were functionalized with a priorly coupled ligand to achieve a specific binding between a desired protein and the particles, as well as their stabilization in aqueous and biologic media. This system showed great potential as a highly specific magnetic purification system while time and labor could be minimized compared to a traditional chromatographic approach.

Publications:
Functionalization of magnetic nanoparticles with high-binding capacity for affinity separation of therapeutic proteins, I.-C. Masthoff, F. David, C. Wittmann, G. Garnweitner, Journal of Nanoparticle Research, 2013, 16, 1-10.

Formation of magnetic nanoparticles studied during the initial synthesis stage, M. Kraken, I.-C. Masthoff, A. Borchers, F.J. Litterst, G. Garnweitner, Hyperfine Interact 2013, 1-7.

# Synthesis of p-type ZnO nanorods on an InP substrate by pulse laser deposition

Dongqi Yu[1], Weihang Yu[1], Nan Zhang[1], Yan Wang[1], Yun Zhou[1], Lizhong Hu[*2], Jingchang Sun[**1]

*1 School of Physics and Electronic Technology, Liaoning Normal University, Dalian 116029, China*

*2 School of Physics and Optoelectronic Technology of Dalian University of Technology, Dalian 116024, China*

**Abstract**

Highly oriented and aligned crystalline ZnO:P nanorod arrays (NRAs) were successfully synthesized by phosphorus diffusion from the InP substrates using pulse laser deposition (PLD) technique. To verify the role of the impurity bound excitons induced by the phosphorus, photoluminescence (PL) spectroscopy was employed to investigate the energy states of acceptor dopants in the ZnO NRAs. Low temperature PL spectrum demonstrated four acceptor-related near-band-edge emission peaks. The acceptor energy level of the phosphorus dopant was estimated to be 125 meV.

Keywords: ZnO, nanorods, dopant phosphorus, PLD.

## I. Introduction

ZnO is a oxide semiconducting material with wide direct band gap of 3.37 eV and large exciton binding energy of 60 meV which can provide more efficient exciton emission at room temperature, has attracted increasing interest due to its potential applications for the next generation short-wavelength optoelectronic devices operating in the near ultraviolet regions. In addition, it also has excellent piezoelectric and photo-catalytic properties. With the development

* Corresponding author. Tel.: +86 411 84707865; fax: +86 411 84709304
E-mail address: useeu@163.com
** Corresponding author. Tel.: +86 411 82159023; fax: +86 411 82158555.
E-mail address: jingchangsun@163.com

of nanoscience and nanotechnology, ZnO nanomaterials are expected to have potential applications in nano-optoelectronics, electro-mechanics. In order to develop nanoscale devices, *p*-type ZnO nanostructures are essential. However, as to GaN, unintentionally doped ZnO is naturally an *n*-type material, *p*-type doping is still a major obstacle to the device application due to the deep acceptor levels, low dopant solubility and self-compensating process [1-3]. Therefore, most efforts have been focused on fabricating *p*-type ZnO materials. Although *p*-type ZnO thin films have been reported by some researchers [4-6], these results are still controversial, and the availability of *p*-type doping ZnO nanostructure is very scarce. The Lack of *p*-type ZnO nanostructure is currently the major factor precluding the realization of a wide range of functional nanodevices based on ZnO. Limpijumnong *et al.* has developed a theoretical model for large-sized-mismatch group-V elements dopants, such as phosphorus, arsenic and antimony, which may act as effective *p*-type dopants for ZnO through an $X_{Zn}$-$2V_{Zn}$ complex structure [7]. This is paving the way to researching group-V *p*-type doping in ZnO. Following this theoretical breakthrough, phosphorus, arsenic, and antimony have been employed in successfully obtaining *p*-type ZnO thin films. Among these *p*-type ZnO materials, phosphorus-doped ZnO (ZnO:P) thin films exhibit a high carrier concentration, reasonable mobility, and low resistivity [8-10]. As for the development of ZnO:P nanostructures, very few reports have been published thus far [11-13]. The dopant phosphorus can be realized by phosphorus diffusion through thermal annealing film accompanied by an InP wafer [11]. Additionally, InP wafer has several fundamental advantages for fabricating ZnO-based devices, such as its conductivity and cleavability.

In this paper, ZnO:P nanorods were deposited on InP wafer by phosphorus diffusion from InP substrate using PLD technique, ZnO seed-layer rather than metal was used as the catalyst for fabricating ZnO nanorods. To verify the information of the dopant phosphorus in the films, the characterizations of low temperature photoluminescence spectroscopy were investigated dependent on temperature model and excitation intensity model. PL spectrum demonstrated four acceptor-related near-band-edge emission peaks. The acceptor energy level of the phosphorus dopant was estimated to be about 125 meV.

## II. Experimental procedures

The ZnO:P nanorods were synthesized on InP substrate by PLD technique at 550 °C. A KrF excimer laser (λ=248nm, 5HZ) was employed to ablate a ZnO ceramic target (5N purity). The deposition chamber was initially evacuated up to a base pressure of $10^{-3}$ Pa, then oxygen (99.999%) was let into the chamber. The ZnO NRs were synthesized under optimized conditions. Phosphorus could diffuse into ZnO when InP substrate temperature was above 450 °C [14]. The morphology of ZnO:P nanorods arrays were investigated by scanning electron microscopy (SEM). X-ray diffraction (XRD) was employed to analyze the crystallinity of nanorods arrays. PL spectroscopy, employing He-Cd laser (λ = 325nm) as the excitation source, was adopted to characterize the optical properties of ZnO:P nanorods.

## III. Results and discussion

The morphologies of ZnO:P nanorods arrays were shown in Fig. 1. Uniform ZnO nanorods with smooth profiles were vertically deposited on the substrate, and well separated from each other. ZnO nanorods showed uniform diameters along the c-axis and exhibited smooth top surface.

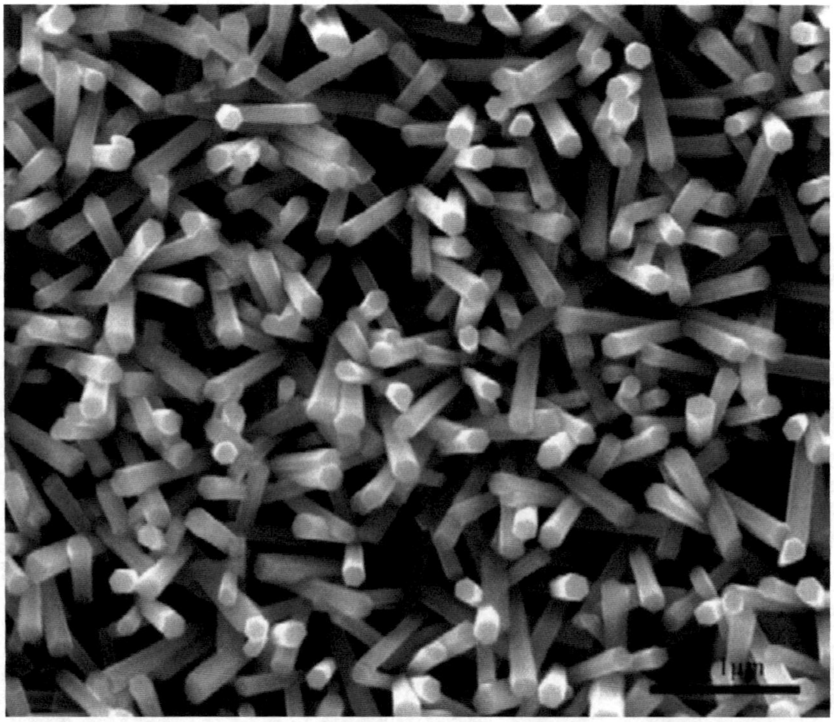

Fig. 1 SEM images ZnO nanorods.

The crystal structure and orientation of ZnO:P nanorods were analyzed by XRD measurements.

Besides the InP substrate (002) diffraction peak located, a dominant diffraction peak corresponding to the (002) plane of ZnO hexagonal wurtzite structure was observed. Enhancement of ZnO (002) peak indicated that the ZnO nanorods arrays exhibit a preferred orientation along *c*-axis, which is normal to the substrate surface. This result is consistent with the SEM image. There was no phosphorus related peak was exhibited indicating that the crystal structure and orientation of ZnO nanorods were not degraded by phosphorus diffusion.

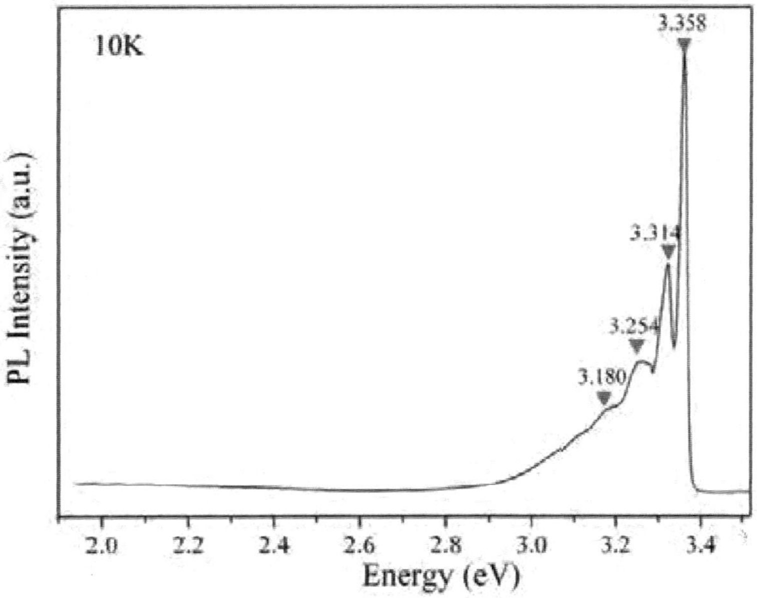


Fig. 2 Low temperature PL spectra of ZnO:P nanorods at 10 K

Low-temperature PL spectrum taken at 10 K was shown in Fig. 2. Four near-band-edge emission peaks emerged at 3.358, 3.314, 3.254, and 3.180 eV. A very weak deep-level emission was observed, indicating that there were very few native donor and acceptor type defects in ZnO nanorods. Therefore, the four emission peaks in Fig. 2 should be attributed to the phosphorus related emissions.

The emission located at 3.358 eV may be related to phosphorus dopants. Equivalent peak corresponding to phosphorus acceptor in ZnO thin films and nanowires were previously reported in the same spectral region [4, 11], therefore it can be assigned to the radiative recombination from neutral-acceptor-bound exciton ($A^oX$). It is well known that ZnO doped by group-V elements has an $E_{FA}$ emission around 3.315 eV [4, 6, 15]. A characteristic feature of $A^oX$ and free electrons to the neutral-acceptor (FA) transitions emission is independent of the PL excitation intensity. Therefore, the emission peak at 3.314 eV could be associated with FA. The difference of energy between $A^oX$ and FA is 44 meV, which agrees well with the reported results [4, 16]. The origin of the emission peak at 3.254 eV was from donor-acceptor pair (DAP) transition. For a DAP transition, the energy of the photon resulting from radiative recombination is given as [4]

$$hn_{DA} = E_{gap} - E_D - E_A + e^2 / 4per \quad \textbf{(1)}$$

where $E_{gap}$, $E_D$, $E_A$, $e$, $\varepsilon$, and $r$ are band-gap energy, donor-binding energy, acceptor-binding energy, elementary electric charge, dielectric constant, and a DAP distance. The number of photo-excited DAPs increases with increasing the excitation intensity, and their average distance $r$ decreases. A weak emission peak was also observed at 3.180 eV, exhibiting separation of 74 meV relative to DAP emission, they should be the first-order longitudinal optical (LO) phonon replicas of the DAP emission, i.e., DAP-1LO. The assignment of PL peaks also accord well with the reported luminescent features in ZnO:P by other growth techniques [4, 12, 11]. The acceptor binding energy $E_A$ could be calculated by the Eq. (1), taking into account that $E_{gap}$ = 3.437 eV [19], $E_D$ = 90 meV [17], and the last term is about 30 ~ 60 meV [7, 10]. As a result, $E_A$ was estimated to be 123 ~ 153 meV. This binding energy of the acceptors $E_A$ is quite reasonable considering Limpijumnong's model predicting a level of 160 meV [7] and the experimental results [4, 11].

Observation of $A^oX$, FA, and DAP indicated that ZnO nanorods grown on InP substrates contained some amount of phosphorus acting as acceptors. The $E_A$ could be calculated by the formula [15]

$$E_A = E_{gap} - E_{FA} + k_B T / 2 \quad \textbf{(2)}$$

where $k_B$ and $T$ are Boltzmann constant and temperature, respectively. Using the values $E_{gap}$ = 3.437 eV, $E_{FA}$ = 3.314 eV, $E_A$ was estimated to be 123 meV at 10 K, this result was in good agreement with the value obtained from Eq. (1).

The temperature dependence of the integrated PL intensity could be described by [17]

$$I = I_o / [1 + B \exp(- E_b^{A^o X} / k_B T)] \quad \textbf{(3)}$$

where $B$ is a fitting parameter, $I_o$ is the integrated PL intensity at 0 K, and $E_b^{A^oX}$ is the binding energy between the acceptor and free exciton. From the fitting to Eq.(3), $E_b^{A^oX}$=15.8 meV was obtained. Using the Haynes rule $E_b^{A^oX} / E_A » 0.1$ for ZnO material [17], $E_A$ was estimated to be 158 meV, the result is in agreement with the values obtained from Eq. (1) and Eq. (2).

## IV. Conclusions

In summary, ZnO:P nanorods were successfully synthesized using PLD method by phosphorus diffusing from InP substrate at 550 °C. The desirable distinct acceptor-bound excitons related emission peaks were appeared at low temperature PL spectra. The results illuminate that the emission peaks at 3.358, 3.314, and 3.254 eV are attributed to A°X, FA, and DAP transition, respectively. The acceptor energy level of the phosphorus dopant determined by experimental data and fitting value agrees well with the corresponding theoretic and experimental values in phosphorus-doped ZnO films and nanowires. These achievements indicate that it maybe a good way to realize ZnO-based nano-devices by phosphorus diffusion doping technique on InP substrate.

## References

[1]Y. Yan, S. B. Zhang, S. T. Pantelides, Phys. Rev. Lett. 86 (2002) 5723.

[2] F. Oba, S. R. Nishitani, S. Isotani, H. Adachi, I. Tananka, J. Appl. Phys. 90 (2001) 824.

[3] C. H. Park, S. B. Zhang, S. H. Wei, Phys. Rev. B 66 (2002) 073202.

[4] D. K. Hwang, H. S. Kim, J. H. Lim, J. Y. Oh, J. H. Yang, S. J. Park, K. K. Kim, D. C. Look, Y. S. Park, Appl. Phys. Lett. 86 (2005) 151917.

[5] T. S. Jeong, M. S. Han, C. J. Youn, Y. S. Park, J. Appl. Phys. 96 (2004) 175.

[6] F. X. Xiu, Z. Yang, L. J. Mandalapu, D. T. Zhao, J. L. Liu, Appl. Phys. Lett. 87 (2005) 152101.

[7] S. Limpijumnong, S.B. Zhang, S. H. Wei, C. H. Park, Phys. Rev. Lett. 92 (2004) 155504.].

[8] F. X. Xiu, Z. Yang, L. J. Mandalapu, J. L. Liu, W. P. Beyermann, Appl. Phys. Lett. 88 (2006) 052106.

[9] V. Vaithianathan, B. T. Lee, S. S. Kim, J. Appl. Phys. 98 (2005) 043519.

[10] X. H. Pan, J. Jiang, Y. J. Zeng, H. P. He, L. P. Zhu, Z. Z. Ye, B. H. Zhao, X. Q. Pan J. Appl.

Phys. 103 (2008) 023708.

[11] C.X. Shan, Z. Liu, S.K. Hark, Applied Physics Letters 92/7 (2008) 073103.

[12] D.S. Kim, J. Fallert, A. Lotnyk, R. Scholz, E. Pippel, S. Senz, H. Kalt, U. Gosele, M. Zacharias, Solid State Communications 143/11-12 (2007) 570.

[13] B. Xiang, P. W. Wang, X. Z. Zhang, S. A. Dayeh, D. P. R. Aplin, C. Soci, D. P. Yu, D. L. Wang, Nano Lett. 7 (2007) 323.

[14] T.W. Kim, Y.S. Yoon, J. Cryst. Growth **212** (2000) 411.

[15] C.H. Zang, D.X. Zhao, Y. Tang, Z. Guo, J.Y. Zhang, D.Z. Shen, Y.C. Liu, Chem. Phys. Lett. **452** (2008) 148.

[16] C.H. Ahn, Y.Y. Kim, S.W. Kang, H.K. Cho, Physica B **401-402** (2007) 370.

[17] J.D. Ye, S.L. Gu, F. Li, S.M. Zhu, R. Zhang, Y. Shi, Y.D. Zheng, X.W. Sun, G.Q. Lo, D.L. Kwong, Appl. Phys. Lett. **90** (2007) 152108.

# MOLECULAR SYNTHESIS AND CHARACTERISATION OF NANOPARTICLES USING IR SPECTROPMETRY

**Engr. Dr. M. N. Idris*** and **Ibrahim Kadafur**
Energy & Process Refining Technology Group
Department of Chemical Engineering
University of Maiduguri
Borno State, Nigeria
Tel: +234 (0) 705 877 1096
Email: idrismn@unimaid.edu.ng, muhinu@hotmail.co.uk

## ABSTRACT

*In nanotechnology, a particle is defined as a small object that behaves as a whole unit with respect to its transport and properties. These particles are further classified according to their diameter. Coarse particles cover a range between 10,000 and 2,500 nanometres. Fine particles are sized between 2,500 and 100 nanometres.*

*There are several methods used in the synthesis of nanoparticles, which includes the reduction of some chemical reagents, attrition and pyrolysis. In attrition, macro- or micro- scale particles are ground in a ball mill, a planetary ball mill, or other size reducing mechanism.*

*In this study, the FT-IR was used to characterize the nanoparticles synthesized. The IR analysis has shown some of the properties of the nanoparticles in terms of wave number, maximum and minimum absorbance and transmittance, nature of bonds likely to be found and the amount of vibrational peaks in the spectrum.*

*It is also observed that the aggregation is one of the major problems of these nanoparticles, a clear indication can be seen from the results obtained using the FT-IR spectrometry; were the nanoparticle aggregation has effect on absorbance, transmittance, number of peaks and even the nature of bonds likely to be found within the samples.*

***Keywords:*** *Synthesis, Characterization FT-IR, Nanoparticles & Aggregation*

---

***Correspondence about this article**: Please for all correspondence on this article, kindly contact the principal author as stated above. Thank you.*

## 1.0 Introduction

Matter can be placed into broad spectrums that are categories according to size. Macroscopic matter is visible with the naked eye while microscopic ones are invisible. Atoms and *most* molecules are microscopic with dimensions less than 1nm. Mesoscopic particles, such as bacteria and cells that have dimensions on the order of micron(s), can be observed with optical

microscopes. Falling into the gap between the microscopic and the mesoscopic particles is another class of matter, called the nanoscopic particles (Poole et al., 2003). Nanotechnology deals with processes that take place on the nanometer (nm) scale, that is, from approximately 1 to 100 nm. Nanoscale materials possess a unique property which requires intensive research studies. More advance studies are occurring in synthesis of isolated nanostructures because of its usefulness industrially. This opens the possibility for creating a new generation of advanced materials with designed properties, not just by changing the chemical composition of the components, but by controlling the size and shape of the components. For instance, the melting point of nanosized metal particles depends upon the size of the particles. The smaller a particle becomes, the more the proportion of surface atoms increases, which in turn increase the reactivity of the materials. As particles decrease in size the number of surface atoms becomes equal to or even exceeds the number of inner-core atoms. For a typical bulk material the surface is negligibly small in comparison to the total volume. Surface atoms are more easily rearranged than those in the centre of the particle, and so the melting process, which depends on destroying the order of the crystal lattice, which get started at a lower temperature. For example, the melting point (m.pt) of gold metal is 1064°C. For a size of 11-12 nm gold particles, it m.pt is about 1000°C, then begins to drop dramatically to 900°C for 5 to 6 nm particles and to 700°C for 2 to 3 nm particles because of the decrease in the particle sizes.
In this research work, effort is made to synthesize nanoparticles using silver materials as the possible materials available to us presently.

### 1.2 Imaging Devices

Micro-organisms such as bacteria can easily be observed with an optical microscope, but with regard to nanoparticles it requires far better resolution. Invention of new microscopes has catalysed rapid advances in the field of nanotechnology. It is now possible to actually see atoms, nanoparticles, and nanoscale materials due to the recent advances in the imaging technologies such as high resolution transmission electron microscopy (TEM), scanning probe microscopy (SPM), also called scanning tunnelling microscopy (STM), atomic force microscopy (AFM), and powder X-ray diffraction (XRD) to mention but a few (Van et al., 1996).

### 1.3 Objective of the Study

The objective of this research work was aim at synthesizing the nanoparticles using infrared (IR) spectroscopy to study the compositions, sizes, structures etc., at different life span of the particles and also using different approach of synthesizing the particles. The characterization of those different samples of nanoparticle were studied in order to find out the effect of aggregation on the particles and also to determine the effect of using another approach in synthesizing the nanoparticles.

### 1.4 Problem Statement

The evolution of modern technology is a continuous phenomenon where innovations and manufacturing keep improving on daily basis. In the current state, nano-materials is playing a significant role in the production of space-craft panel, micro-chips of cell phones, technology vehicles, health technology applications etc., Therefore, continuous studies of the formation of nanoparticles becomes a very important aspect that can never be over-emphasised, because the economic values of these materials are very enormous.

## 2.0 Background on Nanoparticles

In nanotechnology, a particle is defined as a small object that behaves as a whole unit with respect to its transport and properties. They are further classified according to diameter and sizes. Coarse particles cover a range between 10,000 and 2,500 nanometres. Fine particles are sized between 2,500 and 100 nanometres. Ultrafine particles or nanoparticles are sized between 1 and 100 nanometres. The reason for this double name of the same object is that, during the 1970-80's, when the first thorough fundamental studies with 'nanoparticles' were underway in the USA (by Granqvist and Buhrman) and Japan, (within an ERATO Project) they were called 'ultrafine particles' (UFP). However, during the 1990s before the National Nanotechnology Initiative (NTI) was launched in the USA, the new name, 'nanoparticle' had become fashionable. Nanoparticles may or may not exhibit size-related properties that differ significantly from those observed in fine particles or bulk materials. Although the size of most molecules would fit into the above outline, individual molecules are usually not referred to as nanoparticles (Panigrah et al., 2006).

### 2.1 Health and Safety

An important impact of nanoparticles, particularly in the area of outreach, is health and safety. The most common chemical is that silver nanoparticles are reduced from silver nitrate, which is corrosive. In one study, which was determining a process for silver nanoparticle synthesis specifically for educational use, sodium borohydride is used as the reducing agent. These same nanoparticles have to be displayed to a wide range of ages. The nanoparticles can even be handled by students if the nanoparticles are converted to a 'glass' form using polyvinyl alcohol. One method of avoiding safety issues is using low concentrations of reagents. In the sodium borohydride reduction, concentrations are as low as 1 mM for both reagents (silver nitrate and sodium borohydride). Even if the stock solution touches skin, damage will be minimal. Another safety procedure is to force anyone handling the chemicals to use gloves. The nanoparticles are theoretically harmless when released into the environment because they will aggregate into much larger pieces. This does not mean the chemicals can be flushed down the drain; regulations relating to the waste must be followed (Mathew Wynn, 2009).

Waste was separated into two separate streams: gold and silver nanoparticle waste. The waste concentrated on the presence of toxic sodium borohydride. The gold nanoparticle waste was kept separate to prevent the reacting of any unused sodium borohydride from the silver nanoparticles synthesis, even though the molarity of the sodium borohydride was as low as 2 mM (Mathew Wynn, 2009).

Nanoparticles present possible dangers, both medically and environmentally. Most of these are due to the high surface to volume ratio, which can make the particles very reactive or catalytic. They are also able to pass through cell membranes in organisms, and their interactions with biological systems are relatively unknown. A recent study looking at the effects of ZnO nanoparticles on human immune cells has found varying levels of susceptibility to cytotoxicity. There are concerns that pharmaceutical companies, seeking regulatory approval for nano-reformulations of existing medicines, are relying on safety data produced during clinical studies of the earlier, pre-reformulation version of the medicine (Mathew Wynn, 2009).

### 2.2 Synthesis of Nanoparticles

There are several methods for creating nanoparticles, including reduction of some chemical reagents, attrition and pyrolysis. In attrition, macro or micro scale particles are ground in a ball mill, a planetary ball mill, or other size reducing mechanism. The resulting particles are air classified to recover nanoparticles. In pyrolysis, a vaporous precursor (liquid or gas) is forced through an orifice at high pressure and burned. The resulting solid (a version of soot) is air classified to recover oxide particles from by-product gases. Pyrolysis often results in aggregates and agglomerates rather than single primary particles. But here under this research work the nanoparticles synthesis using the reduction method is been used.

#### 2.2.1 Reduction Method of Synthesizing Nanoparticles

Basically three different methods of synthesizing nanoparticle are briefly discussed in this paper; which are the *gold*, *silver* and *copper* nanoparticle synthesis. However, we synthesized only the silver method in the laboratory, because of the unavailability of some useful reagent e.g. hydrochloroauric acid ($HAuCl_4$) to conduct the other synthesis, during the cause of this research. Details of these syntheses can be found in the standard literatures.

### 2.2.2 Synthesis of Silver Nanoparticles

Yellow colloidal silver has been reported upon reaction with ice-cold sodium borohydride and is the basis for the method used in this work described next. The determination of an optimal set of conditions for the synthesis of silver nanoparticles is described in the sections that follow. An excess of sodium borohydride is needed both to reduce the ionic silver and to stabilize the silver nanoparticles that form. The chemical reaction is the sodium borohydride reduction of silver nitrate:

$$AgNO_3 + NaBH_4 \rightarrow Ag + 1/2\ H_2 + ½\ B_2H_6 + NaNO_3 \qquad (2.2)$$

### 2.2.3 Protocol

Reactions were done in clean dry 125-mL Erlenmeyer flasks. A 10-mL volume of 1.0 mM silver nitrate was added drop-wise (about 1 drop/second) to 30mL of 2.0 mM sodium borohydride solution that had been chilled in an ice-bath. The reaction mixture was stirred vigorously on a magnetic stir plate. The solution turned light yellow after the addition of 2 mL of silver nitrate and a brighter yellow when all of the silver nitrate had been added. The entire addition took about three minutes, after which the stirring was stopped and the stir bar removed. The silver nanoparticle is stable at room temperature stored in a transparent vial for as long as several weeks or months. Upon aggregation the colloidal silver solution turns darker yellow, violet and finally gray as aggregation proceeds (Solomon et al., 2007).

### 2.2.4 Application of Silver Nanoparticles

The ability of silver nanostructures to enhance local electric fields has led to their applications as Surface Enhanced Raman Scattering (SERS). The enhancement factor can be as much as 1014 - 1015, which enables the detection of individual molecules absorbed on metal particles. The maximum signal amplification in SER spectroscopy (1015) can be achieved with silver particles of about 50 nm (Curry et al., 2006).

Silver has been known for antibacterial activity since the ancient Greece times. Currently, the investigation of this phenomenon has gained more attention due to the increase of bacterial resistance to antibiotics, caused by their overuse. Silver nanoparticles can be used in medicine to reduce infections in burn treatment, to prevent bacteria colonization on dental materials, stainless steel materials, to eliminate microorganisms on textile fabrics, or used for water treatment. Antimicrobial activity of colloid silver particles is influenced by the dimensions of the particles. The extremely small size of nanoparticles results in the particles having a large surface area relative to their volume. This allows them to easily interact with other particles and increases their antibacterial efficiency (Pearce et al., 2007).

## 3.0 Research Materials and Methodology

**3.1 Materials:** During the characterization of the silver nanoparticles, the following materials are used. They are: (i) sample of nanoparticle, (ii) Fourier transmitter infrared spectrometry (FT-IR), (iii) a computer system, and (iv) a printer.

### 3.2 Methodology

The process of characterizing the sample of nanoparticle is carried out in the following way. The cover of the FT-IR spectrometry was open to bring out the cuvette which was rinse with distilled water, about 1.5mil of a sample was poured into the cuvette, a cover was used to cover it and then the cuvette was place in its holder in the FT-IR spectrometry.

The scanning process was conducted using the computer system which has an application that is been used to characterize such samples. While scanning the sample a platform for naming the sample will pop out, this allows one to name the sample with the name of his choice.

After the scanning is over a chromatograph of transmitters and frequency is displayed which is printed using the printer. On the chromatograph display on the window one can switch the nature of the graph from that of transmitters against frequency to absorbance against frequency which can also be printed out.
The same procedure was followed for the samples only that the naming differs in order to distinguish them from one another.

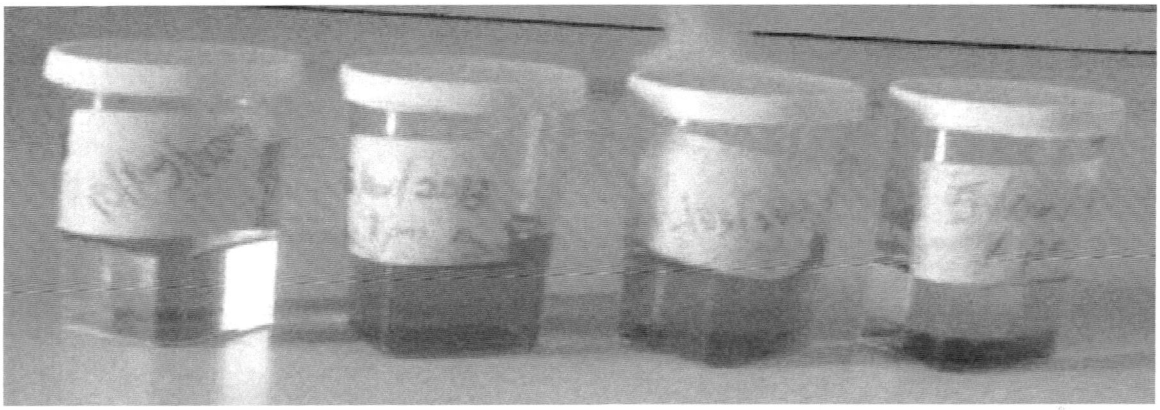

**Figure 1** Laboratory synthesized colloidal silver in various stages of nano-aggregation. From left to right are: (a) clear yellow sol, (b) dark yellow, (c) violet, and (d) grayish, as the aggregation proceeds.

### 3.3 Function of the FT-IR Spectrometry

Fourier Transform Infrared (FT-IR) spectrometry was developed in order to overcome the limitations encountered with dispersive instruments. The main difficulty was the slow scanning process. A method for measuring all of the infrared frequencies simultaneously, rather than individually, was needed. A solution was developed which employed a very simple optical device called an interferometer. The interferometer produces a unique type of signal which has all of the infrared frequencies 'encoded' into it. The signal can be measured very quickly, usually on the order of one second or so. Thus, the time element per sample is reduced to a matter of a few seconds rather than several minutes.

Most interferometers employ a beam splitter which takes the incoming infrared beam and divides it into two optical beams. One beam reflects off of a flat mirror which is fixed in place. The other beam reflects off of a flat mirror which is on a mechanism which allows this mirror to move a very short distance (typically a few millimeters) away from the beam splitter. The two beams reflect off of their respective mirrors and are recombined when they meet back at the beam splitter. Because the path that one beam travels is a fixed length and the other is constantly changing as its mirror moves, the signal which exits the interferometer is the result of these two beams "interfering" with each other. The resulting signal is called an interferogram which has the unique property that every data point (a function of the moving mirror position) which makes up the signal has information about every infrared frequency which comes from the source. This means that as the interferogram is measured; all frequencies are being measured simultaneously. Thus, the use of the interferometer results in extremely fast measurements. Because the analyst requires a frequency spectrum (a plot of the intensity at each individual frequency) in order to make identification, the measured interferogram signal cannot be interpreted directly. A means of "decoding" the individual frequencies is required. This can be accomplished via a well-known mathematical technique called the Fourier transformation. This transformation is performed by the computer which then presents the user with the desired spectral information for analysis.

**Figure 2** The Fourier Transforms Infrared spectrometry

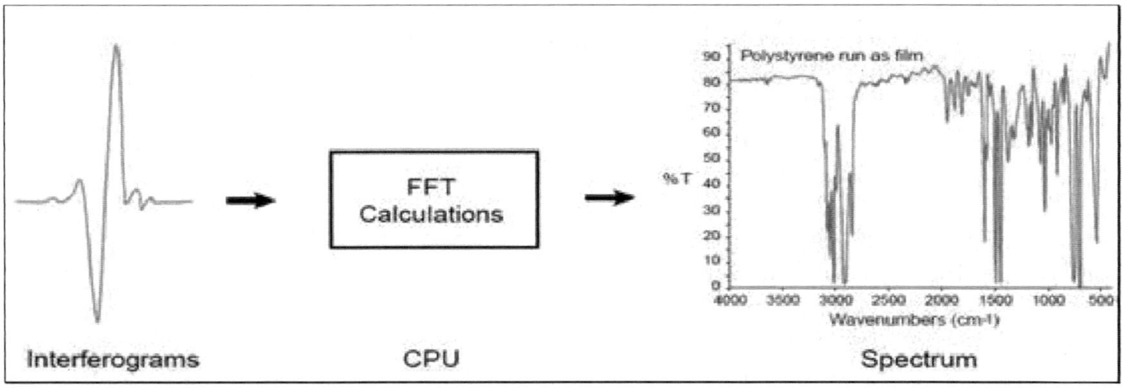


**Figure 3** Depict the path way through which a computer converts the interferogram using the FFT calculations in to spectrum. (Thermo Nicolet 2001)

## 4.0 Experimental Results

The results of the studies carried out on the samples of freshly prepared silver nanoparticles were characterized immediately after the synthesis. The initial synthesis was used as a reference-base to the other samples which where characterize after certain period (hours, minutes or days) of their synthesis. A reference chart obtained from the standard literature was also used to check for the types of bonds existing within the synthesized nano-samples.

### 4.1 Absorption Bands

Figure 4 shown below is the reference chart used in determining the nature of bonds that exist between the various components that make up the nano-samples.

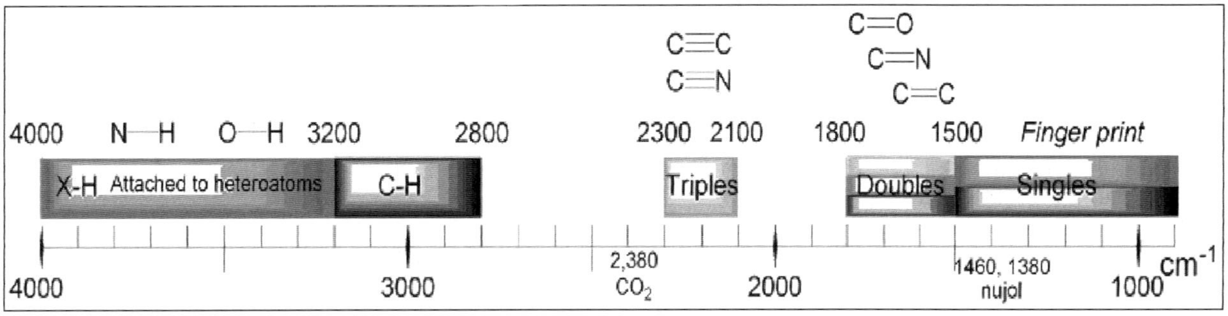


**Figure 4** Absorption band showing the possible bonds at various wave number

The chromatographs of transmittance and absorbance spectrum obtain after characterizing the samples are shown below:

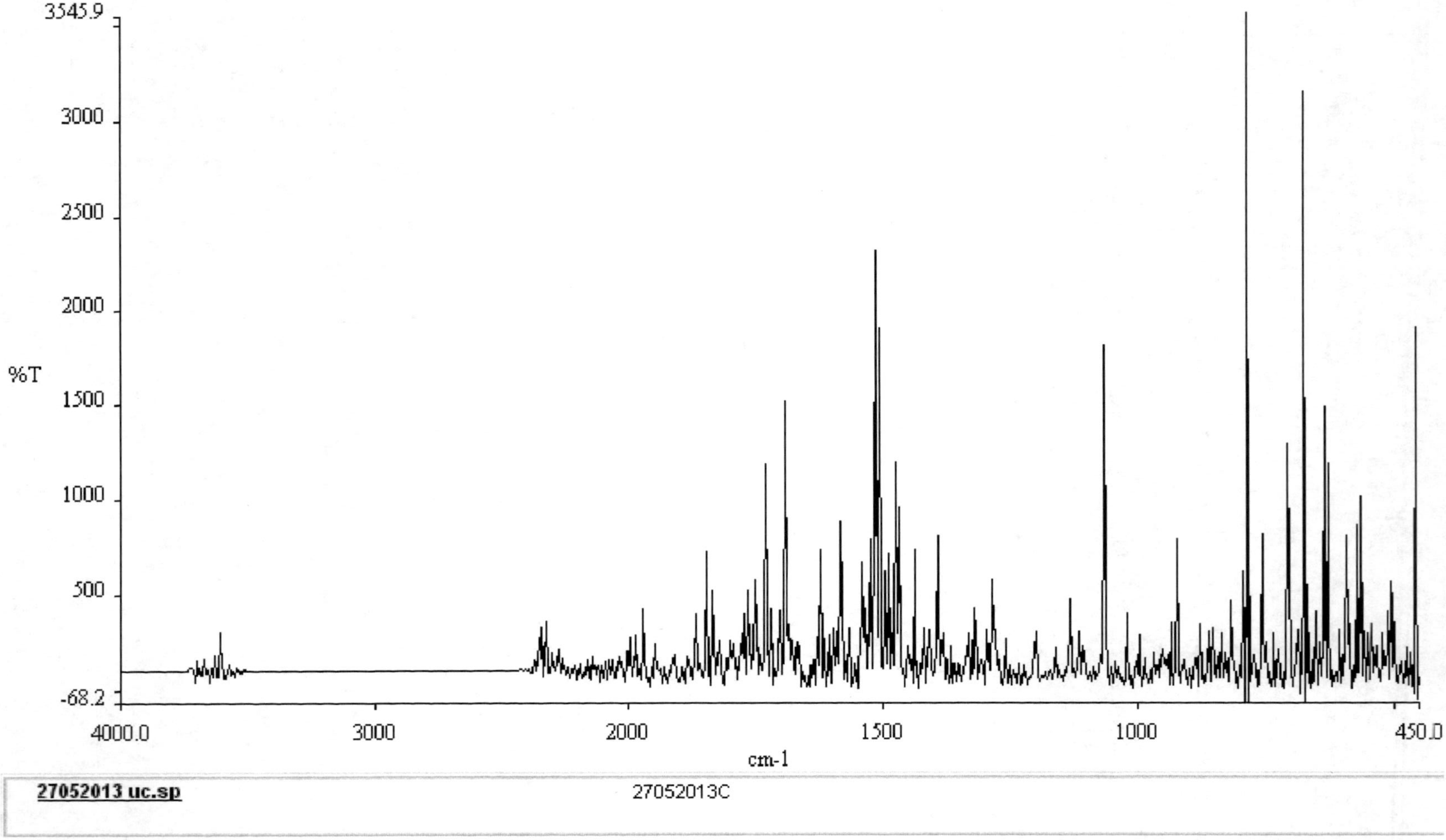


Figure 4.1(a) Chromatograph of T against F of freshly prepared silver nanoparticles of Day 2

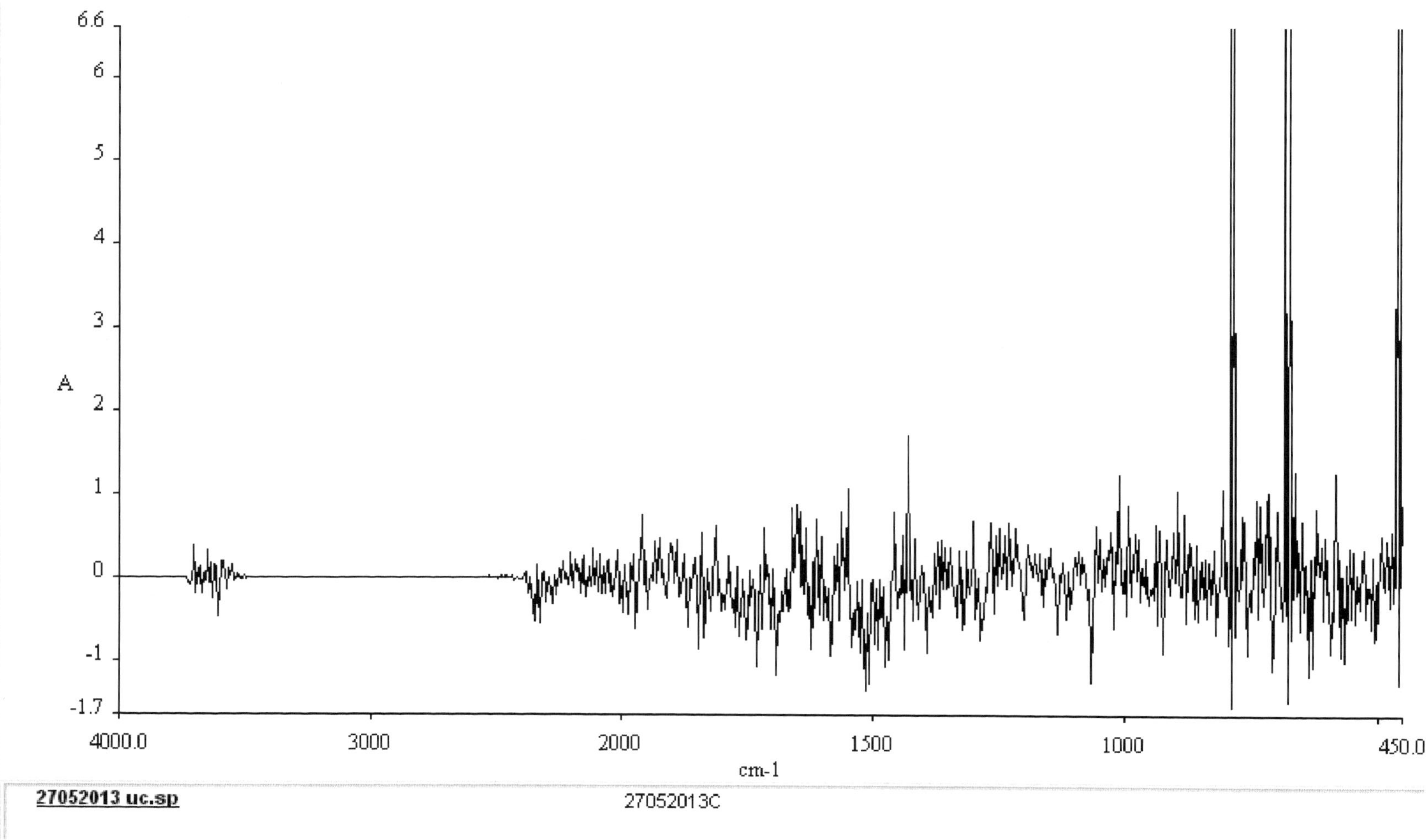


Figure 4.1(b) Chromatograph of A against F of freshly prepared silver nanoparticles of Day 2

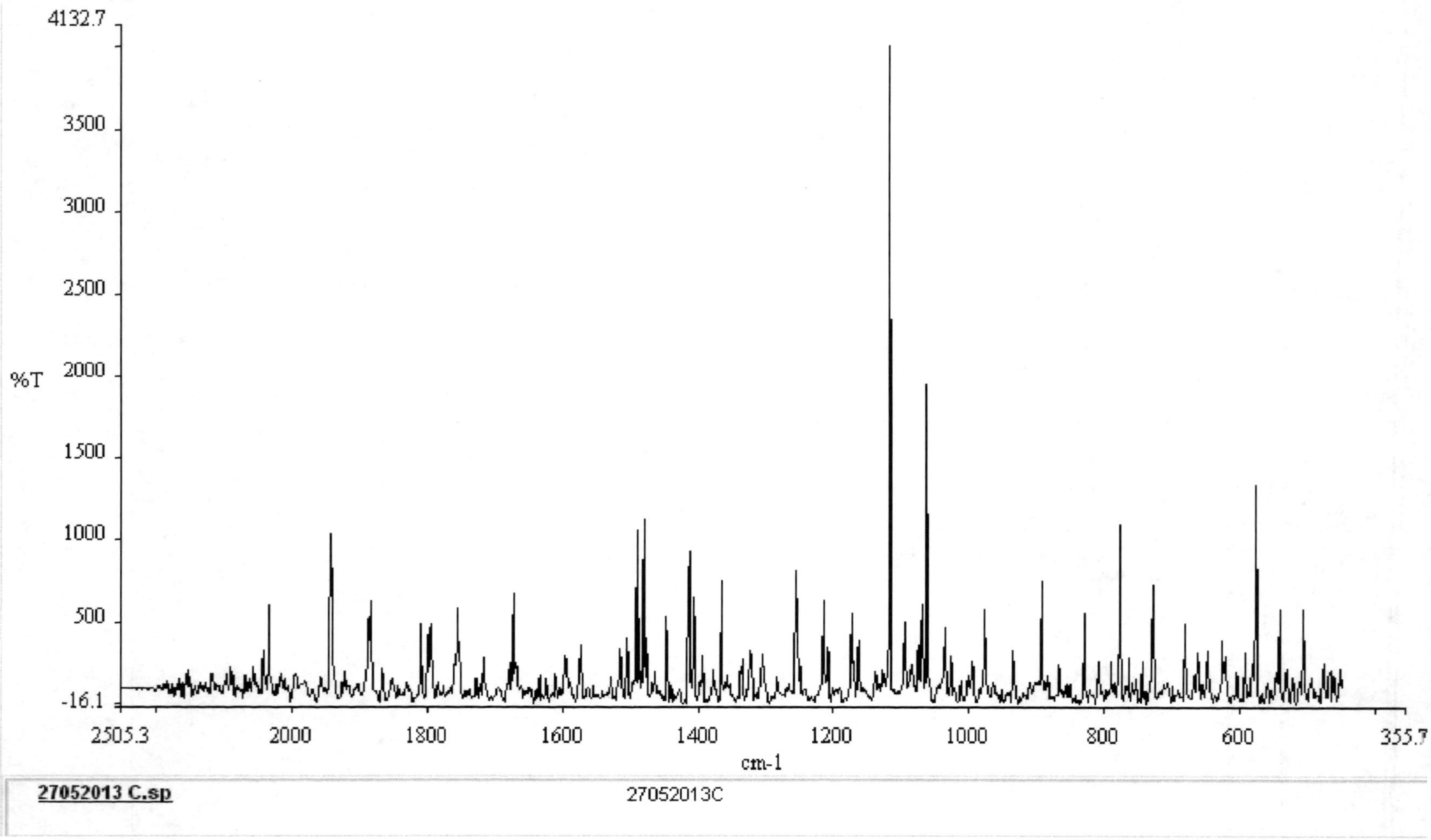


Figure 4.2(a) Chromatograph of T against F of freshly prepared silver nanoparticles using different method of synthesis

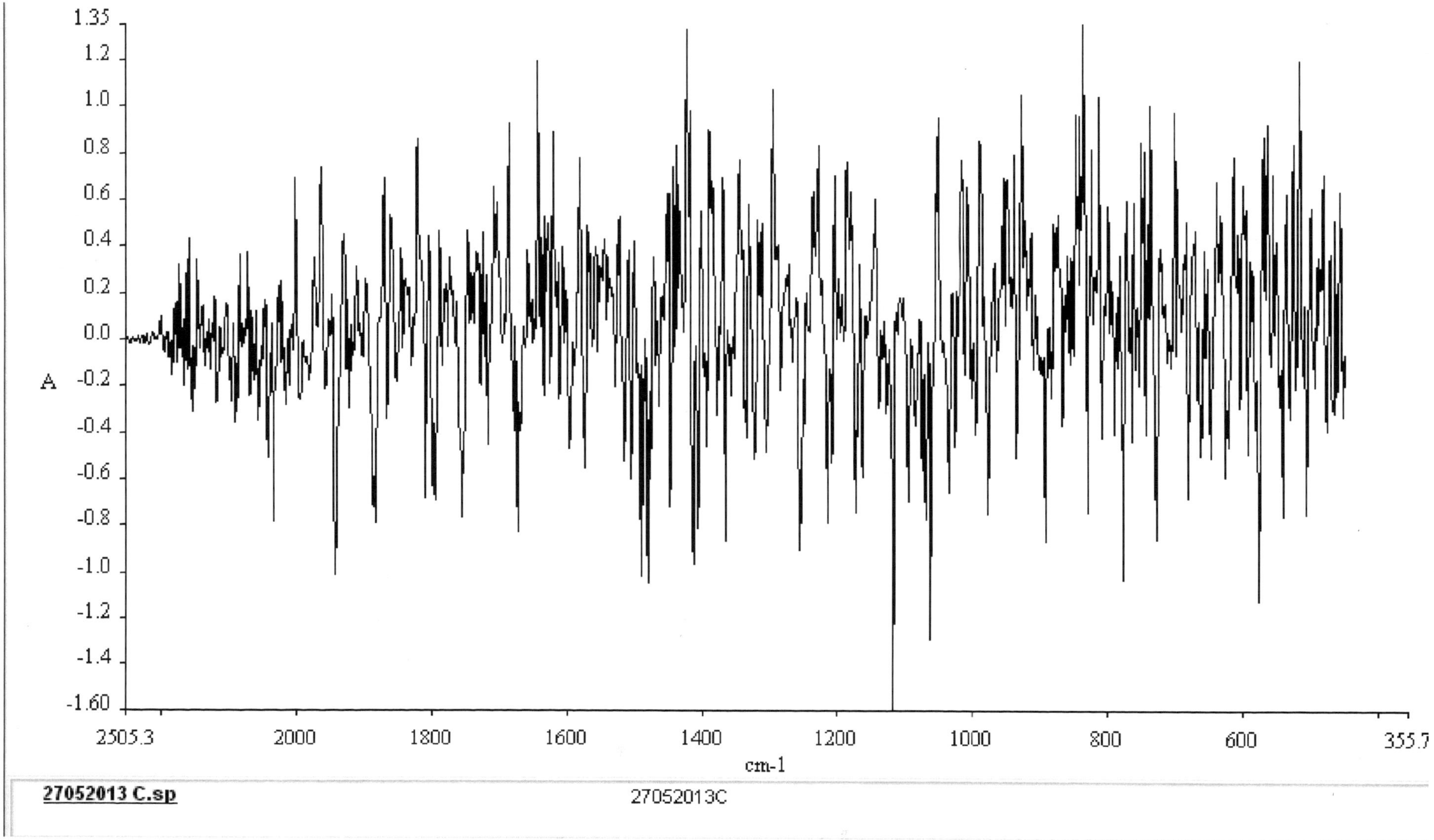


Figure 4.2(b) Chromatograph of T against F of freshly prepared silver nanoparticles using different method of synthesis

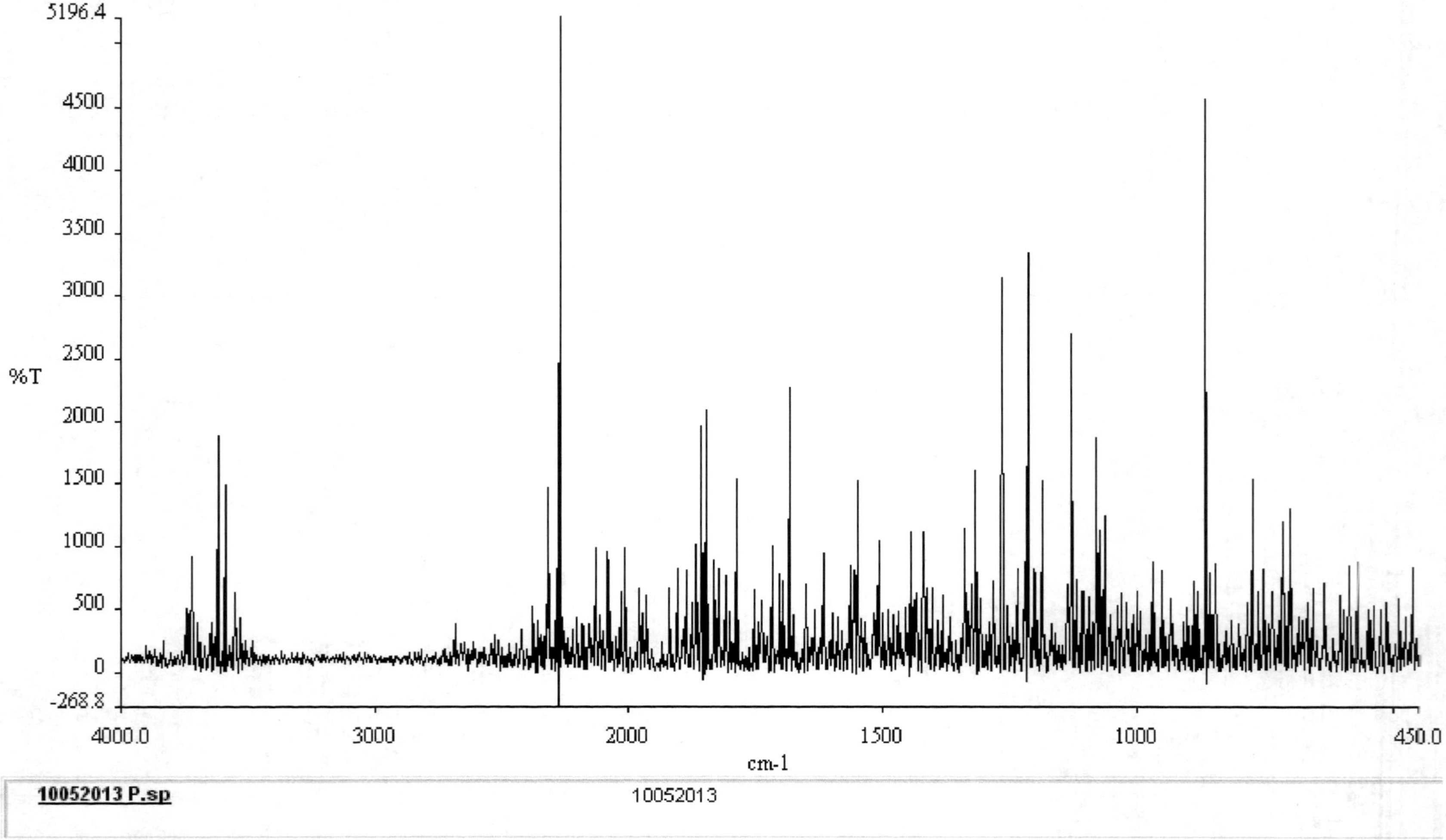


Figure 4.3(a) Chromatograph of T against F of freshly prepared silver nanoparticles of Day 1

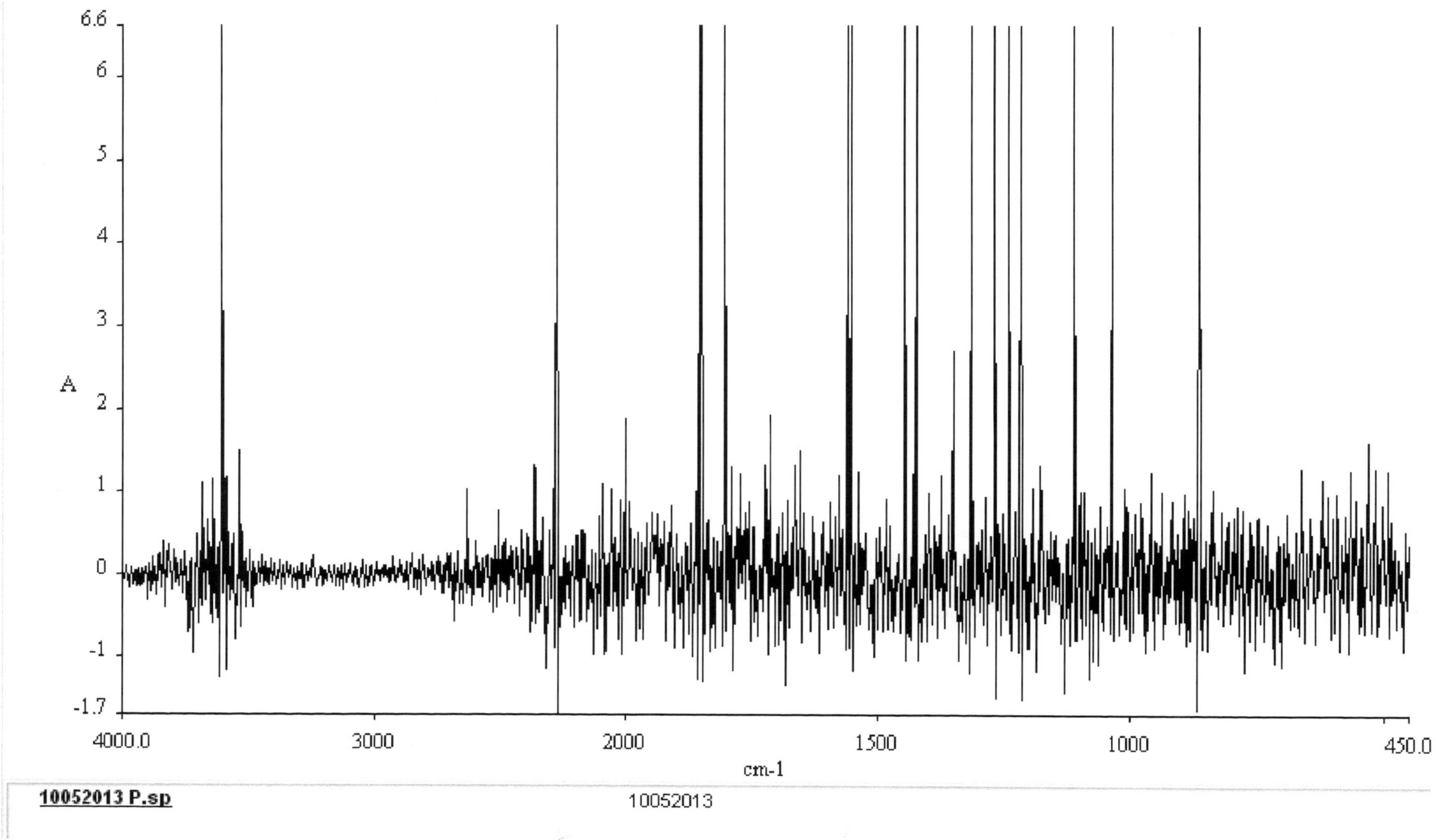


Figure 4.3(b) Chromatograph of A against F of freshly prepared silver nanoparticles of Day 1

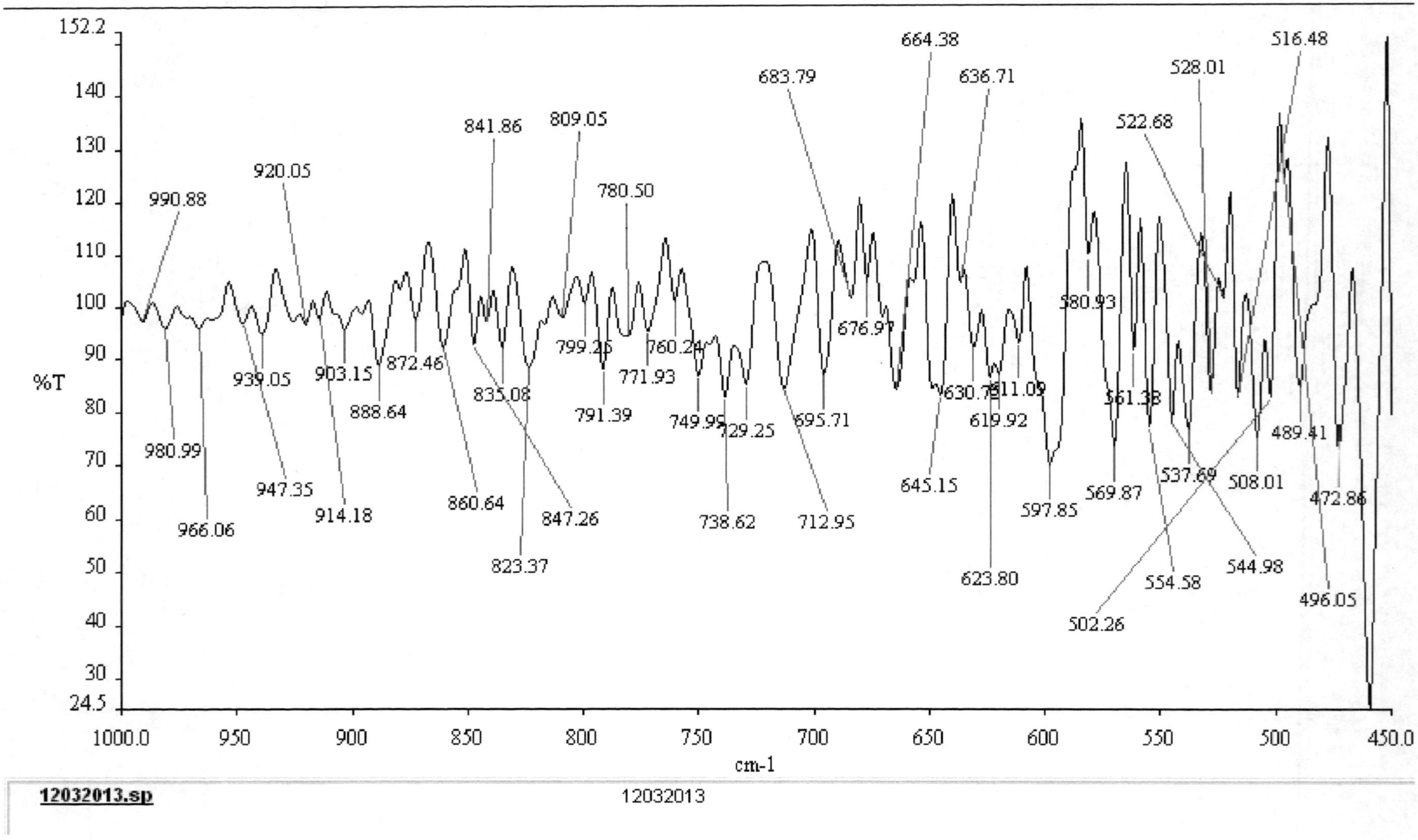


Figure 4.4(a) Chromatograph of T against F of 3 weeks (21-days) prepared silver nanoparticles

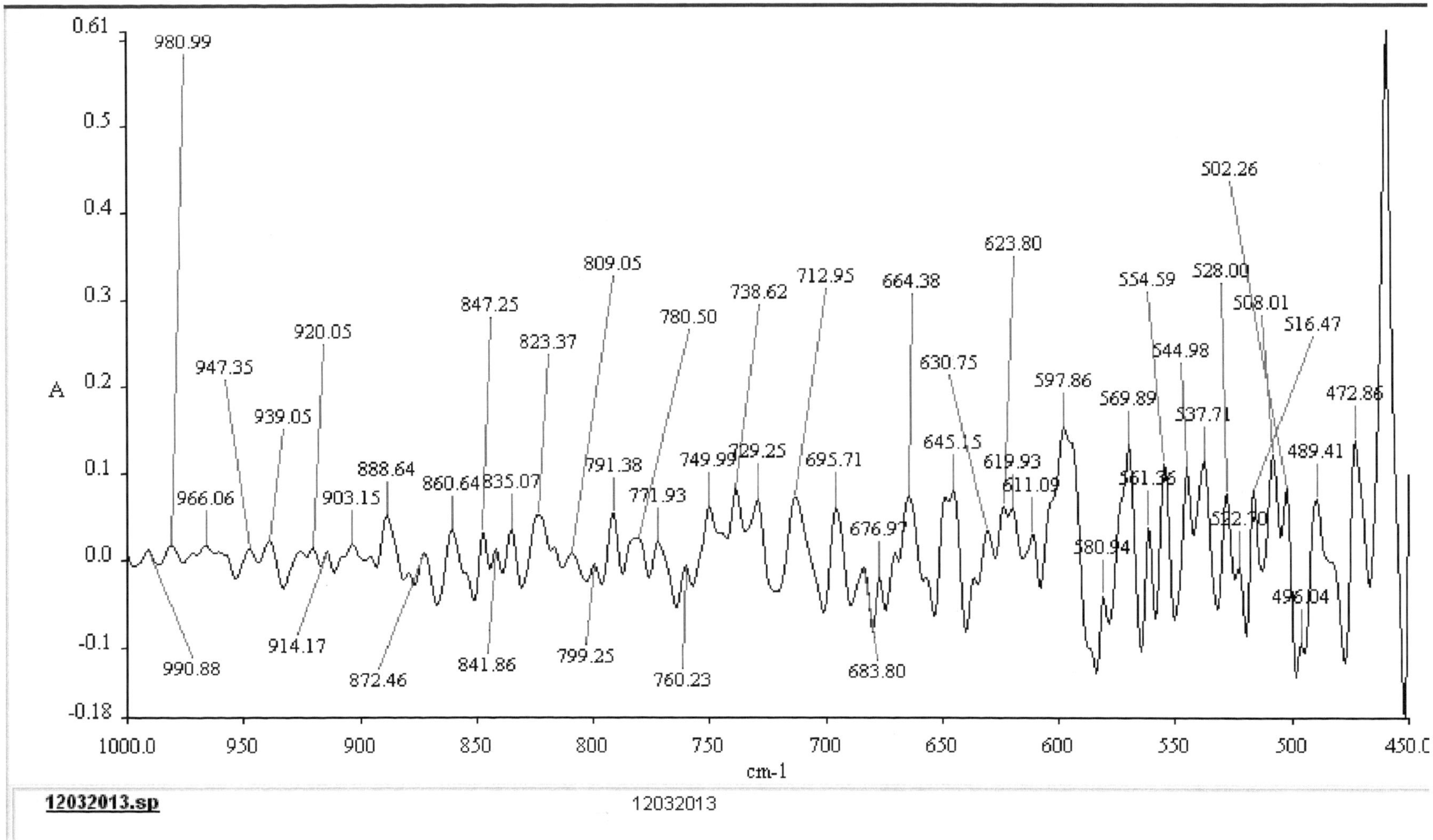


Figure 4.4(b) Chromatograph of A against F of 3weeks (21-days) prepared silver nanoparticles

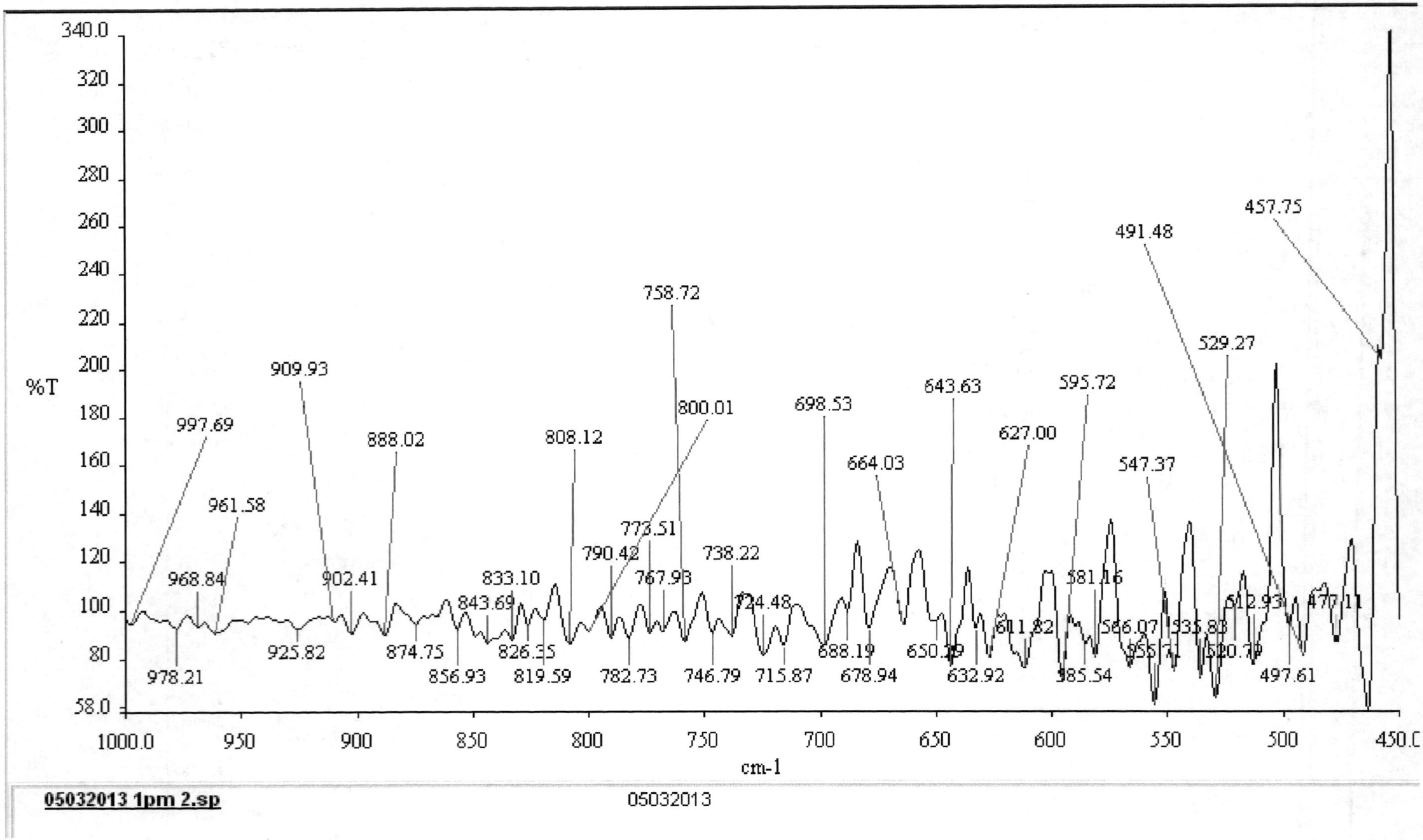


Figure 4.5(a) Chromatograph of T against F of 5weeks (35-days) prepared silver nanoparticles

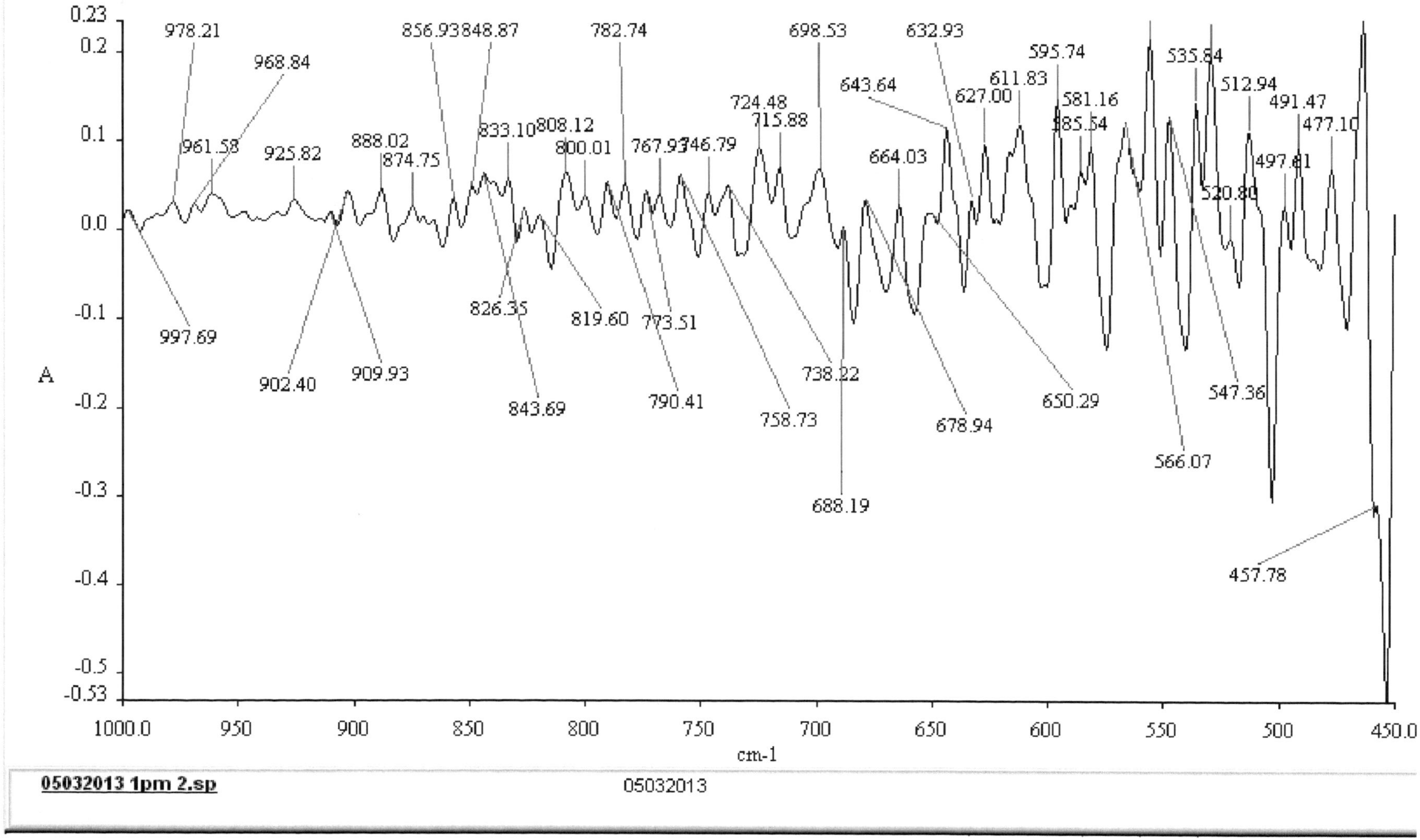


Figure 4.5(b) Chromatograph of A against F of 5weeks (35-days) prepared silver nanoparticles

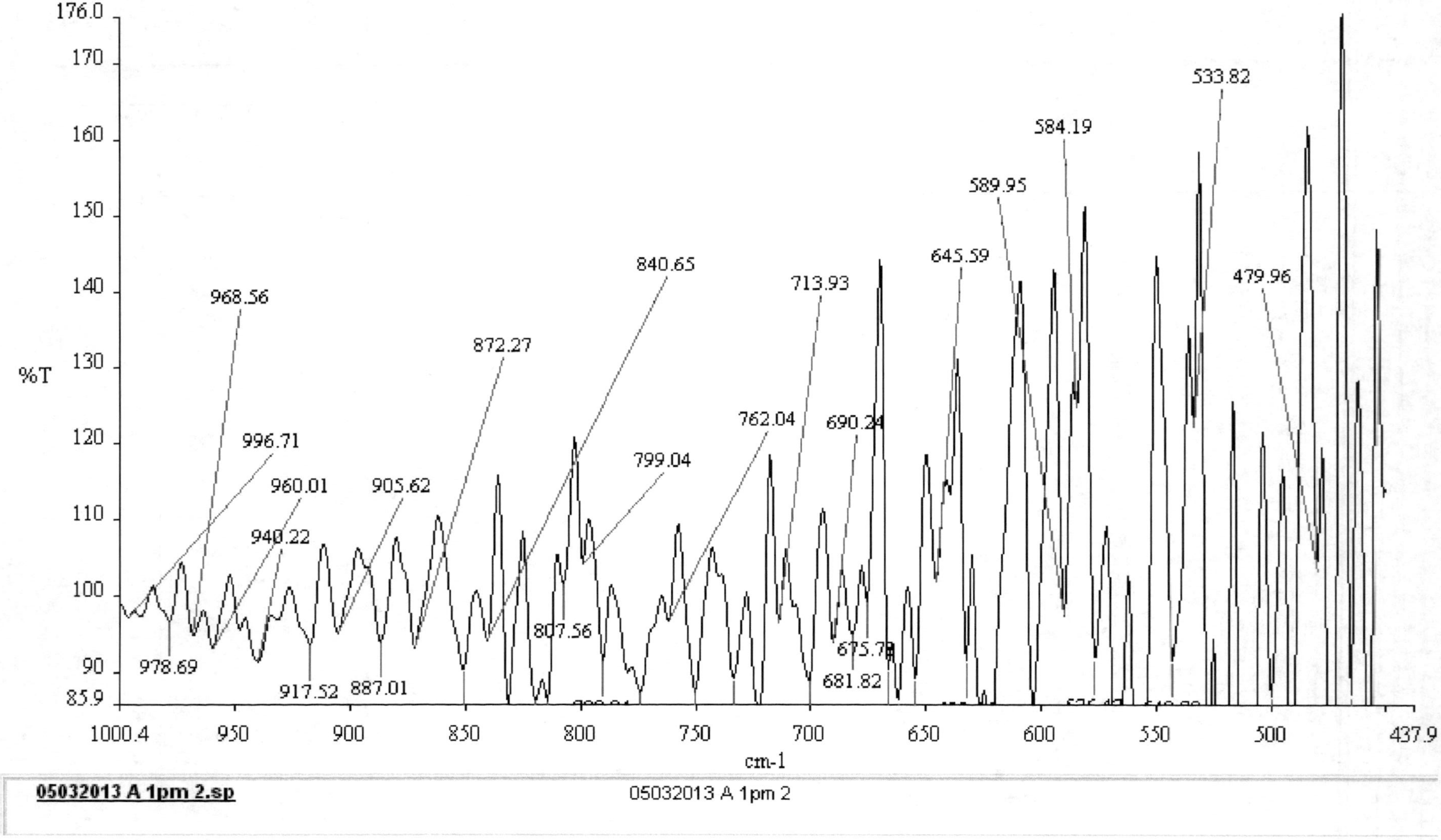


Figure 4.6(a) Chromatograph of T against F of an aggregated silver nanoparticle 1

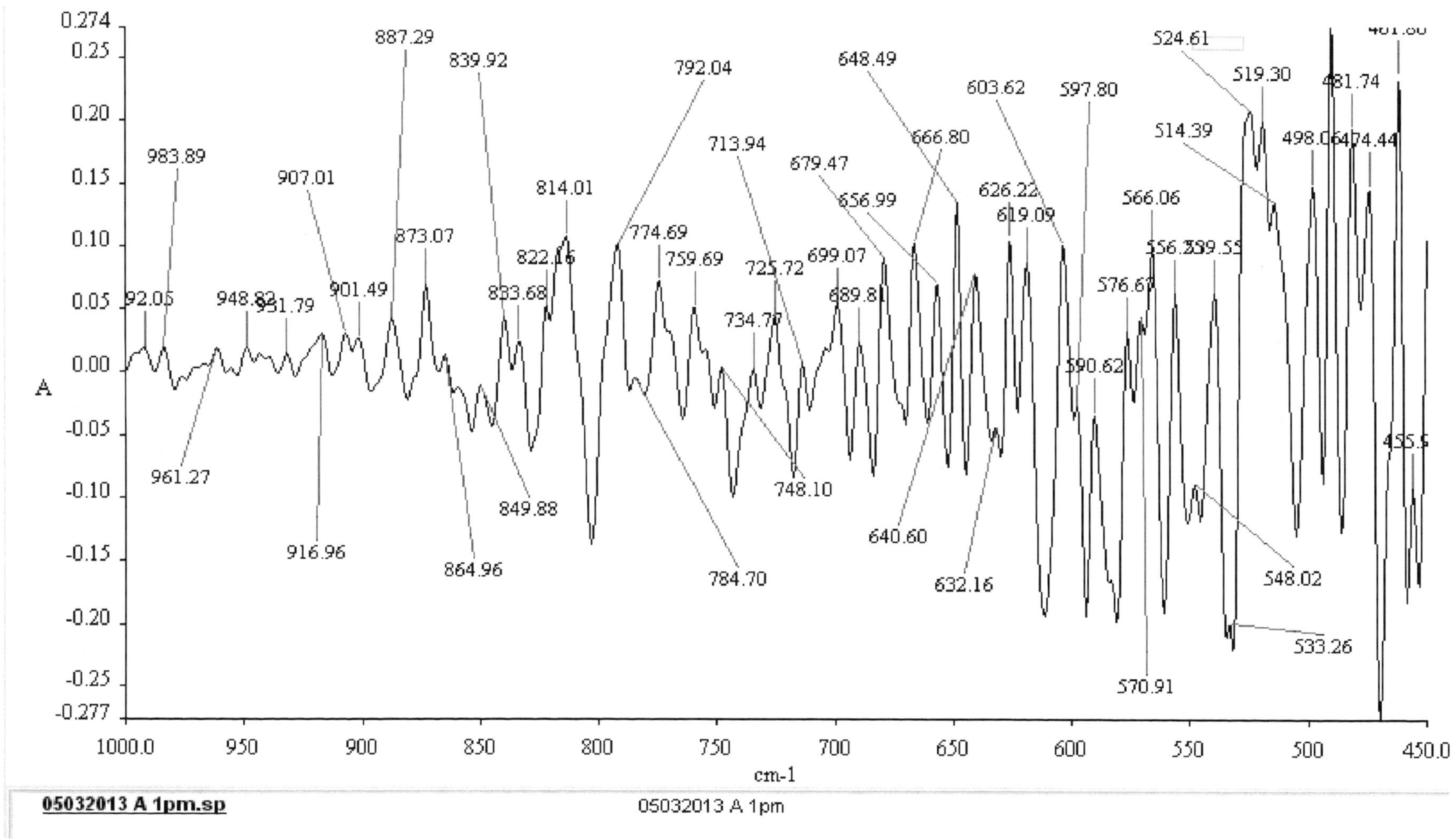


Figure 4.6(b) Chromatograph of A against F of an aggregated silver nanoparticle 2

## 5.0 Discussion of Results

The findings on each of the synthesized samples characterized were discussed in details in this section of this paper. The freshly prepared and characterize sample of day 2 (i.e. 48-hours synthesis) were used as reference to the other samples for comparison. The chat shown in Figure 4 were used to determine (or evaluate) the nature of bonds possible to be available within the sample is also been used.

### 5.1 Freshly Prepared and Characterized Sample of Day 2 (48-days)

From the graphs of figure 4.1(a) and 4.1(b) it can be seen that the sample has a vibration at wave number 4000-450cm$^{-1}$, but the vibration is more prominent from 2500-450 for both the absorbance and transmittance graph. The major goal of any absorption spectrum is to measure how well a sample absorbs light at each wavelength and the graph has shown that this sample absorb very well from wave number of 2500-450cm$^{-1}$ and at a frequency of 4μm to 22.2 μm. The absorption spectrum indicates the IR radiation was absorbed by the sample, whereas that could not be made up of the transmittance spectrum. Absorption is the resonant frequency, which is the frequency of the absorbed radiation matches the transition energy of the bond or group that vibrates. The more the number of peaks in the spectrum the more the material present in the sample and it can be seen clearly that the two spectrums have much numbers of peaks.

The transmittance spectrum graph shows that the sample has a maximum value of 3545.9.7% transmittance and a minimum value of -68.2% transmittance, whereas the absorbance graph has a maximum absorbance at 1.35 and a minimum of -1.60 absorbance.

Based on the analysis carried out, the types of bonds likely to be present are highlighted below:

1. X-H (attached to heteroatom) because we had vibration at wave number between 4700 - 4600 cm-$^{1}$.
2. Triple bonds due to the vibrations between 2300 - 21000 cm-$^{1}$.
3. Doubles and singles (finger print) which is likely to be the most part of the sample, because of much vibration are present within the wave number of 1800 to less than 1000cm-$^{1}$.

### 5.2 Freshly Prepared and Characterized Sample Using Different Method of Synthesis

From Figure 4.2(a) and figure 4.2(b) it can be seen that the sample has vibration at wave number 2505.3 - 355.7cm$^{-}$ for both the absorbance and transmittance graph. Even though the sample was prepared using different method it has shown that it still has many peaks which indicated that there are many materials present but not up to the first one.

The transmittance spectrum graph shows that the sample has a maximum value of 4132.7% transmittance and a minimum value of -16.1% transmittance whereas the absorbance graph has a maximum absorbance at 1.35 and a minimum of -1.60 absorbance.

The chart analyses show the type of bonds that is likely to be present as follows:

1. Triple bonds due to the vibrations between 2300 - 21000 cm-$^{1}$.
2. Doubles and singles (finger print) which is likely to be the most part of the sample because many vibrations are present within that wave number of 1800 to less than 1000cm-$^{1}$.

### 5.3 Freshly Prepared and Characterized Sample of Day 1

This sample was characterized immediately after it synthesis. It has almost similar qualities with the sample of day 2 (24-hours) as it can be seen on Figure 4.3(a) and Figure 4.3(b), only that the nature of peaks and their positions on the spectrum differs. This sample has a maximum value of 5196.4% transmittance and a minimum value of -268.8% transmittance. Whereas for the absorbance, the spectrum has a maximum value of 6.6 and -1.7 absorbances. The size of peaks present in both the transmittance and absorbance spectrum has indicates the present of large number of materials.

About the nature of possible bonds present is also same with the sample in day 2 (48-hours) that is X-H, doubles and finger prints.

## 5.4 Sample Characterized After 3 Weeks (21-days) of Synthesis

Clear differences can be observed between the freshly prepared and those that are characterized days after the synthesis as shown in Figures 4.4(a) and 4.4(b). Vibrations were only present within the wave number of 1000-450cm-$^{1}$, for both the absorbance and transmittance spectrum, which is equivalent to the frequency of 10µm to 22.2µm. The number and size of the peaks are relatively low when compared with the fresh once. The maximum transmittance value of this spectrum is 152.2% transmittance, while the minimum value is 24.5% transmittance, which means no negative transmittance value. Whereas in the case of the absorbance spectrum, the maximum value is 0.61 of absorbance and the minimum is -0.18 of absorbance.

The size and number of the peaks shows that the number of materials in the sample is low which indicated that the particle has start aggregating with each other forming bigger particles and therefore their number decreases.

The fact that all the vibrations are within the wave number of 1000-450cm-$^{1}$ indicates that the only bond likely to be present is the single bond or the finger prints.

## 5.5 Sample Characterized After 5 Weeks (35-days) of Synthesis

Although both the spectrum begins from 1000 to 450cm$^{-1}$, but prominent peaks are only notice at around 700 to 450cm$^{-1}$, as it can be seen in Figures 4.5(a) and 4.5(b). The nature, size and numbers of the peaks has decline again far lesser than those characterized after 3 weeks of synthesis. The transmittance and absorbance value has also drop to 340 to 58% of transmittance and 0.23 to - 0.53 of absorbance which was in thousands for the transmittance and units for the absorbance. The amount of materials (particle) present is this sample is far low due to low number of peaks present and this is a clear indication that the particles have bonded to each other forming larger particles that are no more in nanoscale, in short the sample has almost aggregate only that the colour has not indicate that the sample has aggregate completely. Here too the only possible bonds that could be found are the single bonds (finger) prints anything apart from that could not be found within the sample.

## 5.6 Sample Characterized After It Has Aggregated

The aggregated sample is almost similar with the previous one in terms of their peaks size, number and nature, the only difference is the absorbance and transmittance value which has decline to 176 - 85.9% of transmittance and 0.274 of absorbance as it can be seen in Figures 4.6(a) and 4.6(b). The numbers of peaks are not much which indicates that the number of particles present is low and with the colour of the sample changing to violet is a conclusion that the sample has aggregated. In terms of possible bonds that could be found here is still the single bond or finger print.

## 6.0 Conclusions

The synthesis of nanoparticle is very possible in a laboratory small scale and it doesn't cause much money. The synthesized nanoparticles create a room for further research to be carried out. The use of the FT-IR to characterize the nanoparticles has shown some of the properties of the nanoparticles in terms of wave number, maximum and minimum absorbance and transmittance, nature of bonds likely to be found and the amount of vibrational peaks in the spectrum.

In conclusion, it can be deduce that the freshly synthesized samples have much number of peaks which indicates that the particles present is also much, while the size of the peaks as well, because the maximum transmittance values of fresh samples were in thousands and the minimum values were in negative values. However, the aggregated or almost aggregated samples are in hundred for their maximum values and positive tens for their minimum values. In the case of the absorbance the fresh one's has maximum values in tens and minimum values in negatives tens whereas the aggregated or almost aggregated one's has their for the maximum in positive decimal while the minimum in negative decimals.

In the case of possible bonds that are likely to be found in the sample, for the fresh samples about 2 or 3 difference bonds are likely to be gotten, while for the aggregated or almost aggregated samples only 1 bond is likely to be found.
In addition, the major problem of these nanoparticles as they aggregate is that, it can be seen clearly from the result and discussion that aggregation has effect on absorbance, transmittance, number of peaks and even the nature of bonds likely to be found within the sample. From physical observation, the transformation the nano-samples undergoes from pale yellow down to grey, which indicates aggregation of the synthesized sample. This confirmed that the nanoparticles do aggregates and further studies should be carried out for more results.

**7.0 Terminology**

***Nano***: - Is derived from Greek word meaning dwarf.
***Nanometer***: - This means one-billionth of a meter (1nm =10 E-09m).
***Nanotechnology***: - This involves the design, production and application of structures, devices and systems by controlling the shape and size at the nanometer scale.
***Nanoparticle***: - A nanoparticle (NP) has dimensions in the nanometer range, e.g. between 0.5 and 100 nm.
***Nanocrystal***: - *I*s a solid particle that is a single crystal in the nanometer size range.
***Nanostructured (or nanoscale materials)***: - A nanoscale material is any solid material that has a nanometer dimension; such as three dimension (particles); two dimensions (thin films); one dimension (thin wire).
***Nanoscience***: - Is the study of phenomena and manipulation of materials at atomic, molecular and macromolecular scales, where properties of materials can be quite different from those of larger scales.

## References


Poole, C. P., and Owens, F. J. (2003) '*Introduction to Nanotechnology'*, Wiley Interscience: New Jersey; pp. 35-45

Van Cleef, M., Holt, S. A., Watson, G. S., and Myhra, S., (1996) 'Polystyrene spheres on mica substrates: AFM calibration, tip parameters and scan artifacts', *Journal of Microscopy*, 181 (1), 2-9

Panigrah, S., Kundu, S., Ghosh, S. K., Nath, S., and Pal, T.,(2004) "General method of synthesis for metal nanoparticles", *Journal of Nanoparticle Research*, 6, pp. 411-414.

Klabunde, K., (2001) Ed. *Nanoscale Materials in Chemistry*, Wiley Interscience: New York, p 23

Solomon, S., Bahadory, M., Jeyarajasingam, A., Rutkowsky, S., Boritz, C., and Mulfinger, L., (2007) "Synthesis and study of silver nanoparticles", *Journal of Chemical Education*, 84 (2), 322-325.

Zhu, H., Zhang, C., and Yin, Y., (2005) 'Novel synthesis of copper nanoparticles: influence of the synthesis conditions on the particle size', *Nanotechnology*, 16, 3070-3083.

Yonzon, C. R., Stuart, D. A., Zhang, X., McFarland, A. D. Haynes, C. L., and Van Duyne, R. P., (2005), 'Towards advanced chemical and biological nanosensors', *an overview, Talanta*, 67, 438–448.

Curry, A., Hwang, W. L., and Wax, A., (2006) "Epi-illumination through the microscope objective applied to darkfield imaging and micro spectroscopy of nanoparticles interaction with cells in culture", *Optics Express*, 14 (14), 6535-6542.

Pearce, M. E., Melanko, J. B., and Salem, A. K., (2007) 'Multifunctional nanorods for biomedical applications', *Expert Review, Pharmaceutical Research*, 24 (12), 2335-2352

Cioffi, N., Ditaranto, N., Torsi, L., Picca, R. A., Giglio, E., D., and Zambonin, P., G.,(2005), 'Synthesis, analytical characterization and bioactivity of Ag and Cu nanoparticles embedded in poly-vinyl-methyl-ketone films', *Analytical and Bioanalytical Chemistry*, 2005, 382, 1912–1918.

Capatina, C., (2005) 'The study of copper ruby glass', *Ceramics-Silikaty*, 4, 283- 286.

Ch. Hayashi, Ryozi Uyeda, A. Tasaki. (1997). *Ultra-fine particles: Exploratory science and technology (1997 Translation of the Japan report of the related ERATO Project 1981–86).* Noyes Publications.

Chen K., (2012) Infrared spectroscopy from Wikipedia on 13/06/2013, the free encyclopedia (searched on the Mathew Wynn *Materials Engineering Department California Polytechnic State University, San Luis Obispo* Advised by Dr. Katherine C.

*Introduction to Fourier Transform Infrared spectrometry* by Thermo Nicolet Corporation 2001.

# Preparation of Drug Nanoparticles Stabilized by Polysaccharide Surfactant through Four-Inlet Vortex Mixer

Mingwei Wang, Zhenyu Yuan, Li Li, Jie Wang, Xuhong Guo

Surfactants (Maltose-C12) based on polysaccharide were synthesized by click reaction between reducing end of the Maltose and $NH_2$ group, in the existence of $NaBH_3CN$, Figure 1. Using Maltose-C12 as surfactants, nanoparticles of beta-carotene as model drug were successfully prepared through flash nano-precipitation in a home-made transparent four-inlet vortex mixer. The average diameter is 191 nm with a PDI of 0.105 as shown in Figure 2.

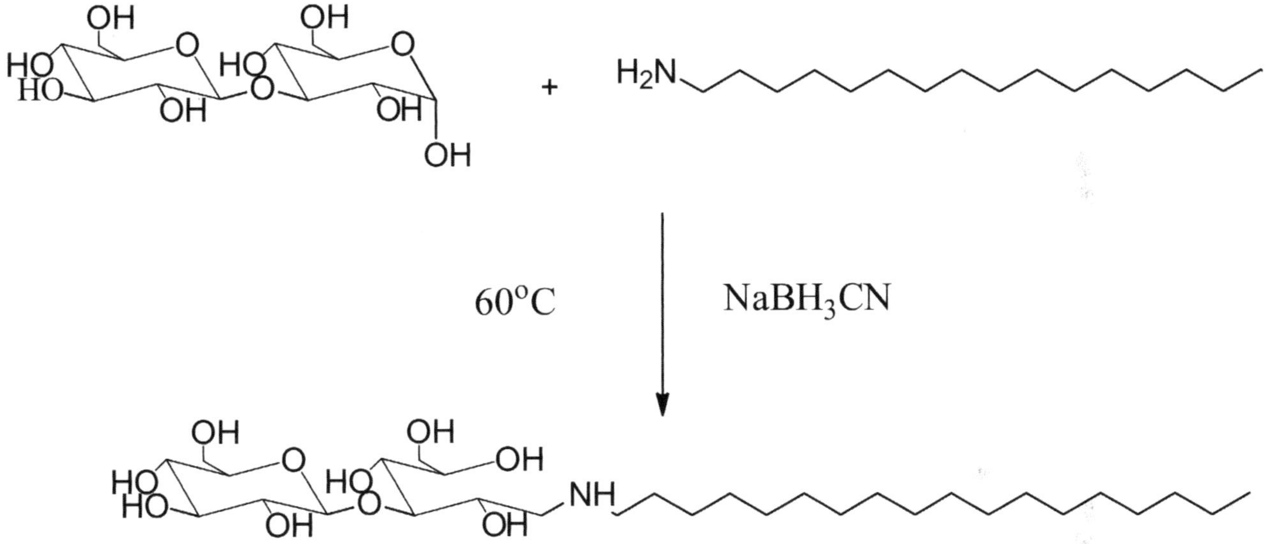


**Figure.1 Reaction of Maltose-C12 synthesis.**

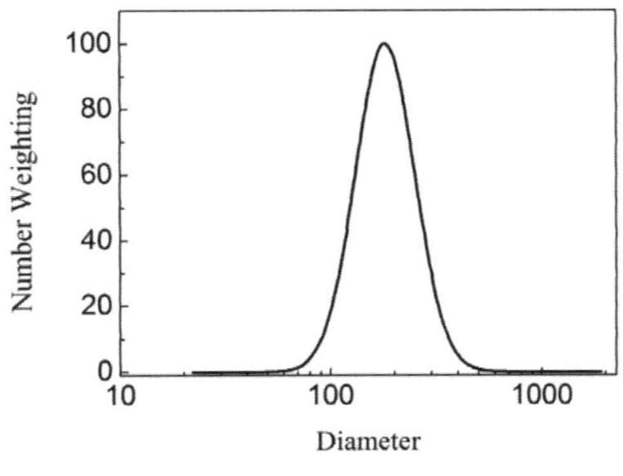


**Figure.2 Size distribution of nanoparticles stabilized by Maltose-C12.**

# 384315 Scale-up of Nanoparticle Manufacturing By Flame Spray Pyrolysis

**Abstract:**

The scaling-up of nanoparticle manufacturing by versatile flame spray pyrolysis (FSP) is investigated numerically and experimentally for production of $ZrO_2$. The process is described with a three-dimensional computational fluid dynamics model that allows extracting the high-temperature particle residence time (HTPRT) that is governed by the dispersion gas to precursor liquid mass flow ratio as well as the flame enthalpy content. The HTPRT is shown to control the primary particle and agglomerate size, morphology and even $ZrO_2$ crystallinity in agreement with experimental data. By keeping the HTPRT constant the production rate for $ZrO_2$ nanoparticles could be scaled up from ~100 to 500 g/h without significantly affecting product particle properties, revealing the HTPRT as a key design parameter for flame aerosol processes.

# 386687 Numerical Analisys of Multicomponent Catalysts Production By Double Flame Spray Pyrolysis (DFSP)

**Abstract:**

The synthesis of nanoparticles by Flame Spray Pyrolysis (FSP) is a newer but already well developed process in terms of product diversity. Many elements of the periodic table can be turned into oxides, salts or even metal NPs in laboratory-scale reactors with production rates of a few grams per hour. In these reactors, low cost metallic precursors (e.g. zirconium(IV) propoxide) dissolved in a liquid fuel (e.g. ethanol, xylene) are supplied into a twin-fluid atomizer. Pure oxygen is utilized as dispersion gas and atomizes the liquid precursor in fine droplets with mass median diameter of ~10 μm. A support flame feed by a mixture of methane and oxygen allow a permanent ignition of the spray. Droplet evaporation, nucleation, cluster formation, coalescence, and finally agglomeration of product nanoparticles are determined as respective process steps in FSP occurring along the spray's trajectory. The combustion of organic solvents and precursors causes a huge release of thermal energy with temperature larger than 2000 K within the flame. Particle agglomerates can be thermophoretically deposited above the flame on temperature-controlled substrates resulting in homogenous films with high porosity (98 %) and tailored thickness. A further step to the FSP process is to apply it to multicomponent catalysts production, which commonly consist of a supporting material (e.g. $TiO_2$) whose surface is coated with a catalyst (e.g. Pt). However some limitations in the synthesis of such nanomaterials arouse because the particle design and distribution of the catalyst on the support can only be hardly controlled. An approach to overcome such limitations is the Double Flame Spray Pyrolysis (DFSP), in which the individual precursors are separately atomized and burned. The fabricated nanoparticles are subsequently mixed in the downstream of the two intersecting jets at a defined temperature. Therefore the objective of the present work is to apply computational fluid dynamics (CFD) to investigate the nanoparticle evolution in the mixing of two jets. These are defined by two nozzle parameters: intersectional nozzle angle and nozzle distance. The former is determined by the angle between the nozzle centerline and the vertical axis, while the latter one determines the displacement of the two nozzle outlets on a horizontal axis. The mathematical model include an Eulerian description of the gas phase mass, energy and momentum balances, coupled with a Lagrangian tracking of the evaporating fuel and precursor droplets. On the converged solution a simple monodispersed population balance model (PBM) is applied, which describes the nucleation and growth, by coagulation and sintering, of the particles. On this work identical precursors and flow rates were investigated for both nozzles. The simulation results show that on lower height above the burner the temperature profiles of the individual jets are still visible because both are not yet well mixed. The fuel and precursor droplet evaporate in this region, and the nanoparticle nucleation and growth process also begins. When approaching the intersection point, both jets unite and form a temperature distribution with a single maximum and forming a single jet. The temperatures are still high enough to allow further coagulation and sintering of nanoparticle agglomerates. Varying the intersection point the flow behavior and temperature profile changes. Accordingly the nanoparticle formation is also affected by the intersection point. Finally, this work was set to determine the viability of the numerical investigations on DFSP systems. Considering the complexity of the multiphase and multiscale system, the numerical model applied showed good results. Furthermore, with the application of CFD simulations, several geometrical configurations and different materials could be efficiently tested, guiding experimental investigations.

# 379404 A Coarse Grain Model for Protein-Nanoparticle Interactions

**Abstract:**

Protein-Nanoparticle (NP) interactions are the key phenomenon in many biological techniques including diagnosis, imaging, drug delivery, catalysis, and biosensors. The distinct small size is a significant reason for the widespread use of NPs in these techniques. For example, substrates of interest could be delivered to biological targets by NPs, which is hard to be achieved by traditional methods. However, adsorbed or tethered proteins on NP surfaces may change their conformations leading to the loss of activity or unexpected toxicity. Understanding the fundamental biophysics behind protein-NP interactions would be the essential of designing and engineering bio-NP systems for better performance.

To address this point, we developed a coarse grain model in this work. This model evolves from a well parameterized and validated coarse grain surface potential for protein-(flat)surface interactions. It is based on the Brooks Go-like protein model that is able to reproduce protein folding mechanisms. The main advantage of this model is that it quantitatively accounts for the hydrophobicity of both residues in a protein and the NP surface. In addition, the curvature of the NP is taken into account so that the protein behavior could be well characterized on NPs of different sizes. To validate this model, an experimental work is referred, which showed protein structure-related information on a nanoparticle. As will be presented, the coarse grain model successfully reproduces the protein GB1 structure on the nanoparticle, which is consistent with evidences produced by experimental methods such as circular dichroism (CD) and fluorescence spectroscopy. Also, the protein adsorption free energy is measured to be on the same order as the result from experiments. In addition, this model has the potential to be applied to predict protein behavior on NP surfaces with various chemical properties and curvatures.

# 390891 Molecular Dynamics Simulations of Peptide-SWCNT Interactions: Role of Curvature

**Abstract:**

Direct electron transfer between the redox active prosthetic group, flavin adenine dinucleotide (FAD), of oxidoreductases and single-walled carbon nanotubes (SWCNT) results in generation 3 biosensors and advanced biofuel cells. Phage display has recently been used to identify a specific dodecapeptide sequence (HWKHPWGAWDTL) that has high affinity for SWCNTs. Using this adsorptive peptide sequence, analogous sequences that share some homology with key enzymes (Glucose oxidase (GOx), Lactate oxidase (LOx) and Laccase (Lac)) used in generation 3 biosensors and advanced enzyme biofuel cells were identified. These high affinity sequences were found to decorate the surfaces of these enzymes, being away from the redox-active co-factor that would otherwise allow direct electron transfer. Molecular dynamics simulations were performed to study the interactions of four discrete dodecapeptides, along with the original phase display dodecapeptide, with a suite of SWCNT of varying diameters (6,0 f=0.47nm; 10,0 f=0.78nm; 15,0 f=1.17nm and 16,0 f=1.25nm) and with graphene in explicit water. Simulations were done using GROMACS parameterized using AMBER for the peptides, OPLS for the SWCNTs and TIP4P for water. From the progression of the radius of gyration, $R_g$, the peptides studied were concertedly adsorbed to both the SWCNT and graphene. Peptide properties based on group contributions, such as hydrophobicity and solubility parameter, did not correlate with the observed adsorption behavior as quantified by $R_g$. The SWCNT diameter was found to influence the adsorption interaction.

# 388600 Enhanced Activity and Stability of Organophosphorus Hydrolase Via Interaction with an Amphiphilic Polymer

**Abstract:**

Neurotoxic organophosphate (OP) compounds are widely used as pesticides and, unfortunately, as chemical warfare agents. A decontamination system based on organophosphorus hydrolase (OPH) is of great interest due to its potential for non-toxic and non-corrosive decontamination of OPs. However, despite its high catalytic activity, OPH has a relatively short shelf-life and low stability under practical working environments, such as pot life, organic solvents, and high temperatures.

Here, a simple approach to enhancing the activity and stability of OPH has been developed based on interactions between the hydrophobic poly(propylene oxide) (PPO) block of the amphiphilic block copolymer Pluronic and the enzyme. Using MALDI-TOF, NMR, Raman and fluorescence spectroscopies, evidence is provided that the PPO block of the polymer can interact with hydrophobic amino acids of OPH near its active site.

The polymer in pre-constructed OPH/Pluronic complexes enhances the stability and activity of the enzyme (i) in aqueous solutions with low enzyme concentration at RT (pot life), (ii) in the presence of an organic solvent (iii) in aqueous solutions at higher temperatures, (iv) after lyophilization, and (v) during long-term storage. PPO-containing block copolymers with various molecular structures have also been demonstrated to enhance OPH stability and its activity. Therefore, this strategy of simply blending enzyme with inexpensive, non-toxic, biocompatible, and commercially available Pluronic provides an efficient route to new formulations for decontaminating organophosphate neurotoxins.

# 378888 Enabling Multi-Enzyme Biocatalysis Using Coaxial Electrospun Hollow Nanofibers: Design and Applications

**Abstract:**

Highly efficient immobilization of multi-enzyme systems involving cofactor regeneration represents one of the greatest challenges in bioprocessing. Multi-enzyme systems that have been reported so far are almost exclusively in discrete spherical capsule formations with single or multicompartments, with a few of examples using porous supports or nanoparticles, which have limitations in some significant applications owing to high diffusional resistance, poor enzyme stability, and usually complicated preparation process. To breakthrough these limitations, the present work proposed a novel hollow nanofiber supported multi-enzyme system through co-axial electrospinning technology. By choosing bio-friendly solution with dissolved multienzymes as core solution and immiscible solvent with dissolved polymer as shell solution of co-axial electrospinning, an in-situ encapsulation of multi-enzyme inside the nano chamber of hollow nanofibers were realized. Compared to the existing multi-enzyme system in the form of particulates, novel hollow nanofiber supported multi-enzyme system is expected to offer several unique advantages: i.e., the woven-membrane formation of electrospun nanofibers enables an easily recycling and operation, facilitating molecular interactions between enzymes and possibly shared cofactor, as well as enhancing stability due to confining multi-enzyme system inside the nano-scale compartments.

To demonstrate the feasibility and advantages of the hollow nanofiber supported artificial cells, two multi-enzymatic systems were investigated. The first one is for bile acid assay, which includes 3α-hydroxysteroid dehydrogenase (3α-HSD), diaphorase (DP) and NAD(H), all of which were encapsulated in-situ inside chamber of hollow nanofiber via co-axial electrospinning. The activity recovery of immobilized 3-α-HSD and diaphorase reached 76% and 82%, respectively. When this nanofiber-supported multi-enzyme system was used for bile acid assay, a good linearity was obtained by varying bile acid concentrations within 0-200 μM, and this linear range was slightly broader than that obtained by using free multi-enzyme system (0-150μM). Investigation on storage stability of the multi-enzyme system showed that its half-life at 4°C and 25°C reached 160 days and 80 days, respectively. The second example is the fabrication of "ready-to-use" glucose test strip with glucose oxidase and horseradish peroxidase encapsulated inside the chamber and chromogenic agents embedded simultaneously in the shell of hollow nanofibers by co-axial electrospinning. The testing strip could be used either as colorimetric sensors in solution or as optical biosensor that can be operated in the "dip-and-read" mode. Both application modes yield a high sensitivity and broad testing range. The nano-confining effect of hollow nanofiber also provided the immobilized multi-enzyme system unique stability mechanism.

# 382547 Confinement Protection Effects of Mesoporous Silica, Inspired By Chaperonin Complexes

**Abstract:**

With the increasing use of biocatalytic enzymes across many fields of industrial manufacturing, more research must be done to stabilise enzymes and thereby increase their utility. The most common challenges in biocatalysis include unsuitable pH, high temperature, biological contaminants, and the presence of organic solvents [1,2]. Immobilisation onto nanostructured materials can combat the inherent instability of enzymes, particularly in the harsh environments encountered in biocatalysis.

Mesoporous silica SBA-15 shows great potential for applications in enzyme immobilisation. SBA-15 is typically formed as rod-shaped microparticles with cylindrical nanopores arrayed hexagonally along their axis. These pores simulate the environment found in the GroEL/ES complex, a pair of chaperone proteins found in bacterial cells that assist in the folding of new proteins [3]. The pores of SBA-15 have a hydrophilic core, similar to the GroEL/ES complex, and exhibit similar electrostatics with respect to protein adsorption. This study investigates the denaturation protection conferred to protein adsorbed to the internal surface of SBA-15 from exposure to proteases and extreme pH.

The model biocatalytic enzymes myoglobin and lysozyme were chosen for this study because of their well-characterised structure and simple activity assays. The SBA-15 used in this study is synthesised at a target pore diameter just larger than that of the enzymes, to lead to tight confinement upon adsorption [4, 5]. The SBA-15 is characterised using scanning electron microscopy (SEM), nitrogen sorption, and small angle x-ray scattering (SAXS). To investigate the protective effects of SBA-15, the enzymes are adsorbed to SBA-15 and these composites are exposed to a range of detrimental pHs. Other confined samples are additionally exposed to protease. Myoglobin activity is tested via peroxidase activity assay, which is measured with UV-Vis spectroscopy. Lysozyme activity is tested via chitinase activity assay, which is measured with fluorescence spectroscopy.

Our results suggest that confinement in tightly-fitting SBA-15 protects the enzymes from both acidic denaturation and proteolysis. More work must be done to investigate further protective effects against high temperature and organic solvents, and further studies will be required to generalise SBA-15's protective effects to enzymes with more useful industrial applications.

***References***

*[1] Sheldon & van Pelt Chem. Soc. Rev. 42, 6223–35 (2013)*

*[2] Tran & Balkus ACS Catal. 1, 956–968 (2011)*

*[3] Mayhew et al. Nature 379, 420–426 (1996)*

*[4] Sang & Coppens Phys. Chem. Chem. Phys. 13, 6689–6698 (2011)*

*[5] Sang et al. Langmuir 27, 13828–37 (2011)*

# 373486 A Combinatorial Strategy Towards a Better Sugar-Powered Enzymatic Fuel Cell

**Abstract:**

Building high-energy-density, green, safe batteries is highly desirable for meeting the rapidly growing needs of the increasing number of portable electronic devices. Enzymatic fuel cells (EFCs) are appealing metal-free bio-inspired batteries that directly convert chemical energy to electricity using enzymes. However, quite a few challenges such as incomplete oxidation of sugars (e.g., glucose and starch), low power output, short lifetime of enzymes, and expensive and non-stable mediators, have to be solved before EFCs can be commercialized in the future.

In this work, we have improved the performance of our EFC through working on several aspects: (1) using a synthetic enzymatic pathway to complete oxidize glucose unit and achieve the highest energy storage density in the EFC; (2) using thermostable free enzymes with free mediators to achieve high a power density; (3) using a novel carbon material to improve the power density; (4) engineering the redox enzyme and changing its preference to an inexpensive and stable cofactor.

EFCs featuring an energy storage density of ~800 Ah/kg, 100% biodegradability, a high level of safety, and rapid refilling, and a high power output of 1 mW/cm2, could be next-generation green power sources, particularly for portable electronic devices (e.g., smartphones, iPads, and GPS devices).

# 388495 Invited: Packing and Assembly of Particles: Nanowires and Beyond

**Abstract:**

Particle shape affects their packing and assembly into larger scale structures. Recent breakthroughs in the modeling, simulation and theory of patchy particles - including both entropic and enthalpic patchiness - allows for the *a priori* design and synthesis of particles optimized for target geometries, including those with highly anisotropic, chiral, or hierarchical features. We discuss recent progress, new understanding, and future trends.

**Forming Aligned Micellar Nanowires by De-wetting on Micro-pillars**

Piotr J. Glazer, Leo Bergen, Laurence Jennings, Arjan J. Houtepen, Eduardo Mendes and Pouyan E. Boukany*

Macroscopic nanowire fabrication with well-ordered arrangement is of great interest in the physical, chemical, and biological fields due to their importance in both fundamental studies and technological applications in functional devices [1-2]. Organic nanowires have desirable properties for producing high-performance nano-devices, because of their low-cost, facile and large-scale synthesis, flexibility, simple processability and tunability by molecular design, in contrast with other. However, they display a disordered alignment with random positions during/after processing on large-scale. Their disorder alignment strongly hinders their application in the future functional devices, where often cross-linking and overlapping of nanowire takes place. Therefore, the ability to align the assembly of organic nanowires into desired architectures without damaging them is a prerequisite for integrating them into functional devices.

Here, we report a low-cost, novel and simple method to form nanowires by aligning ultra long worm-like micelles (see Figure 1) [3]. During the dewetting process the receding waterline slides on the micro-pillars and in so doing orients the worm-like micelles. The alignment speed and directionality is controlled by dragging the micelles solution along the array with a paper tissue that serves as an external capillary force source. In addition, the micelles are successfully decorated with hydrophobic molecules that incorporate into the micelle core or with the hydrophilic nanoparticles such as quantum dots entrapped in the corona. The presented block-copolymer patterning process based on surface dewetting has several advantages over conventional wire formation methods. The PDMS fabrication is a simple and affordable technique for which neither clean-room facility nor specialized training are required. In addition, the structure of the wires can easily be modified either chemically or by simple physical interactions creating new opportunities for specific chemical/biosensing applications.

Figure 1: (Left) Schematic picture of suspended micellar nanowires on micro-patterned surfaces; (Right) Scanning Electron Microscopy (SEM) image of de-wetted PS-PEO nanowires on micro-patterned PDMS stamp.

**References:**

[1] **The art of aligning one-dimensional (1D) nanostructures**, Su, B.; Wu, Y. C.; Jiang, L., *Chem. Soc. Rev.* **2012,** *41*, 7832-7856.

[2] **The large area of laterally ordered nanochannel arrays from DNA combing and imprinting**, Jingjiao Guan, Pouyan E. Boukany, O. Hemminger, L. J. Lee,*Adv. Mat.* **2010**, 22, 3997-4001. DOI: 10.1002/adma.201000136.

[3] **Generating Aligned Micellar Nanowire Arrays by Dewetting of Micro-patterned Surfaces**, P.J. Glazer, L. Bergen, L. Jennings, A. J. Houtepen, E. Mendes and Boukany, P.E.*, *Small*, **2014**, 9, 1657. DOI:. DOI: 10.1002/smll.201303414.

# 389185 The New Liquid Crystals: Self-Assembly and Processing of Nanocylinder Dispersions

**Abstract:**

Nanoscience has enabled the controllable production of materials with an unprecedented range of chemistries, sizes, and shapes including cylindrical nanomaterials. Nanocylinder dispersions undergo self-assembly and rheological behaviors that can be generally be explained in terms of the known behavior of lyotropic liquid crystalline polymer solutions. However, there are several key distinctions. Nanocylinders can have significantly larger aspect ratios, persistence lengths, polydispersities, attractive interactions, and densities than their polymeric cousins. Comparison of various nanocylinder dispersions to each other, and to lyotropic liquid crystalline polymers, highlights opportunities for new insights into the fundamental behavior of colloidal rods. In addition, understanding the interplay between dispersion microstructure and applied shear can enable producing bulk nanocylinder films with controlled morphologies.

Over the last ten years, we have investigated nanocylinder dispersions with a broad range of chemical and physical properties. This talk highlights our work on cellulose nanocrystals in water, carbon nanotubes in aqueous dsDNA, mixtures of silver nanowires and spheres in ethylene glycol, and silica nanorods in dimethyl sulfoxide. We have shown that nanocylinders can assemble into cholesteric, nematic, and smectic phases. We have also shown that these structures can be maintained, enhanced, or erased during processing into solid coatings. Finally, we have raised new questions about what determines biphasic microstructure and rheological behavior at low shear. Answering these questions promises to enhance our fundamental understanding and ability to control the microstructure of bulk materials produced from nanoscale building blocks.

# 388479 Flow-Guided Template Manufacturing of Nanowire-Array Sensors

**Abstract:**

Nanowires hold tremendous potentials to build next-generation sensors, electronics, and optoelectronics. Compared to the thin film planar counterparts, these devices better define the accumulation or depletion of charge carriers to attain ultrahigh sensitivity. However, their wide adoption is seriously impeded from the great challenges on the fabrication, integration, and manufacturing of nanowire-based electronics. We adopt microchannel-based templates in the fabrication of nanowire-based sensing system. A novel flow-guided assembly approach we invented, in which polymer strands are aligned and patterned along microchannels via air blowing to produce highly ordered nanowire arrays with each nanowire centimeter long and having tunable diameter. The polymer nanowires were further converted to carbon, following a procedure which has used for years in the production of carbon nanofibers in industry. Unlike in other template synthesis processes where nanowires are randomly distributed on sensing substrates, our approach pre-defines the position and patterns of nanowires at desired locations (e.g., between the contact micropads). We are now working on its sensing performance on proteins after being built as nanowire field emission transistors. Their small lateral size, high surface-to-volume ratio, and tunable quantum confinement effect demonstrate nonlinear behaviors on the mobility of carriers and ultrahigh sensitivities.

# 383231 Synthesis and Characterization of Silver Nanowire for Printable Conductive Media

**Abstract:**

Screen printing of conductive inks is a promising technologies for applications including solar panel current collectors. In this work, the purpose is to create ink with a low resistance, allowing for the higher collection of solar radiation to increase efficiency. Conductive inks generally contain metal powder, solvent, binder, and rheological agent. Among different forms and types of metal powder, 1D silver nanostructures or silver nanowires have gained attention due to their unique electrical, thermal, and optical properties. We have used a polyol process to synthesis silver nanowire in a scaled up batch reactor. The morphology and microstructure were characterized by electron microscopy and X-ray diffraction. Based on these materials, silver nanowire inks, suitable for screen printing process, have been formulated with various silver and binder contents. The inks show time-dependent thixotropic behavior. Rheological measurements under conditions that resemble the screen printing process have been done on the conductive inks to assess visco-elastic properties induced by flow alignment of the wires. Flow-induced alignment in screen printed samples was characterized by SEM.

# 378721 In Situ Formation of Silver Nanowire Networks on Paper and Plastic Substrates Using a Desktop Inkjet Printer

**Abstract:**

Simple, universally adaptable techniques for fabricating nanostructured films are required to translate lab scale innovations into low-cost solutions for the developing world. Silver nanowires have emerged as an attractive building block for fabricating a wide range of sensors and devices. Herein, we report the in situ formation of *conductive* silver-nanowire networks on paper, thereby, eliminating the need for either cost-intensive ink formulation or substrate preparation or complex post-deposition sintering steps. Reminiscent of the photographic process of 'salt printing', a desktop office printer was used to deposit desired patterns of silver bromide on paper/plastic substrates, which were subsequently overexposed to light from a halogen lamp and then immersed in a photographic developer. Percolating networks of silver nanowires (30-100 nm dia. and 1-2 μm length) that conformally coated the substrate were formed. Thus, films of silver nanowires can be fabricated over large areas at a low-cost by combining two widely used processes—inkjet printing and photographic development. These films show excellent performance characteristics as strain gauges and as SERS (Surface Enhanced Raman Spectroscopy) substrates.

# 390282 Phase Transformation As a Strategy for the Synthesis and Assembly Via Welding of Single-Crystalline Metal Silicide Nanowires

**Abstract:**

A simple and versatile strategy for not only the synthesis, but also the large-scale assembly via welding of single-crystalline metal silicide nanowires was developed. The strategy relies on the solid-state phase transformation of pre-synthesized silicon nanowires that have sharp tips. Experiments performed using $Mg_2Si$ materials system as an illustrative example indicated that both the formation and welding of $Mg_2Si$ nanowires is possible using solid-state phase transformation. The use of sharp tipped nanowires, in conjunction with solid-sate phase transformation, circumvented the formation of multiple $Mg_2Si$ nuclei within the nanowire during phase transformation, and hence the formation of polycrystalline $Mg_2Si$ nanowires. Furthermore, simultaneous synthesis and the large-scale assembly of the nanowires via welding were accomplished by the phase transformation of silica nanoparticle decorated silicon nanowires. Phase transformation of silica nanoparticle decorated silicon nanowires with magnesium foils results in the formation of single-crystalline $Mg_2Si$ nanowires that are bridged together via $Mg_2Si$ bridges. This strategy allows for the large-scale assembly of nanowires in an interface-engineered manner, unlike traditional nanowire assembly strategies that result in the formation of oxides hat have electrically-insulating MgO at the interfaces. Such welding of nanowires is expected to offer the ability to independently tune thermal and electrical transport, and the fabrication of highly efficient thermoelectrics.

# 377258 Nanofiber-Based Dye Sensitized Solar Cells with a Novel Nano-Particle Scattering Layer

**Abstract:**

Dye sensitized solar cells (DSSC) is a very promising low-cost, high efficiency, thin-film solar cells. Here, we introduce a novel technique where incoming solar light can be effectively trapped thereby increasing the cell efficiency.

Polydispersed-size nanocrystallite titanium dioxide ($TiO_2$) with large surface area are fabricated by nozzle-less electrospraying using a precursor solution consisted of titanium isopropoxide (TIP) / polyvinylpyrrolidone (PVP) and ethanol.

The polydispersed $TiO_2$ nanoparticles are applied to dye sensitized solar cell as a novel scattering layer, which function both as dye loading and light trapping. By modifying the electrospraying condition, such as changing the voltage across the rotating electrode and the ground receiving the nanostructures, the mean size and size distribution of TiO2 nanoparticle can be engineered to balance both dye loading with smaller size and light trapping with larger size. After optimization, a high power conversion efficiency (PCE) of 8.71% can be achieved with the polydispersed-size $TiO_2$ nanoparticles as the scattering layer when compared with PCE of 7.17% obtained for the photoanode with the traditional scattering layer made of 200nm monodispersed $TiO_2$ nanoparticles; which represents an impressive 21% increase

# 384513 Hydrophobically Modified Chitosan Coated Liposomes for Gene Delivery

**Abstract:**

Non-viral gene delivery is not as effective as viral gene delivery when it comes to transfection efficiency. The highly studied complexes of DNA with cationic polymers and cationic liposomes have major obstacles of inflammatory toxicity, aggregation and rapid clearance from the blood stream. Encapsulating DNA within neutral liposomes has been comparatively less studied than other strategies of gene delivery. Liposomes are known to effectively deliver cargo and they can be coated with some polymers such as polyethylene glycol (PEG) which can enhance their blood circulation half-life and circumvent the problem of aggregation. The concern with PEG coated liposomes is that the covalently attached PEG to the lipid extends inside the aqueous core of the liposomes, which physically hinders the cargo of liposomes. Thus, to encapsulate condensed DNA complexes which are around 100nm in size, larger liposomes are required which are not efficiently taken up by cells. Employing a different polymer for coating liposomes such as hydrophobically modified chitosan (HMC) which coats only the outside of liposomes allows one to encapsulate bigger complexes in small liposomes. HMC has alkyl groups covalently attached to the chitosan backbone which insert themselves in the bilayer of liposomes and at the right concentrations, coat liposomes effectively. Chitosan being a slightly cationic polymer at neutral pH helps the liposomes from aggregation and increases the blood residence time of the liposomes like PEG. Plasmid DNA was condensed in a core using cationic chitosan and polyethyleneimine (PEI) and then encapsulated inside liposomes by using the thin film hydration method. The effect of different molecular weights and concentration ratios of the chitosan on the DNA condensation was evaluated using dynamic light scattering and transmission electron microscopy analysis. Then these various complexes were evaluated for their in vitro transfection. Further in vivo studies are necessary but nonetheless delivering DNA using coated liposomes can be an effective approach as the major difficulties of toxicity, aggregation and rapid clearance can be avoided. This work was supported by the National Science Foundation.

# 387565 Local Sustained Knockdown for Accelerated Diabetic Wound Healing

**Abstract:**

Diabetic foot ulcers (DFUs) affect more than one million Americans annually, costing nearly 11 billion dollars a year in healthcare spending and leading to over 70,000 lower limb amputations. These wounds are painful, nagging injuries that significantly reduce the quality of life for patients and their families. To date most pharmacological approaches to DFU therapy focuses on the delivery of super-physiological concentrations of recombinant growth factors, which has shown only limited success with serious side-effects. In this work we report the use of layer-by-layer technique to effectively deliver short interfering RNA (siRNA) directly to the wound bed in a diabetic mouse model of chronic wound healing. We targeted the expression of ECM protease MMP-9, known to be chronically upregulated in the wounds of diabetic patients and diabetic wound healing models. It was observed that MMP-9-specific siRNA therapy led to reduced MMP-9 expression and to significant improvements in wound healing, with increased granulation tissue formation and accelerated epithelial closure. This work points to a new focus in localized RNAi therapy, where dysregulated and pathologic gene expression can be directly targeted and corrected.

**Identification of Kinase Inhibitors for Enhanced Polymer-Mediated Transgene Expression**

Matthew Christensen[1], Jacob Elmer[1], Seron Eaton[2], Laura Gonzalez[2], Joshua LaBaer[2], Kaushal Rege[1]

Department of Chemical Engineering[1], Biological Design Institute, Personalized Medicine[2]
Arizona State University

Human cells contain hundreds of kinase enzymes which regulate several cellular processes, including those that can influence endogenous gene and transgene expression. We carried out a kinome-level screen of nearly 200 kinase inhibitors, using which, we identified small molecule inhibitors that enhance polymer-mediated transgene expression in prostate and bladder cancer cells. In particular, our lead targets include kinases associated with cell cycle regulation as well as transcription. Dose response studies revealed optimal concentrations of kinase inhibitors for enhancing transgene expression in different cell lines. Furthermore, these dose response studies demonstrated that the use of different polymer carriers, or different polymer to DNA ratios, led to different levels of transgene expression enhancement induced by the drug. This suggests that different polymer carriers may assist in overcoming particular barriers to gene delivery, in some cases perhaps overlapping with the inhibitor's mechanism.

The role of cancer cell cycle and intracellular trafficking were investigated in order to elucidate mechanisms by which kinase inhibitors enhance transgene expression. Pairwise combination of small molecule inhibitors of epigenetic targets and kinases resulted in significant enhancement of polymer-mediated transgene expression, although the most dramatic enhancements were seen with a cocktail of two different kinase inhibitors and the one epigenetic modulator. The observed enhancement in transgene expression with the addition of each inhibitor suggests that the mechanisms by which these inhibitors gene expression are independent of each other.With combination treatments, minimal effects on cell viability were observed with each drug addition, indicating that inhibiting specific intracellular biomarkers is a powerful strategy for enhancing non-viral transgene expression. Our results on combination kinase inhibition for enhancing transgene expression have transformative implications for biotechnology and medicine.

# 388267 Protease-Activatable Viruses Based on Adeno-Associated Virus for Cardiovascular Disease Therapy

**Abstract:**

Cardiovascular disease (CVD) is the leading cause of death around the globe today. Gene therapy is an emerging technology that may be used to treat and cure a wide range of pathologies, including CVD. However, a major issue most researchers face is targeting specificity of gene vector delivery. One hallmark of cardiovascular disease (as well as most human cancers and neurological disorders) is increased protease activity at the affected site. In particular, matrix metalloproteinase-9 (MMP-9) plays a major role in the progression of CVD. Our lab has recently developed a protease-activatable virus (PAV) based on adeno-associated virus serotype-9 (AAV9) that delivers genetic cargo to diseased cells only upon MMP-9 proteolysis. Briefly, a negatively charged tetra-aspartic acid ($D_4$) peptide "lock" was inserted throughout the capsid to prevent the virus' interactions with its primary cell receptor, galactose. This peptide lock is flanked by two protease-cleavable peptide sequences, which, upon MMP-9 proteolysis, release the $D_4$ lock and allow subsequent cell binding and internalization of the virus. A major advantage of our approach compared to other targeting approaches, such as targeting peptide insertion, is several diseased cells lack specific receptors that identify them as unique. Instead, PAVs recognize the diseased, MMP-rich tissue microenvironment and deliver genetic cargo only to areas that require therapy. We have already created and characterized over a dozen PAV variants that display varying specificity and kinetic activation rates for MMP-9. Our goal is to develop specialized PAVs that recognize the specific biomolecular protease signatures of CVD for targeted therapeutic gene delivery to the myocardium.

# 388698 Gold Nanoparticle Electroporation Enhanced Polyplex Delivery

**Abstract:**

Non-viral delivery strategies have been widely developed to serve as favorable alternative in cell therapy for their low toxicity and immunogenicity. However, they still suffer some unsatisfied performance (e.g., low cellular uptake, high cytotoxicity, and/or low efficiency). We hypothesize that appropriate combinations of some non-viral delivery approaches could help improve their overall therapy performance. In this study, we combined polyplex and electroporation to improve the delivery of DNA and RNA based probes. Specifically, polyplex (i.e., complex of cationic polymers and DNA/RNA probes) are fixed on the surface of gold nanoparticles (AuNPs) to avoid high cytotoxicity of free cationic polymers when dissociated in cell post internalization. AuNPs also serve as homogeneous probe carrier to help produce AuNP-polyplex complex with more uniform in size. Electroporation helps the quick and direct internalization of complex without the slow and inefficient endocytosis route. The presence of AuNPs further enhances electroporation performance from two aspects: free AuNPs help reduce the electroporation solution resistance for its high conductivity, leading to enhancement of the actual pulse strength on cells; transferrin grafting AuNPs (Tf-AuNPs) bound to TfR receptors on cell membrane and served as virtual microelectrodes to localize the electric fields. The delivery efficiency was evaluated with leukemia cells with both DNA plasmids (e.g., pGFP & pLuc) or siRNA molecules (e.g., siRNA & miRNA). Compare to traditional polyplex, AuNPs-polyplex have more homogeneous morphology, as confirmed with AFM and dynamic light scattering measurements. The transfection enhancement to plasmid DNA was shown 1.5~2 folds improvement with no significant increase of toxicity is observed. Similar effectiveness was also achieved in siRNA and miRNA delivery. Such a combination of physical and chemical delivery concept may stimulate further exploration in the delivery of various therapeutic materials for both *in vitro* and *in vivo* applications.

# 389423 Near-Infrared–Actuated Devices for Remotely Controlled Drug Delivery

**Abstract:**

A reservoir that could be remotely triggered to release a drug would enable the patient or physician to achieve on-demand, reproducible, repeated, and tunable dosing. Such a device would allow precise adjustment of dosage to desired effect, with a consequent minimization of toxicity, and could obviate repeated drug administrations or device implantations, enhancing patient compliance. It should exhibit low off-state leakage to minimize basal effects, and tunable on-state release profiles that could be adjusted from pulsatile to sustained in real time. Despite the clear clinical need for a device that meets these criteria, none has been reported to date to our knowledge. To address this deficiency, we developed an implantable reservoir capped by a nanocomposite membrane whose permeability was modulated by irradiation with a near-infrared laser. Irradiated devices could exhibit sustained on-state drug release for at least 3 h, and could reproducibly deliver short pulses over at least 10 cycles, with an on/off ratio of 30. Devices containing aspart, a fast-acting insulin analog, could achieve glycemic control after s.c. implantation in diabetic rats, with reproducible dosing controlled by the intensity and timing of irradiation over a 2-wk period. These devices can be loaded with a wide range of drug types, and therefore represent a platform technology that might be used to address a wide variety of clinical indications.

# 390535 Degradable Lipid Nanoparticles with Predictable in Vivo siRNA Delivery Activity

**Abstract:**

One of the most significant challenges in the development of clinically-viable delivery systems for RNA interference therapeutics is to understand how molecular structures influence delivery efficacy. To this end, we synthesized 1400 degradable lipidoids and evaluated their transfection ability and structure function activity. Here we show that lipidoid nanoparticles mediate potent gene knockdown in hepatocytes and immune cell populations upon IV administration to mice (siRNA $EC_{50}$ values as low as 0.01 mg/kg). Surprisingly, we identify four necessary and sufficient structural and pKa criteria that robustly predict the ability of nanoparticles to mediate greater than 95% protein silencing in vivo. Because these efficacy criteria can be dictated through chemical design, this discovery could eliminate our dependence on time-consuming and expensive cell culture assays and animal testing. In this talk, we will identify promising degradable lipidoids and describe new design criteria that reliably predict *in vivo* siRNA delivery efficacy without any prior biological testing.

# 390203 Poly (propargyl L-glutamate) Based Linear-Dendritic Block Copolymers As pH-Responsive Hydrophobic Drug Carrier

**Abstract:**

To design biocompatible, pH-responsive and long-circulating nanoparticulate drug delivery systems, we have coupled a hydrophobic poly (propargyl L-glutamate) (PPLG) linear polypeptide block to a hydrophilic poly (ester) dendron by highly efficient thiol-maleimide chemistry. The propargyl side-chains of the polypeptide segment were attached to diethylamine moieties through Cu-mediated azide-alkyne 'click' cycloaddition. The synthesized block copolymers with linear-dendritic topology were found to self-assemble and disassemble into stable nanoparticles in a pH-responsive manner over a range of biologically relevant conditions. These sterically stabilized block copolymer assemblies were able to encapsulate both hydrophilic and hydrophobic anticancer agents without compromising their cytotoxic property, and undergo controlled release at the pH of the endosomal compartments within which the nanoparticles accumulate. The dendritic block is designed to present a cluster arrangement of ligands on the surfaces of the vesicles generated by these block copolymers, thus enhancing targeted tumor uptake. Whole-animal fluorescence imaging showed that, when administered systemically to BALB/c mice, Cy 5.5 conjugated linear-dendritic nanoparticles persisted in blood for a period of more than 30h with substantially controlled liver accumulation.

# 373593 PLGA-Peg-Based pH Sensitive Nanoparticles for Targeted and Controlled Drug Delivery: In Vitro Evaluation

**Abstract:**

PLGA-PEG-Based pH Sensitive Nanoparticles for Targeted and Controlled Drug Delivery: In Vitro Evaluation

*Zilan Zhou, Apurva Badkas, and Joo-Youp Lee*

Chemical Engineering Program

Department of Biomedical, Chemical, and Environmental Engineering

University of Cincinnati, Cincinnati, Ohio 45221-0012

In this study, we built a drug delivery system for targeted and controlled drug delivery, which composes both active targeting abilities against HER-2 overexpressed breast cancer cells and pH response to an acidic environment inside and outside the tumor cells. A bi-functional tri-block copolymer, Poly(lactic-co-glycolic acid)-block-poly(L-histidine)-block-poly(ethylene glycol)-azide, (PLGA-Phis-PEG-$N_3$) was prepared and used for the construction of nanoparticles via nanoprecipitation. This tri-block copolymer contained two features. The imidazole ring in the histidine part could be protonated or deprotonated, which endowed the tri-block copolymer with pH sensitivity. The azide group at the end of the polymer chain was coupled with HerceptinÒ through a linker using "click chemistry" for targeted drug delivery.

The pH responsiveness was examined for the drug release kinetics at different pH conditions as well as the particle size change. The SK-BR-3 and MCF-7 breast cancer cell lines were used to examine the efficacy of the drug delivery system. The in vitro cellular uptake of the fluorescence-labeled system was quantified using fluorometry. The cytotoxicity of the PLGA-Phis-PEG-based nanoparticles was examined by MTS assay.

# 390139 Molecular Bioelectronics: Enzyme-Swnt Conjugates for Bioanalytical Biochips

**Abstract:**

Generation-3 (Gen-3) biosensors that are based on direct electron transfer are reagentless [1] and hold the potential for long term in vivo stability to enable fully implantable continual monitoring of various analytes of a wide variety of physiological conditions and pathologies [2]. Among the novel approaches for establishing a Gen-3 biotransduction mechanism, ultrasonic processing [3] of single walled carbon nanotube-enzyme (***e.g.*** Glucose oxidase, GOx) conjugates [4] was investigated as a function of sonication time (5 and 60 min), SWNT functionalities (-OH and –COOH) and tube length (<1 μm and 5-10 μm) [5]. Using the blue shift ($\Delta\lambda_{max}$) in the absorbance maximum of the FAD-apoenzyme, the change in the sum of α-helix and β-sheet in the circular dichroism spectra, and the change in the $k_{cat}/K_M$ and n of the enzyme conjugate expressed relative to the pristine GOx, were found to show only modest change to the enzyme's structure during supramolecular conjugation. Short SWNTs appeared to support conjugate formation with no egress of FAD for the enzyme; medium SWNT appeared to support conjugate formation moderately while the long and non-functionalized SWNT caused partial denaturation leading to the egress of FAD.

Microdisc electrode arrays (MDEA 5037 Pt) were microlithographically fabricated on borosilicate glass from 10 nm TiW adhesion layer and 100 nm e-gun deposited Pt. The Pt layer was patterned into working, counter and reference electrodes and all electrodes covered by $Si_3N_4$ which was windowed to reveal a pattern of microdiscs (37 discs, each f=50 μm, A= ???$cm^2$) over the working electrode and exposed counter and reference electrodes. Biofabrication [6] ***of enzyme amperometric biosensors*** using the process of electropolymerization (Py to PPy at 0.8 V vs. Ag/AgCl) onto the MDEA 5037 in the presence of glucose oxidase (1 mg/ml, DI water, 100 mC/$cm^2$) resulted in GOx, GOx-ferrocene monocarboxylic acid and GOx-SWNT biotransducers that were responsive to glucose and that allowed parallel evaluation of the relative efficacy of Gen-1, Gen-2 and Gen-3 mechanisms of transduction. The bioanalytical performance of sensitivity (S), linear dynamic range (LDR) and detection limit [3(SD(blank)/S] as well as apparent enzyme kinetic parameters of $\mathbf{K}_M$, $I_{max}$, and $k_{cat}$, were obtained from the dose-response curves for mutarotated glucose solutions (0.0 mmol/L to 13.7 mmol/L in PBS 7.2) at RT and compared to solution phase behavior. There was a good linear relationship ($R^2$=0.99) between glucose concentration and response from 2.0 mM – 9.0 mM. The sensitivity was $10^{-3}$ mA/mM and the detection limit was 0.002 mA/mM.

Keywords: Biosensors, Bioimmobilization, Nanotubes, Direct Electron Transfer, Glucose-SWNT

References:

[1] Guiseppi-Elie A, Lei C, Baughman RH. Direct electron transfer of glucose oxidase on carbon nanotubes. Nanotechnology. 2002;13:559.

[2] Kotanen CN, Moussy FG, Carrara S, Guiseppi-Elie A. Implantable Enzyme Amperometric Biosensors. Biosensors and Bioelectronics. 2012:(in press).

[3] Guiseppi-Elie A, Choi S-H, Geckeler KE. Ultrasonic processing of enzymes: Effect on enzymatic activity of glucose oxidase. Journal of Molecular Catalysis B: Enzymatic. 2009;58:118-23.

[4] Guiseppi-Elie A, Choi S-H, Geckeler K, Sivaraman B, Latour R. Ultrasonic Processing of Single-Walled Carbon Nanotube–Glucose Oxidase Conjugates: Interrelation of Bioactivity and Structure. NanoBiotechnology. 2008;4:9-17.

[5] Karunwi O, Guiseppi-Elie A. Supramolecular glucose oxidase-SWNT conjugates formed by ultrasonication: effect of tube length, functionalization and processing time. Journal of Nanobiotechnology. 2013;11:6.

[6] Karunwi O, Wilson AN, Kotanen C, Guiseppi-Elie A. Engineering the Abio-Bio Interface to Enable More than Moore in Functional Bioelectronics. Journal of The Electrochemical Society. 2013;160:B60-B5.

AIChE 2014 Annual Meeting Abstract

Binding between DNA and Carbon Nanotubes Strongly Depends upon Sequence and Chirality

Certain single-stranded DNA sequences are known to recognize their partner single-walled carbon nanotube (CNT). Here, we report on activation energies for the removal of several ssDNA sequences from a few CNT species by a surfactant molecule. We find that DNA sequences systematically have higher activation energy on their CNT recognition partner than on non-partner species. For example, the DNA sequence $(TAT)_4$ has much lower activation energy on the (9,1) CNT than on its partner (6,5) CNT, whereas the DNA sequence $(CCA)_{10}$ binds strongly to its partner (9,1) CNT compared to (6,5) CNT. The (6,5) and (9,1) CNTs have the same diameter but different electronic properties, suggesting that the activation energy difference is related to electronic properties. The activation energies of increasing lengths of closely related sequences from the 11-mer $(TAT)_3TA$ to the 21-mer $(TAT)_7$ on three different CNT species (9,1), (6,5), and (8,3) were measured. For the shorter sequences, the activation energy on the CNT varies periodically with the sequence.

# *Thermodynamics of DNA Strands Encapsulated into Electrically Charged Nanotubes*

*Fernando J.A.L. Cruz* and *José P. B. Mota*
Requimte/CQFB, Dept. Chemistry, Universidade Nova de Lisboa, Caparica, Portugal.
fj.cruz@fct.unl.pt

## I. Introduction

Deoxyribonucleic acid (DNA) and single-walled carbon nanotubes (SWCNTs) are prototypical one-dimensional structures; the former plays a central role in chemical biology and the latter holds promise for nanotechnology applications [1-3]. From the point of view of biological purposes and DNA manipulation, carbon nanotubes have been proposed as templates for DNA encapsulation, intracellular penetration via endocytosis and delivery of biological payloads and ultrafast nucleotide sequencing. While structure in its natural form and environment is well established (*e.g.* B-DNA in aqueous solution), their interactions have been the subject of intense investigation, nonetheless, the corresponding molecular-level phenomena remain rather unexplored. Recently we have shown that, given a sufficiently large hydrophobic nanotube, the confinement of a double-stranded DNA segment, 5'-D(*CP*GP*CP*GP*AP*AP*TP*TP*CP*GP*CP*G)-3', is thermodynamically favourable under physiological environments (134 mM, 310 K, 1 bar), leading to DNA-nanotube hybrids with lower free energy than the unconfined biomolecule [4]. To accommodate itself within the $D$ = 4nm nanopore, the DNA's end-to-end length increased from 3.85 nm up to approximately 4.1 nm, due to a 0.3 nm elastic expansion of the strand *termini*. The canonical Watson-Crick H-bond network was essentially preserved throughout encapsulation, showing that contact between the DNA dodecamer and the hydrophobic carbon walls results in minor rearrangements of the nucleotides H-bonding. A diameter threshold of 3 nm was established below which encapsulation is inhibited.

It is well known that nanotubes can be electrically charged, either using an AFM tip and applying a voltage bias or by chemically doping the solids with p-type dopants to obtain positively charged nanotubes [5, 6]. The effect of charge density upon the energetics and dynamics of confinement needs to be carefully addressed; because DNA's outer surface is negatively charged (phosphate moieties), its interaction with a positively charged solid might lead to the occurrence of encapsulation which is inhibited for hydrophobic pores. We address this issue using molecular dynamics and enhanced sampling algorithms (metadynamics, umbrella sampling) to probe the encapsulation mechanism of an atomistically detailed DNA dodecamer, onto positively charged ($q$ = + 0.05 $e^{-}$/C) SWCNTs of different diameters (3 nm, 4 nm), while employing precise physiological conditions.

## II. Results and Discussion

In contrast with the purely hydrophobic (40,0) topology ($D$ = 3 nm), the existence of an overall positive charge density on the solid indeed favours the encapsulation of the DNA molecule. To probe the thermodynamical stability associated with encapsulation, free-energy landscapes are built using the well-tempered metadynamics scheme [7] and two order parameters relating the distance between centres of mass of DNA and SWCNT, $\xi_1$, and the end-to-end length of the biomolecule, $\xi_2$. The corresponding free-energy maps recorded in Fig.1 show that: *i)* the nanopore endohedral volume ($\xi_1 < 2$) is the thermodynamically preferred region, by comparison with the bulk ($\xi_1 > 2$), *ii)* encapsulated DNA retains its translational mobility, diffusing freely between adjacent free-energy minima contained within the solid and *iii)* whilst DNA maintains a *quasi* B-form end-to-end length within the (51,0) topology ($D$ = 4 nm), the double-strand seems to suffer an elastic contraction when subjected to such a constraining volume as a (40,0) nanotube. The end-to-end length probability distributions, $P(\Omega_2)$, have been determined by independent umbrella sampling calculations and the results are recorded in Fig.2 for both topologies, (40,0) and (51,0), along with the previous curve obtained for a purely hydrophobic (51,0) SWCNT [4]. It now becomes clear that charge density on the solid plays a paramount role upon the encapsulation mechanism; the elastic expansion of the double-strand observed for the (51,0) hydrophobic pore ($L$ = 4.1 nm) is annihilated when the solid becomes electrically charged resulting in a maximum probability DNA end-to-end length of $L$ = 3.73 nm, consistent with the canonical B-DNA form [8]. On the other hand, to accommodate itself within the constricting volume of the (40,0) topology, the DNA molecule undergoes a contraction and exhibits a maximum probability of occurrence of $L$ = 3.54 nm.

A nanoscopic picture of the encapsulated DNA molecule can be produced by calculating the corresponding number density maps, as indicated in Fig.3 and obtained from atomically detailed mass histograms. Fig.3 reveals the existence of a cylindrical exclusion volume centred along the (51,0) main axis, where molecular density is $\rho \approx 0$, which can be attributed to the strong electrostatic attraction between DNA ($PO_4^{3-}$ ions) and the solid, driving the former towards the walls and away from the nanopore center. Entropic effects caused by the pore narrowness of the (40,0)

SWCNT force the DNA molecule to cluster tightly around the nanopore center, where it exhibits the region of highest molecular density.

As far as we are aware these observations are the first of their kind, and they come to pave the way for the design of smart nanotube based devices for *in vivo* DNA encapsulation.

**Acknowledgements.** The authors would like to acknowledge Requimte/CQFB (Universidade Nova de Lisboa, Portugal) for generous CPU time. This work makes use of results produced with the support of the Portuguese National Grid Initiative (more information in https://wiki.ncg.ingrid.pt). F.J.A.L. Cruz gratefully acknowledges financial support from FCT/MCTES (Portugal) through grants SFRH/BPD/45064/2008 and EXCL/QEQ-PRS/0308/2012.

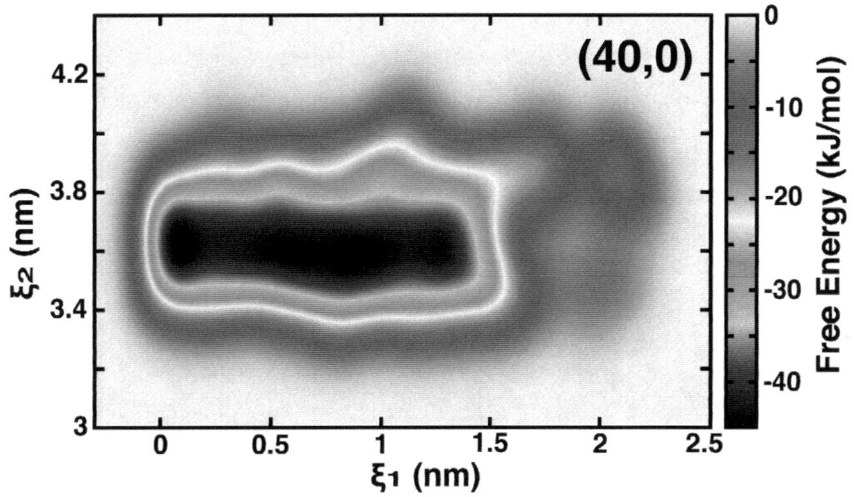

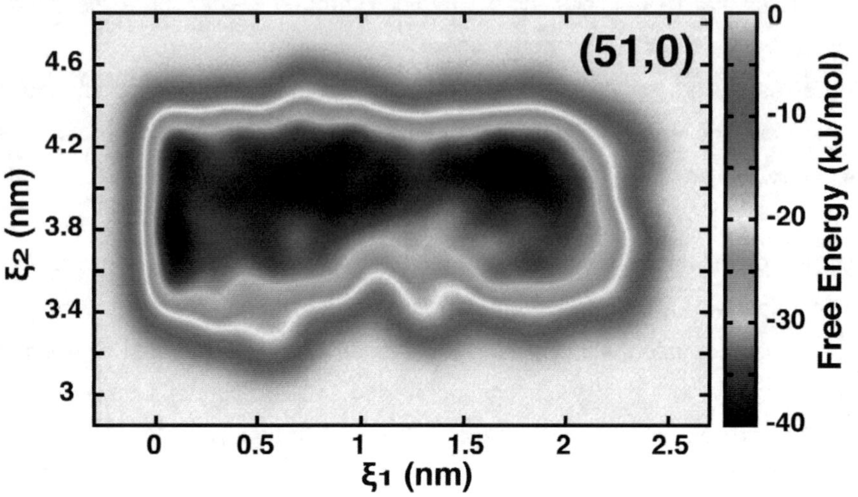


**ꝚE 1 – Free energy landscapes of DNA@SWCNT hybrids.** $\xi_1$ is the distance between centres of mass of the DN
ꞮT, projected along the nanopore main axis, and $\xi_2$ is the absolute distance between (GC) termini on opposite sides
ɜ-strand, equivalent to the DNA end-to-end length. The several free-energy minima along $\xi_1$ demonstrate that the moleɕ
ely mobile to translocate along the nanotube; interestingly, all the $\xi_1$ minima are located along a *quasi*-linear path defined ŀ
ɿ (40,0) and $\xi_2 \approx 4$ nm (51,0) highlighting the enhanced thermodynamical stability corresponding to the canonical B form
1,0) topology.

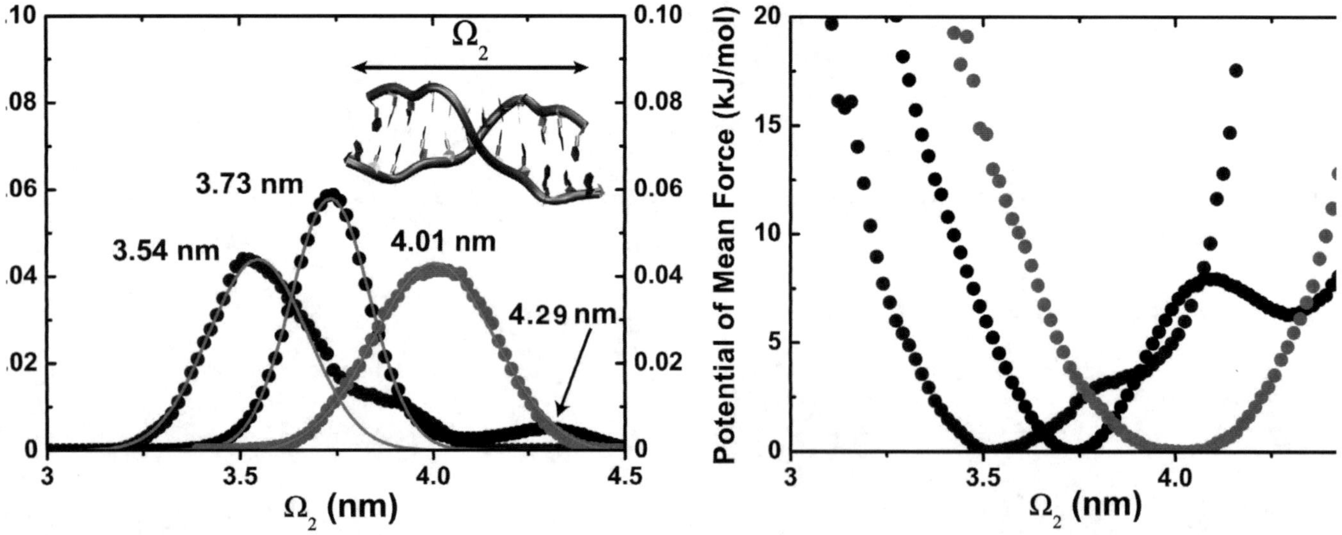


**ꝚE 2 – Potential of mean force and probability distribution profiles.** $\Omega_2$ corresponds to the end-to-end length of
ɔls are umbrella sampling results and red lines are free fittings of data to Gaussian statistics, $P(\Omega_2)=\phi\exp\left[-\frac{1}{2}\left(\Omega_2-\Omega_2^0/\sigma\right)^2\right]$
Ꝺ(40,0) SWCNT, *black)* DNA@(51,0) SWCNT, *green)* DNA@(51,0) hydrophobic SWCNT [4].

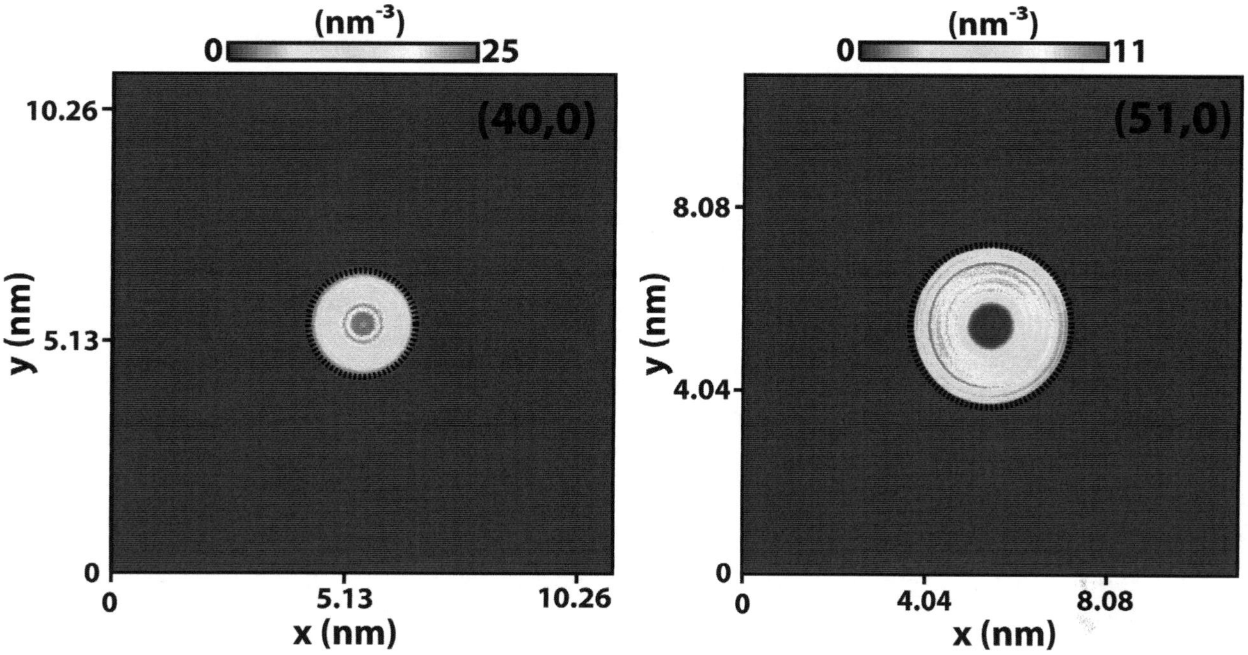


ꓤE 3 – **Number density maps of DNA@SWCNT.** The existence of a cylindrical exclusion volume centred along the ore main axis, $\rho \approx 0$, is the direct consequence of strong electrostatic attractions between the heavily charged phosphate ꞔ e solid. The dashed lines indicate the boundaries of the nanotube.

***References***
[1] H. Kumar *et al.*, Soft Matter **7**, 5898 (2011).
[2] B. M. Venkatesan, and R. Bashir, Nature Nano. **6**, 615 (2011).
[3] A. D. Franklin *et al.*, Nano Lett. **12**, 758 (2012).
[4] F. J. A. L. Cruz, J. J. de Pablo, and J. P. B. Mota, J. Chem. Phys. **(submitted)** (2014).
[5] X. Zhao, and J. K. Johnson, J. Am. Chem. Soc. **129**, 10438 (2007).
[6] W. Lv, Chem. Phys. Lett. **514**, 311 (2011).
[7] A. Barducci, G. Bussi, and M. Parrinello, Phys. Rev. Let. **100**, 020603 (2008).
[8] J. M. Vargason, K. Henderson, and P. S. Ho, Proc. Nat. Acad. Sci. **98**, 7265 (2001).

# Effect of organic acids on nucleation and growth of calcium phosphate crystals on glutamic acid functionalized nanofibers

**Danial Barati, Seyedsina Moeinzadeh, Esmaiel Jabbari**

Department of Chemical Engineering, University of South Carolina, Columbia, SC, USA 29208

**Introduction:** Aligned collagen fibrils nucleated with calcium phosphate (apatite) crystals are the building blocks of cortical bone. The high strength of compact bone stems from the high loading of calcium phosphate crystals (250-300% based on the fibril weight) deposited on the aligned collagen fibers, facilitated by the acidic ECM proteins. ECM proteins that are rich in glutamic acid (Glu) residues nucleate CaP crystals by surface-immobilization on collagen fibrils. Further, it has been shown that the spacing between carboxylate groups matches those between calcium ions in the growing CaP nanocrystals. Considering that carboxylate-rich organic acids account for 5.5% of the organic matter in bone, they should play a key role in controlling crystal growth and size. The objective of this work was to investigate the effect of organic acids on the extent of crystal nucleation, average size and crystal type of Glu-functionalized aligned nanofiber microsheets. Organic acids included citric acid (CA), maleic acid (MA), and salicylic acid (SalA).

**Materials and method:** Cysteine-terminated EEGGC (GLU) peptide was synthesized in the solid phase. The peptide was conjugated to low molecular weight acrylated poly(L-lactide) (PLA) via Michael addition to synthesize the PLA-Glu conjugate. Aligned PLA nanofibers sheets were electrospun from a 10 wt% solution of high molecular weight PLA and PLA-Glu with 1.0 mL/h injection rate, 20 kV electrical potential, and 7.5 cm needle-to-collector distance and the fibers were collected on a rotating wheel at 1200 rpm. The nanofiber sheets were incubated in a modified solution of 10x concentrated SBF supplemented with citric acid (CA), malic acid (MA) and salicylic acid (SalA). The amount of CaP nucleation on the nanofibers was characterized using a QuantiChrom calcium assay. The elemental composition of the samples was analyzed using an Energy Dispersive X-ray Spectrometer (EDS) connected to FESEM at an accelerating voltage of 15 kV.

**Results:** Figure 1a and 1b shows the effect of incubation time and organic acid concentration on CaP content of nanofiber microsheets. The CaP content of nanofibers increased over 16 hours of incubation in the organic acid modified SBF, and did not change significantly with further incubation. The plateau CaP content depended on type of organic acid with citric acid (CA) having the highest CaP content and SalA the lowest. The CaP content was also a strong function of organic acid concentration and reached a maximum value between concentrations of 5-10 mM. The maximum CaP content of the nanofibers was 225±10, 200±10 and 170±10 wt% (based on fiber mass) for citric acid, maleic acid, and salicylic acid, respectively. Further, the type, average size, and calcium to phosphate ratio of the crystals nucleated on the fibers depended on the type of organic acid. For example, based on EDS data, the calcium to phosphorous (Ca:P) ratio increased from 1.77±0.02 to 1.79±0.04 and 1.83±0.04 as the organic acid changed from citric acid to maleic anhydride and salicylic acid, respectively. Results of this work could be used to synthesize fiber-reinforced composites mimicking the nanostructure of compact bone.

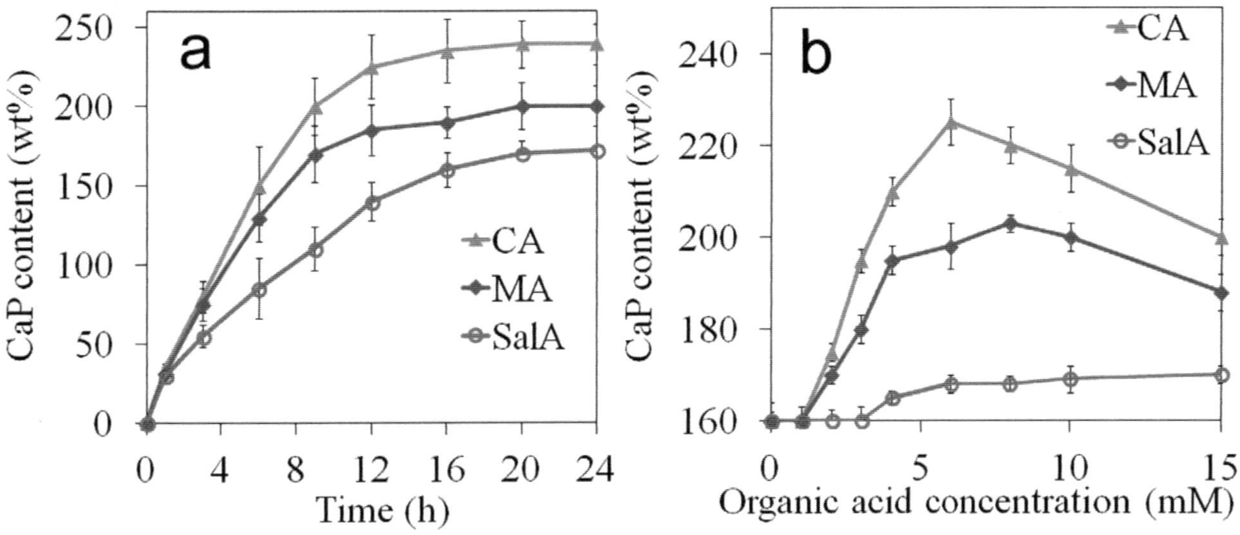


*Figure 1. (a) effect of incubation time and (b) organic acid concentration on CaP content of the nanofibers.*

# 388215 Biophysical Characterization of Immobilized, Self-Assembled Phospholipid Bilayer Based Structures in Sol-Gel Derived Silica

**Abstract:**

The entrapment of Nanolipoprotein Particles (NLPs) and liposomes in transparent, nanoporous silica derived from the precursor Tetramethylorthosilicate (TMOS) was investigated. NLPs are discoidal patches of lipid bilayer that are belted by amphiphilic scaffold proteins and have an average thickness of 5 nm, with a diameter that ranges from 10-15 nm. Liposomes, which are spherical self-assemblies of lipid bilayer, have previously been examined inside of silica gels and have been shown to be somewhat unstable. This is attributed by their size (~150 nm) and altered structure and lipid dynamics upon entrapment within the nanometer scale pores (5-50 nm) of the silica gel. We have demonstrated that NLPs are much more compatible with the nanometer scale size of the porous environment by analysis of lipid phase behavior via fluorescence anisotropy and analysis of scaffold protein conformation via circular dichroism spectroscopy.

NLPs are capable of solubilizing a wide variety of Integral Membrane Proteins (IMPs). NLP immobilization allows for the synthesis of bio-inorganic hybrid nanomaterials that exploit the functionality of IMPs. These types of materials were either previously non feasible or very limited. We have examined temperature-dependant visible light absorption spectra of the IMP bacteriorhodopsin (BR) inside of NLPs as a means of investigating protein functionality. We intend to immobilize and examine BR-NLPs inside of silica gels via various biophysical characterization techniques as a model system for IMPs. The success of this work could lead to the development of novel platforms in several areas, including high-throughput drug screening, chromatography, and biosensors.

# 370168 Computational Modelling of Functionalized Nanoparticles Interacting with Bio-Inspired Membranes and Vesicles

**Abstract:**

Our objective is to design bio-inspired membranes and vesicles composed of different molecular species and investigate their interactions with nanoparticles of various functionalizations. Multi-component membrane and vesicles are designed such that they are composed of different representative amphiphilic molecular species present in biological cell membranes [1.] These individual molecular species can differ from each other due to the dissimilar chemical properties and molecular geometry of their hydrophilic and hydrophobic groups. We use a Molecular Dynamics-based mesoscopic simulation technique called Dissipative Particle Dynamics to simultaneously resolve the structural and dynamical properties of stable mixed bilayer membranes and vesicles. We investigate the factors that control the self-organization of the molecular species within the hybrid bilayer and illustrate its role on the dynamical and mechanical properties of the bilayers. In addition, we explore the effect of the relative concentrations of the molecular species on the stability and properties of the bilayers. We introduce nanoparticles of various functionalizations in solution at a certain distance from the bilayers such that there is no interaction between them in the beginning. We investigate the interactions between the nanoparticles and the bilayers with and without curvature and find out the role of each amphiphilic molecular species on those interactions. The results of our investigations can be used to design effective cell targeting vehicles for applications in drug delivery, sensing and imaging.

[1] E. Koufos, B. Muralidharan and M. Dutt, submitted.

# 389113 Morphological Characterization of Detergent-Mediated Photosystem I (PS I)-Proteoliposome Formation

**Abstract:**

In-vitro functional and structural studies of membrane proteins rely on suitable choice of synthetic membrane mimics (e.g., detergents and liposomes) that maintain the protein function[1]. To this end, we investigate the interaction of two nonionic detergents *n-dodecyl*-β-D-maltoside (DDM) and Triton X-100 (TX-100) with two phospholipids, namely DPhPC (1,2-diphytanoyl-*sn*-glycero-3-phosphocholine) and DPPG (1,2-dipalmitoyl-*sn*-glycero-3-phospho-(1'-*rac*-glycerol)), that commonly constitute the natural thylakoid membrane housing of Photosystem I (PS I), the photosynthetic protein complex. We use optical density and dynamic light scattering (DLS) measurements in conjunction with advanced transmission electron microscopy (TEM) techniques (namely, negative staining and cryo-TEM) to study the detailed mechanistic picture behind processes that drive membrane bilayers to undergo complex structural arrangements during different stages of detergent-mediated solublization.. Specifically, negative staining and cryo-TEM images are used to investigate the solution-phase morphological arrangement of PS I-proteoliposomes. These characterizations provide a fundamental understanding of the role of detergent-mediated protein reconstitution in driving unidirectional arrangements of PS I trimeric complexes within lipid vesicles during PS I-lipid interactions in colloidal solutions. In turn, the aforementioned studies provide valuable insight into mimicking the naturally occurring protein-phospholipid interactions in cell membranes. In future, these bio-mimetic systems shall facilitate easy incorporation of membrane protein-lipid complexes into novel bio-hybrid devices.

1. Dewald A.H., Hodges J.C., Columbus L. Biophys J. 2011; 100(9): 2131–2140

# 378032 Multi-Site Functionalization of Protein Scaffolds for Bimetallic Nanoparticle Templating

**Abstract:**

The use of biological scaffolds to template inorganic material offers unique strategies to synthesize precise composite nanostructures of different sizes and shapes. Proteins are unique biological scaffolds that consist of multiple binding regions or epitope sites that site-specifically associate with conserved amino acid sequences within protein binding partners. These binding regions can be exploited as synthesis sites for multiple inorganic species within the same protein scaffold, resulting in bimetallic inorganic nanostructures. We demonstrate this strategy with the scaffold protein clathrin, which self-assembles into spherical cages. Specifically we design tether peptides that noncovalently associate with distinct clathrin epitope sites while initiating simultaneous synthesis of two inorganic species within the assembled clathrin protein cage. We demonstrate the flexibility and diversity of this unique biotemplating strategy by synthesizing two types of composite structures, silver-gold mixed bimetallic nanoparticles and silver-gold core-shell nanostructures, from a single clathrin template. This noncovalent, Template Engineering Through Epitope Recognition or TEThER strategy can be readily applied to any protein system with known epitope sites to template a variety of bimetallic structures, without the need for chemical or genetic mutations.

# 390627 Nature-Inspired Microfluidic Template As a Substrate for Plasmonic Nanoparticles

**Abstract:**

The sophisticated vasculature of leaves has been replicated to aid in the construction of a microfluidic like device. Their abundance in nature necessitates research into their development into low-cost, renewable platforms for various applications. The presence of functional groups of amines, hydroxyls in the leaf vasculature along with its porous microstructure facilitate formation of metallic nanoparticles from metal salts in situ. This technique of using only the leaf skeleton does not require additional reducing agents to assist nanoparticle formation. The templated nanoparticles within leaves demonstrate excellent catalytic properties, thereby imparting catalytic and plasmonic properties to the leaf itself. Additionally it has been shown that the nanoparticles can be recovered in aqueous medium by simple thermal decomposition of the leaf. The nanoparticles have been further characterized by TEM, SEM, UV-Visible Spectroscopy and Selected area diffraction. FTIR studies were performed on the crushed leaves to verify the presence of functional groups. Everything considered, this facile approach in bio-templating plasmonic nanoparticles has led to the generation of organic-inorganic structures which has potential for diverse applications in sensing, catalysis and medicine.

# 367133 Diffusing Colloidal Probes of Cancer Cells

**Abstract:**

To engineer nanoparticles for novel biosensing, diagnostic imaging, and drug delivery applications, the interactions of nanoparticles with the surface of cells must be characterized to better understand their influence on specific and non-specific cell surface adhesion, internalization pathways, and toxicity. In order to develop models to aid in optimization and design in these systems, direct measurements of colloidal interactions with cell surfaces are required that are both quantitative and sensitive to determine the impact of particle composition and surface chemistry. In this talk, we report the use of diffusing colloidal probes to directly measure cell surface interactions with *kT*-scale sensitivity with newly developed imaging and analysis techniques. Dark field video microscopy allows for label-free imaging of colloidal particles and live cells. With combined real-time particle tracking and cell boundary determination, particle trajectories can be monitored in relation to their distance from the cell surface which allows for direct measurement of equilibrium and non-equilibrium colloid-cell interactions. Using this technique, polyethylene glycol (PEG), bovine serum albumin (BSA), and dextran, hyaluronic acid (HA) functionalized colloidal silica were used to measure specific and non-specific macromolecular interactions with the surface of epithelial breast cancer cells. Ultimately, our results demonstrate how diffusing colloidal probe microscopy can be used to characterize biomolecular and biophysical properties of cell surfaces.

# Enhancement of Cancer Cellular Uptake by the Morphology of Lipid-Based Nanodiscs

*Wafa Aresh[1], Ying Liu[2], Jessica Sine[3], Derek Thayer[3], Anu Puri[3], Yike Huang[4], Yong Wang[4], and Mu-Ping Nieh[1,2,5*]*

[1]*Department of Biomedical Engineering, University of Connecticut, Storrs, CT, USA*

[2]*Department of Chemical & Biomolecular Engineering, University of Connecticut, Storrs, CT, USA*

[3]*Basic Research Laboratory, Center for Cancer Research, National Cancer Institute, Frederick, MD, USA*

[4]*Department of Biomedical Engineering, Pennsylvania State University, University Park, PA, USA*

[5]*Polymer Program, Institute of Materials Science, University of Connecticut, Storrs, CT, USA*

## Abstract

The interaction between cells and particles is recently found morphologically and material dependent. In many cases, "spheres" may not be the best candidates for cellular uptake. In this study, self-assembled nanodiscs and nano- oblate vesicles composed of a zwitterionic long-chain 1,2-dihexadecanoyl-*sn*-3-phosphocholine (DPPC), a zwitterionic short chain1,2-dihexanoyl-*sn*-glycero-3-phosphocholine (DHPC) doped with small quantities of charged long-chain 1,2-dihexahecanoyl-*sn*-glycero-3-phospho-(1'-rac-glycerol) (DPPG) and a PEGylated lipid, 1,2-distearoyl-sn-glycero-3-phosphoethanolamine-N-[methoxy(polyethylene glycol)-2000] (DSPE-PEG2000) were synthesized at the same composition. Both morphologies are uniform and have similar sizes. Small angle neutron scattering data show that the average diameter and thickness of the nanodiscs are 21.0 and 5.4 nm, respectively, and that the outer long- and short- axes of the oblate vesicle yield 34.2 and 12.0 nm, respectively. The cellular uptake of these two types of Nilered-containing nanoparticles was investigated as incubated with the lymphoblast cancer cells (CCRF-CEM) at 37 °C for 2 hours. The result indicates that the cellular uptake of nanodiscs is 3 ~ 5 fold of that of the nanovesicles. Moreover, the fact that the locations of nanoparticles are different in the cells suggests a morphological dependence of cellular uptake mechanism.

*Mu-Ping Nieh. Email: mu-ping.nieh@ims.uconn.edu

# 379128 Mitochondrial Targeting with Triphenylphosphonium (TPP)-Conjugated and TPP-Peg-Conjugated PAMAM Dendrimer Nanocarriers

**Abstract:**

Mitochondria play vital roles in energy production within cells, are key to eukaryotic cell survival, in the generation of reactive oxygen species, in the regulation of calcium homeostasis, metabolism, and of programmed cell death (intrinsic apoptosis). Mitochondrial dysfunction is related to a range of diseases including heart disease, diabetes, cancer, neurodegenerative diseases, neuromuscular disorders, and aging. Their crucial role in apoptosis, metabolism, and energy production make these organelles a crucial target for many mitochondria-targeted drug therapies that can modulate the defects associated with the mitochondria and mitochondrial signaling pathways or induce apoptosis for cancer treatment strategies.

Among the many polymeric nanocarriers, Poly(amido amine) (PAMAM) dendrimers are a promising nanocarriers for the delivery of therapeutics including in the treatment of mitochondrial-related diseases. PAMAM dendrimers are polymeric hyperbranched tree-like structures of nanometer sizes comprising of a central core, repeated branches increasing with each generation, with a large number of surface functional groups that can be used to conjugate therapeutics for controlled release and other functionalities to enhance tissue/cellular/intracellular targeting. Polyethylene glycol (PEG) has also been extensively used to modulate the properties of dendrimers including improving hydrophobicity, decreasing toxicity, protecting therapeutic cargo, improving circulation time of dendrimers, and becoming a flexible linker to targeting moieties, such as mitochondrial targeting groups. Delocalized lipophilic cations (DLCs), such as triphenylphosphonium cations (TPP), are the widely used mitochondrial targeting agents due to their delocalized positive charge, which allows them to pass through plasma and mitochondrial membranes in a potential driven process, showing 100-1000 fold selective accumulation into the cell cytosol and mitochondrial compartments. Therefore, the use of DLCs such as TPP may offer unique strategies to deliver numerous therapeutic agents preferentially to mitochondria within cells, and can thus be very effective in maximizing drug effects while reducing side effects. The combination of DLCs and PEG with drug nanocarriers, such as PAMAM dendrimers, may thus afford unique opportunities for the treatment of mitochondrial-related diseases.

In this work, we report i) the synthesis and characterization of TPP and TPP-PEG conjugated to Generation 4 PAMAM amine terminated dendrimers ($G4NH_2$), which are potential nanocarriers of therapeutic agents, ii) the cellular internalization and colocalization of such nanocarriers in A549 alveolar epithelial adenocarcinomal cell line using flow cytometry and confocal microscopy, respectively, and iii) surface properties of TPP and TPP-PEG conjugated PAMAM dendrimers ($G4NH_2$-TPP and $G4NH_2$-PEG-TPP) and their microstructure. We systematically investigate the rate of internalization and colocalization of the conjugates as a function of TPP density with or without PEG.

# 379607 Effect of Incorporation of Lysolipid on Dipalmitoylphosphatidylcholine Bilayer Membrane: Study on the Permeability Enhancement of Liposomes at the Phase Transition Using Molecular Dynamics Simulations

**Abstract:**

Drug delivery and its release depend greatly on the ability to control the permeability of the vesicle encapsulating the drug. Liposomes, which are spherical nanoparticles composed of one or more phospholipid bilayers, have emerged as a promising delivery system for potent chemotherapeutics. It is well known that the permeability of drug through lipid bilayers exhibits a maximum as the lipid bilayer undergoes the transition from gel to liquid-crystalline phase, which has been studied intensively in order to control the permeability. One of the ways for the controlled drug release is an addition of the lysolipid, which affects the transition properties such as temperature.

In this study, we run fully atomistic molecular dynamics (MD) simulation with flat liposome model comprised of both DiPalmitoyl Phosphatidyl Choline (DPPC) and 0-30% lysolipids of MonoPalmitoyl Phoshatidyl Choline (MPPC). The phase transition temperature is obtained by monitoring the area per lipid and the lateral diffusion coefficient as the temperature is changed. At this phase transition temperature, we examine the structural characteristic of the mixed system by observing the density profile and interfacial tension profile. Furthermore, to clarify the mechanism of enhanced drug release of lysolipid containing liposome, we investigate the most probable distribution of MPPC in the mixed membrane, which could be in the form of either dispersed or assembled configuration.

# 379713 Self-Assembled Polymer Nanomedicines for Synergistic and Synthetic Lethal Drugging of Breast and Ovarian Tumors

**Abstract:**

Synergistic and synthetic lethal drug delivery to solid tumors can be difficult to achieve using traditional pharmaceutical excipients and nanotechnologies. Ideally, these strategies not only require that drug compounds with vastly differing size, charge, and hydrophobicity be co-delivered to tumor cells, but also that delivery (i) targets multiple cell populations within a heterogeneous tumor, (ii) is non-toxic to healthy tissues, and (iii) circumvents known mechanisms of resistance. Here, we describe recent progress in our laboratory utilizing self-assembled polymer drug carriers to address these difficult challenges. (i) We report the development of a nanoscale drug carrier based on Layer-by-Layer (LbL) polymer self-assembly that is capable of simultaneously targeting both cancer stem cells and the hypoxic tumor microenvironment in vitro and in vivo. (ii) We also describe the discovery of a novel DNA damage signal effector whose loss is synthetic lethal in ovarian cancer cells that lack functional tumor suppressor protein, p53, and the development of a series of lipid-like peptide co-polymers capable of co-delivering combinations of small interfering RNA and platinum therapy to ovarian tumors. (iii) We present a novel nanoscale combination therapy of molecularly-targeted inhibitors that horizontally block oncogene-addicted and resistance-associated signaling in triple negative breast and non-small cell lung cancers in vitro and in vivo. These strategies are part of a comprehensive effort to directly address unmet clinical needs in the therapy of breast and ovarian carcinoma.

# 381884 Dendrimer Nanodevices for Pediatric Brain Injury

**Abstract:**

**Introduction**: Maternal-fetal inflammation (MI) and hypoxia-ischemia (HI) are major risk factors for perinatal/neonatal brain injury, and both can lead to neurological diseases including cerebral palsy (CP) and autism. Neuroinflammation, mediated by activated microglia and astrocytes, is implicated in the pathogenesis of both etiologies, resulting in diffuse white and grey matter injury. Targeting activated microglia/astrocytes may offer such an opportunity. However, this is a challenge on multiple levels because (1) transport of drugs and drug delivery vehicles across the blood-brain-barrier (BBB) is difficult to achieve, (2) inflammation and injury are often diffuse, making it difficult for therapeutics to reach target cells even if administered locally, and (3) the role of microglia and extent of BBB disruption as disease progresses in the developing brain is not well understood, especially after a therapy is administered. Recent studies from our laboratory have shown that dendrimer nanomaterials can target activated microglia in the brain upon systemic administration, producing dramatic motor function improvements and neuronal repair in a rabbit model of CP. These studies suggest that nanotechnology-based approaches provide potential platforms for CNS therapy. However, there is a major knowledge gap relating to the effect of changes in extracellular matrix, brain edema, glial cell function and BBB disruption, on the movement, interactions, and cellular uptake of nanoparticles following injury to the brain.

**Materials and Methods**: In this work, we use two clinically relevant, broad etiological mechanisms of neonatal brain injury (HI or MI) to study the effect of dendrimer size and surface functionality, as well as disease etiology and pathophysiology, on the ability of the dendrimer to penetrate a disrupted BBB, diffuse within the brain parenchyma, and selectively uptake in microglia cells. For efficacy, N-acetyl cysteine (NAC) conjugated to OH-dendrimer (D-NAC) was administered systemically, and animals were followed for survival or sacrificed at specified timepoints for quantitative, immunohistochem (IHC), or RT-PCR analysis. For fluorescent-dendrimer conjugates injected systemically, brains were harvested at specified timepoints and used for IHC or for quantitative analysis using spectrophotometry. Blinded neurobehavioral analysis was completed and scored to determine extent of disease and efficacy of dendrimer-drug conjugates compared to free drug (NAC), saline controls, or age-matched healthy controls.

**Results and Discussion**: In both etiologies, we show that systemic administration of a single dose of dendrimer-drug conjugate results in accumulation in activated microglia in the brain of newborn rabbits with MI-induced CP or newborn mice following HI, leading to sustained attenuation of neuroinflammation up to 8 days after injury. BBB breakdown is greater in HI-mediated brain disease compared to MI-induced brain disease. Following systemic administration, neutral dendrimer penetrates a disrupted BBB in both models, and rapidly accumulates in microglia within 4 hours. Cationic dendrimers do not penetrate within the brain parenchyma or localize in microglia cells, and anionic dendrimers show delayed uptake in the brain and in microglia cells. These results demonstrate the importance of dendrimer surface functionality on movement and localization within the brain. Dendrimer was not found in the brain in age-matched healthy controls. Importantly, a strong positive correlation between % injected dose of dendrimer uptake in the brain and disease score based on neurobehavioral testing was found, with sicker animals showing greater uptake, likely due to increased BBB impairment and greater microglia activation representative of a stronger inflammatory response.

**Conclusions:** Pediatric illnesses are often underserved by novel drug delivery technologies, which focus primarily on adults. Therefore, there is great opportunity to bring nanotherapeutic approaches to perinatal and neonatal brain injury, with implications that can be translated to adult brain injury. These findings begin to address the critical knowledge gap in identifying the physicochemical properties of nanoparticle platforms and pathological properties of disease that facilitate brain penetration, diffusion and specific cellular uptake that would promote targeted attenuation of neuroinflammation in the brain.

Abstract for: 2014 AIChE Conference, Atlanta, GA, November 16-21, 2014

# Dry powder aerosols based on mucus-penetrating nanocomposite microparticles for pulmonary antibiotics delivery

**Zimeng Wang**[1], Samantha A. Meenach[1, 2]

[1] Department of Chemical Engineering, [2] Biomedical & Pharmaceutical Sciences, University of Rhode Island

Pulmonary antibiotic delivery is increasingly recommended as maintenance therapy to prolong the interval between pulmonary exacerbations and to slow the progression of lung disease in cystic fibrosis (CF) patients with chronic *Pseudomonas aeruginosa* infection due to its capability to achieve high drug concentrations at the site of infection and to minimize the risk of systemic toxicity and drug resistance. Currently used inhaled formations so far include the nebulization of drugs such as tobramycin which are regarded as inconvenient due to the long administration time and limited portability for chronic drug therapy in daily life of patients. Dry powder formulations of tobramycin have many advantages over nebulizers including faster delivery, easy use, portability, reduced need for cleaning and room temperature storage. However, minimal studies have been completed to control its drug release rate. Besides, the abnormally thick and sticky mucus in respiratory tract of CF patients impairs efficient mucus penetration and limits the range of antibiotics for inhalation.

To overcome the obstacles of pulmonary antibiotic delivery, we developed nanocomposite microparticles (nCμP) for pulmonary delivery of antibiotics in the form of dry powder aerosols. Our system consists of a microparticle carrier comprised of mannitol aimed on targeted delivery with drug-loaded nanoparticles entrapped in the carrier. The drug-loaded nanoparticles contain tobramycin as a model drug and the polymer acetalated dextran which provides a controlled drug release rate. The nanoparticle is coated by a Vitamin E poly (ethylene glycol) (PEG) 5000 layer which has been shown to improve the stability and mucus penetrating rate of nanoparticles. Upon synthesis of the tobramycin-loaded nanoparticles via nanoprecipitation, they were loaded into the mannitol carrier via advanced spray drying techniques. The resulting nCμP were then characterized for their shape, drug loading, thermodynamic stability, mucus-penetrating, and aerosol characteristics. Scanning electron microscopy and dynamic light scattering showed approximately 150 nm nanoparticles entrapped in 1 – 2 μm microparticles with smooth morphology. The system effectively entrapped 12% of the loaded nanoparticles and further studies are underway.

This nCμP system sheds a light on dry powder-based antibiotic delivery due to its features including targeted delivery, rapid mucus penetrating rate, and controlled drug release. It can be applied as a promising alternative of the nebulized antibiotic treatment by providing high efficacy of the delivery, more convenient administration, more flexible storage condition and lower risk of contamination in the device.

Submitted to:
22B02 Bionanotechnology for Gene and Drug Delivery

# 383792 Multi-Modal Gold Nanoparticles for Simultaneous Lineage-Specific Targeted Delivery and Tracking of Chemotherapeutics

**Abstract:**

Au nanoparticles, along with other drug delivery technologies such as micellar and liposomal or polymeric encapsulations, and monoclonal antibody conjugation, have been intensively pursued as candidates to advance nanomedicine. Realizing multiple functionalities in one platform is an important prospect of nanomedicine. We have developed an amine-passivated Au nanoparticle drug delivery platform that allows for controlled payload release in specific cell lineage. Our approach is to use Au nanoparticles as intermediaries to which drug and targeting moieties are simultaneously attached using simple but different chemistries that permit controlled release of drug, while ensuring targeting functionality intact. To demonstrate this, we established robust and scalable synthesis of daunorubicin (DNR)-loaded Au nanoparticles functionalized with erythropoietin mimetic peptide (EMP). In essence, we use Au nanoparticles as a “linker” between drug molecules and EMP. We elucidate payload delivery to erythroid progenitors in murine bone marrow via flow cytometry and fluorescence microscopy. Exploiting the recovery of quenched DNR fluorescence upon release from Au nanoparticle surface as a reporter, cell lineage-targeted payload delivery is demonstrated. The drug-Au-EMP platform allows for versatile choice of drugs as well as targeting moieties, and could potentially improve therapeutic index in breadth of existing drugs.

## Controlled uptake of therapeutic protein nanoparticles

Proteins have significant potential to treat diseases due to their specific mechanisms of action compared to small molecule drugs. The relatively large size of proteins, however, gives them poor permeability and selectivity across the cell membrane. Additionally, protein structure is susceptible to degradation by proteases and denaturation by changes in pH, temperature, and salt concentration *in vivo*. Protein nanoparticles offer a means to turn soluble protein into a colloidal drug carrier capable of enhanced cellular uptake, improved stability, and protection of protein cargo. It is known that nanoparticle size, surface charge, hydrophobicity, and the absorbed protein corona all affect the ability of nanoparticle to be internalized and trafficked. These properties are controlled by selection of the carrier protein used to fabricate the nanoparticles.

In this work, a therapeutic anti-inflammatory protein, AvrA, was desolvated with a carrier protein, eGFP, to control protein nanoparticle uptake. Nanoparticle size was controlled by tuning temperature, protein concentration, type of crosslinker, and crosslinker concentration in the desolvation process. Epithelial cells internalized 3.7 times more AvrA in protein nanoparticles compared with soluble protein; macrophages internalized 420 times more. We determined that the nanoparticles are internalized via receptor-mediated endocytosis and have minimal cytotoxicity. We will discuss the role of the carrier protein and how it can control the physiochemical properties of the nanoparticle, thereby controlling endocytic uptake. Delivered AvrA retained its immunological activity by ameliorating inflammatory response induced by TNF-α. Collectively, this data demonstrates the ability of protein nanoparticles as functional delivery vehicles. Future work will improve cellular uptake and direct nanoparticles toward specific endocytic routes by modifying the nanoparticle properties. Applications of this study will enable controlled cellular uptake of a variety of other proteins therapeutics by altering protein nanoparticle properties.

# 390398 Improved Entrapment of N-Acetyl Cysteine (NAC) in Nanoparticles of Poly[Lactic-Co-Glycolic Acid] (PLGA) Prepared By Double Emulsion Method By Using Bovine Serum Albumin (BSA)

**Abstract:**

NAC is a drug with many actual and potential applications as a result of its antioxidant properties [1]. However, low oral bioavailability (4 to 10 %) and its strong hydrophilic character allowing easy excretion by renal pathway, impose limits to systemic concentration of NAC and, therefore, to therapeutic activity [2]. Encapsulation and slow release of NAC from a polymeric matrix could extend both residence time and time-averaged concentration in the body.

The double emulsion method has been used to entrap hydrophilic compounds within hydrophobic polymeric matrices such as PLGA; however, the resulting capture efficiency of the drug tends to be extremely low. The use of BSA has been reported to improve entrapment efficiency [3]. The general purpose of this study was to evaluate the use of BSA as a medium to improve the encapsulation efficiency of NAC in nanoparticles of PLGA made by the double emulsion method.

Methodology:

PLGA and a surfactant were dissolved into Dichloromethane (DCM). An aqueous solution of BSA and NAC was then emulsified into the DCM/surfactant/PLGA solution by ultrasound. The resultant emulsion was subsequently emulsified into an aqueous solution of Pluronic F127 by ultrasound. After DCM evaporation, PLGA nanoparticles containing NAC were formed and characterized by particle size, zeta potential, drug efficiency entrapment and drug release kinetics. The effect of PLGA and surfactant concentrations in DCM in the first emulsion, of BSA and NAC concentrations on water in the first emulsion, and of organic/aqueous phase ratio in the second emulsion on particle characteristics was assessed.

Results:

BSA improves the entrapment efficiency of NAC in nanoparticles of PLGA. Concentration of polymer in DCM or NAC in water affect both entrapment efficiency and particle size. The presence of a surfactant is important in order to stabilize the first emulsion.

Overall, for best conditions evaluated, nanoparticles showed a particle size between 200 and 350 nm and entrapment efficiency around 8%. *In vitro* drug release studies showed a burst release within the first hour, followed by a reduction in the release rate.

Conclusion:

Results suggest that bovine serum albumin improves retention of hydrophilic compounds like N-Acetylcysteine into hydrophobic polymeric matrices such as PLGA.

References:

[1] M. Zafarullah, W. Q. Li, J. Sylvester, and M. Ahmad, "Cellular and Molecular Life Sciences Molecular mechanisms of N -acetylcysteine actions," *C. Cell. Mol. Life Sci.*, vol. 60, pp. 6–20, 2003.

[2] K. Suzuki, "Anti-oxidants for therapeutic use: why are only a few drugs in clinical use?," *Adv. Drug Deliv. Rev.*, vol. 61, no. 4, pp. 287–9, Apr. 2009.

[3] D. Chitkara and N. Kumar, "BSA-PLGA-Based Core-Shell Nanoparticles as Carrier System for Water-Soluble Drugs.," *Pharm. Res.*, no. 12, Jun. 2013.

# 382154 DNA Nanotubes and Nanotapes Formed Via Self-Assembly of ssDNA-Amphiphiles

**Abstract:**

DNA is a popular material for constructing complex, multi-dimensional nanostructures due to its ability to organize via Watson-Crick base-pairing in a precise and predictable manner. Popular approaches to create nanostructures from DNA include DNA origami, DNA tile assembly, and DNA brick assembly, each of which rely on numerous strands of single-stranded DNA (ssDNA) with different and complimentary sequences to direct the DNA assembly process. These approaches and others have been used to create numerous structures including three-dimensional polyhedra, two-dimensional lattices, and nanoribbons and nanotubes. An alternate approach to direct the assembly of ssDNA is to conjugate a hydrophobic moiety (i.e., polymer or lipid) to the ssDNA to form an amphiphilic molecule that spontaneously self-assembles when added to aqueous solutions. It was recently shown by our group that the type of molecular spacer linking a ssDNA headgroup to a lipid-like molecule affects the amphiphile's assembly and produces spherical micelles or bilayer nanotapes (flat or twisted) depending on the type of spacer used in the amphiphile design.

In this work we explore the impact of the ssDNA length and secondary structure on the self-assembly of ssDNA-amphiphiles using circular dichroism and cryo-transmission electron microscopy (cryoTEM). We found that ssDNA-amphiphiles with 10, 25, and 40 nucleotide ssDNA sequences form a variety of structures including spherical micelles, twisted or helical nanotapes, and nanotubes. Interestingly, our cryoTEM images capture for the first time ssDNA-amphiphile structures that are transitioning between nanotapes and nanotubes, which suggests an assembly mechanism that includes transitions from twisted nanotapes to helical nanotapes to ssDNA nanotubes. The ability to form structures of great interest, such as nanotubes, from a single ssDNA sequence in a simple way to the common DNA assembly techniques could be useful for DNA nanotechnology applications that can trade complexity in supramolecular structure for ease-of-design and a robust assembly process.

# 388310 Targeted Binding of Nanolipoprotein Particles to Phase Separated Lipid Domains

**Abstract:**

Targeted binding of functional proteins and/or biomolecules to supported lipid bilayers has a variety of applications, including development of nano-array technologies and biosensors. Use of Integral Membrane Proteins (IMPs) in such systems is difficult, as incorporation into planar bilayers is not straightforward and can result in protein inactivation. We intend to overcome this limitation by use of Nanolipoprotein Particles (NLPs), which are able to solubilize IMPs. NLPs are discoidal patches of lipid bilayer that have their hydrophobic faces covered by amphiphilic scaffold proteins. The scaffold proteins can bear polyhistadine tags, which are capable of chelating to $Cu^{2+}$ metal ions. Phase specific, Iminodiacetic acid functionalized lipids are also capable of chelating $Cu^{2+}$, providing a mechanism for targeted binding of NLPs. We investigate this binding via fluorescence microscopy and characterize interaction on supported bilayers and giant unilamellar vesicles (GUVs). The thermodynamics (enthalpy of lipid mixing and steric pressure of protein crowding) and morphology of binding are also examined.

Self-assembly of Functionalized Virus Capsids

Viral capsids have been the subject of a significant amount of research in a variety of fields. Their potential relevance to biomedical science arises from their ability to encapsulate therapeutic agents and deliver them to targeted cellular sites. The capsids can be decorated with polymers to target specific sites or promote controlled aggregation for creating novel materials. We used coarse-grained molecular dynamics to simulate the aggregation dynamics of Cow Pea Mosaic Virus capsids grafted with Polyethylene Oxide chains, and the role of the polymer length, polymer grafting density and volume fraction on the aggregate morphology.

# 370156 Design of Sterically Stable Bio-Inspired Hybrid Soft Biomaterials Via Computational Modelling

**Abstract:**

Our objective is to investigate the design of sterically stabilized hybrid biomaterials through the self-assembly of the end-functionalized amphiphilic molecular species such as glycolipids. Individual molecular species are composed of a hydrophilic head group and two hydrophobic tails. The species can differ in terms of specific chemical properties of the polar head groups and the hydrocarbon tail groups. The stable hybrid vesicles can be formed through the self-assembly of amphiphilic molecules in the presence of a suitable solvent. We use a Molecular Dynamics-based mesoscopic simulation technique called Dissipative Particle Dynamics to simultaneously resolve the aggregation dynamics, structure and morphology of the hybrid aggregate. In addition, we investigate the factors that control the self-organization of the molecular species within the hybrid soft biomaterials. We characterize the effects of functional groups grafted to hydrophilic segment of the lipid species on the stability, the aggregation kinetics, the equilibrium structure and morphology of the hybrid soft materials. The results of our investigations can be used for the design and prediction of novel hybrid soft materials such as sterically stabilized vehicles for drug delivery.

# 394855 Programming Colloidal Phase Transitions with DNA Strand Displacement

**Abstract:**

Interactions induced by the transient bridging of complementary DNA strands grafted to colloidal particles can direct the self-assembly of nanoparticle-based materials. Such DNA-grafted nanoparticles have been called 'programmable atom-equivalents': Like atoms, they form three-dimensional crystals, but unlike atoms, the particles themselves carry information (the sequences of the grafted strands), which can be used to 'program' the equilibrium crystal structures. We present experimental results showing that the programmability of these colloids can be generalized to the full temperature-dependent phase diagram, and not just the structures themselves. We add additional information to the buffer, in the form of soluble DNA strands designed to compete with the grafted strands through strand displacement. Employing only two displacement reactions, we program phase behavior not found in atomic systems or other DNA-grafted colloids, including arbitrarily wide gas-solid coexistence, crystals that melt upon cooling, and even reversible transitions between distinct crystal phases. Our findings outline a new direction in self-assembly, in which additional information supplied to the buffer programs equilibrium pathways between many different target structures within a closed system.

# 358347 Ibuprofen Loading and Release in Micelles Formed By Amphiphilic Peptide $(AF)_6H_5K_{15}$: A Coarse-Grained Molecular Dynamics Simulation

**Abstract:**

Amphiphilic peptides have been recognized as good candidates for drug delivery due to their self-assembly nature. We report a molecular dynamics simulation study for drug (ibuprofen) loading and release in micelles formed by peptide $(AF)_6H_5K_{15}$ (FA32). The peptide, drug, water and counterions are represented by the MARTINI coarse-grained model. With 5 μs simulation duration, ibuprofen is loaded into the micelle core and covered by peptide residues. Upon protonating Histidine residues and along with applying an external force on solvent to mimic stirring effect, ibuprofen is observed to be released. This suggests that the drug can be released from the micellar core when it reaches the vicinity of tumor cells with an acidic pH. The microscopic insight from molecular simulation is helpful to better understand the mechanisms of drug loading and release, and could facilitate the development of new amphiphilic peptides for high-efficacy drug delivery.

# Measuring How the Local Glass Transition Temperature Shifts Across a Glassy-Rubbery Polymer-Polymer Interface

Roman R. Baglay and Connie B. Roth
Department of Physics, Emory University, Atlanta, GA, USA

Contact information:
Roman Baglay: rbaglay@emory.edu
Connie Roth: cbroth@emory.edu

Recent advances in block copolymer synthesis have led to new routes to forming nanostructured polymer blends putting rubbery, glassy, and crystalline phases in intimate contact. These new multicomponent block copolymers can be designed to self-assemble into a range of morphologies including unique intertwined co-continuous phases with domain sizes easily less than 100 nm in size. Efforts to understand the unique properties of these new nanostructured materials have been limited to measurements of macroscopic properties correlated with the observed microstructure. Here, we use individually assembled multilayer films with specific layer thicknesses to act as a model system mimicking the local morphology of such nanostructured blends. A fluorescence technique is used to directly measure the local glass transition temperature ($T_g$) profile as a function of distance from a glassy-rubbery polymer-polymer interface.

We focus on a polystyrene (PS) / poly(*n*-butyl methacrylate) (PnBMA) weakly immiscible system with a 7 nm interfacial width [1] that creates an 80 K $T_g$ difference across the interface from one bulk $T_g$ value to another. Multilayer films of high molecular weight PS and PnBMA are constructed where an ultrathin (10-15 nm thick) pyrene-labeled layer is inserted into the multilayer structure at different positions ($z$) from the glassy-rubbery interface. Temperature-dependent fluorescence intensity is collected to determine the local $T_g(z)$ at a given position z from the interface. Using a series of different samples, we are able to map the $T_g(z)$ profile across this glassy-rubbery interface. Our work reveals a gradient in the local $T_g(z)$ that spans ~400 nm across the interface from one bulk $T_g$ value ($T_g^{PnBMA} = 20$ °C) to another ($T_g^{PS} = 100$ °C). The results demonstrate that local properties can be highly coupled across polymer-polymer domains within nanostructured blends, easily spanning entire domain sizes. This is in contrast to how the glass transition of polymer blends are traditionally treated, assumed to depend only on the local composition within a few nanometers [2], such that neighboring domains, even in nanostructured polymer blends, are presumed to have independent $T_g$s.

Surprisingly we find that the local $T_g(z)$ gradient across the glassy-rubbery PS/PnBMA interface is highly asymmetric, extending much further into the glass PS side. This observation is consistent with recent theoretical predictions from a limited mobility model for glassy dynamics in confined systems [3]. For decades $T_g$ in confined systems has been studied with the hopes of uncovering the length scales that impact the glass transition. In ultrathin confined films, free surface and substrate perturbations induce gradients in local $T_g$ spanning several tens of nanometers indicating that cooperative dynamics are coupled in some fashion across multiple cooperatively rearranging regions (CRRs) [4,5]. However, understanding length scales of local $T_g$ gradients near a free surface has been hampered by limitations of how to treat the enhanced mobility at the free surface theoretically. In our study, a glassy-rubbery polymer-polymer interface is used to establish an 80 K local $T_g$ gradient from one well-defined $T_g$ value to another, thereby introducing a well-defined system with $T_g$ asymptotics on both sides of the glassy and rubbery polymer-polymer interface.


[1] Rauscher, P. M.; Pye, J. E.; Baglay, R. R.; Roth, C. B. *Macromolecules* **2013**, *46*, 9806–9817.

[2] Lodge, T. P.; McLeish, T. C. B. *Macromolecules* **2000**, *33*, 5278–5284.

[3] Tito, N. B.; Lipson, J. E. G.; Milner, S. T. *Soft Matter* **2013**, *9*, 9403–9413.

[4] Ellison, C. J.; Torkelson, J. M. *Nature Materials* **2003**, *2*, 695–700.

[5] Pye, J. E.; Rohald, K. A.; Baker, E. A.; Roth, C. B. *Macromolecules* **2010**, *43*, 8296–8303.

# 383404 Tuning the Wall Thickness of Templated Polystyrene Nanotubes Produced from Melt Infiltration

**Abstract:**

In this work, we demonstrate methods of varying the wall thickness of polystyrene nanotubes produced from template synthesis by utilizing classical theories of liquids spreading on flat and curved substrates. We find that factors which affect the surface energy of the polymer and substrate such as the annealing temperature, annealing atmosphere and polymer molecular weight influence the thickness of the nanotubes which are achieved. Results from this study are in qualitative agreement with what is predicted from classical theories. Nanotubes with wall thicknesses ranging from 18 to 40 nm were achieved by infiltrating different molecular weight polystyrene. Higher molecular weight samples possessed larger wall thicknesses compared to lower molecular weight samples. Thermal characterization with differential scanning calorimetry measurements revealed significant Tg-confinement behavior as a function of wall thickness. Post infiltration processing such as annealing was also found to affect the thickness. Supported tubes annealed at high temperatures possessed thinner wall diameters compared to those annealed at lower temperatures. Additional results suggest that atmosphere can have an effect on the wall thickness. Alteration of the surface energy through changes to factors such as the temperature, polymer molecular weight and atmosphere affects the equilibrium tube thickness and may allow for dimensional control over the nanostructures produced from template synthesis.

The ability to tune the dimensions of nanomaterials is critical for applications which rely upon the unique properties which arise due to the greater effect of free surfaces, interfacial interactions and small diffusion lengths at the nanoscale. Confinement in one or more dimensions can lead to enhancements in the mechanical strength, electrical, optical, thermal properties of materials and benefit areas involving drug delivery, medical devices, interconnects, sensing, catalysis, filtration and nanocomposites. The ability to adjust material properties using the length scale of these structures imparts additional requirements e.g. the ability to achieve uniform size and shapes and the means by which one can tune their dimensions.

Template synthesis is often employed because it allows for dimensional tunability and can be used to infiltrate a wide range of organic and inorganic materials. Anodic aluminum oxide templates (AAO) can be produced using a variety of different electrolytes and conditions and yield pores tens to several hundred nanometers in size. Infiltration into the templates results in negative replication of the size, shape and length of the pore morphology. This is useful for creating nanostructures with custom exterior dimensions and shapes; however, the ability to tune the characteristics of internal features such as the wall thickness of nanotubes is very limited. Control over the wall thickness of nanotubes is typically carried out using solution based techniques to infiltrate polymer into the membrane. The nanotube wall thickness can be varied by changing the molecular weight of the polymer, solution concentration, type of solvent and size of the template used. However, the uniformity and the reproducibility of nanostructures produced using solution based approaches can be very unpredictable. Holes or undulations are often observed as a result of the solvent evaporation process or the evolution of Rayleigh instabilities. Furthermore, extensive drying of the samples must be carried out because residual solvent may influence the morphology of the nanostructures or influence their properties. Melt infiltration generally yields more uniform and reproducible nanostructures and is simpler because it is solvent-less but little is known or has been done regarding the wall thickness tunability of nanotubes produced in this manner. Classical theories regarding the spreading of low energy fluids on flat and curved substrates and recent experimental work suggest that the thickness of polymer films or tubes supported in AAO templates can be modified by annealing under different conditions.

According to classical wetting theories the propensity for a low energy liquid (e.g. a polymer melt) to spread across a high energy surface is characterized by the spreading coefficient which accounts for the difference in energy between an exposed substrate versus one that has been wetted by a liquid film. A polymer liquid will wet the substrate when the total system energy can be lowered be minimized. This process can occur very rapidly and is often described as a "precursor film". When exposed to a template membrane under complete wetting conditions, e.g. high temperatures, the polymer coats the interior and forms nanotubes. The thickness of the precursor film reaches an equilibrium value which is determined by the balance between long range Van der Waals forces and surface energies. This prevents the film from spreading without bounds into an infinitely thin layer of material. Changes to the surface energy can shift the equilibrium value and result in changes to the equilibrium thickness. Factors such as the temperature and atmosphere are known to influence the surface energy of a polymer melt and can be used to modify the wall thickness of template nanotubes. Likewise, the molecular weight of the polymer or the functionality of the substrate can also affect the surface energy of the system.

Various molecular weights of polystyrene were melt infiltrated into AAO templates under the same thermal conditions. Nanotubes with wall thicknesses ranging from 18 to 40 nm were achieved. Thicker nanotubes were created by infiltrating higher molecular weight samples whereas thinner tubes were obtained from lower molecular weights. This is in agreement with classical theory; the surface energy of a polymer increases with molecular weight and increases the equilibrium thickness. The infiltrated templates were also measured using differential scanning calorimetry to determine the change in Tg-confinement behavior as a function of thickness. Tg reductions as large as 20 °C from the bulk Tg were observed for the thinnest tubes whereas bulk-like Tg behavior was obtained for the thickest tubes. It was also discovered that wall thickness can be modified by annealing at different temperatures after infiltration. For the same molecular weight polymer, nanotubes annealed at 160 °C possessed a lower wall thickness than those annealed at 130 °C. These results suggest that it is possible to tailor the wall thickness of a polymer sample by modifying the annealing conditions without having to use different molecular weight polymers or modifying the surface energies of the substrate. These results further suggest that other factors such as the atmosphere and the surface energy of the substrate may affect the thickness and that factors such as the amount of polymer available for infiltration will not based upon what is predicted from classical spreading behavior.

# 362494 Depletion, Bridging, and Ordering in Bare and Grafted Nanorod-Polymer Mixtures

**Abstract:**

We present a detailed study of polymer-nanorod composites. Using integral equation theory we have determined the structure and phase behavior non-interacting, attractive, and grafted nanorods in polymer melt. Depletion-induced phase separation is predicted between polymer and rods of nearly all aspect ratios. Miscibility is achieved by introducing a small amount of attractive interaction between polymer and nanoparticles. At large polymer-nanoparticle attraction strengths a few polymer molecules bridge several nanoparticles giving rise to another phase separated regime. Large diameter nanorods tend to be bridged my more than a single polymer layer (telebridging). This creates a finite miscibility window below the depletion temperature and above the bridging temperature.

Small changes in nanorod aspect ratio induce a drastic narrowing of the miscibility window, particularly for thin nanorods. However, as the rod radius of gyration approaches the radius of gyration of the polymer, the phase boundaries are nearly independent of aspect ratio. Our work confirms previous experimental findings that depletion-driven phase separation is enhanced with increasing size of nanospheres. However, we predict that nanorods display a subtle, and non-montonic dependence of the depletion line on the rod thickness. In the bridging regime, larger nanospheres show enhanced bridging while nanorods display a non-monotonic behavior. The width of the miscibility window is, therefore, a non-monotonic function of the nanorod thickness. Close to both bridging-driven, and depletion-driven phase separation, the attractive force between rods increases with rod thickness, while remaining independent of rod length. However, this attractive force operates at different inter-particle separations in the two regimes. Our work with grafted nanorods (with various graft lengths) illustrates that even short polymer grafts at moderate grafting densities can prevent the entropic depletion-driven phase separation. On the other hand, grafted nanoparticles are likely to form ordered, microphase separated structures within the polymer melt.

# 384265 Ordered Polymer-Grafted Nanoparticle Networks for Ionic Conductivity

**Abstract:**

Ionic liquids have been used effectively to improve ionic conductivity of diblock copolymers in past works. We suggest that polymer-grafted nanoparticles with sulfonated end groups can be used as a new way of controlling particle dispersion and ion aggregation within ionic liquids. In this talk, the structures of synthesized ionically functionalized polymer-grafted iron oxide nanoparticles will be examined and discussed through SAXS and TEM analysis.

# 390373 Preparation of Conductive Polymer Nanofibers with a Coaxial Nozzle Electrospinning Method

**Abstract:**

Conductive polymer is a polymeric compound which shows electrical conductivity, and is applied to various fields because of its high electrical conductivity, high flexibility, and lightweight characteristics. Further, nonwoven microfabrics of the conductive polymers are expected to be utilized for the highly sensitive sensors in medical and environmental fields. However, because of their infusible and insoluble nature, process of the conductive polymer nonwoven microfabrics must become a multistep. In recent years, the electrospinning of microfibers has been employed due to its simplicity, versatility, and scale-up opportunities. In the electrospinning, coaxial nozzles are often used as the discharge port for the formation of the fine structured materials. In this study, we developed a single step processing method for the formation of nonwoven conductive polymer fabrics which uses electrospinning method with a coaxial two fluid nozzle of a monomer and an oxidant solution.

3,4-ethylenedioxythiophene (EDOT) was chosen as a monomer due to the facts that poly-EDOT (PEDOT) has a good conductivity, environmental stability, and harmless to the human body. Iron( ) *p*-toluenesulfonatehexahydrate (FeTS), one of the most efficient oxidants for EDOT polymerization and known to produce intrinsically doped and electrically conductive PEDOT, was chosen as a oxidant. Before the electrospinning experiments, appropriate solvent for the synthesizing the PEDOT was examined with a bulk polymerization of EDOT, and acetonitrile was chosen for the solvent. Cellulose acetate (CA) and Poly(methyl methacrylate) (PMMA) were used as matrices for the conductive polymer microfibers. Effects of spinning conditions on the properties of conductive polymer were examined.

The monomer solution was prepared by dissolving the CA powder and the liquid EDOT into acetone or acetonitrile. The oxidant solution was prepared by dissolving the PMMA powder and the FeTS powder into acetone or acetonitrile. Both solutions were filled into grass syringes, and placed in automatic syringe pumps. The grass syringe filled with the monomer solution was connected to the outer tube of the coaxial nozzle, while that of the oxidant solution was connected to the inner tube. Both solutions were fed simultaneously to the coaxial nozzle, and electrospunned to the aluminum foil counter electrode to obtain the PEDOT nonwoven microfabrics. The PEDOT microfibers were characterized with scanning electron microscopy (SEM), mass spectrometric analysis (MALDI-TOF-MS) and four-point probes conductivity measurements.

With the coaxial nozzle electrospinning, nonwoven microfabrics were successfully obtained. Average diameter of PEDOT microfibers were varied from 490.6 nm (PMMA 15 wt%, CA 1 wt%) to 800 nm (PMMA 15wt%, CA 10 wt%) with the change in the CA concentration. This fact indicates that the fiber diameter could be controlled by the concentration of the matrix solution. Conductivity of thus formed microfabrics was the order of $10^{-3}$ S/cm, which is the same with the PEDOTS obtained by the bulk polymerization. We are now working on the improvement of the conductivity.

# 383834 Novel Pressure-Responsive Shape Memory Polymers

**Abstract:**

Shape memory polymers (SMPs) are stimuli-responsive materials that can change their shape upon application of an external stimulus (e.g., heat and light). They have been extensively exploited for a wide spectrum of technological applications ranging from smart biomedical devices to aerospace morphing structures. Most of the currently available SMPs are thermoresponsive and they suffer from slow response speed and heat-demanding programming and recovery steps. Although pressure is an easily adjustable process variable like temperature, pressure-responsive SMPs are largely unexplored. Here we demonstrate a new class of pressure-sensitive SMPs that enable the instantaneous recovery of the memorized shapes by applying a very small pressure. By combining these novel SMPs with self-assembled colloidal crystals, we have developed tunable macroporous photonic crystal films showing large optical stop band shifts triggered by a small pressure change. Rewritable photonic crystal microstructures, such as optical waveguides, can be easily printed on the pressure-responsive SMP films. We have also demonstrated that nanoporous smart coatings exhibiting tunable anti-glare properties can be fabricated by using the new SMPs. These pressure-responsive SMPs could find important applications in anti-counterfeiting and biometric recognition markets.

# 370773 Synthesis of Polymer Brush in Oil-Phase through Photo-Emulsion Polymerization

**Abstract:**

Oil-dispersed spherical polymer brush consisting of a hydrophilic core and lipophilic shell was synthesized through a novel "grafting from" procedure. Poly(acrylic acid) (PAA) nanoparticles with an average diameter of 100 nm were firstly prepared in toluene by inverse emulsion polymerization. At the end of polymerization, photo initiator 2[p-(2-hydroxy-2-methylpropiophenone)]-ethylene glycol- methacrylate (HMEM) was slowly added to copolymerize at the surface of PAA nanoparticles. After the addition of N-vinyl carbazole monomer, poly(N-vinyl carbazole) (PVK) chains finally generated from the surface of PAA core to form a brush-like structure via photo-emulsion polymerization under the irradiation of UV light. Dynamic light scattering (DLS) results display an increase in hydrodynamic diameters after photo-emulsion polymerization. In addition, the characteristic peaks of PVK were clearly observed in UV-vis spectra and photoluminescence spectra, indicating PVK chains were grafted onto PAA matrix. Transmission electron microscopy (TEM) shows the spherical shape of PAA-PVK brushes with relatively narrow size distribution. Compared with pure PAA core, the stability of PVK grafted PAA brush was significantly improved in organic solvent. This kind of novel core-shell structure is promising in the field of fluorescence labeling carrier and protein encapsulation in non-aqueous media.

# 387876 Thermodynamic Interactions and Tunable Properties of Thermoplastic Elastomers Derived from Vegetable Oils

**Abstract:**

Linear ABA triblock copolymers, such as poly(styrene-*b*-butadiene-*b*-styrene) and poly(styrene-*b*-isoprene-*b*-styrene), are among the most important and widely used thermoplastic elastomers (TPEs). However, those TPEs are derived from fossil fuels. Due to the finite availability of fossil fuels and the environmental impact of fossil fuels manufacturing, there has recently been an urgent need to develop alternative polymeric materials from sustainable sources. In this study, lauryl acrylate (LAc) and stearyl acrylate (SAc), derived from vegetable oils, were polymerized with reversible addition-fragmentation chain transfer polymerization, and subsequently extended for the preparation of poly(styrene-*b*-(LAc-*co*-SAc)-*b*-styrene) triblock copolymers. The thermal and mechanical properties of polyacrylate random copolymers were readily tuned by variation of the acrylate composition and resulting distribution of side-chain lengths. The order-disorder transition temperatures of the triblock copolymers were surprisingly independent of the mid-block composition, implying the Flory-Huggins interaction parameter is also independent of the acrylate side-chain length. Structural analysis revealed the non-equilibrium spherical morphologies of the triblock copolymers, which transformed to cylindrical microstructures under large amplitude oscillatory shear with the cylinders aligned transverse to the shear flow direction. Tensile testing experiments indicated that these triblock copolymers exhibit elastomeric behavior at room temperature.

# 379939 Invited Talk: Using Light to Locally Tune the Properties of Polymers

**Abstract:**

This talk will describe two related projects that utilize generic light sources to induce localized changes to the physical and mechanical properties of polymers in a strategic manner. The first project utilizes photopolymerization to change the properties of the walls of microfluidic channels. Poly(dimethylsiloxane) is one of the most popular materials for microfluidics, but swells in the presence of most organic solvents and is therefore limited to handling aqueous fluids. We show that it is possible to diffuse photoinitiator into the walls of the channel to localize photopolymerization that results in a thin shell that is impervious to organic solvents due to crosslinking. The remainder of the polymer that composes the microchannel retains the desirable properties of siloxanes. The second project utilizes uniform light from an IR heat lamp to induce self-folding of polymer sheets. Regions of black ink patterned on sheets of pre-strained polymer absorb the light and heat up due to the photothermal effect. The heat causes the polymer beneath the ink to relax and therefore shrink, which can induce self-folding. Self-folding is similar to origami but is done without direct human intervention. The appeal of our approach is that it utilizes two dimensional patterns to create a variety of complex three dimensional objects. In essence, it allows shape memory polymers to be programmed using a variety of two dimensional patterning techniques and triggered using only light. The talk will discuss the fundamental science that enables both of these projects, which are united by the importance of localized changes in polymer properties initiated by exposure to light.

# Neurotransmitter Detection Using Corona Phase Molecular Recognition on Fluorescent Single-Walled Carbon Nanotube Sensors

Temporal and spatial changes in neurotransmitter concentrations are central to information processing in neural networks. Therefore, biosensors for neurotransmitters are essential tools for neuroscience. In this work, we applied a new technique, corona phase molecular recognition (CoPhMoRe), to identify adsorbed polymer phases on fluorescent single-walled carbon nanotubes (SWCNTs) that allow for the selective detection of specific neurotransmitters, including dopamine. We functionalized and suspended SWCNTs with a library of different polymers ($n = 30$) containing phospholipids, nucleic acids, and amphiphilic polymers to study how neurotransmitters modulate the resulting band gap, near-infrared (nIR) fluorescence of the SWCNT. We identified several corona phases that enable the selective detection of neurotransmitters. Catecholamines such as dopamine increased the fluorescence of specific single-stranded DNA- and RNA-wrapped SWCNTs by 58–80% upon addition of 100 μM dopamine depending on the SWCNT chirality (n,m). In solution, the limit of detection was 11 nM [Kd = 433 nM for $(GT)_{15}$ DNA-wrapped SWCNTs]. Mechanistic studies revealed that this turn-on response is due to an increase in fluorescence quantum yield and not covalent modification of the SWCNT or scavenging of reactive oxygen species. When immobilized on a surface, the fluorescence intensity of a single DNA- or RNA-wrapped SWCNT is enhanced by a factor of up to $5.39 \pm 1.44$, whereby fluorescence signals are reversible. Our findings indicate that certain DNA/RNA coronae act as conformational switches on SWCNTs, which reversibly modulate the SWCNT fluorescence. These findings suggest that our polymer–SWCNT constructs can act as fluorescent neurotransmitter sensors in the tissue-compatible nIR optical window, which may find applications in neuroscience.

**Title:**
Photo-switchable quantum dots based on reversible FRET and azobenzene DNA

**Abstract:**

Super-resolution fluorescence microscopy has become a powerful tool to resolve biological structures smaller than the diffraction limit of light microscopy (~200nm). Nonetheless most super-resolution techniques (STORM/PALM, STED) use photo-switchable fluorescent probes (i.e., dyes and fluorescent proteins) that are limited in brightness and photostability, which reduces potential image resolution. Here, we describe photo-switchable quantum dots (QDs) with enhanced brightness and stability, as candidates for super-resolution imaging techniques. These QDs are composed of one green QD, one gold nanoparticle (AuNP), and complimentary single stranded DNA (ssDNA) conjugated to each particle. The DNA is modified with azobenzene groups that are responsive to light of different wavelengths. Under visible light, the azobenzene groups adopt a *trans* configuration and the two pieces of ssDNA hybridize, which tethers the two nanoparticles, quenching the QD fluorescence through Förster resonance energy transfer (FRET), a highly distance dependent mechanism (i.e., $\alpha$ $1/r^6$, in this case, a few nanometers). The azobenzene groups switch to the *cis* form when excited by UV light, which disrupts the DNA double helix, and the two particles diffuse from each other, thus restoring QD fluorescence. This design displays stochasticity in the on/off switching process, which is a key requirement for single-molecule localization-based super-resolution imaging techniques (i.e., STOMR/PALM), and is absent in previous light responsive QDs.

# 373375 Selective Intracellular Labeling Using Microfluidic Electroporation-Delivered Quantum Dots

**Abstract:**

Quantum dots (QDs) are luminescent semiconductor nanocrystals that have extraordinary brightness and exceptional photostability. Due to these superiorities, QDs have been widely used as a replacement of organic dyes as fluorophores for studying cell biology, particularly in long-term observation and single particle tracking. However, QDs without specific coating are unable to penetrate the cell membrane due to their size and surface properties. Currently, QDs with cell penetrating coating are mostly used for delivering, which cause the encapsulation of QDs in endosomes. Microinjection is a way to overcome this restriction, but it requires single cell manipulation and complicated operation which results in a very limited throughput. Our studies revealed that electroporation deliveries QDs into cells with high efficiency and these QDs are available for subcellular targeting and tracking.

In our project, we used electroporation as an efficient method for QDs delivery and intracellular targeting in microfluidic devices. Electroporation has been commonly used for gene transfer and drug delivery in vitro and in vivo. Microfluidic chips provide us convenient platform for culturing, operating and viewing cells, as well as conducting in vitro electroporation. Several groups have reported the delivery of QDs by electroporation, but no systematical research has been conducted. Furthermore, no study of intracellular targeting with electroporation-delivered QDs has been conducted. We demonstrated that electroporation deliveries QDs into cells independent of endocytosis, and their distribution within the cytoplasm is homogeneous. We also systematically studied, characterized and optimized the microfluidic electroporation conditions for QDs delivery and cell viability. With cytosolic dispersed QDs that have been conjugated with anti-kinesin antibody, we were able to visualize the dynamic motion of intracellular kinesin, which is a linear motor protein supports transportation of cellular cargo. We believe this is the first tracking of endogenous molecules in living cells with microfluidic electroporation-delivered QDs. Hence, our results indicate that electroporation may serve as a powerful tool to delivery antibody coated QDs into cells for single molecular tracking within living cells.

# 384406 Biochemical Ratiometric Sensing By Single Chirality Carbon Nanotubes in Living Tissues

**Abstract:**

Single walled carbon nanotubes (SWNT) are well positioned to act as single particle sensors in living organisms as they do not photobleach and fluoresce in the near infrared (NIR) in which tissues are relatively transparent. However, lack of specificity of SWNT fluorescence response to biochemical compounds has limited their applications as sensors *in vivo*. Here we show that SWNT localized inside leaves enable a more selective ratiometric sensing of plant free radicals and pollutants like nitric oxide and hydrogen peroxide. Using a novel method for rapid and efficient coating exchange of single chirality SDS-SWNT, we synthesized the first SWNT ratiometric sensor with 6,5 and 7,6 chiralities coated with single stranded DNA and Polyvinyl alcohol. These ratiometric platform was designed for 7,6 SWNT NIR fluorescence to quench in the presence of the free radicals while being unresponsive in the 6,5 SWNT chirality. Thus creating a multimodal NIR spectral signature that increases selectivity to biochemicals generated by plants or pollutants in the environment. Spatial and temporal patterns of the ratio sensor nIR fluorescence in the leaf lamina in response to exogenously and endogenously generated nitric oxide and hydrogen peroxide were monitored *in vitro* and imaged *in vivo* in real time. This nanobionic approach of interfacing nanosensors with leaves will lead to the development of cost effective, selective, and stable plant biochemical detectors.

# 375633 Si Nanocrystal Surface Chemistry: Room Temperature Hydrosilylation and Thermally-Induced Thiolation

**Abstract:**

Silicon (Si) nanocrystals are heavy-metal free, light-emitting nanomaterials that could be used in applications such as biomedical imaging and light emitting devices. They exhibit stable, bright, size-tunable photo- and electroluminescence. To prevent surface corrosion and degradation, Si nanocrystals are typically passivated with covalently bound ligand monolayers to eliminate surface dangling bonds and provide dispersibility in solvents. One common approach to ligand passivation is via hydrosilylation between surface Si-H species and terminal alkenes. Hydrosilylation requires either high temperature or photon irradiation, which can cause unwanted side reactions in certain cases, such as derivatization with bio-active molecules or other temperature/light sensitive ligands. We have discovered a path to room temperature hydrosilylation of Si nanocrystals that does not require light irradiation or added catalyst. The hydrosilylation reaction is promoted by polar ester or carboxylic acid groups, which facilitate direct nucleophilic attack of the highly curved Si surface with unsaturated C-C bonds, leading to covalent Si-C linkages to the hydrocarbon capping ligands. As a proof of concept, the mild hydrosilylation conditions were utilized for styrene passivation of Si nanocrystals without thermally-induced self-polymerization of styrene as a side reaction. Besides Si-C bonds, we have found that Si-S bonds can also be formed to passivate Si nanocrystals, through thermally-induced thiolation. In contrast to a Si-C passivated Si nanocrystals, which are stable in water for several months, Si-S passivated Si nanocrystals are extremely sensitive to moisture.

# 387242 Detection of a Peanut Allergen, Ara h1, on a Novel, Biodegradable Biosensor Platform Using Surface Enhanced Raman Spectroscopy

**Abstract:**

Food allergy is one of the major health issues related to food as the consequences can be lethal for susceptible populations. Peanuts (*Arachis hypogea*) are one of the leading agents of food allergy. Among the eight (*Ara h1* to h8) identified peanut allergen proteins, *Ara h1* has been found to affect up to 100% of the allergic patients [1]. Since allergens should always be monitored for cross-contamination, development of rapid, on-site diagnostic tools is very beneficial for the food industry [2]. Due to their high speed, ease of use and high degree of automation, biosensors present an alternative solution to the allergen detection problem. For this reason, a nano-biosensor that utilizes Surface Enhanced Raman Spectroscopy (SERS) was investigated to detect and identify *Ara h1* allergen protein. SERS makes use of noble metals, mostly gold and silver, with a rough feature to create intensified Raman signals, which gives molecular level information of analytes. Roughness is generally created by either nanoparticles or regular nano-patterns. Here we utilize this highly specific technique to detect *Ara h1*. The novelty of the biosensor comes from the unique biomaterial that it is made of, zein. Zein is a corn prolamin, which has shown to form free-standing films and capable of replicationg micro and nano-scale features [3,4]. In this research, biosensors were obtained by direct transfer of 3D structures (pyramids) with a gold coating onto zein films [5]. By using Raman spectroscopy, the signature of *Ara h1* protein was identified and a calibration curve was obtained for different concentrations. We propose that in the future SERS technique could be used to identify and monitor allergen proteins with the help of an environmentally friendly platform of zein.

## References

[1]. Burks, A.; Williams, L. W.; Connaughton, C.; Cockrell, G.; O'Brien, T. J.; Helm, R. M. Identification and characterization of a second major peanut allergen, *Ara h* II, with use of the sera of patients with atopic dermatitis and positive peanut challenge. *J. Allergy Clin. Immunol.,* **1992**, *90*, 962-969.

[2]. McGrath, T.; Elliott, C.; Fodey, T. Biosensors for the analysis of microbiological and chemical contaminants in food. *Analytical and bioanalytical chemistry,* **2012**, *403*, 75-92.

[3] Luecha, J.; Hsiao, A.; Brodsky, S.; Liu, G. L.; Kokini, J. L. Green microfluidic devices made of corn proteins. *Lab on a Chip*, **2011**, *11, 3419-3425.*

[4] Altunakar, B.; Luecha, J.; Kokini, J. In In Fabrication of biodegradable zein films by using soft lithography; *Proceeding Nanotechnology Conference, Anaheim*, **2010**, *2, 253-256.*

[5] Hsiao, A.; Gezer P.G.; Kokini, J.L.; Liu G.L. Green Nanophotonic Sensor Made by Direct Transfer of Three-Dimensional Metallic Nanostructures onto Corn Protein Film, *J. Nanotechnology* **2014**, *submitted*

# 389692 Spatiotemporal Intracellular Nitric Oxide Signaling Captured Using Internalized, Near Infrared Fluorescent Carbon Nanotube Nanosensors

**Abstract:**

Nitric oxide (NO), a biologically important reactive nitrogen species, is implicated in numerous pathologies but is also vital to the proper function of the vascular, nervous and immune systems. These paradoxical roles of NO have yet to be fully explained, but are hypothesized to be governed by local NO concentrations within tissues and cells. This article describes a state-of-the-art technology, using an intracellular near infrared fluorescent single-walled carbon nanotube (SWCNT) sensors, from which spatial and temporal detection of NO can be monitored at a single cell level with a resolution superior to any existing method. This sensor allows for the evaluation of intracellular NO gradients and dynamics for the first time with results suggesting the presence of local intracellular sources and sinks of NO linked to cellular machinery involved in signaling. With this assay, intracellular transient phenomena involved in disease and healthy cell processes can now be documented and characterized.

# 390530 A Facile Colorimetric Sensor of Ionizing Radiation Using Polypeptide Mediated Nanoparticle Formation

**Abstract:**

Widespread use of ionizing radiation, including X-rays and gamma rays, in food technology, imaging and cancer therapy necessitates precise control over dose, which in turn motivates the development of sensors that can detect various energy levels. A variety of dosimeters, including one-dimensional ion-chambers have been explored for different applications that require ionizing radiation. However, most available sensors suffer from drawbacks including portability and conformability. A potential solution to this problem lies in the development of liquid-phase dosimeters and sensors. We have developed a polypeptide-based sensor which changes color upon irradiation with ionizing radiation. In this approach, exposure to ionizing radiation results in radiolysis of water, resulting in the formaiton of free radicals. These, in concert of polypeptides, result in reduction of colorless metal (e.g. gold) salts to colored plasmonic nanoparticle dispersions. Polypeptides are hypothesized to act as both, templating as well as stabilizing agents in nanoparticle formation. The effect of radiation dose as well as kinetics of nanoparticle formation were investigated. The nanoparticles formed were extensively characterized using TEM, DLS and elemental analysis studies. Our studies on polypeptide-templated plasmonic metal nanoparticle formation indicate the possibility of developing a new class of dosimeters which can be used to detect ionizing radiation using a straightforward colorimetric method.

# 390162 Self-Assembled Polymeric Nanoparticles Encapsulating Hydrophobic Curcumin to Attenuate Morphine Tolerance

**Abstract:**

Self-Assembled Polymeric Nanoparticles Encapsulating Hydrophobic Curcumin to Attenuate Morphine Tolerance

*Magdalena Szymusiak[1], Hao Shen[1], Xiaoyu Hu[2], Zaijie Jim Wang[2,*], and Ying Liu[1,2,*]*

[1] Department of Chemical Engineering, University of Illinois at Chicago, Chicago, IL 60607, USA

[2] Department of Biopharmaceutical Sciences, University of Illinois at Chicago, Chicago, IL 60607, USA

* To whom correspondence should be addressed: liuying@uic.edu, zjwang@uic.edu

**Abstract**

Curcumin [1,7-bis-(4-hydroxy-3-methoxyphenyl)-1,6-heptadiene-3,5-dione] is the active constituent of *Curcuma longa* - a yellow-colored traditional herb. It was recently found to have a wide spectrum of pharmacological activities, including antioxidant, cancer chemopreventive, anti-inflammatory, neuro-protective and antinociceptive actions. However, curcumin's bioavailability is limited by low solubility, high rate of metabolism, and rapid clearance from the body. We have formulated this hydrophobic compound into stable nanoparticle suspensions (nano-curcumin) to overcome these difficulties. Employing the curcumin nanoformulations as the platform, we also discovered that curcumin has the potential to alleviate morphine tolerance. The three types of stable polymeric nanoparticles: (1) poly(lactic-co-glycolic acid) (PLGA), (2) poly(ethylene glycol)-*b*-poly(lactic acid) (PEG-*b*-PLA), and (3) the hybrid of the two were generated using flash nanoprecipitation integrated with spray or freeze drying. The process is scalable and reproducible, and the optimized formulations have high drug loading (50%), small particle size (100-200nm), and controlled surface properties. Mice behavioral studies (tail-flick and hot-plate tests) were conducted to verify the effects of nano-curcumin on attenuating morphine tolerance. Significant analgesia was observed in mice during both tail-flick and hot-plate tests using orally administrated nano-curcumin following subcutaneous injections of morphine. However, unformulated curcumin at the same dose showed no effect. Furthermore, the initial pharmacokinetic analysis revealed that nano-curcumin increased the oral bioavailability of the compound in blood plasma more than 80 fold.

**Keywords:** curcumin, polymeric nanoparticles, flash nanoprecipitation, oral delivery, drug tolerance

## Abstract

The rapidly diminishing number of effective antibiotics in a physicians' arsenal will have a drastic impact on how well infectious diseases and complications are able to be managed in the clinic. According to the Center for Disease Control's 2013 report on antibiotic resistance in the United States, at least 2 million people acquire serious infections from antibiotic resistant bacteria each year, and over 23,000 die as a direct result. Even when alternative treatments exist, patients with antibiotic resistant infections have significantly higher mortality rates, and survivors often have increased hospital stays and long-term complications. These infections cost an estimated $20 billion in excess direct healthcare costs to treat.[1] This study explored the development and optimization of a polymersome nanocarrier formed from a biodegradable diblock copolymer to overcome this resistance. The polymersomes were designed with silver nanoparticles embedded in the particle's hydrophobic membrane bilayer, and an antibiotic solution encapsulated in the particle's aqueous core. This provides a dual-mechanism, concentrated, and localized treatment. Here, these silver nanoparticle-embedded polymersomes (AgPs) were loaded with ampicillin and subsequently tested for bactericidal function against *Escherichia coli* that had been transformed with a gene for ampicillin resistance (*bla*). The AgPs killed the antibiotic-resistant bacteria, whereas free antibiotic, encapsulated antibiotic without the addition of the silver nanoparticles, and AgPs without the addition of ampicillin did not kill the bacteria.


## Introduction

Antibiotic drugs have been extensively used since their discovery in the 1940s to treat patients suffering from a wide variety of infectious diseases. When utilized correctly, these drugs are extremely effective at reducing patient illness lengths and mortality rates, which makes them invaluable in the clinic. Unfortunately, however, antibiotics have been used so prevalently over the last 70 years that the bacterial organisms they were designed to kill have begun to evolve and adapt, rendering these drugs ineffective.[2,3] Infections caused by Gram negative bacteria are particularly difficult to treat because their robust outer lipopolysaccharide membrane helps to impede the influx of antibiotics into the cell.[4] Of additional concern is the appearance of bacterial strains that are resistant to multiple types of antibiotics (known as multi-drug resistant or MDR strains). Recently, scientists and clinicians have even begun to notice that some of these strains have become so antibiotic-resistant that no available drug can successfully treat the infections they cause.[1] Thus, the need to develop alternative treatment strategies has never been more urgent.

## Materials and Methods

A diblock copolymer of methoxypoly(ethelyne glycol) (mPEG) and poly(*D)*-(*L)*-lactic acid (PDLLA) was utilized for the polymersome synthesis. The mPEG block has been documented to confer a "stealth" property *in vivo* in order to help prevent the particles from being cleared prematurely by the immune system.[5] In addition, the racemic mixture of *D*- to *L*-lactides in the PDLLA block has been optimized to generate polymersomes whose release rate is sensitive to changes in temperature.[6] First, hydrophobic silver nanoparticles (5 ± 2nm) were resuspended in organic solvent, and were ultrasonicated to prevent aggregation. The copolymer was added to the mixture, which was again ultrasonicated until the copolymer was completely dissolved. This nanoparticle/polymer solution was then injected through an atomizer into a saline solution (with or without ampicillin). Finally, the entire polymersome solution was transferred to a dialization tube and allowed to dialyze against pure saline for 48 hours to remove all traces of the organic solvent.

*Escherichia coli* (*E. coli*) is a Gram-negative, rod-shaped bacterium which has been extensively investigated in the laboratory for over 60 years, making it one of the most widely studied prokaryotic organisms and thus ideal for a proof-of-concept application. First, *E. coli* (strain K-12) was transformed with a plasmid containing the *bla* gene encoding for the enzyme TEM-1 ß-lactamase using the heat-shock method. TEM-1 is the most common ß-lactamase found in enterobacteriaceae, and confers resistance to narrow-spectrum cephalosporins, cefamandole, cefoperazone, and all the penicillins except temocillin.[7] After the transformation, the *E. coli* solution was streaked for inoculation on a tryptic soy agar + ampicillin plate. For each experimental trial, a single bacterial colony was selected and grown overnight in Lysogeny broth (LB) on a shaking incubator set at 200 rpm and 37ºC.

## Results and Discussion

Results show that the ampicillin-loaded AgPs display significant bactericidal action against *E. coli* with a genetic resistance to ampicillin. At treatment concentrations above 15%, the AgPs completely inhibited all bacterial growth. At sub-lethal concentrations, the treatment still delayed bacterial proliferation time. For every 2.5% increase in treatment concentration, the time for the bacteria to reach exponential and stationary phase was delayed ~2-3 hours (Figure 1 and Table 1). Transmission electron microscopy confirmed interaction between the bacteria and AgPs (Figure 2). Free ampicillin, PBS loaded AgPs, and ampicillin loaded Ps displayed no inhibition of bacterial growth at any concentration (graphs omitted for brevity).

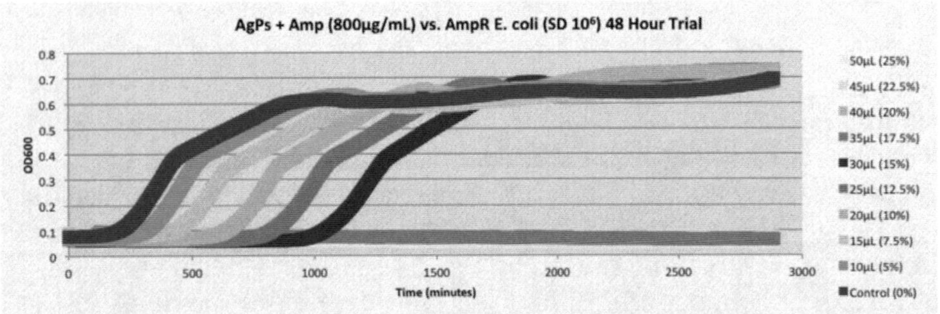


**Bacterial Growth Time from Figure 1**

| Treatment Concentration | Time to Exponential Phase | Time to Stationary Phase |
|---|---|---|
| **0µL (Control)** | 2.50 hrs | 15.17 hrs |
| **10µL** | 4.20 hrs | 17.77 hrs |
| **15µL** | 6.53 hrs | 19.07 hrs |
| **20µL** | 9.50 hrs | 23.03 hrs |
| **25µL** | 12.67 hrs | 27.13 hrs |
| **30µL** | 15.77 hrs | 30.57 hrs |
| **35µL** | DNR | DNR |
| **40µL** | DNR | DNR |
| **45µL** | DNR | DNR |
| **50µL** | DNR | DNR |

**Figure and Table 1. AgPs were synthesized from an 800µg/mL ampicillin solution and tested against the ampR *E. coli*. 100µL of LB containing $10^6$ CFU/mL bacteria were seeded into wells with various concentrations of AgPs and PBS (total well volume of 200µL). Bacterial growth was monitored for 48 hours by measuring the $OD_{600}$ of the wells every 2 minutes.**

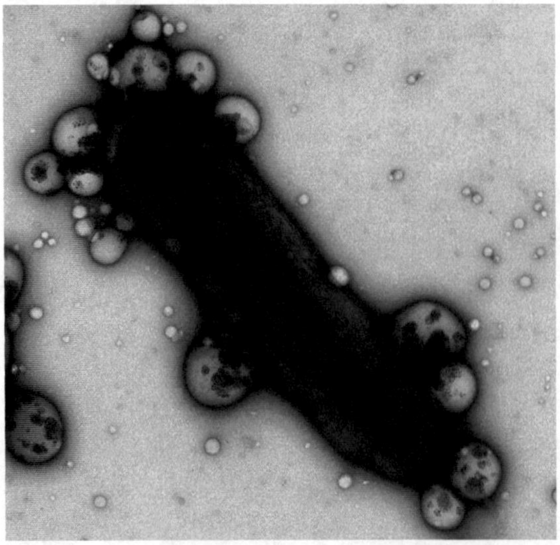

**Figure 2. TEM micrograph (x12,000) shows interaction between the *E. coli* and AgPs. Prior to imaging, the solution was fixed with 4% paraformaldehyde and stained with 1.5% uranyl acetate.**

## Conclusions

Prevention of access of the drug is critical to antibiotic resistance, especially in Gram negative bacteria because of their robust outer lipopolysaccharide membrane. Here, we show that the combined, synergistic delivery of silver nanoparticles and ampicillin enhanced bacterial cell death, even when the gene for ampicillin resistance was present. The specific mechanisms of action for the observed results are currently under investigation by the authors. This work has many important implications. A long-lasting, dual-mechanism, concentrated treatment in one particle able to overcome antibiotic resistance would be extremely beneficial in the clinic, both as a systemic treatment and as a coating for implanted medical devices. The designed particle is also easily customizable, allowing for the potential of personalized infection treatments.

## Acknowledgements

This work was supported by the IGERT Nanomedicine Science and Technology program at Northeastern University, NSF/DGE-096843.

## References

[1] Antibiotic Resistance Threats in the United States, 2013. Centers for Disease Control and Prevention.

[2] Taubes G. The bacteria fight back. Science. 2008;321:356.

[3] Rice, L. B. Emerging issues in the management of infections caused by multidrug-resistant gram-negative bacteria. *Cleve Clin. J. Med.* **74**, S12–S20 (2007).

[4] Pages JM, James CE, Winterhalter M. The porin and the permeating antibiotic: a selective diffusion barrier in Gram-negative bacteria. Nature Reviews Microbiology. 2008;6:893.

[5] Scott, MD, Murad KL, Koumpouras F, Talbot M, Eaton JW. Chemical camouflage of antigenic determinants: Stealth erythrocytes. *Proc. Natl. Acad. Sci.* 1997;94:566-7571.

[6] Santin M. Strategies in Regenerative Medicine: Integrating Biology with Materials Design. Springer. 2009:62.

[7] Matagne A, Lamotte-Brasseur J, Frere JM. Catalytic properties of class A beta-lactamases: efficiency and diversity. *Biochem. J.* 1998;330(2):581-598.

THE
UNIVERSITY
OF RHODE ISLAND
COLLEGE OF
ENGINEERING

# Tumor-Penetrating Acetalated Dextran Nanoparticles Capable of Tandem Delivery of Agents for the Treatment of Lung Cancer

Qihua Sun[1], Samantha A. Meenach[1,2]

[1]Department of Chemical Engineering, [2]Department of Biomedical & Pharmaceutical Sciences
University of Rhode Island, Kingston, RI 02881 USA

November 20, 2014

THINK BIG WE DO™

## Background Information: Significance

**Impact of Non-Small Cell Lung Cancer (NSCLC)**
**Five-year survival rates of patients diagnosed with NSCLC**

| Stage | 5 Year Survival Rate |
|---|---|
| 0 | Too few cases to evaluate |
| I | 54% |
| II | 35% |
| IIIA | 10 - 15% |
| IIIB | < 5% |
| IV | < 2% |

***These dismal statistics show the importance in improved treatment for patients diagnosed with NSCLC throughout the course of their disease***

THINK BIG WE DO™

Resource from the National Cancer Institute's Physician Data Query System, July, 2002

THE
UNIVERSITY
OF RHODE ISLAND

## Background Information: Anticancer Agents

| Anticancer Agent | Paclitaxel (PTX) | Imatinib mesylate (IM) |
|---|---|---|
| Trade Name | **Abraxane** (paclitaxel protein-bound nanometer particles for injectable suspension) | **Gleevec** (also known as STI571) |
| Treatment Use | First-line treatment of locally advanced or metastatic NSCLC, in combination with carboplatin | Chronic myelogenous leukemia, several types of GI tumors |
| Current Research | Nanocarriers loaded with Paclitaxel and its combination with other anticancer agents. | Inhibition of lung cancer cell growth and its synergistic effect with other anticancer agents |

**Paclitaxel**

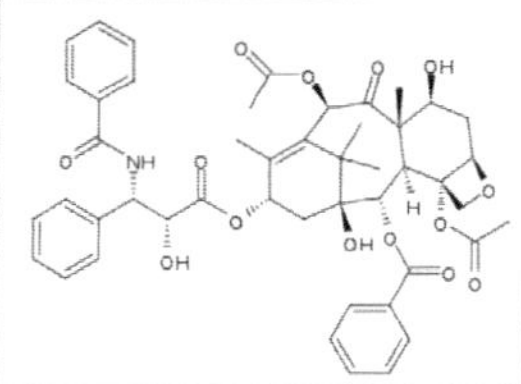

**Imatinib**

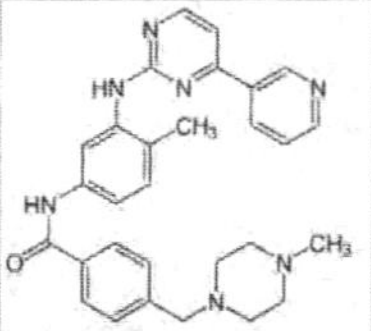

THINK BIG WE DO™

THE UNIVERSITY OF RHODE ISLAND

Resource from http://abraxane.com/hcp/advanced-non-small-cell-lung-cancer/advanced-nsclc/epidemiology/
http://www.nature.com/scitable/topicpage/gleevec-the-breakthrough-in-cancer-treatment-565
Zhang, Peilin, et al., Molecular cancer 2.1 (2003): 1.

## Background Information: Acetalated Dextran (Ac-Dex)

- Ac-Dex = polymer used in nanoparticle synthesis
- Acid-responsive, biodegradable material
- Easy to synthesize

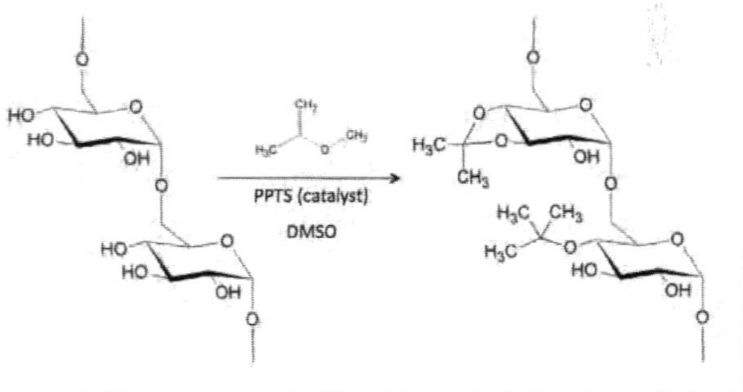


**Comparison Between Ac-Dex and PLGA (FDA-approved polymer):**

| Acetalated Dextran (Ac-Dex) | Poly (lactic-co-glycolic acid) (PLGA) |
|---|---|
| No significant difference in toxicity between the two materials in both HeLa cells and RAW macrophages | |
| pH-neutral byproduct of dextran + an alcohol + acetone | Acidic byproducts that may alter the local pH |
| Acid-sensitive = a broader range of tunable degradation rates | Degradation does not conform to a simple model and depends on many other parameters |

THINK BIG WE DO™

THE UNIVERSITY OF RHODE ISLAND

Uhrich, Kathryn E., et al., Chemical reviews 99.11 (1999): 3181-3198.
Meenach, Samantha A., et al., Molecular pharmaceutics 9.2 (2012): 290-298.
Bachelder, Eric M., et al., Journal of the American Chemical Society 130.32 (2008): 10494-10495.

Background Information

## Tumor-Penetrating Peptide: iRGD

- iRGD = CRGDKRGPCD
- Enhances tumor penetration
  - Targeting occurs via homing to integrins overexpressed in cancer tissue
  - Penetration occurs after peptide cleavage and interaction with neuropilin-1

- **Advantages of iRGD:**
  - No need for the drug or particle to be conjugate with iRGD
  - Can be applied to many systems: small molecule, nanoparticles and monoclonal antibodies
  - Co-injection of iRGD results in significant increased effect in anticancer drug in *in vitro* studies

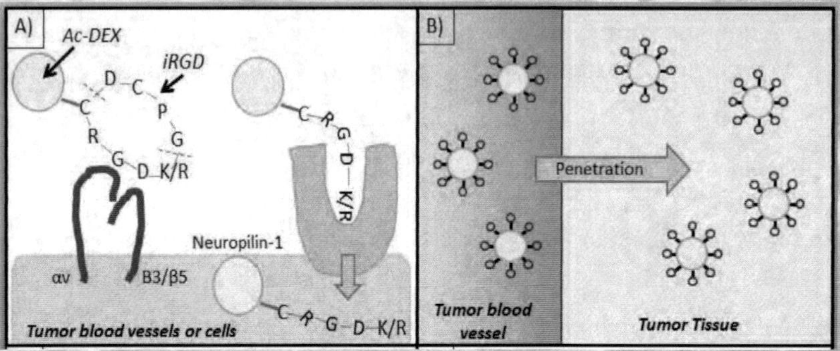




Sugahara, Kazuki N., et al., science 328.5981 (2010): 1031-1035.
Sugahara, Kazuki N., et al., Cancer cell 16.6 (2009): 510-520.



OBJECTIVES

**Overall:**

To develop and optimize peptide-conjugated Ac-Dex nanoparticles capable of encapsulating paclitaxel and imatinib to enhance the targeting and treatment of non-small cell lung cancer.

**Specific:**

- To synthesize acetalated dextran-based iRGD-conjugated nanoparticles loaded with paclitaxel and/or imatinib mesylate
- To evaluate and optimize the physicochemical properties the peptide-conjugated Ac-Dex nanoparticles
- To assess the efficiency of the proposed nanoparticle system in the treatment of non-small cell lung cancer using *in vitro* methods

## Methods and Results: Synthesis and Characterization of Ac-Dex

**Schematic of Synthesis of Ac-Dex**

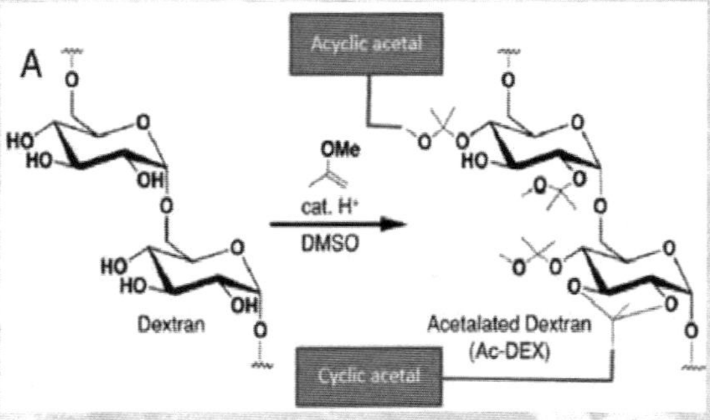


**$^1$H-NMR spectrum of Ac-Dex**

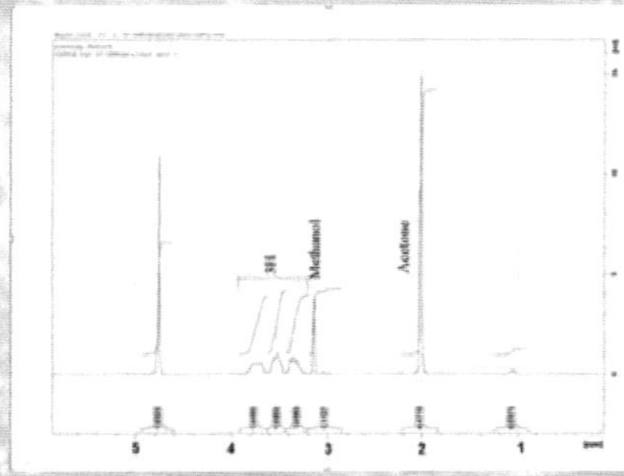


**Solid state NMR used to evaluate:**

- Cyclic acetal conversion (CAC)
- Proportion of converted hydroxyl groups

**CAC = 70%**
**-OH Conversion = 80%**

***CAC and –OH conversion values appropriate for Ac-Dex degradation within 48 hours (fast time)***

THINK BIG WE DO

THE UNIVERSITY OF RHODE ISLAND


Broaders, Kyle E., et al., Proceedings of the National Academy of Sciences 106.14 (2009): 5497-5502.
Bachelder, Eric M., et al., Journal of the American Chemical Society 130.32 (2008): 10494-10495.


## Methods and Results: Nanoparticle Synthesis and Characterization

**Nanoparticle Synthesis**

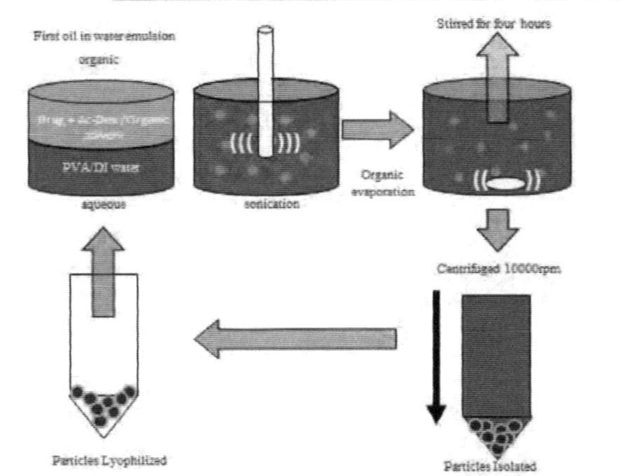


| Phase | 0.1% Paclitaxel | 10% Imatinib |
|---|---|---|
| Organic | 0.05 mg PTX + 50 mg Ac-Dex + 1 ml DCM | 5 mg IM + 50 mg Ac-Dex + 1 ml DCM |
| Water | 6 ml 3% PVA | |

**Scanning Electron Micrographs**

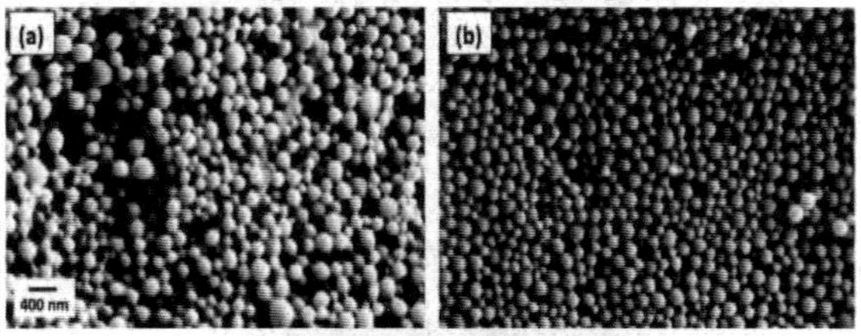

**(a) 0.1% PTX, (b) 10% Imatinib, and (c) Blank acetalated dextran nanoparticles**

***SEM images confirm successful particle formation → particles exhibit smooth, round morphology***

THINK BIG WE DO

THE UNIVERSITY OF RHODE ISLAND

## Methods and Results: Nanoparticle Synthesis and Characterization

- Dynamic light scattering (DLS) → size, polydispersity index (PDI), and zeta potential
- Ultra performance liquid chromatography (UPLC) → drug encapsulation and loading

| Particle System | Diameter (nm) | PDI | Zeta Potential (mV) | Encapsulation Efficiency (%) | Drug Loading (μg drug/mg NP) |
|---|---|---|---|---|---|
| **0.1% PTX** | 232 ± 13 | 0.21 ± 0.08 | -0.40 ± 0.02 | 50 ± 17 | 0.5 ± 0.17 |
| **10% Imatinib** | 282 ± 2 | 0.20 ± 0.02 | -0.30 ± 0.03 | 14 ± 3 | 14 ± 3 |
| **Blank** | 255 ± 25 | 0.19 ± 0.02 | -2.43 ± 0.13 | ---- | ---- |

- ***Size of nanoparticles → appropriate for nanoparticle targeting***
- ***Low polydispersity index → uniform size distribution***
- ***Neutral zeta potential***
- ***Appropriate EE and drug loading for effective drug delivery***

THINK BIG WE DO™

THE UNIVERSITY OF RHODE ISLAND

## Methods and Results: Drug Release

**Release Analysis**: 1 mg/ml NPs in PBS with surfactant, centrifugation at each time point

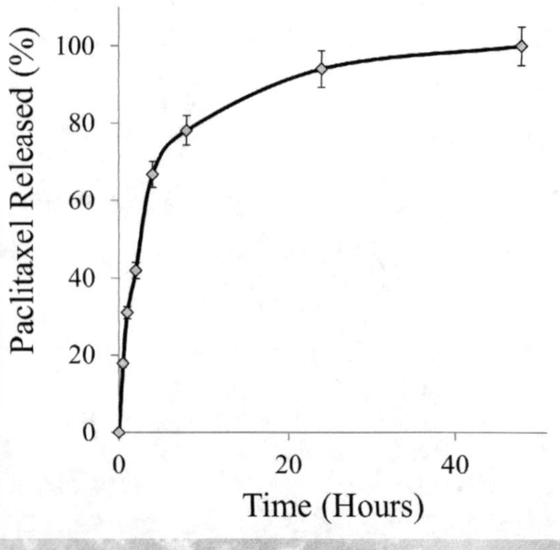


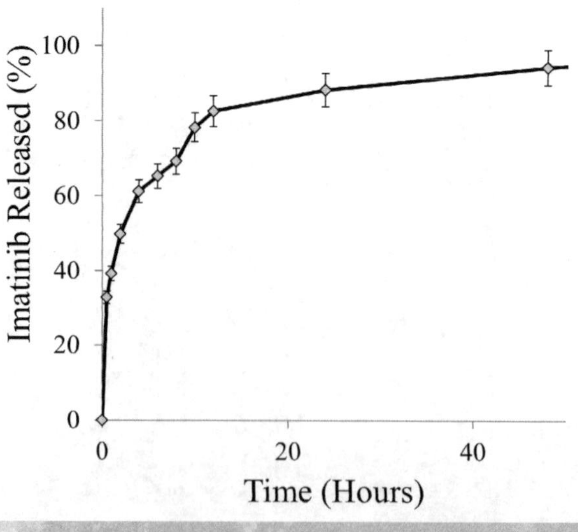


***PTX and IM-loaded nanoparticles exhibited sustained release over 24 hours***

THINK BIG WE DO™

THE UNIVERSITY OF RHODE ISLAND

## Methods and Results: *In Vitro* Biological Characterization

**Viability Assay/Dose-Response Curve:**

- A549 cells (lung adenocarcinoma)
- Exposure to PTX in DMSO or drug-loaded nanoparticles
- Viability → resazurin assay

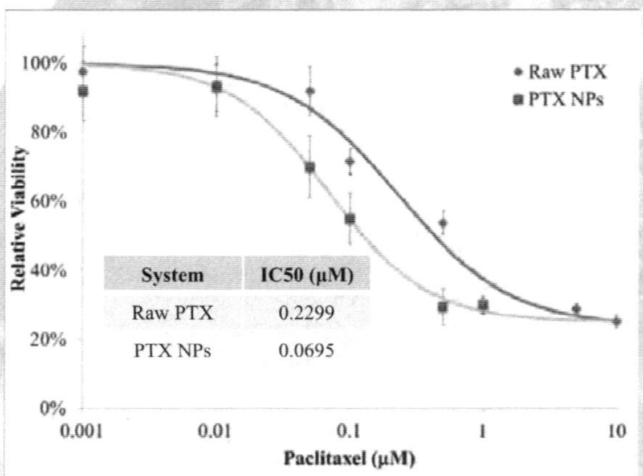


***PTX-loaded nanoparticles resulted in lower IC50 value → more effective than raw PTX in vitro***

**Nanoparticle Uptake Evaluation:**

- A549 cells (lung adenocarcinoma)
- Exposure to PTX-loaded nanoparticle (FITC encapsulation allows for fluorescence)
- Fluorescence imaging

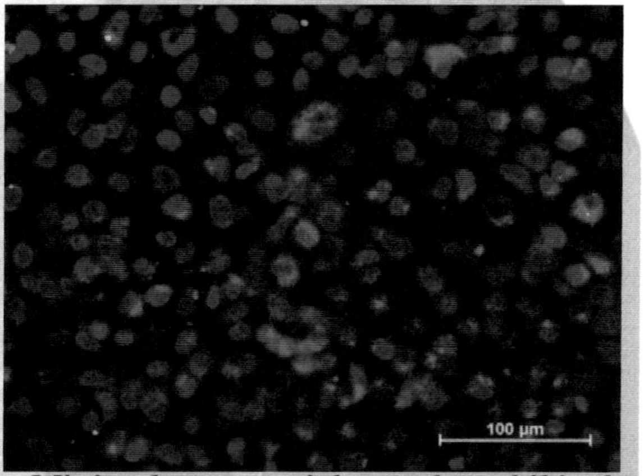


***Minimal nanoparticle uptake exhibited for non-iRGD NPs***

THINK BIG WE DO™

THE UNIVERSITY OF RHODE ISLAND

## Methods and Results: Generation of 3D Multicellular Spheroids

- A549 cells (lung adenocarcinoma)
- 20 μg/ml collagen (rat tail type I) added to media
- Formed in non-adherence round bottom 96 well plates
- 1500 rcf for 15 minutes

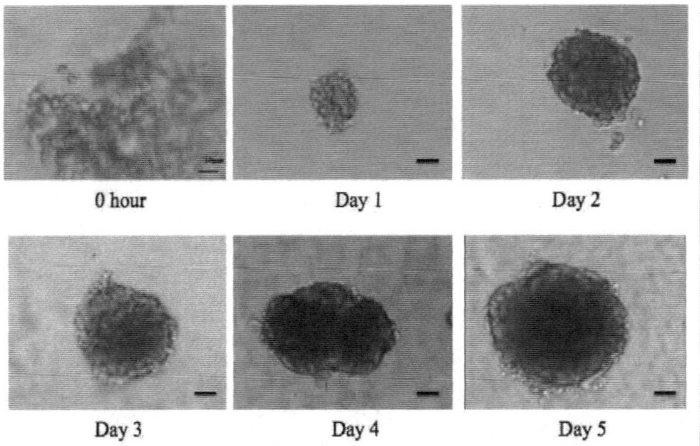


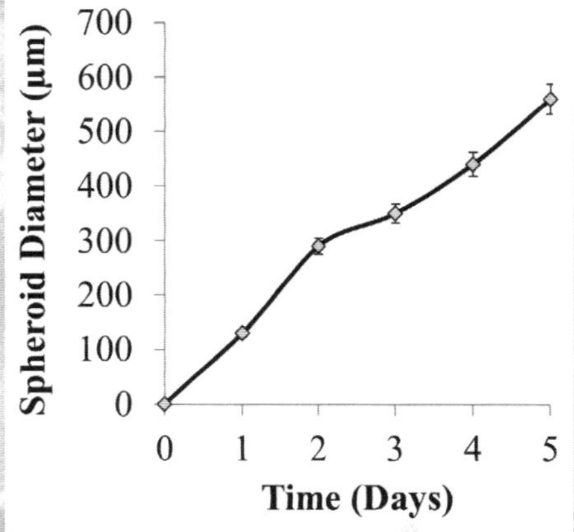


***3D spheroid formation of A549 lung adenocarcinoma successful***

***Will be used for future studies***

THINK BIG WE DO™

THE UNIVERSITY OF RHODE ISLAND

## Methods and Results: iRGD-Nanoparticle Conjugation and Affinity to A549 Cells

**iRGD-Nanoparticle Conjugation:**

- Alkoxyamine-terminated iRGD allows for conjugation with Ac-Dex
- 5 mg iRGD + 10 mg Ac-Dex NPs for 24 hours

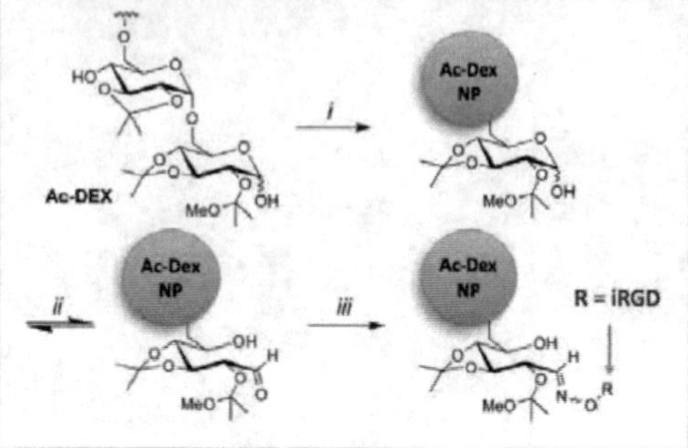


**80% of iRGD conjugated to NPs**

***iRGD was successfully conjugated to Ac-Dex nanoparticles***

**iRGD Affinity to A549 Cells**

- A549 cells seeded into 96 well plates
- Cells exposed to differing concentrations of fluorescently-labeled iRGD
- Fluorescence measured after washing

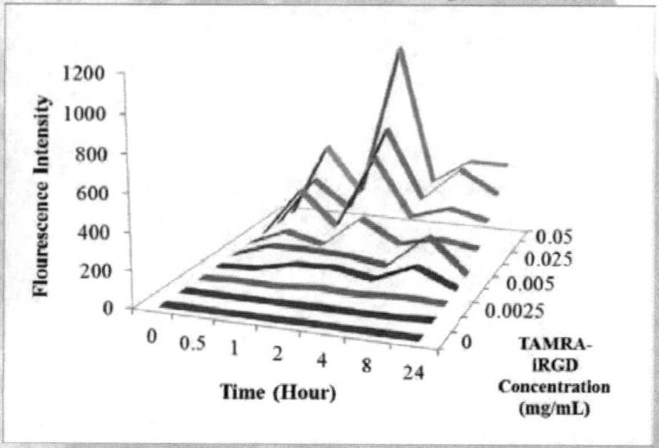


***iRGD exhibited greatest affinity for A549 cells at 4 hours***

THINK BIG WE DO™

THE UNIVERSITY OF RHODE ISLAND

## Conclusions

- Two kinds of drug-loaded nanoparticles which were composed of 0.1% paclitaxel (PTX) and 10% imatinib were successfully synthesized
- Physical and chemical characteristics of nanoparticles were evaluated for the application for potential the treatment of lung cancer
- *In vitro* cell studies showed that drug-loaded nanoparticles increases the efficacy of chemotherapeutic drugs (in particular paclitaxel) for A549 lung carcinoma cells
- Drug-loaded Ac-Dex nanoparticles could potentially be actively targeted to non-small cell lung cancer tumors and vasculature and then penetrate into the tumors cells with the help of tumor-penetrating peptide, iRGD

THINK BIG WE DO™

THE UNIVERSITY OF RHODE ISLAND

## Future work

- Optimization of nanoparticles size and encapsulation efficiency to achieve a better effect during *in vitro* studies
- Investigation of the methods of loading two drugs into one nanoparticle system to reduce side effects and allow for dual delivery
- Optimizing and evaluating the conjugation of iRGD onto blank and drug-loaded nanoparticle systems
- Investigation into the ability of iRGD-conjugated nanoparticles to increase the efficacy of non-iRGD nanoparticles against A549 cells on both 2D monolayer and 3D spheroid cultures
- Evaluating the ability of iRGD-conjugated nanoparticles to penetrate 3D multicellular spheroids

THINK BIG WE DO™

THE UNIVERSITY OF RHODE ISLAND

## Acknowledgements


- Meenach Research Group
  - Zimeng Wang
  - Elisa Torrico
  - Nick Fraunfelter
  - Stephanie Geiger
- Core Analytical Facilities:
  - RI-INBRE
  - RIN$^2$
- Funding:
  - RI-INBRE (2P20GM103430)
  - URI Council for Research

THINK BIG WE DO™

THE UNIVERSITY OF RHODE ISLAND

# Design and Synthesis of Aminoglycoside Antibiotic-derived Polymeric Micelles for Drug Delivery to Cancer Cells

Bhavani Miryala[a], Taraka Sai Pavan Grandhi[b], Matthew D. Christensen[a], Yanqing Tian[c], and Kaushal Rege[a,b]*

[a]Chemical Engineering, School for Engineering of Matter, Transport, and Energy, Arizona State University, Tempe, AZ 85287-6106, USA
[b]Harrington Biomedical Engineering, Arizona State University, Tempe, AZ 85287, USA
• [c] Center for Biosignatures Discovery Automation, Biodesign Institute, Arizona State University, Tempe, Arizona, USA.

## Abstract

In the constant increasing interest in developing drug delivery systems, polymeric micelles that are self-assembled core-shell structures has emerged as promising strategies. The possibility of structural variations of polymeric micelles makes them particularly well suited for drug delivery purposes. Nanoscale micelles, composed of an aminoglycoside polymer core and polyethylene glycol corona, were derived from the self-assembly of poly(ethylene glycol)-grafted aminoglycoside polymer (mPEG-PAG) in aqueous media. The mPEG-PAG polymer was synthesized by conjugating succinyl methoxy PEG to the free amines on aminoglycoside polymer. The synthesized polymer was well characterized using $^1$H NMR and FTIR spectroscopy for structural information and gel permeation chromatography for molecular weight. The critical micelle concentration of mPEG-PAG micelle, determined using pyrene fluorescence assay, was approximately 0.001 mg/mL, and the hydrodynamic diameter, determined using dynamic light scattering, was 30-40 nm. A zeta potential value of +1-6 mV indicated slightly cationic character consistent with aminoglycoside properties. Doxorubicin-loaded mPEG-PAG nanomicelles were prepared by solvent evaporation method; doxorubicin loading of 14.5 wt % and entrapment efficiency of 66-wt % was observed in mPEG-PAG micelles.The $\pi$-$\pi$ interaction between the DOX and the polymer backbone was studied by UV absorbance and Fluorescence methods. $IC_{50}$ values of Dox-loaded mPEG-PAG micelles indicated their efficacy for inducing death in several cancer cell lines. Taken together, our results indicate that these novel mPEG-PAG micelles are potent nanocarriers for drug delivery and can be particularly attractive in anti-cancer applications.

# 384596 Size Effect of Hollow Gold Nanoparticles on Photo-Activated Drug Release of a Liposomal Carrier

**Abstract:**

Liposome drug delivery systems are faced by the quandary of minimizing non-specific drug release and initiating fast release at the site of interest. A novel strategy is to separate the mechanism of drug release from drug retention by using an external agent to trigger drug release. Near infra-red (NIR) light is an appropriate choice as it is physiologically friendly with minimal thermal injury to normal tissues and light penetration of several cm. We can synthesize Hollow Gold Nanoparticles (HGN) to absorb NIR light over the range from 700-900 nm by manipulating the diameter to shell thickness ratio. NIR picosecond pulsed laser light absorbed by liposome encapsulated or tethered HGN is rapidly converted into thermal energy; the large temperature gradients lead to the formation of transient vapor nanobubbles in aqueous solution. The collapse of the nanobubbles ruptures cell, endosome, or liposome membranes; thereby releasing their contents with minimal damage to the contents or surroundings. We find smaller (10 – 15 nm) HGN show higher specific absorbance of NIR light than larger (30-50 nm) HGN, which leads to more efficient membrane rupture and delivery of liposome contents. We have systematically synthesized monodisperse HGN of sizes from 10 – 40 nm and determined the relative efficacy of laser power and duration on the rate of contents release from 200 nm liposomes. A fluorescent dye was encapsulated within a liposome as a model agent to study release kinetics triggered by a picosecond pulsed NIR laser irradiation of the HGN coupled to the liposome. Subsequently, an anticancer drug was encapsulated within the liposomes to deliver the drug to prostate cancer cell in vitro; near-complete cell killing was observed. The study provides design parameters for engineering drug delivery systems that ultimately will be apply to tumor targeting therapeutics.

# 385320 Single Lipoplex Fluorescent Nanoparticle Tracking Analysis and Applications for the Study of Exosomes

**Abstract:**

The study of exosomes, 40-100nm lipid bodies released by virtually all cells in the body, has seen tremendous growth in recent years. Long considered artifacts of cellular processing, new results have revealed that these bodies not only carry important cargo, but have the ability to interact specifically with surrounding cells and elicit functional changes in the target. Recent research has identified exosomes as key vehicles of intercellular communication, and characterization of the pathways and machinery they utilize to transfer information between cells is a rapidly expanding research field. Beyond cellular communication a number of studies have generated interest in exosomes as biomarkers for a variety of human diseases including cancers, Alzheimer's disease, HIV, and kidney disorders. The success of liposome encapsulated therapeutics has also led to proposals for producing therapeutic exosomes, wherein a desired cargo will be completely produced, enclosed in an exosome, and separated from a culture supernatant. With the explosion of interest in this field, the development of reliable techniques for isolating and characterizing exosomes is necessary and relevant. To date more than four thousand proteins and two thousand mRNAs have been identified in exosomes, however many studies have relied on the examination of hetergenous bulk exosome populations to produce results, leading to obscuration of markers expressed at low levels or in specific smaller subpopulations. Moving forward, the ability to identify specific proteins and RNAs at the single exosome level will be required, along with reliable methods of assessing separation schemes for isolating specific subpopulations from bulk exosome preparations. This presentation examines the application of nanoparticle tracking analysis (NTA) to identify and track single fluorescently labeled nanoparticles for both characterization and isolation purposes.

Nanoparticle Tracking Analysis (NTA) is a widely used technique for determining the size and number concentration of various submicron particles, including catalyst particles, viruses, protein aggregates, in addition to exosomes. NTA allows for the visualization and sizing of single nanoparticles in suspension by observing their Brownian motion from light scattering methods. However, impure preparations and lack of specific signaling can make this analysis prone to errors from contaminating particles in the exosome preparation. NTA can also track fluorescently labeled nanoparticles in the same manner, and this method in conjunction with an appropriate labeling method can provide a confirmation of particle identity. Herein we demonstrate the effect of fluoroprobe loading in self assembling lipoplex (anionic nucleic acid – cationic lipid complexes) particles on fluorescent NTA size and concentration measurements when compared to light scattering measurements of the same particles. Understanding the difference between these two methods and the associated phenomena is critical to proper application of NTA to the study of fluorescently labeled exosomes. In addition the lipoplex particles employed in this work have the ability to bind exosomes, and we demonstrate a method using NTA for determining the exosome binding capacity of these particles, which will find application in future techniques for separating and characterizing specific subpopulations of exosomes. The results of this presentation provide a method for determining the lower limit of fluorescence sensitivity in an NTA system, a value which must be known for effective fluorescent labeling of exosomes, an understanding of how measured results will differ between fluorescent and scattering observations, and a method of quantifying the exosome binding ability of particles for separations.

# 388549 Electronic Platform to Quantify Cellular Mechanisms Associated with Carbon Nanotubes Exposure in Real-Time

**Abstract:**

Despite the intriguing potential of carbon nanotubes (CNTs) for biomedical applications, their implementation is currently being hindered by many uncertainties regarding their toxicity and fate inside the biological systems. Several reports have shown that CNTs toxicity can be attributed to their length, surface chemistry, aggregation and metal impurities, just to name a few. However, due to the various types of CNTs being synthesized every day along with the different surface functionalization techniques, there is no fundamental understanding of the toxicological and pharmacological profiles of cellular systems exposed to CNTs with different physico-chemical properties. In this research, we provide a comprehensive analysis of the cellular behavior of human lung cells post exposure to CNTs using a combination of electronic and cell-based techniques. Our analyses rely on a non-invasive electronic cell impedance sensing (ECIS) platform to provide real-time measurements of cell adhesion, cell-cell interactions as well as changes in cellular morphology upon exposure to CNTs with user-defined physical and chemical properties. The approach is complemented by standard microscopy techniques as well as conventional *in vitro* cellular-based assays, such as cell cycle analysis, cell proliferation and viability, in order to derive structure-function relations associated with the cytotoxic and apoptotic events induced by exposure to CNTs. Our results correlated the different physico-chemical properties of CNTs with the various biological responses and provided mechanistic insights into their Cyto- and genotoxic effects. Out study help define a novel platform for nanomaterials toxicity analysis and thus could expand the biomedical applications of CNTs and other nanomaterials.

# 372737 Viscosity Increase in Concentrated Mab Solutions Due to Large Transient Clusters


**Abstract:**

This work addresses a fundamental challenge in controlling properties and stability in concentrated protein solutions, which is both of basic scientific interest and industrial importance. Proteins in densely packed environments are abundant in our bodies and in commercially produced biopharmaceutical products. In addition to product stability during lyophilization and long-term storage, successful drug product formulation is essential for the processing and drug delivery stages of a product's life cycle. In particular, an increase in a therapeutic's viscosity could introduce challenges to material handling, such as during filling, and also adversely limit the variety of available drug delivery methods. Recent reports on monoclonal antibodies (mAbs) have highlighted their tendency to form small, strongly bound, reversible clusters, which produce an undesirable significant increase in viscosity at elevated concentrations (Biophys. J. 105:720, 2013). Clusters in fluids are important structural features formed by the balance of different interaction potential features that are especially prominent in concentrated protein solutions. Several types of short-range attractive interactions contribute to effective inter-protein interactions that may serve as the origin of clustering. While previous studies have discussed clustering in mAb solutions via electrostatic driven interactions, here we highlight a more hydrophobic mAb where the addition of salt induces the formation of large transient clusters. Cluster formation becomes more extensive with increasing salt concentration and reduced temperature. At the lowest temperatures studied, signs of slow kinetic arrest are present. However, despite the formation of a gel, the local dynamics of individual proteins in these transient clusters measured by neutron spin echo are surprisingly found to be very fast. This is dramatically different from that of globular proteins and previously studied mAb systems. Despite the fast motion of these mAbs, the short range interaction is still strong enough to produce a very large weakly linked protein network causing the increase of solution viscosity or gelation. These results highlight the difference between transient and strongly bound dynamic clusters and the sensitivity of mAb therapeutic material properties to the origin of inter-protein interactions and the purity of product formulation.

# 382768 Protein Confinement in Mesoporous Materials

**Abstract:**

Challenges in biotechnology and therapeutics have brought considerable interest to protein immobilization in porous hosts[1]. Hence, it is necessary to develop a more thorough understanding of the fundamental behavior of protein confinement. While pore surface properties such as charge and hydrophobicity are known to dramatically affect protein stability[2], our previous studies on confinement effects in nanoporous particles have shown that geometric properties such as surface curvature play a significant role as well[3]. These studies have shown that high concave surface curvature can stabilize the native protein structure and protect enzymes from extreme environmental conditions. Here we expand our investigation to include additional confinement geometries including the Ia3d (KIT-6), Im3m (SBA-16), and Fm3m (FDU-12) cubic space groups while maintaining consistent surface properties. This allows for further elucidation of the fundamental dependence of protein behavior on protein confinement geometry.

References:

1. Siefker J, Karande P, Coppens M-O. Packaging Biological Cargoes in Mesoporous Materials: Opportunities for Drug Delivery. Expert Opin Drug Deliv. 2014 (submitted)
2. Yuan L, Yu Q, Li D, Chen H. Surface Modification to Control Protein/Surface Interactions. Macromolecular Bioscience. 2011;11(8):1031
3. Sang L-C, Coppens M-O. Effects of surface curvature and surface chemistry on the structure and activity of proteins adsorbed in nanopores. Phys Chem Chem Phys. 2011;13(14):6689

# AUTHOR INDEX


Adigun, Oluwamayowa ....... 287
Aichele, Clint P. ....... 157
Ajji, Abdellah ....... 13
Akcora, Pinar ....... 390
Alloul, Ouassima ....... 145, 170
Almquist, Benjamin D. ....... 347
Anani, Tareq ....... 247, 294
Anderson, Daniel G. ....... 352
Anderson, Kelly M. ....... 127
Anderson, Kimberly W. ....... 88, 190, 242, 254
Ansaloni, Luca ....... 9
Ao, Geyou ....... 151, 228
Aravagiri, Kannan K. ....... 364
Archer, Lynden A. ....... 44
Aresh, Wafa ....... 370
Armani, Andrea M. ....... 93
Arora, Jaspreet S. ....... 346
Arya, Gaurav ....... 191
Avouris, Phaedon ....... 85
Aydin, Fikret ....... 365, 382
Badkas, Apurva ....... 354
Baglay, Roman R. ....... 385
Bailey, Christina ....... 188
Balbuena, Perla B. ....... 131, 132
Baldwin, Scott ....... 419
Banfield, Ryan ....... 342
Banta, Scott A. ....... 46
Bao, Yuping ....... 296
Barati, Danial ....... 362
Barkey, Dale P. ....... 342
Barton, Paul I. ....... 179
Baschetti, Marco Giacinti ....... 9
Baskin, Artem ....... 106
Bauer, Gernot ....... 196
Bedar, Amita ....... 140
Beetge, Jenni M. ....... 131
Beleño, José A. ....... 378
Bermudez, Harry ....... 78
Bevan, Michael A. ....... 369
Bhardwaj, Garima ....... 238
Bhattacharya, Soumya ....... 251
Bielski, Elizabeth ....... 371
Blankschtein, Daniel ....... 82
Bo, Zhenyu ....... 282
Bogart, Timothy ....... 400
Boghossian, Ardemis A. ....... 402
Bognet, Brice ....... 87
Bohorquez, Ana ....... 268
Boltyanskiy, Rostislav ....... 16
Bonzongo, Jean-Claude J. ....... 149, 208
Bothun, Geoffrey D. ....... 45
Boukany, Pouyan E. ....... 339
Braband, Vanessa Wood ....... 233
Brazel, Christopher S. ....... 237, 293
Breaux, Caleb ....... 4
Briggs, Nicholas ....... 157
Brooks III, Charles L. ....... 332
Brown, Matthew ....... 371
Burgos, Juan C. ....... 131
Burnett, Daniel J. ....... 83
Burns, Mark A. ....... 103
Cabral, Vladimir ....... 371
Cai, Fang ....... 214
Cambre, Sofie ....... 152
Camesano, Terri A. ....... 188
Cao, Xijia ....... 214
Cao, Zhenning ....... 398
Carpenter, Corinne ....... 164
Castleberry, Steven ....... 347
Chakraborty, Saptarshi ....... 333
Chalmers, Jeffrey J. ....... 419
Champion, Julie A. ....... 377
Chang, Hsueh-Chia ....... 95
Chang, John ....... 403
Charubin, Kamil ....... 205
Chen, Fengqiu ....... 271
Chen, Hao ....... 230
Chen, Jeane ....... 91, 250
Chen, Juan ....... 341
Chen, Yantao ....... 166
Cheng, Dangguo ....... 271
Cheung, Kei Bo ....... 230
Chi, Miaofang ....... 289
Chinesta, Francisco ....... 87
Chistiakova, Maria V. ....... 93
Chmelka, Bradley F. ....... 172
Choi, Jae Hyung ....... 372
Choi, Young S. ....... 247, 294
Chong, Leebyn ....... 381
Chopra, Maulick ....... 197
Choudhury, Pabitra ....... 249
Christensen, Matthew ....... 92, 348, 416
Chu, Xiaolei ....... 365
Chun, ByeongJae (Ben) ....... 207
Chung, Pil Seung ....... 146
Cisneros, Elizabeth ....... 172
Clar, Justin G. ....... 149, 208
Coffel, Joel ....... 295
Coletta, Elyse ....... 3
Conrad, Jacinta C. ....... 84
Coppens, Marc-Olivier ....... 279, 336, 422
Cosgrove, Terence ....... 48
Creighton, Megan A. ....... 86, 143
Crossley, Steven ....... 157
Cruz, Fernando J.A.L. ....... 357
Cruz-Acuña, Melissa ....... 297
Da Rocha, Sandro R.P. ....... 371
Dai, Leilei ....... 214
Damasceno, Pablo F. ....... 177
Danecker, Fabian ....... 196

# AUTHOR INDEX


Das, Saikat ..... 15
Das, Saroj Kumar ..... 140
Das, Sriya ..... 147
Datta, Dibyadyuti ..... 346
Dauthal, Preeti ..... 159
David, Allan ..... 245, 247, 294
Davis, Todd P. ..... 293
Davis, Virginia ..... 61, 340
De Miranda, Mariané Silva ..... 172
Deighan, Clayton ..... 419
Demokritou, Philip ..... 175
Dhakal, Subas ..... 210
Dickey, Michael D. ..... 395
Dimitrakopoulos, Christos ..... 14, 85
Dimitratos, Nikolaos ..... 114
Dinu, Cerasela Zoica ..... 90, 186, 420
Dix, Sean T. ..... 284
Dixit, Ninad ..... 2
Dobson, Jon ..... 297
Dodd, Paul ..... 177
Dong, Chenbo ..... 90, 186, 420
Dongqi, Yu ..... 300
Doorn, Stephen ..... 152
Dreaden, Erik ..... 373
Dufresne, Eric R. ..... 16
Duncan, Gregg ..... 369
Duque, Juan G. ..... 152
Dutt, Meenakshi ..... 365, 381, 382
Dziubla, Thomas D. ..... 190
Eaton, Seron ..... 92, 348
Egusa, Shunji ..... 376
Elabd, Yossef A. ..... 9
Eldawud, Reem ..... 90, 186, 420
Elmer, Jacob ..... 92, 348
Epps III, Thomas H. ..... 5
Epstein, Arthur ..... 236
Ercan, Batur ..... 28, 30
Esfandiari, Leyla ..... 109
Essien, Michael ..... 3
Fagan, Jeffrey A. ..... 150, 244
Fairhurst, David ..... 48
Falconer, John L. ..... 113
Falus, Peter ..... 89, 421
Fan, Qirui ..... 397
Fan, Tianju ..... 189, 235, 236
Fan, Xiaobin ..... 64
Fang, Yin ..... 392
Farmer, Damon ..... 85
Fatemi, Ali ..... 374
Filler, Michael A. ..... 289, 291
Fink, Alke ..... 255
Finn, M.G. ..... 47
Fischer, Paige ..... 104
Forde, Michael ..... 114
Frank, Curtis W. ..... 3
Freer, Alexander ..... 287
Fritsching, Udo ..... 331
Furukawa, Riku ..... 136
Gädke, Johannes ..... 298
Gallagher, James R. ..... 282
Gao, Fan ..... 107
Garcia, Armando R. ..... 83
Garlapalli, Ravinder ..... 15
Garnweitner, Georg ..... 298
Gascon, Jorge ..... 155
Geilich, Benjamin M ..... 405
Geng, Tao ..... 398
Genzer, Jan ..... 395
George, Cassie ..... 282
Getman, Rachel B. ..... 284
Gezer, Pervin Gizem ..... 401
Giraldo, Juan Pablo ..... 399
Gkikas, Manos ..... 334
Glotzer, Sharon C. ..... 177, 290, 338
Godfrin, P. Douglas ..... 89, 421
Godoy, Luiz ..... 402
Godoy, Rubén D. ..... 378
Gomez, Enrique D. ..... 394
Gomez-Ballesteros, Jose L. ..... 132
Gong, Xun ..... 402
Gonzales, Arthur ..... 238
Gonzalez, Laura ..... 92, 348
Goswami, Joyanta ..... 61
Goulas, Aristeidis ..... 111
Grady, Brian P. ..... 176
Graham, David E. ..... 185
Grandhi, Taraka Sai Pavan ..... 416
Green, Micah ..... 58, 147
Greenlee, Lauren F. ..... 153
Grill, Alfred ..... 85
Grillo, Fabio ..... 111
Gröhn, Arto J. ..... 330
Gross, Joachim ..... 196
Grzybowski, Bartosz A. ..... 211
Gu, Zhiyong ..... 107
Gualdron, Diego A. Gomez ..... 131
Guan, Jingjiao ..... 17, 96
Guenther, Caitlin ..... 349
Guerrero, Carlos Arturo ..... 378
Guiseppi-Elie, Anthony ..... 333, 355
Guo, Jiang ..... 130
Guo, John Zhanhu ..... 170
Guo, Xuhong ..... 329, 393
Guo, Zhanhu ..... 130, 145, 171
Gurung, Navaraj ..... 233
Gustitus, Sarah ..... 208
Haag, DelRae ..... 168
Hallinan Jr., Daniel T. ..... 232
Hammond, Paula T. ..... 347, 353, 373
Han, Rebecca ..... 148

# AUTHOR INDEX


Han, Wei ........ 230
Hanley, Alan ........ 245
Haroz, Erik ........ 150, 152
Harris, Kathleen R. ........ 250
Harris, Michael T. ........ 287
Harwell, Jeffrey ........ 157
He, Jie ........ 214
He, Qian ........ 114
He, Qingliang ........ 130, 171
He, Shuai ........ 84
He, Steve ........ 3
Headrick, Sierra ........ 249
Hedden, Ronald ........ 58
Heilshorn, Sarah C. ........ 367
Hemmati, Shohreh ........ 342
Henderson, Clifford L. ........ 4, 181
Herberle, Frederick A. ........ 205
Hessel, Colin ........ 400
Hickey, Daniel J. ........ 30
Hilmer, Andrew J. ........ 396
Hilt, J. Zach ........ 190
Hilt, James Z. ........ 88, 242, 254
Hilt, Silvia L. ........ 364
Hirahashi, Ryoto ........ 136
Ho, Michelle L. ........ 349
Hodapp, Maximilian J ........ 331
Hong, Yang-Ki ........ 212
Hosokawa, Akane ........ 108
Hu, Hanqiong ........ 16
Hu, Lin ........ 164, 167
Hu, Xiaoyu ........ 404
Hu, Zhongqiao ........ 384
Huang, Aaron ........ 334
Huang, Danting ........ 96
Huang, Shuyan ........ 350
Huang, Su ........ 231
Huang, Yike ........ 370
Huang, Yinlun ........ 173
Hudson, Benjamin C. ........ 213
Huggins, Kelly ........ 367
Hui, Ho Yee ........ 291
Hurt, Robert ........ 86
Hurt, Robert H. ........ 143
Hutchings, Graham J. ........ 114
Hwang, Hyea Hennim ........ 372
Hwang, Sang Youp ........ 158
Idris Sr., Muhammad Nuru ........ 307
Irin, Fahmida ........ 147
Islam, Syed ........ 15
Ivanov, Ilia N. ........ 185
Iverson, Nicole ........ 399, 402
Iwamura, Shinichiro ........ 136
Jabbari, Esmaiel ........ 362
Jagota, Anand ........ 356
Jain, Rishabh ........ 148, 399
James, Asha Liza ........ 140
Jang, Gyoung gug ........ 185
Jang, Seung Soon ........ 207, 372
Jasuja, Kabeer ........ 140
Jennings, G. Kane ........ 7
Jhon, Myung S. ........ 146, 187
Ji, Xiaoyuan ........ 335
Jiang, Chengjun ........ 110
Jiang, Jianwen ........ 194, 384
Jiang, Jing ........ 24
Jiang, Peng ........ 25, 392
Jiang, Wenjuan ........ 180
John, Vijay T. ........ 346
Johnston, Michael ........ 374
Joseph, Eugene ........ 2
Joshi, Pushkaraj ........ 343
Jung, Sukwon ........ 229
Kamaloo, Elaheh ........ 188
Kamat, Prashant V. ........ 139
Kang, Jeon Woong ........ 334
Kang, Yijin ........ 184
Kang, Yongmin ........ 344
Kannan, Rangaramanujam ........ 374
Kannan, Sujatha ........ 374
Kapteijn, Freek ........ 155
Karra, Vyshnavi ........ 381
Karunwi, Olukayode ........ 333, 355
Katsaras, John ........ 205
Katz, Alexander ........ 279
Kazbial, Andrew ........ 81, 144
Kelley, Elizabeth G. ........ 5
Kempler, Paul A. ........ 7
Kesava, Sameer Vajjala ........ 394
Kesavan, Lokesh ........ 114
Khademhosseini, Ali ........ 43
Khomami, Bamin ........ 366
Khripin, Constantine ........ 151
Kidder, Michelle K ........ 185
Kiely, Chris K. ........ 114
Kim, Jeehwan ........ 14
Kim, Minkyu ........ 334
Kim, Young Kyoung ........ 372
Kirkland, Brett ........ 96
Kloxin, April M. ........ 77
Kobayashi, Daisuke ........ 108, 391
Kohane, Daniel S. ........ 351
Kokini, Jozef ........ 401
Kokkoli, Efrosini ........ 379
Kong, Yi Wen ........ 373
Kooijman, Dirkjan ........ 111
Korgel, Brian A. ........ 285, 400
Kornblum, Noga ........ 231
Kothari, Rohit ........ 198
Kotov, Nicholas ........ 290
Kraken, Mathias ........ 298

# AUTHOR INDEX


Král, Petr ........ 106
Krull, Rainer ........ 298
Kruse, Anastasia M. ........ 88, 242, 254
Kruss, Sebastian ........ 396
Kumar, Bijandra ........ 106
Kumar, Nirbhay ........ 346
Kumar, Sunil ........ 206
Kung, Harold H. ........ 168
Kwak, Seonyeong ........ 399
Kwansa, Albert L. ........ 248
LaBaer, Joshua ........ 92, 348
Lam, Michael ........ 349
Lam, Stephanie ........ 244
Landry, Markita ........ 396, 399
Lane, Alan ........ 212
Lange, Alexander ........ 196
Lapotko, Dmitri ........ 418
Larson, Andrew C. ........ 250
Lattuada, Marco ........ 182, 255
Lawson, Richard A. ........ 4, 181
Lee, Jaegeun ........ 133
Lee, Joo-Youp ........ 354
Lee, Keewon ........ 372
Lee, Kun-Hong ........ 133
Lee, Ming ........ 296
Leung, Wallace Woon-Fong ........ 345
Lewandowski, Robert J. ........ 250
Li, Dongmei (Katie) ........ 128
Li, Fanxing ........ 156
Li, Lei ........ 81, 144
Li, Li ........ 329, 393
Li, Luping ........ 192, 288
Li, Ming ........ 205
Li, Wei ........ 347
Li, Yang ........ 64
Li, Ying ........ 243
Li, Zhiting ........ 81, 144
Li, Zhiyang ........ 107
Liang, Siwei ........ 8
Liang, Xinhua ........ 110, 270
Lima, Barbara ........ 396
Lin, Pin Ann ........ 132
Lin, Wen-Chi ........ 103
Ling, Kevin ........ 377
Litterst, Jochen ........ 298
Liu, G Logan ........ 401
Liu, Haining ........ 76
Liu, Haitao ........ 81
Liu, Lianjun ........ 243
Liu, Luting ........ 28
Liu, Tao ........ 17
Liu, Xiaochi ........ 393
Liu, Yidong ........ 189, 235, 236
Liu, Ying ........ 205, 370, 395, 404
Liu, Yun ........ 89, 421
Longo, Marjorie L. ........ 364, 380
Lu, Chang ........ 398
Lu, Shih-Yuan ........ 169
Lubers, Alia M. ........ 127
Ludovice, Peter ........ 4, 181
Luo, Sida ........ 17
Lv, Hao ........ 214
Lynch, Michele ........ 336
Lynd, Nathaniel A ........ 8
Ma, Anson W. K. ........ 87
Ma, Guanghui ........ 335
Machado, Alyne ........ 165
Macher, Thomas ........ 296
Magboo, Ramiro ........ 11
Majewska, Magdalena M. ........ 138
Majumder, Mainak ........ 59, 141
Maldonado-Camargo, Lorena ........ 246, 297
Mantri, Anirudh ........ 389
Mao, Shifan ........ 3
Markovic, Nenad ........ 184
Maroudas, Dimitrios ........ 164, 165, 167
Marquardt, Drew ........ 205
Marson, Ryan L. ........ 290
Martin, Stephen M. ........ 2
Masthoff, Ingke-C. ........ 298
Mathias, Ronita ........ 242
Mays, Robin ........ 395
Mazumder, Sonal ........ 197, 241
McArdle, Timothy J. ........ 85
McIntyre, Carl ........ 209
Meenach, Samantha A. ........ 88, 254, 408
Mefford, O. Thompson ........ 269
Mehta, Shalin B ........ 59, 141
Meier, Henry F. ........ 331
Meierhofer, Florian ........ 331
Meng, Fanxu ........ 174
Menkhaus, Todd ........ 233
Mensah, Lawrence ........ 353
Meyer III, Harry M. ........ 185
Mi, Gujie ........ 251
Miller, Daniel J. ........ 8
Miller, Jeffrey T. ........ 282
Min, Yong ........ 235, 236
Min, Yong G. ........ 189
Mineart, Kenneth ........ 1
Minelli, Matteo ........ 9
Miryala, Bhavani ........ 416
Mishra, Manoj ........ 374
Mittal, Jeetain ........ 356
Mo, Shenbin ........ 189, 235
Moayeri, Ali ........ 13
Moeinzadeh, Seyedsina ........ 362
Molla, Shahnawaz ........ 105
Monaharan, Vinothan ........ 383
Monbouquette, Harold G. ........ 109

# AUTHOR INDEX


Monnier, Christophe....255
Moon, Ji-Won....185
Morton, Stephen....353, 373
Mostowfi, Farshid....105
Mota, José P.B....357
Mu, Bin....396
Muhich, Christopher L....127
Mukai, Shin R....136
Mukherjee, Dibyendu....366
Mukhopadhyay, Debabrata....402
Mukhopadhyay, Mausumi....159
Muniz, Andre R....164, 165
Munoz, Enrique....80
Murphy, Ryan P....5
Murray, Christopher B....184
Murthy, Chaitanya....191
Musin, Ildar....291
Nabar, Gauri M....6
Naderi, Majid....83
Nagarajan, Ramanathan....334
Nagpure, Suraj....15
Nance, Elizabeth....374
Nangia, Shikha....180
Nation, Benjamin....181
Nayar, Suprabha....251
Nelson, Justin....396
Ni, Siyu....28
Nieh, Mu-Ping....205, 370
Nigra, Michael M....279
Nikles, David E....293
Niroomand, Hanieh....366
Njoroge, Ian G....7
Noriler, Dirceu....331
Notestein, Justin M....282
Nuxoll, Eric....295
Nykaza, Jacob....9
Oakey, John....104
Ober, Christopher K....24
Ogino, Isao....136
Oguntoye, Moses....234
Ogunyankin, Maria O....418
Oh, Eugene....133
Oldenbourg, Rudolf....59, 141
Olsen, Bradley D....334
Osuji, Chinedum....231
Osuji, Chinedum O....16
Otake, Katsuto....108, 391
Panthani, Matthew....400
Papavassiliou, Dimitrios V....62
Pape, Alicia....2
Paravastu, Anant K....213
Parikh, Atul N....364
Park, Hongsik....14
Park, Jihoon....212
Parmar, Shravan Kumar....343
Parra-Vasquez, Nicholas....152
Parviz, Dorsa....58, 147
Patanwala, Huseini....87
Patel, Rajankumar....270
Paulaitis, Michael....419
Pearce, Timothy R....379
Pelligra, Candice....231
Peng, Zhenmeng....158
Pereira, Ericson....209
Pesika, Noshir S....234
Peters, Andrew J....181
Peterson, Amy M....11
Pfeiffer, Dirk....85
Phillips, Blayne....25
Picher, Matthieu....132
Porambo, Michael....374
Porcar, Lionel....89, 421
Pratsinis, Sotiris E....330
Pratt, Lawrence R....234
Puri, Anu....370
Pushpavanam, Karthik....368, 403
Qin, Guan....84
Quadir, Mohiuddin....353
Radich, James G....139
Rajagopalan, kartik Kumar....341, 350
Ramasubramaniam, Ashwin....164
Ramaswamy, Sivaraman....179
Rankin, Stephen E....15
Rasch, Michael....400
Rege, Kaushal....92, 348, 368, 403, 416
Rentz, Nikki S....153
Repnin, Nikita....106
Resasco, Daniel....157
Reuel, Nigel F....396
Rikima, Yuusuke....136
Rima, Simonetta....182
Rinaldi, Carlos....246, 268, 297
Risbud, Subhash H....364
Roberts, Christopher B....154
Robertson, Megan L....394
Rodrigues, Flávio A....172
Rogers, William Benjamin....383
Rojanasakul, Yon....90, 186, 420
Roth, Connie B....385
Saha, Shibely....128
Salas, Ruth J. Lancheros....378
Salehi-Khojin, Amin....106
Sambasivam, Abhinanden....210
Sangodkar, Rahul P....172
Santhanam, Venugopal....343
Santra, Sanjitarani....368
Sapareto, Stephen....403
Sauntharajah, Yogen....376
Saxena, Shilpi....280
Schmidt, Jacob....109

# AUTHOR INDEX


Schneiderman, Steven.......233
Sebastiano, Michael.......365
Sen, Fatih.......402
Sen, Selda.......402
Shah, Rhythm R.......237, 293
Shang, Zeyu.......270
Shankar, Akshaya.......356
Shankar, Uma K.......389
Sharma, Aditya.......241
Sharma, Renu.......132
Shea, Lonnie D.......250
Shen, Hao.......404
Sheng, Yuan.......283
Sherwood, Jennifer.......296
Shi, Di.......251
Shimada, Yuichiro.......108, 391
Shin, Nae Chul.......289
Shono, Atsushi.......108, 391
Shopsowitz, Kevin.......353
Shukla, Anita.......79
Siefert, Sönke.......6
Siefker, Justin.......422
Silvera-Batista, Carlos.......208
Sine, Jessica.......370
Singer, Jonathan P.......16
Singh, Dilpreet.......175
Singh, Meenesh R.......8
Slouka, Zdenek.......95
Smith, Robert.......146
Song, Hao.......173
Sotiriou, Georgios A.......175
Souva, Matthew S.......6
Spakowitz, Andrew J.......3
Spontak, Richard.......1
Sreevatsan, Srinand.......246
Sresht, Vishnu.......82
Stair, Peter C.......282
Stallings, Danielle S.......248
Stamenkovic, Vojislav.......184
Stephanopoulos, George.......179
Stolzoff, Michelle.......26
Strano, Michael S.......82, 148, 396, 399, 402
Striolo, Alberto.......176
Strzalka, Joseph.......15
Style, Robert W.......16
Su, Ren.......114
Su, Zhiguo.......335
Subbaiyan, Navaneetha.......152
Suh, Junghae.......349
Sullivan, Millicent O.......5
Sun, Chen.......398
Sun, Gongchen.......95
Sun, Jinsheng.......214
Sun, Linlin.......28, 30, 251
Sun, Qihua.......408
Sureshkumar, Radhakrishna.......210
Suthiwangcharoen, Nisaraporn.......334
Suttipong, Manaswee.......176
Szymusiak, Magdalena.......404
Talebi, Mojgan.......111
Talluri, Suvarna.......233
Tan, Anthony.......387
Tang, Wei.......189, 235, 236
Teng, Xiaowei.......342
Thayer, Derek.......370
Thompson, Michael.......24
Thota, Naresh.......194, 384
Tian, Yanqing.......416
Timko, Brian P.......351
Tkacz, Rachel.......59, 141
Tong, Songzhao.......189, 235, 236
Torkelson, John M.......387
Tripathy, Mukta.......195, 389
Turner, C. Heath.......76
Tvrdy, Kevin.......148
Ugaz, Victor M.......174
Ulissi, Zachary.......82, 148, 402
Uppaladadium, Geetartha.......382
Uruma, Takashi.......391
Vaddiraju, Sreeram.......344
Vahdat, Vahid.......81, 144
Valdesueiro, David.......111
Van Anders, Greg.......290
Van De Ven, Anne L.......405
Van Krieken, Finn.......86, 143
Van Norman, Staci A.......113
Van Ommen, J. Ruud.......111, 113
VanDer Ende, Emma.......396
Vasiraju, Venkata.......344
Vemuri, Sesha Hari.......146, 187
Vengsarkar, Pranav S.......154
Venkataramani, Deepika.......157
Visi, Akanksha.......241
Vo, Minh.......62
Vora, Sahil.......87
Voss, John C.......364
Wagner, Norman J.......89, 421
Wan, Chao.......271
Wang, Chun-Chieh.......169
Wang, Jian.......189, 235
Wang, Jie.......329
Wang, Kathleen.......188
Wang, Mian.......12
Wang, Mingwei.......329
Wang, Ping.......335
Wang, Shengnian.......341, 350
Wang, Shu.......394
Wang, Xiaohan.......393
Wang, Yiran.......130
Wang, Yong.......370

# AUTHOR INDEX


Wang, Zaijie Jim ........ 404
Wang, Zhibin ........ 17, 96
Wang, Zhong Lin (ZL.) ........ 286
Wang, Zimeng ........ 375
Waterman, Kellie ........ 188
Watkins, James J. ........ 198
Webster, Thomas J. ........ 12, 26, 28, 30, 238, 251, 405
Wegner, Karsten ........ 330
Wei, Huige ........ 130, 171
Wei, Shuai ........ 332
Wei, Suying ........ 130, 145, 170, 171
Weimer, Alan W. ........ 113, 127
Wenning, Brandon ........ 24
Weston, Javen ........ 157
White, Sarah B. ........ 250
Whitehead, Kathryn A. ........ 352
Widgeon, Scarlett ........ 172
Winans, Randall ........ 6
Winter, H. Henning ........ 198
Winter, Jessica O. ........ 6, 397
Wogan, Gerald ........ 402
Wohlleben, Wendel ........ 175
Wong, Justin C. ........ 193, 292
Wong, Min Hao ........ 399
Wu, Qingliu ........ 15
Wu, Wenzhuo ........ 286
Wyant, Spencer ........ 164
Wydra, Robert J. ........ 190
Wyslouzil, Barbara E. ........ 6
Xia, Bingzhao ........ 104
Xia, Junfei ........ 17, 96
Xia, Yan ........ 205
Xiao, Jie ........ 173
Xiao, Zhiwei ........ 17
Xu, Cheng ........ 192, 288
Xu, Jun ........ 393
Xu, Rui ........ 154
Xu, Xia ........ 212
Xu, Yisheng ........ 393
Yaffe, Michael ........ 373
Yamagami, Momona ........ 349
Yan, Xingru ........ 130, 171
Yang, Cuixian ........ 229
Yang, Guang ........ 232
Yang, Lijun ........ 345
Yang, Lin ........ 371
Yang, Peidong ........ 184
Yaolin, Yaolin ........ 296
Yap, Woon Teck ........ 250
Yasaei, Poya ........ 106
Yeom, Jihyeon ........ 178
Yeung, King Lun ........ 230
Yi, Hyunmin ........ 229
Yingling, Yaroslava G. ........ 248
Youn, Sejin ........ 208
Yu, Yixuan ........ 400
Yu, Ziniu ........ 58
Yuan, Chunqiu ........ 189, 235, 236
Yuan, Tianyu ........ 149
Yuan, Zhenyu ........ 329
Zarraga, Isidro (Dan) ........ 89, 421
Zarzar, Jonathan ........ 89, 421
Zasadzinski, Joseph A. ........ 418
Zeng, Hua Chun ........ 135, 283
Zeno, Wade F. ........ 364, 380
Zhan, Guowu ........ 135
Zhan, Xiaoli ........ 271
Zhang, Changlin ........ 158
Zhang, Fan ........ 374
Zhang, Fang ........ 175
Zhang, Fengbao ........ 64
Zhang, Guoliang ........ 64
Zhang, Jingqing ........ 82, 396
Zhang, Junshe ........ 156
Zhang, Peipei ........ 96
Zhang, Qianyi ........ 243
Zhang, Rui ........ 393
Zhang, Songping ........ 335
Zhang, Xi ........ 145, 170, 171
Zhang, Y.-H. Percival ........ 337
Zhang, Zhi ........ 374
Zhang, Zhongtao ........ 76
Zhao, Chunyan ........ 189, 235
Zhao, Cunyu ........ 243
Zhao, Huilei ........ 243
Zhao, Qingshan ........ 64
Zhao, Yang ........ 149, 192, 288
Zhao, Zheng ........ 157
Zheng, Ming ........ 150, 151, 228
Zhong, Qian ........ 371
Zhou, Feng ........ 81, 144
Zhou, Yao ........ 183
Zhou, Yunlong ........ 290
Zhou, Zilan ........ 354
Zhu, Zhiguang ........ 337
Ziegler, Kirk J. ........ 149, 192, 193, 208, 288, 292
Zu, Yingbo ........ 341, 350